SENSORS
AND
ACTUATORS

Control Systems

Instrumentation

SENSORS AND ACTUATORS

Control Systems

Instrumentation

CLARENCE W. de SILVA

CRC Press
Taylor & Francis Group
Boca Raton   London   New York

CRC Press is an imprint of the
Taylor & Francis Group, an informa business

CRC Press
Taylor & Francis Group
6000 Broken Sound Parkway NW, Suite 300
Boca Raton, FL 33487-2742

© 2007 by Taylor & Francis Group, LLC
CRC Press is an imprint of Taylor & Francis Group, an Informa business

International Standard Book Number-10: 1-4200-4483-4 (Hardcover)
International Standard Book Number-13: 978-1-4200-4483-6 (Hardcover)

Library of Congress Cataloging-in-Publication Data

De Silva, Clarence W.
  Sensors and actuators : control system instrumentation / Clarence W. de Silva.
     p. cm.
  "A CRC title."
  Includes bibliographical references and index.
  ISBN-13: 978-1-4200-4483-6 (alk. paper)
  ISBN-10: 1-4200-4483-4 (alk. paper)
  1. Automatic control. 2. Detectors. 3. Actuators. I. Title.

TJ213.D386 2007
670.42'7--dc22   2006024039

Visit the Taylor & Francis Web site at
http://www.taylorandfrancis.com

and the CRC Press Web site at
http://www.crcpress.com

Dedication

Dedicated to Charmaine, CJ, and Cheryl, since their "senses" have developed and since they have become rather "active."

But as artificers do not work with perfect accuracy, it comes to pass that mechanics is so distinguished from geometry that what is perfectly accurate is called geometrical; what is less so, is called mechanical. However, the errors are not in the art, but in the artificers.

Sir Isaac Newton, Principia Mathematica,
Cambridge University, May 8, 1686

Preface

This is an introductory book on the subject of control system instrumentation, with an emphasis on sensors, transducers, and actuators. Specifically, the book deals with "instrumenting" a control system through the incorporation of suitable sensors, actuators, and associated interface hardware. It will serve as both a textbook for engineering students and a reference book for practicing professionals. As a textbook, it is suitable for courses in control system instrumentation; control sensors and actuators; and mechatronics; or a second course in feedback control systems. The book has adequate material for two 14-week courses, one at the junior (third-year undergraduate) or senior (fourth-year undergraduate) level and the other at the first-year graduate level. In view of the practical considerations, design issues, and industrial techniques that are presented throughout the book, and in view of the simplified and snap-shot style presentation of more advanced theory and concepts, the book will serve as a useful reference tool for engineers, technicians, project managers, and other practicing professionals in industry and in research laboratories, in the fields of control engineering, mechanical engineering, electrical and computer engineering, manufacturing engineering, aerospace engineering and mechatronics.

A control system is a dynamic system that contains a controller as an integral part. The purpose of the controller is to generate control signals, which will drive the process to be controlled (the plant) in the desired manner. Actuators are needed to perform control actions as well as to directly drive or operate the plant. Sensors and transducers are necessary to measure output signals (process responses) for feedback control; to measure input signals for feedforward control; to measure process variables for system monitoring, diagnosis and supervisory control; and for a variety of other purposes. Since many different types and levels of signals are present in a control system, signal modification (including signal conditioning and signal conversion) is indeed a crucial function associated with any control system. In particular, signal modification is an important consideration in component interfacing. It is clear that a course in control system instrumentation should deal with sensors and transducers, actuators, signal modification, and component interconnection. Specifically, the course should address the identification of control system components with respect to functions, operation and interaction, and proper selection and interfacing of these components for various control applications. Parameter selection (including system tuning) is an important step as well. Design is a necessary and integral part of control system instrumentation, for it is design that enables us to build a control system that meets the performance requirements—starting, perhaps, with basic components such as sensors, actuators, controllers, compensators, and signal modification devices. The book addresses all these issues, starting from the basics and systematically leading to advanced concepts and applications.

The approach taken in the book is to treat the basic types of control sensors and actuators in separate chapters, but without losing sight of the fact that various components in a control system have to function as an interdependent and interconnected (integrated) group in accomplishing the specific control objectives. Operating principles, modeling, design considerations, ratings, performance specifications, and applications of the individual components are discussed. Component integration and design considerations are addressed as well. To maintain clarity and focus and to maximize the usefulness of the book, the material is presented in a manner that will be useful to anyone with a basic engineering background, be it electrical, mechanical, mechatronic, aerospace, control, manufacturing, chemical, civil, or computer. Case studies, worked examples, and exercises are provided throughout the book, drawing from such application systems as robotic manipulators, industrial machinery, ground transit vehicles, aircraft, thermal and fluid process plants, and digital computer components. It is impossible to discuss every available control system component in a book of this nature; for example, thick volumes have been written on measurement devices alone. In this book, some types of sensors and actuators are studied in great detail, while some others are treated superficially. Once students are exposed to an in-depth study of some components, it should be relatively easy for them to extend the same concepts and the same study approach to other components that are functionally or physically similar. Augmenting their traditional role, the problems at the end of each chapter serve as a valuable source of information not found in the main text. In fact, the student is strongly advised to read all the problems carefully in addition to the main text. Complete solutions to the end-of-chapter problems are provided in a solutions manual, which is available to instructors who adopt the book.

About 10 years after my book *Control Sensors and Actuators* (Prentice-Hall, 1989) was published, I received many requests for a revised and updated version of the book. The revision was undertaken in the year 2000 during a sabbatical leave. As a result of my simultaneous involvement in the development of undergraduate and graduate curricula in mechatronics and in view of substantial new and enhanced material that I was able to gather, the project quickly grew into one in mechatronics and led to the publication of the monumental 1300-page textbook: *Mechatronics—An Integrated Approach* (Taylor & Francis, CRC Press, 2005). In meeting the original goal, however, the present book was subsequently developed as a condensed version of the book on mechatronics, while focusing on control sensors and actuators. The manuscript for the original book evolved from the notes developed by me for an undergraduate course entitled "Instrumentation and Design of Control Systems" and for a graduate course entitled "Control System Instrumentation" at Carnegie Mellon University. The undergraduate course was a popular senior elective taken by approximately half of the senior mechanical engineering class. The graduate course was offered for students in electrical and computer engineering, mechanical engineering, and chemical engineering. The prerequisites for both courses were a conventional introductory course in feedback controls and the consent of the instructor. During the development of the material for that book, a deliberate attempt was made to cover a major part of the syllabuses for the two courses: "Analog and Digital Control System Synthesis," and "Computer Controlled Experimentation," offered in the Department of Mechanical Engineering at the Massachusetts Institute of Technology. At the University of British Columbia, the original material was further developed, revised, and enhanced for teaching courses in mechatronics and control sensors and actuators. The material in the book has acquired an application orientation through my industrial experience in the subject at places such as IBM Corporation, Westinghouse Electric Corporation, Bruel and Kjaer, and NASA's Lewis and Langley Research Centers.

The material presented in the book will serve as a firm foundation, for subsequent building up of expertise in the subject—perhaps in an industrial setting or in an academic research laboratory—with further knowledge of control hardware and analytical skills (along with the essential hands-on experience) gained during the process. Undoubtedly, for best results, a course in control sensors and actuators, mechatronics, or control system instrumentation should be accompanied by a laboratory component and class projects.

Main Features of the Book

The following are the main features of the book, which will distinguish it from other available books on the subject:

- The material is presented in a progressive manner, first giving introductory material and then systematically leading to more advanced concepts and applications, in each chapter.
- The material is presented in an integrated and unified manner so that users with a variety of engineering backgrounds (mechanical, electrical, computer, control, aerospace, manufacturing, chemical, and material) will be able to follow and equally benefit from it.
- Practical procedures and applications are introduced in the beginning and then uniformly integrated throughout the book.
- Key issues presented in the book are summarized in boxes and in point form, at various places in each chapter, for easy reference, recollection, and for use in Power-Point presentations.
- Many worked examples and case studies are included throughout the book.
- Numerous problems and exercises, most of which are based on practical situations and applications, are given at the end of each chapter.
- References and reading suggestions are given at the end of the book, for further information and study.
- A solutions manual is available for the convenience of the instructors.

Clarence W. de Silva
Vancouver, Canada

Acknowledgments

Many individuals have assisted in the preparation of this book, but it is not practical to acknowledge all such assistance here. First, I wish to recognize the contributions, both direct and indirect, of my graduate students, research associates, and technical staff. Special mention should be made of Jason Zhang, my research engineer. I am particularly grateful to Michael Slaughter, senior editor, engineering, CRC Press, for his interest, enthusiasm, constant encouragement, and support, throughout the project. Other staff of CRC Press and its affiliates, particularly, Dr. S. Vinithan, Glenon Butler, Jessica Vakili, and Liz Spangenberger deserve special mention here. Finally, I wish to acknowledge the advice and support of various authorities in the field, particularly, Professor Devendra Garg of Duke University, Professor Mo Jamshidi of the University of Texas (San Antonio), Professors Marcelo Ang, Ben Chen, Tong-Heng Lee, Jim A.N. Poo, and Kok-Kiong Tan of the National University of Singapore, Professor Arthur Murphy (DuPont Fellow Emeritus), Professor Max Meng of the Chinese University of Hong Kong, Professor Grantham Pang of the University of Hong Kong, Dr. Daniel Repperger of U.S. Air Force Research Laboratory, and Professor David N. Wormley of the Pennsylvania State University. My wife and children deserve much appreciation for their support and understanding during the production of the book.

Author

Dr. Clarence W. de Silva, P.E., Fellow ASME, and Fellow IEEE, is a professor of mechanical engineering at the University of British Columbia, Vancouver, Canada and has occupied the NSERC-BC Packers Research Chair in industrial automation since 1988. Before this, he has served as a faculty member at Carnegie Mellon University (1978–1987) and as a Fulbright visiting professor at the University of Cambridge (1987–1988). Dr. de Silva has earned PhD degrees, one from Massachusetts Institute of Technology (1978) and the other from the University of Cambridge, England (1998). Dr. de Silva has also occupied the Mobil Endowed Chair Professorship in the Department of Electrical and Computer Engineering at the National University of Singapore (2000). He has served as a consultant to several companies including IBM and Westinghouse in the United States, and has led the development of six industrial machines.

Dr. de Silva is a recipient of the Henry M. Paynter outstanding investigator award from the Dynamic Systems and Control Division of ASME, Killam research prize, outstanding engineering educator award of the Institute of Electrical and Electronics Engineers (IEEE), Canada, education award of the Dynamic Systems and Control Division of the American Society of Mechanical Engineers (ASME), lifetime achievement award of the World Automation Congress, IEEE third millennium medal, meritorious achievement award of the Association of Professional Engineers of BC, the outstanding contribution award of the Systems, Man, and Cybernetics Society of IEEE, outstanding chapter chair award of the IEEE Vancouver Section, outstanding chapter award of the IEEE Control Systems Society, and the outstanding large chapter award of the IEEE Industry Applications Society. He has authored 16 technical books including *Mechatronics—An Integrated Approach* (CRC Press, Taylor & Francis, 2005); *Soft Computing and Intelligent Systems Design—Theory, Tools, and Applications* (with F. Karry, Addison Wesley, 2004); *Vibration—Fundamentals and Practice* (CRC Press, Taylor & Francis, 2nd Edition, 2007), *Intelligent Control—Fuzzy Logic Applications* (CRC Press, 1995), *Control Sensors and Actuators* (Prentice Hall, 1989), 12 edited volumes, about 165 journal papers, and a similar number of conference papers and book chapters.

Dr. de Silva has served on the editorial boards of twelve international journals, in particular as the editor-in-chief of the *International Journal of Control and Intelligent Systems*, editor-in-chief of the *International Journal of Knowledge-Based Intelligent Engineering Systems*, senior technical editor of *Measurements and Control*, and regional editor, North America, of *Engineering Applications of Artificial Intelligence – the International Journal of Intelligent Real-Time Automation*. He is a Lilly Fellow at Carnegie-Mellon University, NASA-ASEE Fellow, Senior Fulbright Fellow to Cambridge University, Fellow of the Advanced Systems Institute of British Columbia, Killam Fellow, and Fellow of the Canadian Academy of Engineering. Research and development activities of Dr. de Silva are primarily centered in the areas of process automation, robotics, mechatronics, intelligent control, and sensors and actuators.

Contents

1

Control, Instrumentation, and Design

This chapter introduces the subject of instrumentation, as related to control engineering. Various architectures of a control system are outlined and the important role that is played by sensors and actuators in them is indicated. Important issues in the practice of control system instrumentation are identified. How instrumentation, modeling, and design of a control system are interrelated is highlighted. The steps of instrumenting a control system with sensors, transducers, actuators, and associated hardware are stated. The organization of the book is summarized.

1.1 Introduction

This is an introductory book on the subject of control system instrumentation, with emphasis on sensors, transducers, and actuators. Specifically, the book deals with instrumenting a control system through the incorporation of suitable sensors, actuators, and associated hardware.

A control system is a dynamic system that contains a controller as an integral part. The purpose of the controller is to generate control signals, which will drive the process to be controlled (the plant) in the desired manner. Actuators are needed to perform the control actions as well as to drive the plant directly. Sensors and transducers are necessary to measure output signals (process responses) and to measure input signals for feedforward control; to measure process variables for system monitoring, diagnosis, and supervisory control; and for a variety of other purposes. Since many different types and levels of signals are present in a control system, signal modification (including signal conditioning and signal conversion) is indeed a crucial function associated with any control system. In particular, signal modification is an important consideration in component interfacing. It is clear that the subject of control system instrumentation should deal with sensors, transducers, actuators, signal modification, and component interconnection. In particular, the subject should address the identification of control system components with respect to functions, operation and interaction, and proper selection and interfacing of these components for various control applications. Parameter selection (including system tuning) is an important step as well. Design is a necessary part of control system instrumentation, for it is design that enables us to build a control system that meets the performance requirements—starting, perhaps, with basic components such as sensors, actuators, controllers, compensators, and signal modification devices.

Control engineers should be able to identify or select components, particularly sensors and actuators, for a control system, model and analyze individual components and the overall systems, and choose parameter values to perform the intended functions of the particular system in accordance with some specifications. Several applications and their use of sensors and actuators are noted in Table 1.1. Identification, analysis, selection

TABLE 1.1

Sensors and Actuators Used in Some Common Engineering Applications

Process	Typical Sensors	Typical Actuators
Aircraft	Displacement, speed, acceleration, elevation, heading, force, pressure, temperature, fluid flow, voltage, current, global positioning system (GPS)	DC motors, stepper motors, relays, valve actuators, pumps, heat sources, jet engines
Automobile	Displacement, speed, force, pressure, temperature, fluid flow, fluid level, voltage, current	DC motors, stepper motors, valve actuators, pumps, heat sources
Home heating system	Temperature, pressure, fluid flow	Motors, pumps, heat sources
Milling machine	Displacement, speed, force, acoustics, temperature, voltage, current	DC motors, AC motors
Robot	Optical image, displacement, speed, force, torque, voltage, current	DC motors, stepper motors, AC motors, hydraulic actuators
Wood Drying Kiln	Temperature, relative humidity, moisture content, air flow	AC motors, DC motors, pumps, heat sources

matching and interfacing of components, and tuning of the integrated system (i.e., adjusting parameters to obtain the required response from the system) are essential tasks in the instrumentation and design of a control system. The book addresses these issues, starting from the basics and systematically leading to advanced concepts and applications.

1.2   Control Engineering

The purpose of a controller is to make a plant (i.e., the system to be controlled) behave in a desired manner. The overall system that includes at least the plant and the controller is called the control system. The system can be quite complex and may be subjected to known or unknown excitations (i.e., inputs), as in the case of an aircraft (see Figure 1.1).

A schematic diagram of a control system is shown in Figure 1.2. The physical dynamic system (e.g., a mechanical system) the response of which (e.g., vibrations) needs to be controlled is called the plant or process. The device that generates the signal (or command)

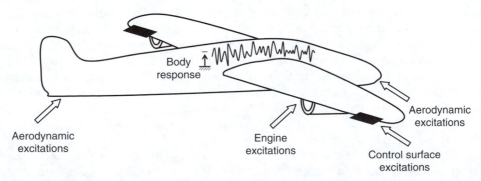

FIGURE 1.1
Aircraft is a complex control system.

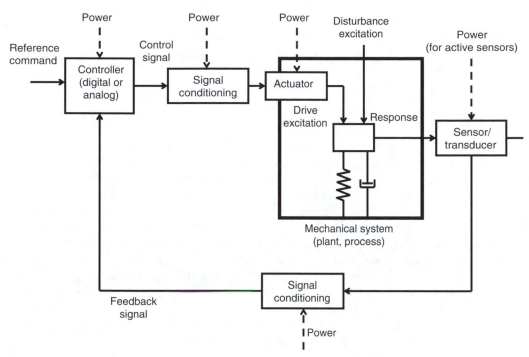

FIGURE 1.2
Schematic diagram of a feedback control system.

according to some scheme (or control law) and controls the response of the plant is called the controller. The plant and the controller are the two essential components of a control system. Certain command signals or inputs are applied to the controller and the plant is expected to behave in a desirable manner, under control. In feedback control, the plant has to be monitored and its response needs to be measured using sensors and transducers, for feeding back into the controller. Then, the controller compares the sensed signal with a desired response as specified externally and uses the error to generate a proper control signal.

In Figure 1.2, we have identified several discrete blocks, depending on various functions that take place in a typical control system. In a practical control system, this type of clear demarcation of components might be difficult; one piece of hardware might perform several functions, or more than one distinct unit of equipment might be associated with one function. Nevertheless, Figure 1.2 is useful in understanding the architecture of a general control system. This is an analog control system because the associated signals depend on the continuous time variable; no signal sampling or data encoding is involved in the system.

A good control system should possess the following performance characteristics:

1. Sufficiently stable response (stability): Specifically, the response of the system to an initial-condition excitation should decay back to the initial steady state (asymptotic stability). The response to a bounded input should also be bounded (bounded-input-bounded-output—BIBO stability).

2. Sufficiently fast response (speed of response or bandwidth): The system should react quickly to a control input.

3. Low sensitivity to noise, external disturbances, modeling errors, and parameter variations (sensitivity and robustness).

4. High sensitivity to control inputs (input sensitivity).

5. Low error; for example, tracking error and steady-state error (accuracy).

6. Reduced coupling among system variables (cross sensitivity or dynamic coupling).

1.2.1 Instrumentation and Design

In some situations, each function or operation within a control system can be associated with one or more physical devices, components, or pieces of equipment and in other situations one hardware unit may accomplish several of the control system functions. In the present context, by instrumentation we mean the identification of these instruments or hardware components with respect to their functions, operation, and interaction with each other and the proper selection, interfacing, and tuning of these components for a given application; in short, instrumenting a control system. A simplified schematic example of an instrumented control system is shown in Figure 1.3.

By design, we mean the process of selecting suitable equipments to accomplish various functions in the control system; developing the system architecture; matching and interfacing these devices; and selecting the parameter values, depending on the system characteristics, to achieve the desired objectives of the overall control system (i.e., to meet design specifications), preferably in an optimal manner and according to some performance criterion. Design may be included as an instrumentation objective. In particular, there can be many designs that meet a given set of performance requirements (see Chapter 3). Identification of key design parameters, modeling of various components, and analysis are often useful in the design process. Modeling (both analytical and experimental) is important in analyzing, designing, and evaluation of a control system.

Identification of the hardware components (perhaps commercially available off-the-shelf items) corresponding to each functional block in Figure 1.2 is one of the first steps in the instrumentation of a control system. For example, in process control applications off-the shelf analog proportional-integral-derivative (PID) controllers may be used. These controllers for process control applications traditionally have knobs or dials for control

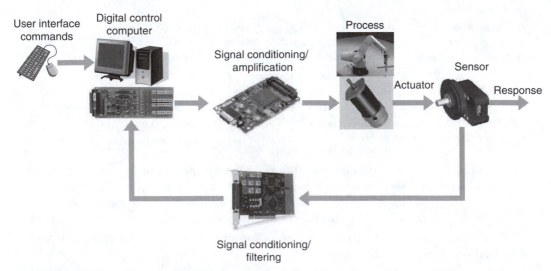

FIGURE 1.3
An instrumented feedback control system.

parameter settings; that is, proportional band or gain, reset rate (in repeats of the proportional action per unit time), and rate time constant. The operating bandwidth (operating frequency range) of these control devices is specified (see Chapter 3). Various control modes—on/off, proportional, integral, and derivative or combinations—are provided by the same control box.

Actuating devices (i.e., actuators) include stepper motors, dc motors, ac motors, solenoids, valves, and relays (see Chapter 6 and Chapter 7), which are also commercially available to various specifications. An actuator may be directly connected to the driven load, and this is known as the direct-drive arrangement. More commonly, however, a transmission device (see Chapter 8) may be needed to convert the actuator motion into a desired load motion and for proper matching of the actuator with the driven load. Potentiometers, differential transformers, resolvers, synchros, gyros, strain gauges, tachometers, piezoelectric devices, fluid flow sensors, pressure gauges, thermocouples, thermistors, and resistance temperature detectors (RTDs) are examples of sensors used to measure process response for monitoring its performance and possible feedback (see Chapter 4 and chapter 5) for control. An important factor that we must consider in any practical control system is noise, including external disturbances. Noise may represent actual contamination of signals or the presence of other unknowns, uncertainties, and errors, such as parameter variations and modeling errors. Furthermore, weak signals have to be amplified, and the form of the signal might have to be modified at various points of interaction. Charge amplifiers, lock-in amplifiers, power amplifiers, switching amplifiers, linear amplifiers, pulse-width modulated (PWM) amplifiers, tracking filters, low-pass filters, high-pass filters, band-pass filters, and band-reject filters or notch filters are some of the signal-conditioning devices used in analog control systems (see Chapter 2). Additional components, such as power supplies and surge-protection units, are often needed in control, but they are not indicated in Figure 1.2 because they are only indirectly related to control functions. Relays and other switching and transmission devices, and modulators and demodulators may also be included.

1.2.2 Modeling and Design

A design may use excessive safety factors and worst-case specifications (e.g., for mechanical loads and electrical loads). This will not provide an optimal design or may not lead to the most efficient performance. Design for optimal performance may not necessarily lead to the most economical (least costly) design, however. When arriving at a truly optimal design, an objective function that takes into account all important factors (performance, quality, cost, speed, ease of operation, safety, environmental impact, etc.) has to be optimized. Multiple objective functions may be necessary and an integrated design approach (i.e., mechatronic design), which simultaneously takes into account the entire system, is desirable. A complete design process should generate the necessary details of a system for its construction or assembly. Of course, in the beginning of the design process, the desired system does not exist. In this context, a model of the anticipated system, particularly an analytical model or a computer model, can be very useful, economical, and time efficient. In view of the complexity of a design process, particularly when striving for an optimal design, it is valuable to incorporate system modeling as a tool for design iteration.

Modeling and design can go hand in hand, in an iterative manner. In the beginning, by knowing some information about the system (e.g., intended functions, performance specifications, past experience, and knowledge of related systems) and using the design objectives, it will be possible to develop a model with sufficient (low to moderate) details and complexity. By analyzing and carrying out computer simulations of the model, it will be possible to generate useful information that will guide the design process (e.g., generation of a preliminary design). In this manner, design decisions can be made, and the model can be refined using the available (improved) design.

1.3 Control System Architectures

Some useful terminologies related to a control system, as introduced in the previous section, are summarized as follows:

- Plant or process: System to be controlled
- Inputs: Excitations (known, unknown) to the system
- Outputs: Responses of the system
- Sensors: Devices that measure system variables (excitations, responses, etc.)
- Actuators: Devices that drive various parts of the system
- Controller: Device that generates control signal
- Control law: Relation or scheme according to which control signal is generated
- Control system: At least the plant and the controller (may include sensors, signal conditioning, and other components as well)
- Feedback control: Control signal is determined according to plant response
- Open-loop control: Plant response is not used to determine the control action
- Feed-forward control: Control signal is determined according to plant excitation

The significance of sensors and actuators for a control system is implicit here. This importance holds regardless of the specific control system architecture that is implemented in a given application. We now outline several architectures of control system implementation while indicating the presence of sensors and actuators in them.

   If the plant is stable and is completely and accurately known, and if the inputs to the plant can be precisely generated (by the controller) and applied, then accurate control might be possible even without feedback control. Under these circumstances, a measurement system is not needed (or at least not needed for feedback) and thus, we have an open-loop control system. In open-loop control, we do not use current information on system response to determine the control signals. In other words, there is no feedback. The structure of an open-loop control system is shown in Figure 1.4. Note that a sensor is not explicitly indicated in this open-loop architecture. However, sensors may be employed within an open-loop system to monitor the applied input, the resulting response, and possible disturbance inputs even though feedback control is not used.

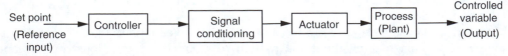

FIGURE 1.4
An open-loop control system.

1.3.1 Feedback Control with PID Action

In a feedback control system, as shown in Figure 1.2 and Figure 1.3, the control loop has to be closed, by measuring the system response and employing that information to generate control signals to correct any output errors. Hence, feedback control is also known as closed-loop control.

A control law is a relationship between the controller output and the plant input. Common control modes are

- On/off (bang–bang) control
- Proportional (P) control
- Proportional control combined with reset (i.e., integral, I) and rate (i.e., derivative, D) actions (i.e., multimode or multiterm control)

Control laws for commonly used control actions are given in Table 1.2. Some advantages and disadvantages of each control action are also indicated. Many control systems employ three-mode controllers or three-term controllers (i.e., PID controllers), which can provide the combined action of proportional, integral, and derivative modes. The control law for PID control is given by

$$c = k_p \left(e + \tau_d \dot{e} + \frac{1}{\tau_i} \int e \, dt \right) \tag{1.1a}$$

or in the transfer function form

$$\frac{c}{e} = k_p \left(1 + \tau_d s + \frac{1}{\tau_i s} \right) \tag{1.1b}$$

where e is the error signal (controller input), c is the control or actuating signal (controller output or plant input), k_p is the proportional gain, τ_d is the derivative time constant, and τ_i is the integral time constant.

TABLE 1.2

Comparison of Some Common Control Actions

Control Action	Control Law	Advantages	Disadvantages
On/off	$\frac{c_{max}}{2}[\text{sgn}(e) + 1]$	Simple Inexpensive	Continuous chatter Mechanical problems Poor accuracy
Proportional	$k_p e$	Simple Fast response	Offset error (Steady-state error) Poor stability
Reset (Integral)	$\frac{1}{\tau_i} \int e \, dt$	Eliminates offset Filters out noise	Low bandwidth (Slow response) Reset windup Instability problems
Rate (Derivative)	$\tau_d \dfrac{de}{dt}$	High bandwidth (Fast response) Improves stability	Insensitive to DC error Allows high-frequency noise Amplifies noise Difficult analog implementation

Another parameter, which is frequently used in process control, is the integral rate. This is defined as

$$r_i = \frac{1}{\tau_i}. \qquad\qquad (1.2)$$

The parameters k_p, τ_d, and τ_i or r_i are used in controller tuning.

The proportional action provides the necessary speed of response and adequate signal level to drive a plant. In addition, increased proportional action has the tendency to reduce steady-state error. A shortcoming of increased proportional action is the degradation of stability. Derivative action (or rate action) provides stability that is necessary for satisfactory performance of a control system. In the time domain, this is explained by the fact that the derivative action tends to oppose sudden changes (large rates) in the system response. Derivative control has its shortcomings, however. For example, if the error signal that drives the controller is constant, the derivative action will be zero and it has no effect on the system response. In particular, derivative control cannot reduce steady-state error in a system. In addition, derivative control increases the system bandwidth, which has the desirable effect of increasing the speed of response (and tracking capability) of a control system but has the drawback of allowing and amplifying high-frequency disturbance inputs and noise components. Hence, derivative action is not practically implemented in its pure analytic form, but rather as a lead circuit.

The presence of an offset (i.e., steady-state error) in the output may be inevitable when proportional control alone is used for a system that has finite dc gain. When there is an offset, one way to make the actual steady-state value equal to the desired value would be to change the set point (i.e., input value) in proportion to the desired change. This is known as manual reset. Note that this method does not actually remove the offset but rather changes the output value. Another way to bring the steady-state error to zero would be to make the dc gain infinity. This can be achieved by introducing an integral term (which has the transfer function $1/s$) in the forward path of the control system (because $1/s \to \infty$ when $s = 0$; i.e., at zero frequency). This is known as integral control or reset control or automatic reset. An alternative explanation for the behavior of the integral action is that an integrator can provide a constant output even when the error input is zero because the initial value of the integrated (accumulated) error signal will be maintained even after the error reaches zero. Integral control is known as reset control because it can reduce the offset to zero and can counteract external disturbances including load changes. A further advantage of integral control is its low-pass-filter action, which filters out high-frequency noise. However, since integral action cannot respond to sudden changes (rates) quickly, it has a destabilizing effect. For this reason integral control is practically implemented in conjunction with proportional control, in the form of proportional and integral (PI) control (a two-term controller), or also including derivative (D) control as PID control (a three-term controller).

1.3.2 Digital Control

In digital control, a digital computer serves as the controller. Virtually any control law may be programmed into the control computer. Control computers have to be fast and dedicated machines for real-time operations where processing has to be synchronized with plant operation and actuation requirements. This requires a real-time operating system. Apart from these requirements, control computers are basically not different from general-purpose digital computers. They consist of a processor to perform computations and to oversee data transfer; memory for program and data storage during processing; mass-storage devices to store information that is not immediately needed; and input or output devices to read in and send out information. Digital control systems might use digital instruments and additional processors for actuating, signal-conditioning, or measuring functions. For example, a stepper motor that responds with incremental motion steps when driven by pulse signals can be considered as a digital actuator (see Chapter 6). Furthermore, it usually contains digital logic circuitry in its drive system. Similarly, a two-position solenoid is a digital (binary) actuator. Digital flow control may be accomplished using a digital control valve. A typical digital valve consists of a bank of orifices, each sized in proportion to a place value of a binary word (2^i, $i = 0, 1, 2, \ldots, n$). Each orifice is actuated by a separate rapid-acting on/off solenoid. In this manner, many digital combinations of flow values can be obtained. Direct digital measurement of displacements and velocities can be made using shaft encoders (see Chapter 5). These are digital transducers that generate coded outputs (e.g., in binary or gray-scale representation) or pulse signals that can be coded using counting circuitry. Such outputs can be read in by the control computer with relative ease. Frequency counters also generate digital signals that can be fed directly into a digital controller. When measured signals are in the analog form, an analog front-end is necessary to interface the transducer and the digital controller. Input/output interface cards that can take both analog and digital signals are available with digital controllers.

Analog measurements and reference signals have to be sampled and encoded before digital processing within the controller. Digital processing can be effectively used for signal conditioning as well. Alternatively, digital signal processing (DSP) chips can function as digital controllers. However, analog signals have to be preconditioned using analog circuitry before digitizing in order to eliminate or minimize problems due to aliasing distortion (high-frequency components above half the sampling frequency appearing as low-frequency components) and leakage (error due to signal truncation) as well as to improve the signal level and filter out extraneous noise (see Chapter 3). The drive system of a plant typically takes in analog signals. Often, the digital output from the controller has to be converted into analog form for this reason. Both analog-to-digital conversion (ADC) and digital-to-analog conversion (DAC) can be interpreted as signal-conditioning (modification) procedures (see Chapter 2). If more than one output signal is measured, each signal will have to be conditioned and processed separately. Ideally, this will require separate conditioning and processing hardware for each signal channel. A less expensive (but slower) alternative would be to time-share this expensive equipment by using a multiplexer. This device will pick one channel of data from a bank of data channels in a sequential manner and connect it to a common input device.

The current practice of using dedicated, microprocessor-based, and often decentralized (i.e., distributed) digital control systems in industrial applications can be rationalized in terms of the major advantages of digital control. The following are some of the important considerations.

1. Digital control is less susceptible to noise or parameter variation in instrumentation because data can be represented, generated, transmitted, and processed as binary words, with bits possessing two identifiable states.
2. Very high accuracy and speed are possible through digital processing. Hardware implementation is usually faster than software implementation.
3. Digital control systems can handle repetitive tasks extremely well, through programming.

4. Complex control laws and signal-conditioning methods that might be impractical to implement using analog devices can be programmed.

5. High reliability in operation can be achieved by minimizing analog hardware
   components and through decentralization using dedicated microprocessors for
   various control tasks.

6. Large amounts of data can be stored using compact, high-density data-storage
   methods.

7. Data can be stored or maintained for very long periods of time without drift
   and without getting affected by adverse environmental conditions.

8. Fast data transmission is possible over long distances without introducing
   excessive dynamic delays and attenuation, as in analog systems.

9. Digital control has easy and fast data retrieval capabilities.

10. Digital processing uses low operational voltages (e.g., 0 to 12 V DC).

11. Digital control is cost-effective.

1.3.3 Feed-Forward Control

Many control systems have inputs that do not participate in feedback control. In other
words, these inputs are not compared with feedback (measurement) signals to generate
control signals. Some of these inputs might be important variables in the plant
(i.e., process) itself. Others might be undesirable inputs, such as external disturbances,
which are unwanted yet unavoidable. Generally, the performance of a control system can
be improved by measuring these (unknown) inputs and somehow using the information
to generate control signals.

   In feedforward control, unknown inputs are measured and that information, along
with desired inputs, is used to generate control signals that can reduce errors due to these
unknown inputs or variations in them. The reason for calling this method feedforward
control stems from the fact that the associated measurement and control (and compensation) take place in the forward path of the control system. Note that in feedback control,
unknown outputs are measured and compared with known (desired) inputs to generate
control signals. Both feedback and feedforward schemes may be used in the same control
system.

   A block diagram of a typical control system that uses feedforward control is shown in
Figure 1.5. In this system, in addition to feedback control, a feedforward control scheme
is used to reduce the effects of a disturbance input that enters the plant. The disturbance
input is measured and fed into the controller. The controller uses this information to
modify the control action so as to compensate for the disturbance input, anticipating its
effect.

   As a practical example, consider the natural gas home heating system shown in Figure
1.6a. A simplified block diagram of the system is shown in Figure 1.6b. In conventional
feedback control, the room temperature is measured and its deviation from the desired
temperature (set point) is used to adjust the natural gas flow into the furnace. On/off
control through a thermostat is used in most such applications. Even if proportional
or three-mode (PID) control is employed, it is not easy to steadily maintain the room
temperature at the desired value if there are large changes in other (unknown) inputs to
the system, such as water flow rate through the furnace, temperature of water
entering the furnace, and outdoor temperature. Better results can be obtained by measuring these disturbance inputs and using that information in generating the control

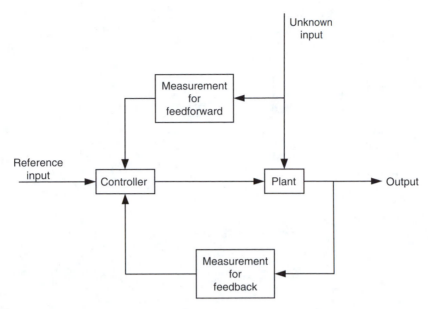

FIGURE 1.5
A system with feedback and feedforward control.

action. This is feedforward control. Note that in the absence of feedforward control, any changes in the inputs w_1, w_2, and w_3 in Figure 1.6 would be detected only through their effect on the feedback signal (i.e., room temperature). Hence, the subsequent corrective action can considerably lag behind the cause (i.e., changes in w_i). This delay will lead to large errors and possible instability problems. With feedforward control, information on the disturbance input w_i will be available to the controller immediately, and its effect on the system response can be anticipated, thereby speeding up the control action and also improving the response accuracy. Faster action and improved accuracy are two very desirable effects of feedforward control.

1.3.4 Programmable Logic Controllers

A programmable logic controller (PLC) is essentially a digital-computer-like system that can properly sequence a complex task, consisting of many discrete operations and involving several devices, which needs to be carried out in a sequential manner. PLCs are rugged computers typically used in factories and process plants, to connect input devices such as switches to output devices such as valves, at high speed at appropriate times in a task, as governed by a program. Internally, a PLC performs basic computer functions such as logic, sequencing, timing, and counting. It can carry out simpler computations and control tasks such as PID control. Such control operations are called continuous-state control, where process variables are continuously monitored and made to stay close to desired values. There is another important class of controls, known as discrete-state control, where the control objective is for the process to follow a required sequence of states (or steps). In each state, however, some form of continuous-state control might be operated, but it is not relevant to the discrete-state control task. PLCs are particularly intended for accomplishing discrete-state control tasks.

There are many control systems and industrial tasks that involve the execution of a sequence of steps, depending on the state of some elements in the system and on some

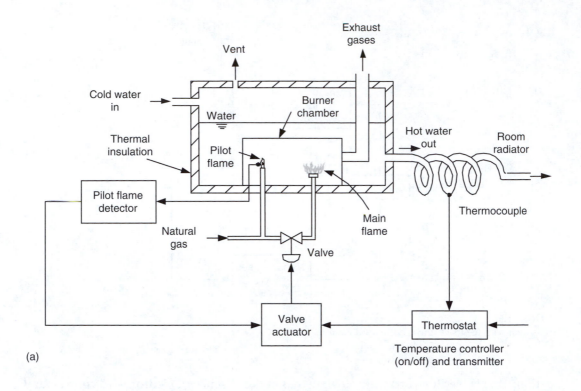

(a)

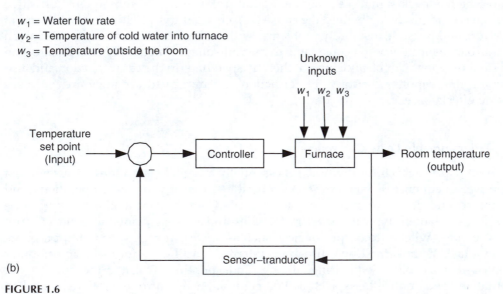

w_1 = Water flow rate
w_2 = Temperature of cold water into furnace
w_3 = Temperature outside the room

(b)

FIGURE 1.6
(a) A natural gas home heating system. (b) A block diagram representation of the system.

external input states. For example, consider an operation of turbine blade manufacture. The discrete steps in this operation might be:

1. Move the cylindrical steel billets into furnace.
2. Heat the billets.

3. When a billet is properly heated, move it to the forging machine and fix it.

4. Forge the billet into shape.

5. Perform surface finishing operations to get the required aerofoil shape.

6. When the surface finish is satisfactory, machine the blade root.

Note that the entire task involves a sequence of events where each event depends on the completion of the previous event. In addition, it may be necessary for each event to start and end at specified time instants. Such time sequencing would be important for coordinating the operation with other activities and perhaps for proper execution of each operation step. For example, activities of the parts-handling robot have to be coordinated with the schedules of the forging machine and milling machine. Furthermore, the billets have to be heated for a specified time, and the machining operation cannot be rushed without compromising product quality, tool failure rate, safety, and so on. Note that the task of each step in the discrete sequence might be carried out under continuous-state control. For example, the milling machine would operate using several direct digital control (DDC) loops (say, PID control loops), but discrete-state control is not concerned with this except for the starting point and the end point of each task.

A process operation might consist of a set of two-state (on/off) actions. A PLC can handle the sequencing of these actions in a proper order and at correct times. Examples of such tasks include sequencing the production line operations, starting a complex process plant, and activating the local controllers in a distributed control environment. In the early days of industrial control, solenoid-operated electromechanical relays, mechanical timers, and drum controllers were used to sequence such operations. An advantage of using a PLC is that the devices in a plant can be permanently wired, and the plant operation can be modified or restructured by software means (by properly programming the PLC) without requiring hardware modifications and reconnection.

A PLC operates according to some logic sequence programmed into it. Connected to a PLC are a set of input devices (e.g., pushbuttons, limit switches, and analog sensors such as RTD temperature sensors, diaphragm-type pressure sensors, piezoelectric accelerometers, and strain-gauge load sensors) and a set of output devices (e.g., actuators such as dc motors, solenoids, and hydraulic rams, warning signal indicators such as lights, alphanumeric light emitting diode (LED) displays and bells, valves, and continuous control elements such as PID controllers). Each such device is assumed to be a two-state device (taking the logical value 0 or 1). Now, depending on the condition of each input device and according to the programmed-in logic, the PLC will activate the proper state (e.g., on/off) of each output device. Hence, the PLC performs a switching function. Unlike the older generation of sequencing controllers, in the case of a PLC, the logic that determines the state of each output device is processed using software and not by hardware elements such as hardware relays. Hardware switching takes place at the output port, however, for turning on/off the output devices controlled by the PLC.

1.3.4.1 PLC Hardware

As noted before, a PLC is a digital computer that is dedicated to perform discrete-state control tasks. A typical PLC consists of a microprocessor, RAM and ROM memory units, and interface hardware, all interconnected through a suitable bus structure. In addition, there will be a keyboard, a display screen, and other common peripherals. A basic PLC system can be expanded by adding expansion modules (memory, I/O modules, etc.) into the system rack.

A PLC can be programmed using a keyboard or touchscreen. An already developed program could be transferred into the PLC memory from another computer or a peripheral mass-storage medium such as hard disc. As noted before, the primary function of a PLC is to switch (energize or de-energize) the output devices connected to it, in a proper sequence, depending on the states of the input devices and according to the logic dictated by the program. A schematic representation of a PLC is shown in Figure 1.7. Note the sensors and actuators in the PLC.

In addition to turning on and off the discrete output components in a correct sequence at proper times, a PLC can perform other useful operations. In particular, it can perform simple arithmetic operations such as addition, subtraction, multiplication, and division on input data. It is also capable of performing counting and timing operations, usually as part of its normal functional requirements. Conversion between binary and binary-coded decimal (BCD) might be required for displaying digits on an LED panel and for interfacing the PLC with other digital hardware (e.g., digital input devices and digital output devices). For example, a PLC can be programmed to make temperature and load measurements, display them on an LED panel, make some computations on these (input) values, and provide a warning signal (output) depending on the result.

The capabilities of a PLC can be determined by such parameters as the number of input devices (e.g., 16) and the number of output devices (e.g., 12) it can handle, the number of program steps (e.g., 2000), and the speed at which a program can be executed (e.g., 1 M steps/s). Other factors such as the size and nature of memory and the nature of timers and counters in the PLC, signal voltage levels, and choices of outputs, are all important factors.

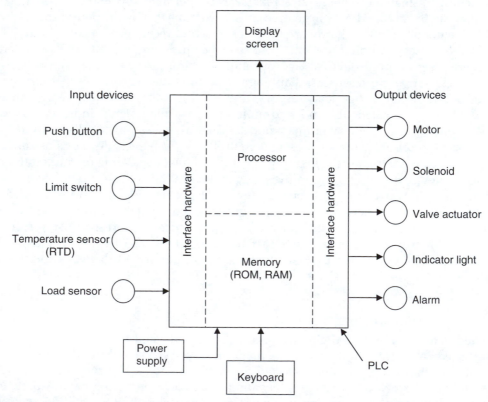

FIGURE 1.7
Schematic representation of a PLC.

1.3.5 Distributed Control

For complex processes with a large number of input or output variables (e.g., a chemical plant and a nuclear power plant) and with systems that have various and stringent operating requirements (e.g., the space shuttle), centralized DDC is quite difficult to implement. Some form of distributed control is appropriate in large systems such as manufacturing workcells, factories, and multicomponent process plants. A distributed control system (DCS) will have many users who would need to use the resources simultaneously and, perhaps, would wish to communicate with each other as well. Moreover, the plant will need access to shared and public resources and means of remote monitoring and supervision. Furthermore, different types of devices from a variety of suppliers with different specifications, data types, and levels may have to be interconnected. A communication network with switching nodes and multiple routes is needed for this purpose.

In order to achieve connectivity between different types of devices with different origins, it is desirable to use a standardized bus that is supported by all major suppliers of the needed devices. The Foundation Fieldbus or Industrial Ethernet may be adopted for this purpose. Fieldbus is a standardized bus for a plant, which may consist of an interconnected system of devices. It provides connectivity between different types of devices with different origins. In addition, it provides access to shared and public resources. Furthermore, it can provide means for remote monitoring and supervision.

A suitable architecture for networking an industrial plant is shown in Figure 1.8. The industrial plant in this case consists of many process devices (PD), one or more PLCs and a DSC or a supervisory controller. The PDs will have direct I/O with their own components, while possessing connectivity through the plant network. Similarly, a PLC may have direct connectivity with a group of devices as well as networked connectivity with other devices. The DSC will supervise, manage, coordinate, and control the overall plant.

1.3.5.1 A Networked Application

A machine that we developed for head removal of salmon is shown in Figure 1.9. The conveyor, driven by an ac motor, indexes the fish in an intermittent manner. Image of each fish, obtained using a charge-coupled device (CCD) camera, is processed to determine the geometric features, which in turn establish the proper cutting location. A two-axis hydraulic drive positions the cutter accordingly, and the cutting blade is operated using a pneumatic actuator. Position sensing of the hydraulic manipulator is carried out using linear magnetostrictive displacement transducers, which have a resolution of 0.025 mm

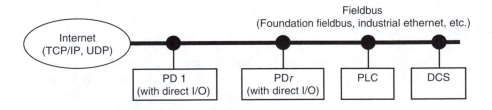

PD = Process device
PLC = Programmable logic controller
DCS = Distributed control system (Supervisory controller)

FIGURE 1.8
A networked industrial plant.

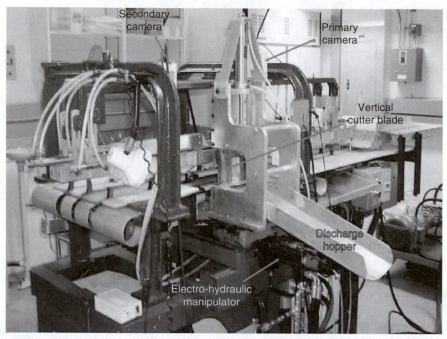

FIGURE 1.9
An intelligent iron butcher.

when used with a 12-bit ADC. A set of six gage-pressure transducers are installed to measure the fluid pressure in the head and rod sides of each hydraulic cylinder and also in the supply lines. A high-level imaging system determines the cutting quality, according to which adjustments may be made online, to the parameters of the control system to improve the process performance. The control system has a hierarchical structure with conventional direct control at the component level (low level) and an intelligent monitoring and supervisory control system at an upper level.

The primary vision module of the machine is responsible for fast and accurate detection of the gill position of a fish, on the basis of an image of the fish captured by the primary CCD camera. This module is located in the machine host and comprises a CCD camera for image acquisition, an ultrasonic sensor for thickness measurement of fish, a trigger switch for detecting a fish on the conveyor, GPB-1 image processing board for frame grabbing and image analysis, and a PCL-I/O card for digital data communication with the control computer of the electrohydraulic manipulator. This vision module is capable of reliably detecting and computing the cutting locations in approximately 300 to 400 ms. The secondary vision module is responsible for acquisition and processing of visual information pertaining to the quality of processed fish that leaves the cutter assembly. This module functions as an intelligent sensor in providing high-level information feedback into the control computer. The hardware associated with this module are a CCD camera at the exit end for grabbing images of processed fish, and a GPB-1 image processing board for visual data analysis. The CCD camera acquires images of processed fish under the direct control of the host computer, which determines the proper instance to trigger the camera by timing the duration of the cutting operation. The image is then transferred to the image buffer in the GPB board for further processing. In this case, however, image processing is accomplished to extract high-level information, such as the quality of processed fish.

With the objective of monitoring and controlling industrial processes from remote locations, we have developed a universal network architecture, both hardware and

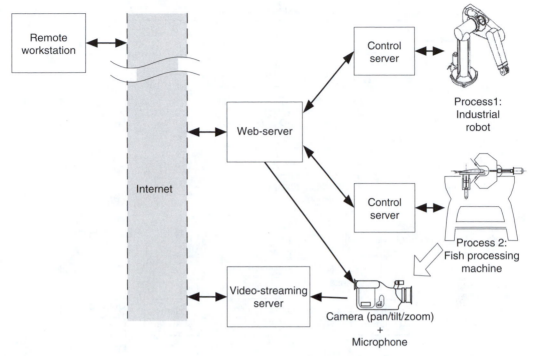

FIGURE 1.10
Network hardware architecture.

software. The developed infrastructure is designed to perform optimally with Fast Ethernet (100Base-T) backbone where each network device needs only a low-cost network interface card (NIC). Figure 1.10 shows a simplified hardware architecture, which networks two machines (a fish-processing machine and an industrial robot). Each machine is directly connected to its individual control server, which handles networked communication between the process and the web-server, data acquisition, sending of control signals to the process, and the execution of low-level control laws. The control server of the fish-processing machine contains one or more data acquisition boards, which have ADC, DAC, digital I/O, and frame grabbers for image processing.

Video cameras and microphones are placed at strategic locations to capture live audio and video signals allowing the remote user to view and listen to a process facility, and to communicate with local research personnel. The camera selected in the present application is the Panasonic Model KXDP702 color camera with built-in pan, tilt, and 21× zoom (PTZ), which can be controlled through a standard RS-232C communication protocol. Multiple cameras can be connected in a daisy-chained manner to the video-streaming server. For capturing and encoding the audio–video (AV) feed from the camera, the Winnov Videum 1000 PCI board is installed in the video-streaming server. It can capture video signals at a maximum resolution of 640 × 480 at 30 fps, with a hardware compression that significantly reduces computational overheads of the video-streaming server. Each of the AV capture boards can support only one AV input. Hence, multiple boards have to be installed.

1.3.6 Hierarchical Control

A favorite distributed control architecture is provided by hierarchical control. Here, distribution of control is available both geographically and functionally. A hierarchical structure can facilitate efficient control and communication in a complex control system.

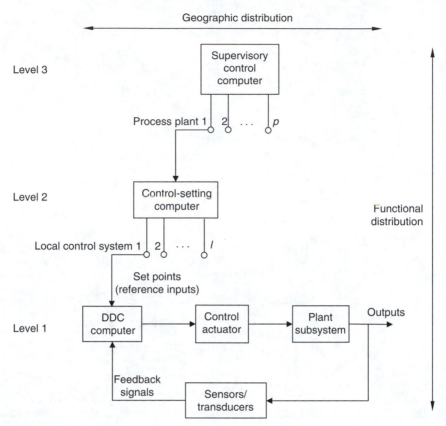

FIGURE 1.11
A three-layer hierarchical control system.

An example for a three-level hierarchy is shown in Figure 1.11. Management decisions, supervisory control, and coordination between plants in the overall facility are provided by the supervisory control computer, which is at the highest level (level 3) of the hierarchy. The next lower level (intermediate level) generates control settings (or reference inputs) for each control region (subsystem) in the corresponding plant. Set points and reference signals are inputs to the computer DDC, which control each control region. The computers in the hierarchical system communicate using a suitable communication network. Information transfer in both directions (up and down) should be possible for best performance and flexibility. In master-slave distributed control, only downloading of information is available.

As another illustration, a three-level hierarchy of an intelligent mechatronic (electro-mechanical) system (IMS) is shown in Figure 1.12. The bottom level consists of electromechanical components with component-level sensing. Furthermore, actuation and direct feedback control are carried out at this level. The intermediate level uses intelligent preprocessors for abstraction of the information generated by the component-level sensors. The sensors and their intelligent preprocessors together perform tasks of intelligent sensing. The state of performance of the system components may be evaluated by this means, and component tuning and component-group control may be carried out as a result. The top level of the hierarchy performs task-level activities including planning, scheduling, monitoring of the system performance, and overall supervisory control. Resources such as materials and expertise may be provided at this level and a human–machine interface would be available. Knowledge-based decision making is carried out at both intermediate and top levels. The resolution of the information that is involved

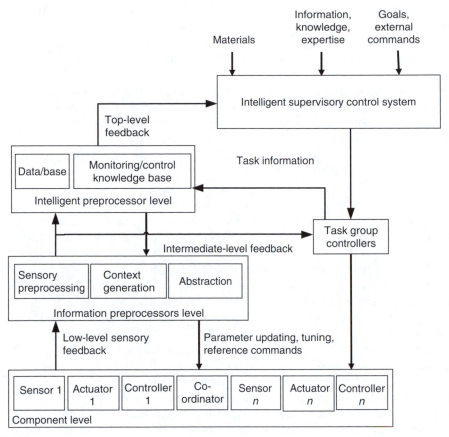

FIGURE 1.12
A hierarchical control and communications structure for an intelligent mechatronic system (IMS).

will generally decrease as the hierarchical level increases, whereas the level of intelligence that would be needed in decision making will increase.

Within the overall system, the communication protocol provides a standard interface between various components such as sensors, actuators, signal conditioners, and controllers, and also with the system environment. The protocol will not only allow highly flexible implementations, but will also enable the system to use distributed intelligence to perform preprocessing and information understanding. The communication protocol should be based on an application-level standard. In essence, it should outline what components can communicate with each other and with the environment, without defining the physical data link and network levels. The communication protocol should allow for different component types and different data abstractions to be interchanged within the same framework. It should also allow for the information from geographically removed locations to be communicated to the control and communication system of the IMS.

1.4   Organization of the Book

The book consists of eight chapters. The chapters are devoted to presenting the fundamentals, analytical concepts, modeling and design issues, technical details, and applications of sensors and actuators within the framework of control system instrumentation.

The book uniformly incorporates the underlying fundamentals as analytical methods, modeling approaches, and design techniques in a systematic manner throughout the main chapters. The practical application of the concepts, approaches, and tools presented in the introductory chapters is demonstrated through numerous illustrative examples and a comprehensive set of case studies.

Chapter 1 introduces the field of control engineering with the focus on instrumentation using sensors and actuators. The relevance of modeling and design in the context of instrumentation is indicated. Common control system architectures are described and the role played by sensors and actuators in these architectures is highlighted. This introductory chapter sets the tone for the study, which spans the remaining seven chapters. Relevant publications in the field are listed.

Chapter 2 presents component interconnection and signal conditioning, which is in fact a significant unifying subject within control system instrumentation. Impedance considerations of component interconnection and matching are studied. Amplification, filtering, ADC, DAC, bridge circuits, and other signal conversion and conditioning techniques and devices are discussed.

Chapter 3 covers performance analysis of a device, component, or instrument within a control system. Methods of performance specification are addressed, in both time domain and frequency domain. Common instrument ratings that are used in industry and generally in engineering practice are discussed. Related analytical methods are given. Instrument bandwidth considerations are highlighted, and a design approach based on component bandwidth is presented. Errors in digital devices, particularly resulting from signal sampling, are discussed from analytical and practical points of view.

Chapter 4 presents important types, characteristics, and operating principles of analog sensors. Particular attention is given to sensors that are commonly used in control engineering practice. Motion sensors, force, torque and tactile sensors, optical sensors, ultrasonic sensors, temperature sensors, pressure sensors, and flow sensors are discussed. Analytical basis, selection criteria, and application areas are indicated.

Chapter 5 discusses common types of digital transducers. Unlike analog sensors, digital transducers generate pulses, counts, or digital outputs. These devices have clear advantages, particularly when used in computer-based digital systems. They do possess quantization errors, which are unavoidable in digital representation of an analog quantity. Related issues of accuracy and resolution are addressed.

Chapter 6 studies stepper motors, which are an important class of actuators. These actuators produce incremental motions. Under satisfactory operating conditions, they have the advantage or the ability to generate a specified motion profile in an open-loop manner without requiring motion sensing and feedback control. However, under some conditions of loading and response, motion steps may be missed. Consequently, it is appropriate to use sensing and feedback control when complex motion trajectories need to be followed under nonuniform and extreme loading conditions.

Chapter 7 presents continuous-drive actuators such as dc motors, ac motors, hydraulic actuators, and pneumatic actuators. Common varieties of actuators under each category are discussed. Operating principles, analytical methods, design considerations, selection methods, drive systems, and control techniques are described. Advantages and drawbacks of various types of actuators on the basis of the nature and the needs of an application are discussed. The subject of fluidics is introduced. Practical examples are given.

Chapter 8 concerns mechanical components that may be employed to connect an actuator to a mechanical load. These transmission devices serve as means of component matching as well, for proper actuation of a mechanical load.

Problems

1.1 a. What are open-loop control systems and what are feedback control systems? Give one example of each case.

b. A simple mass-spring-damper system (simple oscillator) is excited by an external force $f(t)$. Its displacement response y (see Figure P1.1a) is given by the differential equation

$$m\ddot{y} + b\dot{y} + ky = f(t).$$

A block diagram representation of this system is shown in Figure P1.1b. Is this a feedback control system? Explain and justify your answer.

1.2 You are asked to design a control system to turn on lights in an art gallery at night, provided there are people inside the gallery. Explain a suitable control system, identifying the open-loop and feedback functions, if any, and describing the control system components.

1.3 Into what classification of control system components: actuators, signal modification devices, controllers, and measuring devices would you put the following devices?

a. Stepping motor

b. Proportional-plus-integration circuit

c. Power amplifier

d. ADC

e. DAC

f. Optical incremental encoder

g. Process computer

h. FFT analyzer

i. Digital signal processor (DSP).

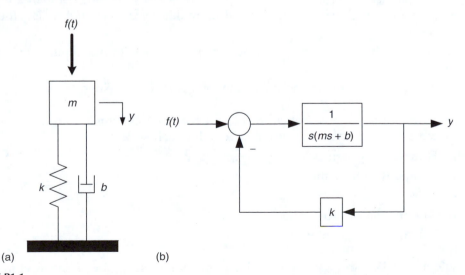

(a)   (b)

FIGURE P1.1
(a) A mechanical system representing a simple oscillator. (b) A block diagram representation of the simple oscillator.

1.4   a. Discuss possible sources of error that can make either open-loop control or feedforward control meaningless in some applications.

b. How would you correct the situation?

1.5   Compare analog control and direct digital control (DDC) for motion control in high-speed applications of industrial manipulators. Give some advantages and disadvantages of each control method for this application.

1.6   A soft-drink bottling plant uses an automated bottle-filling system. Describe the operation of such a system, indicating various components in the control system and their functions. Typical components would include a conveyor belt; a motor for the conveyor, with start/stop controls; a measuring cylinder, with an inlet valve, an exit valve, and level sensors; valve actuators; and an alignment sensor for the bottle and the measuring cylinder.

1.7   Consider the natural gas home heating system shown in Figure 1.6. Describe the functions of various components in the system and classify them into the functional groups: controller, actuator, sensor, and signal modification device. Explain the operation of the overall system and suggest possible improvements to obtain more stable and accurate temperature control.

1.8   In each of the following examples, indicate at least one (unknown) input that should be measured and used for feedforward control to improve the accuracy of the control system.

a. A servo system for positioning a mechanical load. The servo motor is a field-controlled dc motor, with position feedback using a potentiometer and velocity feedback using a tachometer.

b. An electric heating system for a pipeline carrying liquid. The exit temperature of the liquid is measured using a thermocouple and is used to adjust the power of the heater.

c. A room heating system. Room temperature is measured and compared with the set point. If it is low, the valve of a steam radiator is opened; if it is high, the valve is shut.

d. An assembly robot that grips a delicate part to pick it up without damaging the part.

e. A welding robot that tracks the seam of a part to be welded.

1.9   A typical input variable is identified for each of the following examples of dynamic systems. Give at least one output variable for each system.

a. Human body: neuroelectric pulses

b. Company: information

c. Power plant: fuel rate

d. Automobile: steering wheel movement

e. Robot: voltage to joint motor.

1.10   Hierarchical control has been applied in many industries, including steel mills, oil refineries, chemical plants, glass works, and automated manufacturing. Most applications have been limited to two or three levels of hierarchy, however. The lower levels usually consist of tight servo loops, with bandwidths in the order of 1 kHz. The upper levels typically control production planning and scheduling events measured in units of days or weeks.

A five-level hierarchy for a flexible manufacturing facility is as follows: The lowest level (level 1) handles servo control of robotic manipulator joints and machine tool degrees of freedom. The second level performs activities such as coordinate transformation in machine tools, which are required in generating control commands for various servo loops. The third level converts task commands into motion trajectories (of manipulator end effector, machine tool bit, etc.) expressed in world coordinates. The fourth level converts complex and general task commands into simple task commands. The top level (level 5) performs supervisory control tasks for various machine tools and material-handling devices, including coordination, scheduling, and definition of basic moves. Suppose that this facility is used as a flexible manufacturing workcell for turbine blade production. Estimate the event duration at the lowest level and the control bandwidth (in hertz) at the highest level for this type of application.

1.11 According to some observers in the process control industry, early brands of analog control hardware had a product life of about 20 years. New hardware controllers can become obsolete in a couple of years, even before their development costs are recovered. As a control instrumentation engineer responsible for developing an off-the-shelf process controller, what features would you incorporate into the controller to correct this problem to a great extent?

1.12 The PLC is a sequential control device, which can sequentially and repeatedly activate a series of output devices (e.g., motors, valves, alarms, and signal lights) on the basis of the states of a series of input devices (e.g., switches, two-state sensors). Show how a programmable controller and a vision system consisting of a solid-state camera and a simple image processor (say, with an edge-detection algorithm) could be used for sorting fruits on the basis of quality and size for packaging and pricing.

1.13 Measuring devices (sensors, transducers) are useful in measuring outputs of a process for feedback control.

a. Give other situations in which signal measurement would be important.

b. List at least five different sensors used in an automobile engine.

1.14 One way to classify controllers is to consider their sophistication and physical complexity separately. For instance, we can use an x–y plane with the x-axis denoting the physical complexity and the y-axis denoting the controller sophistication. In this graphical representation, simple open-loop on/off controllers (say, opening and closing a valve) would have a very low controller sophistication value and an artificial-intelligence (AI)-based intelligent controller would have a high controller sophistication value. Moreover, a passive device is considered to have less physical complexity than an active device. Hence, a passive spring-operated device (e.g., a relief valve) would occupy a position very close to the origin of the x–y plane and an intelligent machine (e.g., sophisticated robot) would occupy a position diagonally far from the origin. Consider five control devices of your choice. Mark the locations that you expect them to occupy (in relative terms) on this classification plane.

1.15 You are a control engineer who has been assigned the task of designing and instrumenting a control system. In the final project report, you have to describe the steps of establishing the design/performance specifications for the system, selecting and sizing sensors, transducers, actuators, drive systems, controllers, signal conditioning and interface hardware, and software for the instrumentation and component integration of this system. Keeping this in mind, write a project proposal giving the following information:

1. Select a process (plant) as the system to be developed. Describe the plant indicating the purpose of the plant, how the plant operates, what is the system boundary (physical or imaginary), what are the important inputs (e.g., voltages, torques, heat transfer rates, flow rates), response variables (e.g., displacements, velocities, temperatures, pressures, currents, voltages), and what are important plant parameters (e.g., mass, stiffness, resistance, inductance, conductivity, fluid capacity). You may use sketches.

2. Indicate the performance requirements (or operating specifications) for the plant (i.e., how the plant should behave under control). You may use any available information on such requirements as accuracy, resolution, speed, linearity, stability, and operating bandwidth.

3. Give any constraints related to cost, size, weight, environment (e.g., operating temperature, humidity, dust-free or clean room conditions, lighting, and wash-down needs), and so on.

4. Indicate the type and the nature of the sensors and transducers present in the plant and what additional sensors and transducers might be needed to properly operate and control the system.

5. Indicate the type and the nature of the actuators and drive systems present in the plant and which of these actuators have to be controlled. If you need to add new actuators (including control actuators) and drive systems, indicate such requirements in detail.

6. Mention what types of signal modification and interfacing hardware would be needed (i.e., filters, amplifiers, modulators, demodulators, ADC, DAC, and other data acquisition and control needs). Describe the purpose of these devices. Indicate any software (e.g., driver software) that may be needed along with this hardware.

7. Indicate the nature and operation of the controllers in the system. State whether these controllers are adequate for your system. If you intend to add new controllers briefly give their nature, characteristics, objectives, and so on (e.g., analog, digital, linear, nonlinear, hardware, software, control bandwidth).

8. Describe how the users and operators interact with the system, and the nature of the user interface requirements (e.g., graphic user interface or GUI).

The following plants or systems may be considered:

1. A hybrid electric vehicle
2. A household robot
3. A smart camera
4. A smart airbag system for an automobile
5. Rover mobile robot for Mars exploration, developed by NASA
6. An automated guided vehicle (AGV) for a manufacturing plant
7. A flight simulator

8. A hard disc drive for a personal computer
9. A packaging and labeling system for a grocery item
10. A vibration testing system (electrodynamic or hydraulic)
11. An active orthotic device to be worn by a person to assist a disabled or weak hand (which has some sensation, but not fully functional)

2

Component Interconnection and Signal Conditioning

A control system is typically a mixed system, which consists of more than one type of components properly interconnected and integrated. In particular, mechanical, electrical, electronic, and computer hardware are integrated to form a control system. It follows that component interconnection is an important topic in the field of control engineering. When two components are interconnected, signals flow through them. The nature and type of the signals that are present at the interface of two components depends on the nature and type of the components. For example, when a motor is coupled with a load through a gear (transmission) unit, mechanical power flows at the interfaces of these components. Therefore, we are particularly interested in such signals as angular velocity and torque. In particular, these signals would be modified or conditioned as they are transmitted through the gear transmission. Similarly, when a motor is connected to its electronic drive system, command signals of motor control, typically available as voltages, would be converted into appropriate currents for energizing the motor windings so as to generate the necessary torque. Again, signal conditioning or conversion is important here. In general, then, signal conditioning is important in the context of component interconnection and integration, and becomes an important subject in the study of control engineering.

This chapter addresses interconnection of components such as sensors, signal conditioning circuitry, actuators, and power transmission devices in a control system. Desirable impedance characteristics for such components are discussed. Signal modification plays a crucial role in component interconnection or interfacing. When two devices are interfaced, it is essential to guarantee that a signal leaving one device and entering the other will do so at proper signal levels (i.e., the values of voltage, current, speed, force, power, etc.), in the proper form (i.e., electrical, mechanical, analog, digital, modulated, demodulated, etc.), and without distortion (specifically, loading problems, nonlinearities, and noise have to be eliminated, and in this context impedance considerations become important). Particularly for transmission, a signal should be properly modified by amplification, modulation, digitizing, and so on, so that the signal/noise ratio of the transmitted signal is sufficiently large at the receiver. The significance of signal modification is clear from these observations.

The tasks of signal modification may include signal conditioning (e.g., amplification and analog and digital filtering), signal conversion [e.g., analog-to-digital conversion (ADC), digital-to-analog conversion (DAC), voltage-to-frequency conversion, and frequency-to-voltage conversion], modulation [e.g., amplitude modulation (AM), frequency modulation (FM), phase modulation (PM), pulse-width modulation (PWM), pulse-frequency modulation (PFM), and pulse-code modulation (PCM)], and demodulation (i.e., the reverse process of modulation). In addition, many other types of useful signal modification operations can be identified. For example, sample-and-hold circuits (S/H) are used in digital data acquisition systems. Devices such as analog and digital multiplexers and comparators are needed in many applications of data acquisition and processing. Phase shifting, curve shaping, offsetting, and linearization can also be classified as signal modification.

This chapter describes signal conditioning and modification operations that are useful in control applications. The operational amplifier (op-amp) is introduced as a basic element in signal conditioning and impedance matching circuitry for electronic systems. Various types of signal conditioning and modification devices such as amplifiers, filters, modulators, demodulators, bridge circuits, ADCs, and DACs are discussed.

2.1 Component Interconnection

A control system can consist of a wide variety of components that are interconnected to perform the intended functions. When two or more components are interconnected, the behavior of the individual components in the integrated system can deviate significantly from that when each component operates independently. Matching of components in a multicomponent system, particularly with respect to their impedance characteristics, should be done carefully to improve the system performance and accuracy. In this chapter, we first study basic concepts of impedance and component matching. The concepts presented here are applicable to many types of components in a general control system. Discussions and developments given here can be quite general. Nevertheless, specific hardware components and designs are considered as examples in relation to component interfacing and signal conditioning.

2.2 Impedance Characteristics

When components such as sensors and transducers, control boards, process (i.e., plant) equipment, and signal-conditioning hardware are interconnected, it is necessary to match impedances properly at each interface to realize their rated performance levels. One adverse effect of improper impedance matching is the loading effect. For example, in a measuring system, the measuring instrument can distort the signal that is measured. The resulting error can far exceed other types of measurement error. Both electrical and mechanical loading are possible. Electrical loading errors result from connecting an output unit such as a measuring device that has a low input impedance to an input device such as a signal source. Mechanical loading errors can result in an input device because of inertia, friction, and other resistive forces generated by an interconnected output component (i.e., a mechanical load).

Impedance can be interpreted either in the traditional electrical sense or in the mechanical sense, depending on the type of signals involved. For example, a heavy accelerometer can introduce an additional dynamic (mechanical) load, which will modify the actual acceleration at the monitoring location. Similarly, a voltmeter can modify the currents (and voltages) in a circuit, and a thermocouple junction can modify the temperature that is measured as a result of the heat transfer into the junction. In mechanical and electrical systems, loading errors can appear as phase distortions as well. Digital hardware also can produce loading errors. For example, an ADC board can load the amplifier output from a strain gage bridge circuit, thereby affecting digitized data.

Another adverse effect of improper impedance consideration is inadequate output signal levels, which make the output functions such as signal processing and transmission, component driving, and actuation of a final control element or plant very difficult. In the context of sensor–transducer technology, it should be noted here that many types of transducers (e.g., piezoelectric accelerometers, impedance heads, and microphones) have high output impedances in the order of 1000 MΩ (megohm; 1 M$\Omega = 1 \times 10^6$ Ω). These devices generate low output signals, and they would require conditioning to step up the signal level. Impedance-matching amplifiers, which have high input impedances and low output impedances (a few ohms), are used for this purpose (e.g., charge amplifiers are used in conjunction with piezoelectric sensors). A device with a high input impedance has the further advantage that it usually consumes less power (i.e., v^2/R is low) for a given input voltage. The fact that a low input impedance device extracts a high level of power from the preceding output device may be interpreted as the reason for loading error.

2.2.1 Cascade Connection of Devices

Consider a standard two-port electrical device. The output impedance Z_o of such a device is defined as the ratio of the open-circuit (i.e., no-load) voltage at the output port to the short-circuit current at the output port. Open-circuit voltage at output is the output voltage present when there is no current flowing at the output port. This is the case if the output port is not connected to a load (impedance). As soon as a load is connected at the output of the device, a current flows through it, and the output voltage drops to a value less than that of the open-circuit voltage. To measure the open-circuit voltage, the rated input voltage is applied at the input port and maintained constant, and the output voltage is measured using a voltmeter that has a high (input) impedance. To measure the short-circuit current, a very low-impedance ammeter is connected at the output port.

The input impedance Z_i is defined as the ratio of the rated input voltage to the corresponding current through the input terminals while the output terminals are maintained in open circuit.

Note that these definitions are associated with electrical devices. A generalization is possible by interpreting voltage and velocity as "across variables", and current and force as "through variables". Then, mechanical mobility should be used in place of electrical impedance, in the associated analysis.

Using these definitions, input impedance Z_i and output impedance Z_o can be represented schematically as in Figure 2.1a. Note that v_o is the open-circuit output voltage. When a load is connected at the output port, the voltage across the load will be different from v_o. This is caused by the presence of a current through Z_o. In the frequency domain, v_i and v_o are represented by their respective *Fourier spectra*. The corresponding transfer relation can be expressed in terms of the complex frequency response (transfer) function $G(j\omega)$ under open-circuit (no-load) conditions:

$$v_o = G v_i. \tag{2.1}$$

Now consider two devices connected in cascade, as shown in Figure 2.1b. It can be easily verified that the following relations apply:

$$v_{o1} = G_1 v_i, \tag{2.2}$$

$$v_{i2} = \frac{Z_{i2}}{Z_{o1} + Z_{i2}} v_{o1}, \quad \text{and} \tag{2.3}$$

$$v_o = G_2 v_{i2}. \tag{2.4}$$

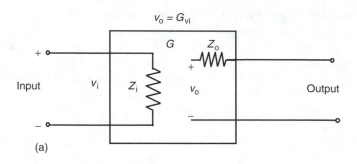

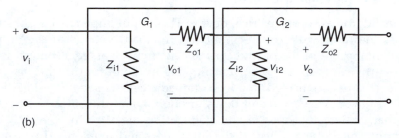

FIGURE 2.1
(a) Schematic representation of input impedance and output impedance. (b) Cascade connection of two two-port devices.

These relations can be combined to give the overall input/output relation

$$v_o = \frac{Z_{i2}}{Z_{o1} + Z_{i2}} G_2 G_1 v_i. \tag{2.5}$$

We see from Equation 2.5 that the overall frequency transfer function differs from the ideally expected product $(G_2 G_1)$ by the factor

$$\frac{Z_{i2}}{Z_{o1} + Z_{i2}} = \frac{1}{(Z_{o1}/Z_{i2}) + 1}. \tag{2.6}$$

Note that cascading has distorted the frequency response characteristics of the two devices. If $Z_{o1}/Z_{i2} \ll 1$, this deviation becomes insignificant. From this observation, it can be concluded that when frequency response characteristics (i.e., dynamic characteristics) are important in a cascaded device, cascading should be done such that the output impedance of the first device is much smaller than the input impedance of the second device.

Example 2.1
A lag network used as the compensation element of a control system is shown in Figure 2.2a. Show that its transfer function is given by

$$\frac{v_o}{v_i} = \frac{Z_2}{R_1 + Z_2},$$

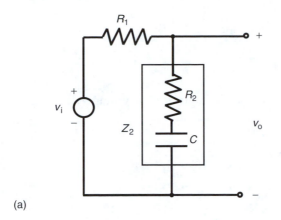

(a)

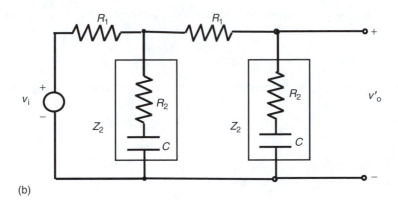

(b)

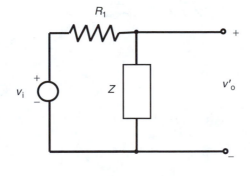

(c)

FIGURE 2.2
(a) A single circuit module. (b) Cascade connection of two modules. (c) An equivalent circuit for (b).

where

$$Z_2 = R_2 + \frac{1}{Cs}.$$

What are the input and output impedances of this circuit? Moreover, if two such lag circuits are cascaded as shown in Figure 2.2b, what is the overall transfer function? How would you bring this transfer function close to the ideal result

$$\left\{\frac{Z_2}{R_1 + Z_2}\right\}^2 ?$$

Solution

To solve this problem, first note that in Figure 2.2a, voltage drop across the element $R_2 + 1/(Cs)$ is

$$v_o = \frac{(R_2 + (1/Cs))}{\{R_1 + R_2 + (1/Cs)\}} v_i.$$

Hence,

$$\frac{v_o}{v_i} = \frac{Z_2}{R_1 + Z_2}.$$

Now, input impedance Z_i is derived by using input current $i = v_i/(R_1 + Z_2)$ as

$$Z_i = \frac{v_i}{i} = R_1 + Z_2,$$

and output impedance Z_o is derived by using short-circuit current $i_{sc} = v_i/R_1$ as

$$Z_o = \frac{v_o}{i_{sc}} = \frac{Z_2/(R_1 + Z_2)v_i}{v_i/R_1} = \frac{R_1 Z_2}{R_1 + Z_2}. \tag{i}$$

Next, consider the equivalent circuit shown in Figure 2.2c. Since Z is formed by connecting Z_2 and $(R_1 + Z_2)$ in parallel, we have

$$\frac{1}{Z} = \frac{1}{Z_2} + \frac{1}{R_1 + Z_2}. \tag{ii}$$

Voltage drop across Z is

$$v_o' = \frac{Z}{R_1 + Z} v_i. \tag{iii}$$

Now apply the single-circuit module result (i) to the second circuit stage in Figure 2.2b. Thus,

$$v_o = \frac{Z_2}{R_1 + Z_2} v_o'.$$

Substituting Equation iii, we get

$$v_o = \frac{Z_2}{(R_1 + Z_2)} \frac{Z}{(R_1 + Z)} v_i.$$

The overall transfer function for the cascaded circuit is

$$G = \frac{v_o}{v_i} = \frac{Z_2}{(R_1 + Z_2)} \frac{Z}{(R_1 + Z)} = \frac{Z_2}{(R_1 + R_2)} \frac{1}{(R_1/Z + 1)}.$$

Now, substituting Equation ii we get

$$G = \left[\frac{Z_2}{R_1 + Z_2}\right]^2 \frac{1}{1 + R_1 Z_2/(R_1 + Z_2)^2}.$$

We observe that the ideal transfer function is approached by making $R_1 Z_2/(R_1 + Z_2)^2$ small compared with unity.

2.2.2 Impedance Matching

When two electrical components are interconnected, current (and energy) flows between the two components and changes the original (unconnected) conditions. This is known as the (electrical) loading effect, and it has to be minimized. At the same time, adequate power and current would be needed for signal communication, conditioning, display, and so on. Both situations can be accommodated through proper matching of impedances when the two components are connected. Usually, an impedance-matching amplifier (i.e., an impedance transformer) would be needed between the two components.

From the analysis given in the preceding section, it is clear that the signal-conditioning circuitry should have a considerably large input impedance in comparison with the output impedance of the sensor–transducer unit to reduce loading errors. The problem is quite serious in measuring devices such as piezoelectric sensors, which have very high output impedances. In such cases, the input impedance of the signal-conditioning unit might be inadequate to reduce loading effects; also, the output signal level of these high-impedance sensors is quite low for signal transmission, processing, actuation, and control. The solution for this problem is to introduce several stages of amplifier circuitry between the output of the first hardware unit (e.g., sensor) and the input of the second hardware unit (e.g., data acquisition unit). The first stage of such an interfacing device is typically an impedance-matching amplifier that has high input impedance, low output impedance, and almost unity gain. The last stage is typically a stable high-gain amplifier stage to step up the signal level. Impedance-matching amplifiers are, in fact, op-amps with feedback.

When connecting a device to a signal source, loading problems can be reduced by making sure that the device has a high input impedance. Unfortunately, this will also reduce the level (amplitude, power) of the signal received by the device. In fact, a high-impedance device may reflect back some harmonics of the source signal. A termination resistance might be connected in parallel with the device to reduce this problem.

In many data acquisition systems, output impedance of the output amplifier is made equal to the transmission line impedance. When maximum power amplification is desired, conjugate matching is recommended. In this case, input and output impedances of the matching amplifier are made equal to the complex conjugates of the source and load impedances, respectively.

Example 2.2
Consider a dc power supply of voltage v_s and output impedance (resistance) R_s. It is used to power a load of resistance R_l, as shown in Figure 2.3. What should be the relationship between R_s and R_l, if the objective is to maximize the power absorbed by the load?

Solution
The current through the circuit is

$$i_l = \frac{v_s}{R_l + R_s}.$$

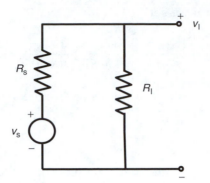

FIGURE 2.3
A load powered by a dc power supply.

Accordingly, the voltage across the load is

$$v_l = i_l R_l = \frac{v_s R_l}{R_l + R_s}.$$

The power absorbed by the load is

$$p_l = i_l v_l = \frac{v_s^2 R_l}{[R_l + R_s]^2}. \tag{i}$$

For maximum power, we need

$$\frac{dp_l}{dR_l} = 0. \tag{ii}$$

We differentiate the RHS expression of Equation i with respect to R_l to satisfy Equation ii. This gives the requirement for maximum power as

$$R_l = R_s.$$

2.2.3 Impedance Matching in Mechanical Systems

The concepts of impedance matching can be extended to mechanical systems and to mixed systems (e.g., electromechanical systems) in a straightforward manner. The procedure follows from the familiar electromechanical analogies. As a specific application, consider a mechanical load driven by a motor. Often, direct driving is not practical owing to the limitations of the speed–torque characteristics of the available motors. By including a suitable gear transmission between the motor and the load, it is possible to modify the speed–torque characteristics of the drive system as felt by the load. This is a process of impedance matching.

Example 2.3

Consider the mechanical system where a torque source (motor) of torque T and moment of inertia J_m is used to drive a purely inertial load of moment of inertia J_L, as shown in Figure 2.4a. What is the resulting angular acceleration $\ddot{\theta}$ of the system? Neglect the flexibility of the connecting shaft.

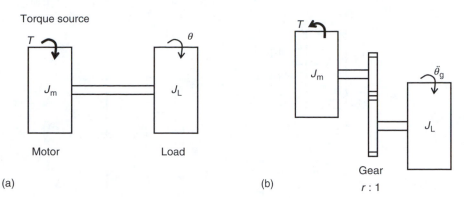

FIGURE 2.4
An inertial load driven by a motor. (a) Without gear transmission. (b) With a gear transmission.

Now suppose that the load is connected to the same torque source through an ideal (loss free) gear of motor-to-load speed ratio r:1, as shown in Figure 2.4b. What is the resulting acceleration $\ddot{\theta}_g$ of the load?

Obtain an expression for the normalized load acceleration $a = \ddot{\theta}_g / \ddot{\theta}$ in terms of r and $p = J_L / J_m$. Sketch a vs. r for $p = 0.1, 1.0,$ and 10.0. Determine the value of r in terms of p that will maximize the load acceleration a.

Comment on the results obtained in this problem.

Solution
For the unit without the gear transmission, Newton's second law gives

$$(J_m + J_L)\ddot{\theta} = T.$$

Hence,

$$\ddot{\theta} = \frac{T}{J_m + J_L}. \tag{i}$$

For the unit with the gear transmission, see the free-body diagram shown in Figure 2.5, in the case of a loss-free (i.e., 100% efficient) gear transmission.

Newton's second law gives

$$J_m r \ddot{\theta}_g = T - \frac{T_g}{r} \tag{ii}$$

and

$$J_L \ddot{\theta}_g = T_g, \tag{iii}$$

where T_g is the gear torque on the load inertia. Eliminate T_g in Equation ii and Equation iii. We get

$$\ddot{\theta}_g = \frac{rT}{(r^2 J_m + J_L)}. \tag{iv}$$

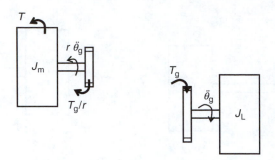

FIGURE 2.5
Free-body diagram.

Divide Equation iv by Equation i.

$$\frac{\ddot{\theta}_g}{\ddot{\theta}} = a = \frac{r(J_m + J_L)}{(r^2 J_m + J_L)} = \frac{r(1 + J_L/J_m)}{(r^2 + J_L/J_m)}$$

or

$$a = \frac{r(1+p)}{(r^2+p)}, \qquad\qquad\qquad\qquad\qquad\text{(v)}$$

where $p = J_L/J_m$.

From Equation v note that for $r = 0$, $a = 0$, and for $r \to \infty$, $a \to 0$. Peak value of a is obtained through differentiation:

$$\frac{\partial a}{\partial r} = \frac{(1+p)[(r^2+p) - r \times 2r]}{(r^2+p)^2} = 0.$$

We get, by taking the positive root,

$$r_p = \sqrt{p}, \qquad\qquad\qquad\qquad\qquad\qquad\text{(vi)}$$

where r_p is the value of r corresponding to the peak of a. The peak value of a is obtained by substituting Equation vi in Equation v. Thus,

$$a_p = \frac{1+p}{2\sqrt{p}}. \qquad\qquad\qquad\qquad\qquad\text{(vii)}$$

Also, note from Equation v that when $r = 1$, we have $a = r = 1$ for any value of p. Hence, all curves in Equation v should pass through the point (1,1).

The relation v is sketched in Figure 2.6 for $p = 0.1$, 1.0, and 10.0. The peak values are tabulated in the following table.

p	r_p	a_p
0.1	0.316	1.74
1.0	1.0	1.0
10.0	3.16	1.74

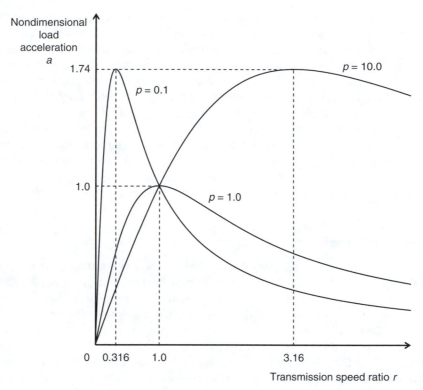

FIGURE 2.6
Normalized acceleration vs. speed ratio.

Note from Figure 2.6 that the transmission speed ratio can be chosen, depending on the inertia ratio, to maximize the load acceleration. In particular, we can state the following:

1. When $J_L = J_m$, pick a direct-drive system (no gear transmission; i.e., $r = 1$).
2. When $J_L < J_m$, pick a speed-up gear at the peak value of $r = \sqrt{J_L/J_m}$.
3. When $J_L > J_m$, pick a speed-down gear at the peak value of r.

2.3 Amplifiers

Voltages, velocities, and temperatures are "across variables" since they are present across an element. Currents, forces, and heat transfer rates are "through variables" since they flow through an element. The level of an electrical signal can be represented by variables such as voltage, current, and power. Analogous across variables, through variables, and power variables can be defined for other types of signals (e.g., mechanical variables velocity, force, and power) as well. Signal levels at various interface locations of components in a control system have to be properly adjusted for satisfactory performance of these components and of the overall system. For example, input to an actuator should possess adequate power to drive the actuator. A signal should maintain its signal level above some threshold during transmission, so that errors due to signal weakening would not be excessive. Signals applied to digital devices must remain within the specified logic levels.

Many types of sensors produce weak signals that have to be upgraded before they could be fed into a monitoring system, data processor, controller, or data logger.

Signal amplification concerns proper adjustment of signal level for performing a specific task. Amplifiers are used to accomplish signal amplification. An amplifier is an active device, which needs an external power source to operate. Even though various active circuits, amplifiers in particular, are commonly produced in the monolithic form using an original integrated-circuit (IC) layout to accomplish a particular amplification task, it is convenient to study their performance using discrete circuit models, with the op-amp as the basic building block. Of course, op-amps are themselves available as monolithic IC packages. They are widely used as the basic building blocks in producing other types of amplifiers, and in turn for modeling and analyzing these various kinds of amplifiers. For these reasons, our discussion on amplifiers will evolve on the op-amp.

2.3.1 Operational Amplifier

The origin of the op-amp dates back to the 1940s when the vacuum tube op-amp was introduced. Op-amp got its name because originally it was used almost exclusively to perform mathematical operations; for example, in analog computers. Subsequently, in the 1950s, the transistorized op-amp was developed. It used discrete elements such as bipolar junction transistors and resistors. Still it was too large, consumed too much power, and was too expensive for widespread use in general applications. This situation changed in the late 1960s when the IC op-amp was developed in the monolithic form, as a single IC chip. Today, the IC op-amp, which consists of a large number of circuit elements on a substrate of typically a single silicon crystal (the monolithic form), is a valuable component in almost any signal modification device. Bipolar complementary metal oxide semiconductor (bipolar-CMOS) op-amps in various plastic packages and pin configurations are commonly available.

An op-amp could be manufactured in the discrete-element form using, say, 10 bipolar junction transistors and as many discrete resistors or alternatively (and preferably) in the modern monolithic form as an IC chip that may be equivalent to over 100 discrete elements. In any form, the device has an input impedance Z_i, an output impedance Z_o, and a gain K. Hence, a schematic model for an op-amp can be given as in Figure 2.7a. Op-amp packages are available in several forms. Very common is the eight-pin dual in-line package (DIP) or V package, as shown in Figure 2.7b. The assignment of the pins (i.e., pin configuration or pin-out) is as shown in the figure, which should be compared with Figure 2.7a. Note the counterclockwise numbering sequence starting with the top left pin next to the semicircular notch (or dot). This convention of numbering is standard for any type of IC package, not just op-amp packages. Other packages include 8-pin metal-can package or T package, which has a circular shape instead of the rectangular shape of the previous package, and the 14-pin rectangular "Quad" package, which contains four op-amps (with a total of eight input pins, four output pins, and two power supply pins). The conventional symbol of an op-amp is shown in Figure 2.7c. Typically, there are five terminals (pins or lead connections) to an op-amp. Specifically, there are two input leads (a positive or noninverting lead with voltage v_{ip} and a negative or inverting lead with voltage v_{in}), an output lead (voltage v_o), and two bipolar power supply leads ($+v_s$ or v_{CC} or collector supply and $-v_s$ or v_{EE} or emitter supply). The typical supply voltage is ± 22 V. Normally, some of the pins may not be connected; for example, pins 1, 5, and 8 in Figure 2.7b.

Note from Figure 2.7a that under open-loop (i.e., no feedback) conditions,

$$v_o = K v_i, \tag{2.7}$$

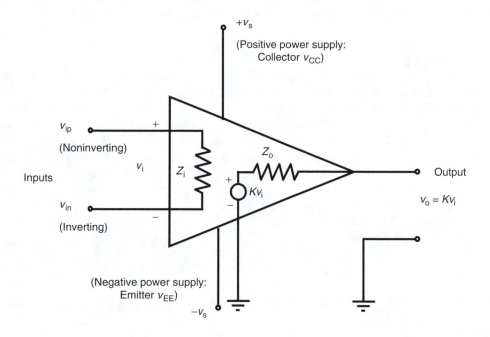

(a)

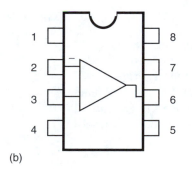

(b)

Pin designations:

1 Offset null
2 Inverting input
3 Noninverting input
4 Negative power supply v_{EE}
5 Offset null
6 Output
7 Positive power supply v_{CC}
8 NC (not connected)

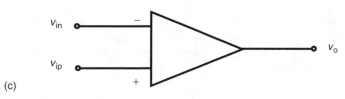

(c)

FIGURE 2.7
Operational amplifier. (a) A schematic model. (b) Eight-pin dual in-line package (DIP). (c) Conventional circuit symbol.

where the input voltage v_i is the differential input voltage defined as the algebraic difference between the voltages at the positive and negative leads. Thus,

$$v_i = v_{ip} - v_{in}. \tag{2.8}$$

The open-loop voltage gain K is very high (10^5 to 10^9) for a typical op-amp. Furthermore, the input impedance Z_i could be as high as 10 MΩ (typical is 2 MΩ) and the output impedance is low, of the order 10 Ω and may reach about 75 Ω for some op-amps. Since v_o is typically 1 to 15 V, from Equation 2.7 it follows that $v_i \cong 0$ since K is very large. Hence, from Equation 2.8, we have $v_{ip} \cong v_{in}$. In other words, the voltages at the two input leads are nearly equal. Now, if we apply a large voltage differential v_i (say, 10 V) at the input, then according to Equation 2.7, the output voltage should be extremely high. This never happens in practice, however, since the device saturates quickly beyond moderate output voltages (of the order 15 V).

From Equation 2.7 and Equation 2.8, it is clear that if the negative input lead is grounded (i.e., $v_{in} = 0$), then,

$$v_o = Kv_{ip}, \tag{2.9}$$

and if the positive input lead is grounded (i.e., $v_{ip} = 0$),

$$v_o = -Kv_{in}. \tag{2.10}$$

This is the reason why v_{ip} is termed *noninverting* input and v_{in} is termed *inverting input*.

Example 2.4

Consider an op-amp with an open-loop gain of 1×10^5. If the saturation voltage is 15 V, determine the output voltage in the following cases:

 a. 5 μV at the positive lead and 2 μV at the negative lead
 b. -5 μV at the positive lead and 2 μV at the negative lead
 c. 5 μV at the positive lead and -2 μV at the negative lead
 d. -5 μV at the positive lead and -2 μV at the negative lead
 e. 1 V at the positive lead and the negative lead is grounded
 f. 1 V at the negative lead and the positive lead is grounded

Solution

This problem can be solved using Equation 2.7 and Equation 2.8. The results are given in Table 2.1. Note that in the last two cases the output will saturate and Equation 2.7 will no longer hold.

Field effect transistors (FET), for example, metal oxide semiconductor field effect transistors (MOSFET), are commonly used in the IC form of an op-amp. The MOSFET type has advantages over many other types; for example, higher input impedance and more stable output (almost equal to the power supply voltage) at saturation, making the MOSFET op-amps preferable over bipolar junction transistor op-amps in many applications.

In analyzing op-amp circuits under unsaturated conditions, we use the following two characteristics of an op-amp:

 1. Voltages of the two input leads should be (almost) equal.
 2. Currents through each of the two input leads should be (almost) zero.

TABLE 2.1

Solution to Example 2.4

v_{ip}	v_{in}	v_i	v_o
5 μV	2 μV	3 μV	0.3 V
−5 μV	2 μV	−7 μV	−0.7 V
5 μV	−2 μV	7 μV	0.7 V
−5 μV	−2 μV	−3 μV	−0.3 V
1 V	0	1 V	15 V
0	1 V	−1 V	−15 V

As explained earlier, the first property is credited to high open-loop gain, and the second property to high input impedance in an op-amp. We shall repeatedly use these two properties to obtain input/output equations for amplifier systems.

2.3.1.1 Use of Feedback in Op-Amps

Op-amp is a very versatile device, primarily owing to its very high input impedance, low output impedance, and very high gain. However, it cannot be used without modification as an amplifier because it is not very stable in the form shown in Figure 2.7. The two main factors that contribute to this problem are frequency response and drift stated in another way, op-amp gain K does not remain constant; it can vary with the frequency of the input signal (i.e., the frequency response function is not flat in the operating range) and also with time (i.e., drift). The frequency response problem arises because of circuit dynamics of an op-amp. This problem is usually not severe unless the device is operated at very high frequencies. The drift problem arises as a result of the sensitivity of gain K to environmental factors such as temperature, light, humidity, and vibration and also as a result of the variation of K due to aging. Drift in an op-amp can be significant and steps should be taken to eliminate that problem.

It is virtually impossible to avoid the drift in gain and the frequency response error in an op-amp. However, an ingenious way has been found to remove the effect of these two problems at the amplifier output. Since gain K is very large, by using feedback we can virtually eliminate its effect at the amplifier output. This closed-loop form of an op-amp has the advantage that the characteristics and the accuracy of the output of the overall circuit depend on the passive components (e.g., resistors and capacitors) in it, which can be provided at high precision, and not the parameters of the op-amp itself. The closed-loop form is preferred in almost every application; in particular, voltage follower and charge amplifier are devices that use the properties of high Z_i, low Z_o, and high K of an op-amp along with feedback through a high-precision resistor, to eliminate errors due to nonconstant K. In summary, op-amp is not very useful in its open-loop form, particularly because gain K is not steady. However, since K is very large, the problem can be removed by using feedback. It is this closed-loop form that is commonly used in practical applications of an op-amp.

In addition to the unsteady nature of gain, there are other sources of error that contribute to the less-than-ideal performance of an op-amp circuit. Noteworthy are

1. The offset current present at the input leads due to bias currents that are needed to operate the solid-state circuitry.
2. The offset voltage that might be present at the output even when the input leads are open.

3. The unequal gains corresponding to the two input leads (i.e., the inverting gain not equal to the noninverting gain).

Such problems can produce nonlinear behavior in op-amp circuits, and they can be reduced by proper circuit design and through the use of compensating circuit elements.

2.3.2 Voltage, Current, and Power Amplifiers

Any type of amplifier can be constructed from scratch in the monolithic form as an IC chip, or in the discrete form as a circuit containing several discrete elements such as discrete bipolar junction transistors or discrete FETs, discrete diodes, and discrete resistors. But, almost all types of amplifiers can also be built using op-amp as the basic building block. Since we are already familiar with op-amps and since op-amps are extensively used in general amplifier circuitry, we prefer to use the latter approach, which uses discrete op-amps for building general amplifiers. Furthermore, modeling, analysis, and design of a general amplifier may be performed on this basis.

If an electronic amplifier performs a voltage amplification function, it is termed a *voltage amplifier*. These amplifiers are so common that, the term "amplifier" is often used to denote a voltage amplifier. A voltage amplifier can be modeled as

$$v_o = K_v v_i, \tag{2.11}$$

where v_o is the output voltage, v_i is the input voltage, and K_v is the voltage gain. Voltage amplifiers are used to achieve voltage compatibility (or level shifting) in circuits.

Current amplifiers are used to achieve current compatibility in electronic circuits. A current amplifier may be modeled by

$$i_o = K_i i_i, \tag{2.12}$$

where i_o is the output current, i_i is the input current, and K_i is the current gain.

A voltage follower has $K_v = 1$ and, hence, it may be considered as a current amplifier. Besides, it provides impedance compatibility and acts as a buffer between a low-current (high-impedance) output device (signal source or the device that provides the signal) and a high-current (low-impedance) input device (signal receiver or the device that receives the signal) that are interconnected. Hence, the name buffer amplifier or impedance transformer is sometimes used for a current amplifier with unity voltage gain.

If the objective of signal amplification is to upgrade the associated power level, then a power amplifier should be used for that purpose. A simple model for a power amplifier is

$$p_o = K_p P_i, \tag{2.13}$$

where p_o is the output power, p_i is the input power, and K_p is the power gain.

It is easy to see from Equation 2.11 through Equation 2.13 that

$$K_p = K_v K_i. \tag{2.14}$$

Note that all three types of amplification could be achieved simultaneously from the same amplifier. Furthermore, a current amplifier with unity voltage gain (e.g., a voltage follower) is a power amplifier as well. Usually, voltage amplifiers and current amplifiers are used in the first stages of a signal path (e.g., sensing, data acquisition, and signal generation), where signal levels and power levels are relatively low, while power amplifiers are typically used in the final stages (e.g., final control, actuation, recording, display), where high signal levels and power levels are usually required.

Figure 2.8a gives an op-amp circuit for a voltage amplifier. Note the feedback resistor R_f which serves the purposes of stabilizing the op-amp and providing an accurate voltage gain. The negative lead is grounded through an accurately known resistor R. To determine the voltage gain, recall that the voltages at the two input leads of an op-amp should be equal (in the ideal case). The input voltage v_i is applied to the positive lead of the op-amp. Then the voltage at point A should also be equal to v_i. Next, recall that the current through the input lead of an op-amp is ideally zero. Hence, by writing the current balance equation for the node point A we have,

$$\frac{v_o - v_i}{R_f} = \frac{v_i}{R}.$$

This gives the amplifier equation

$$v_o = \left(1 + \frac{R_f}{R}\right)v_i. \tag{2.15}$$

Hence, the voltage gain is given by

$$K_v = 1 + \frac{R_f}{R}. \tag{2.16}$$

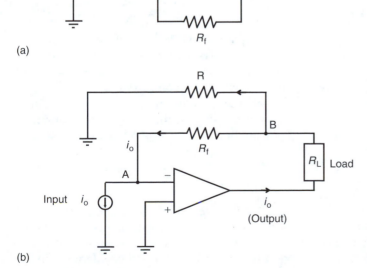

(a)

(b)

FIGURE 2.8
(a) A voltage amplifier. (b) A current amplifier.

Note that K_v depends on R and R_f, and not on the op-amp gain. Hence, the voltage gain can be accurately determined by selecting the two passive elements (resistors) R and R_f precisely. Also, note that the output voltage has the same sign as the input voltage. Hence, this is a *noninverting amplifier*. If the voltages are of the opposite sign, we have an *inverting amplifier*.

A current amplifier is shown in Figure 2.8b. The input current i_i is applied to the negative lead of the op-amp as shown, and the positive lead is grounded. There is a feedback resistor R_f connected to the negative lead through the load R_L. The resistor R_f provides a path for the input current since the op-amp takes in virtually zero current. There is a second resistor R through which the output is grounded. This resistor is needed for current amplification. To analyze the amplifier, use the fact that the voltage at point A (i.e., at the negative lead) should be zero because the positive lead of the op-amp is grounded (zero voltage). Furthermore, the entire input current i_i passes through the resistor R_f as shown. Hence, the voltage at point B is $R_f i_i$. Consequently, current through the resistor R is $R_f i_i / R$, which is positive in the direction shown. It follows that the output current i_o is given by

$$i_o = i_i + \frac{R_f}{R} i_i$$

or

$$i_o = \left(1 + \frac{R_f}{R}\right) i_i. \tag{2.17}$$

The current gain of the amplifier is

$$K_i = 1 + \frac{R_f}{R}. \tag{2.18}$$

As before, the amplifier gain can be accurately set using the high-precision resistors R and R_f.

2.3.3 Instrumentation Amplifiers

An instrumentation amplifier is typically a special-purpose voltage amplifier dedicated to instrumentation applications. Examples include amplifiers used for producing the output from a bridge circuit (bridge amplifier) and amplifiers used with various sensors and transducers. An important characteristic of an instrumentation amplifier is the adjustable-gain capability. The gain value can be adjusted manually in most instrumentation amplifiers. In more sophisticated instrumentation amplifiers, the gain is programmable and can be set by means of digital logic. Instrumentation amplifiers are normally used with low-voltage signals.

2.3.3.1 *Differential Amplifier*

Usually, an instrumentation amplifier is also a *differential amplifier* (sometimes termed difference amplifier). Note that in a differential amplifier both input leads are used for signal input, whereas in a single-ended amplifier one of the leads is grounded and only one lead is used for signal input. Ground-loop noise can be a serious problem in single-ended amplifiers. Ground-loop noise can be effectively eliminated using a differential amplifier because noise loops are formed with both inputs of the amplifier and, hence, these noise signals are subtracted at the amplifier output. Since the noise level is almost the same for both inputs, it is canceled out. Any other noise (e.g., 60 Hz line noise) that might enter both inputs with the same intensity will also be canceled out at the output of a differential amplifier.

A basic differential amplifier that uses a single op-amp is shown in Figure 2.9a. The input/output equation for this amplifier can be obtained in the usual manner. For instance, since current through an op-amp is negligible, the current balance at point B gives

$$\frac{v_{i2} - v_B}{R} = \frac{v_B}{R_f},$$

(i)

where v_B is the voltage at B. Similarly, current balance at point A gives

$$\frac{v_o - v_A}{R_f} = \frac{v_A - v_{i1}}{R}.$$

(ii)

Now we use the property

$$v_A = v_B$$

(iii)

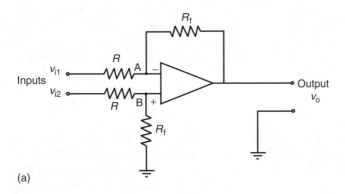

(a)

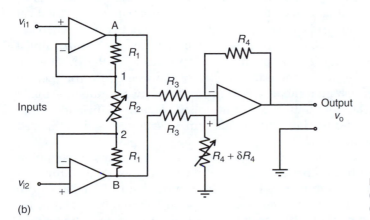

(b)

FIGURE 2.9
(a) A basic differential amplifier.
(b) A basic instrumentation amplifier.

for an op-amp, to eliminate v_A and v_B from Equation i and Equation ii. This gives

$$\frac{v_{i2}}{(1 + R/R_f)} = \frac{(v_o R/R_f + v_{i1})}{(1 + R/R_f)}$$

or

$$v_o = \frac{R_f}{R}(v_{i2} - v_{i1}). \tag{2.19}$$

Two things are clear from Equation 2.19. First, the amplifier output is proportional to the difference and not the absolute value of the two inputs v_{i1} and v_{i2}. Second, voltage gain of the amplifier is R_f/R. This is known as the differential gain. It is clear that the differential gain can be accurately set by using high-precision resistors R and R_f.

The basic differential amplifier, shown in Figure 2.9a and discussed earlier, is an important component of an instrumentation amplifier. In addition, an instrumentation amplifier should possess the capability of adjustable gain. Furthermore, it is desirable to have a very high input impedance and very low output impedance at each input lead. It is desirable for an instrumentation amplifier to possess a higher and more stable gain, and also a higher input impedance than a basic differential amplifier. An instrumentation amplifier that possesses these basic requirements may be fabricated in the monolithic IC form as a single package. Alternatively, it may be built using three differential amplifiers and high-precision resistors, as shown in Figure 2.9b. The amplifier gain can be adjusted using the fine-tunable resistor R_2. Impedance requirements are provided by two voltage-follower type amplifiers, one for each input, as shown. The variable resistance δR_4 is necessary to compensate for errors due to unequal common-mode gain. Let us first consider this aspect and then obtain an equation for the instrumentation amplifier.

2.3.3.2 Common Mode

The voltage that is common to both input leads of a differential amplifier is known as the common-mode voltage. This is equal to the smaller of the two input voltages. If the two inputs are equal, then the common-mode voltage is obviously equal to each one of the two inputs. When $v_{i1} = v_{i2}$, ideally, the output voltage v_o should be zero. In other words, ideally, any common-mode signals are rejected by a differential amplifier. But, since commercial op-amps are not ideal and since they usually do not have exactly identical gains with respect to the two input leads, the output voltage v_o will not be zero when the two inputs are identical. This common-mode error can be compensated for by providing a variable resistor with fine resolution at one of the two input leads of the differential amplifier. Hence, in Figure 2.9b, to compensate for the common-mode error (i.e., to achieve a satisfactory level of common-mode rejection), first the two inputs are made equal and then δR_4 is varied carefully until the output voltage level is sufficiently small (minimum). Usually, δR_4 that is required to achieve this compensation is small compared with the nominal feedback resistance R_4.

Since ideally $\delta R_4 = 0$, we can neglect δR_4 in the derivation of the instrumentation amplifier equation. Now, note from a basic property of an op-amp with no saturation (specifically, the voltages at the two input leads have to be almost identical) that in Figure 2.9b, the voltage at point 2 should be v_{i2} and the voltage at point 1 should be v_{i1}. Next, we use the property that the current through each input lead of an op-amp is negligible.

Accordingly, current through the circuit path B → 2 → 1 → A has to be the same. This gives the current continuity equations

$$\frac{v_B - v_{i2}}{R_1} = \frac{v_{i2} - v_{i1}}{R_2} = \frac{v_{i1} - v_A}{R_1},$$

where v_A and v_B are the voltages at points A and B, respectively. Hence, we get the following two equations

$$v_B = v_{i2} + \frac{R_1}{R_2}(v_{i2} - v_{i1})$$

and

$$v_A = v_{i1} - \frac{R_1}{R_2}(v_{i2} - v_{i1}).$$

Now, by subtracting the second equation from the first, we have the equation for the first stage of the amplifier. Thus,

$$v_B - v_A = \left(1 + \frac{2R_1}{R_2}\right)(v_{i2} - v_{i1}). \qquad \text{(i)}$$

Next from the previous result (see Equation 2.19) for a differential amplifier, we have (with $\delta R_4 = 0$)

$$v_o = \frac{R_4}{R_3}(v_B - v_A). \qquad \text{(ii)}$$

Note that only the resistor R_2 is varied to adjust the gain (differential gain) of the amplifier. In Figure 2.9b, the two input op-amps (the voltage-follower op-amps) do not have to be exactly identical as long as the resistors R_1 and R_2 are chosen to be accurate. This is so because the op-amp parameters such as open-loop gain and input impedance do not enter into the amplifier equations, provided that their values are sufficiently high, as noted earlier.

2.3.4 Amplifier Performance Ratings

Main factors that affect the performance of an amplifier are

1. Stability
2. Speed of response (bandwidth, slew rate)
3. Unmodeled signals

We have already discussed the significance of some of these factors.

The level of stability of an amplifier, in the conventional sense, is governed by the dynamics of the amplifier circuitry, and may be represented by a time constant. But the most important consideration for an amplifier is the parameter variation due to aging, temperature, and other environmental factors. Parameter variation is also classified as a stability issue, in the context of devices such as amplifiers, because it pertains to the steadiness of the response when the input is maintained steady. Of particular importance is the temperature drift. This may be specified as a drift in the output signal per unity change in temperature (e.g., $\mu V/°C$).

The speed of response of an amplifier dictates the ability of the amplifier to faithfully respond to transient inputs. Conventional time-domain parameters such as rise time may be used to represent this. Alternatively, in the frequency domain, speed of response may be represented by a bandwidth parameter. For example, the frequency range over which the frequency response function is considered constant (flat) may be taken as a measure of bandwidth. Since there is some nonlinearity in any amplifier, bandwidth can depend on the signal level itself. Specifically, small-signal bandwidth refers to the bandwidth that is determined using small input signal amplitudes.

Another measure of the speed of response is the slew rate, which is defined as the largest possible rate of change of the amplifier output for a particular frequency of operation. Since for a given input amplitude, the output amplitude depends on the amplifier gain, slew rate is usually defined for unity gain.

Ideally, for a linear device, the frequency response function (transfer function) does not depend on the output amplitude (i.e., the product of the dc gain and the input amplitude). But, for a device that has a limited slew rate, the bandwidth (or the maximum operating frequency at which output distortions may be neglected) will depend on the output amplitude. The larger the output amplitude, the smaller the bandwidth for a given slew rate limit. A bandwidth parameter that is usually specified for a commercial op-amp is the gain-bandwidth product (GBP). This is the product of the open-loop gain and the bandwidth of the op-amp. For example, for an op-amp with GBP = 15 MHz and an open-loop gain of 100 dB (i.e., 10^5), the bandwidth is $15 \times 10^6/10^5\,Hz = 150\,Hz$. Clearly, this bandwidth value is rather low. Since, the gain of an op-amp with feedback is significantly lower than 100 dB, its effective bandwidth is much higher than that of an open-loop op-amp.

Example 2.5
Obtain a relationship between the slew rate and the bandwidth for a slew rate-limited device. An amplifier has a slew rate of 1 V/μs. Determine the bandwidth of this amplifier when operating at an output amplitude of 5 V.

Solution
Clearly, the amplitude of the rate of change signal divided by the amplitude of the output signal gives an estimate of the output frequency. Consider a sinusoidal output voltage given by

$$v_o = a \sin 2\pi\, ft. \tag{2.20}$$

The rate of change of output is

$$\frac{dv_o}{dt} = 2\pi\, fa \cos 2\pi\, ft.$$

Hence, the maximum rate of change of output is $2\pi fa$. Since this corresponds to the slew rate when f is the maximum allowable frequency, we have

$$s = 2\pi f_b a, \tag{2.21}$$

where s is the slew rate, f_b is the bandwidth, and a is the output amplitude.

Now, with $s = 1$ V/μs and $a = 5$ V, we get

$$f_b = \frac{1}{2\pi} \times \frac{1}{1 \times 10^{-8}} \times \frac{1}{5} \text{Hz}$$
$$= 31.8 \text{ kHz}.$$

∎

We have noted that stability problems and frequency response errors are prevalent in the open-loop form of an op-amp. These problems can be eliminated using feedback because the effect of the open-loop transfer function on the closed-loop transfer function is negligible if the open-loop gain is very large, which is the case for an op-amp.

Unmodeled signals can be a major source of amplifier error, and these signals include

1. Bias currents
2. Offset signals
3. Common-mode output voltage
4. Internal noise

In analyzing op-amps, we assume that the current through the input leads is zero. This is not strictly true because bias currents for the transistors within the amplifier circuit have to flow through these leads. As a result, the output signal of the amplifier will deviate slightly from the ideal value.

Another assumption that we make in analyzing op-amps is that the voltage is equal at the two input leads. In practice, however, offset currents and voltages are present at the input leads, due to minute discrepancies inherent to the internal circuits within an op-amp.

2.3.4.1 Common-Mode Rejection Ratio

Common-mode error in a differential amplifier was discussed earlier. We note that ideally the common-mode input voltage (the voltage common to both input leads) should have no effect on the output voltage of a differential amplifier. But, since any practical amplifier has some imbalances in the internal circuitry (e.g., gain with respect to one input lead is not equal to the gain with respect to the other input lead and, furthermore, bias signals are needed for operation of the internal circuitry), there will be an error voltage at the output, which depends on the common-mode input. Common-mode rejection ratio (CMRR) of a differential amplifier is defined as

$$\text{CMRR} = \frac{K v_{cm}}{v_{ocm}}, \tag{2.22}$$

where K is the gain of the differential amplifier (i.e., differential gain), v_{cm} is the common-mode voltage (i.e., voltage common to both input leads), and v_{ocm} is the common-mode output voltage (i.e., output voltage due to common-mode input voltage). Note that ideally $v_{ocm} = 0$ and CMRR should be infinity. It follows that the larger the CMRR the better the differential amplifier performance.

The three types of unmodeled signals mentioned earlier can be considered as noise. In addition, there are other types of noise signals that degrade the performance of an amplifier. For example, ground-loop noise can enter the output signal. Furthermore, stray capacitances and other types of unmodeled circuit effects can generate internal noise. Usually in amplifier analysis, unmodeled signals (including noise) can be represented by a noise voltage source at one of the input leads. Effects of unmodeled signals can be reduced by using suitably connected compensating circuitry, including variable resistors that can be adjusted to eliminate the effect of unmodeled signals at the amplifier output (e.g., see δR_4 in Figure 2.9b). Some useful information about op-amps is summarized in Box 2.1.

Box 2.1   Op-Amps

Ideal Op-Amp Properties

- Infinite open-loop differential gain
- Infinite input impedance
- Zero output impedance
- Infinite bandwidth
- Zero output for zero differential input

Ideal Analysis Assumptions

- Voltages at the two input leads are equal.
- Current through either input lead is zero.

Definitions

- Open-loop gain $= \left| \dfrac{\text{Output voltage}}{\text{Voltage difference at input leads}} \right|$, with no feedback.

- Input impedance $= \dfrac{\text{Voltage between an input lead and ground}}{\text{Current through that lead}}$, with other

  input lead grounded and the output in open circuit.

- Output impedance

  $= \dfrac{\text{Voltage between output lead and ground in open circuit}}{\text{Current through that lead}}$, with normal

  input conditions.

- Bandwidth is the frequency range in which the frequency response is flat (gain is constant).
- GBP $=$ Open-loop gain $\times$ Bandwidth at that gain
- Input bias current is the average DC current through one input lead.
- Input offset current is the difference in the two input bias currents.
- Differential input voltage is the voltage at one input lead with the other grounded when the output voltage is zero.

- Common-mode gain

  $= \dfrac{\text{Output voltage when input leads are at the same voltage}}{\text{Common input voltage}}$

- Common-mode rejection ratio (CMRR) $= \dfrac{\text{Open loop differential gain}}{\text{Common-mode gain}}$
- Slew rate is the rate of change of output of a unity-gain op-amp, for a step input.

2.3.4.2 AC-Coupled Amplifiers

The dc component of a signal can be blocked off by connecting the signal through a capacitor (note that the impedance of a capacitor is $1/(j\omega C)$ and hence, at zero frequency there will be an infinite impedance). If the input lead of a device has a series capacitor, we say that the input is AC coupled and if the output lead has a series capacitor, then the output is AC coupled. Typically, an AC-coupled amplifier has a series capacitor both at the input lead and the output lead. Hence, its frequency response function will have a high-pass characteristic; in particular, the dc components will be filtered out. Errors due to bias currents and offset signals are negligible for an AC-coupled amplifier. Furthermore, in an AC-coupled amplifier, stability problems are not very serious.

2.3.5 Ground-Loop Noise

In instruments that handle low-level signals (e.g., sensors such as accelerometers, signal-conditioning circuitry such as strain gage bridges, and sophisticated and delicate electronic components such as computer disk drives and automobile control modules), electrical noise can cause excessive error, unless proper corrective actions are taken. One form of noise is caused by fluctuating magnetic fields due to nearby ac power lines or electric machinery. This is commonly known as *electromagnetic interference* (EMI). This problem can be avoided by removing the source of EMI, so that fluctuating external magnetic fields and currents are not present near the affected instrument. Another solution would be to use fiber optic (optically coupled) signal transmission, so that there is no noise conduction along with the transmitted signal from the source to the subject instrument. In the case of hard-wired transmission, if the two signal leads (positive and negative or hot and neutral) are twisted or if shielded cables are used, the induced noise voltages become equal in the two leads, which cancel each other.

   Proper grounding practices are important to mitigate unnecessary electrical noise problems and, more importantly, to avoid electrical safety hazards. A standard single-phase ac outlet (120 V, 60 Hz) has three terminals, one carrying power (hot), the second neutral, and the third connected to earth ground (which is maintained at zero potential rather uniformly from point to point in the power network). Correspondingly, the power plug of an instrument should have three prongs. The shorter flat prong is connected to a black wire (hot) and the longer flat prong is connected to a white wire (neutral). The round prong is connected to a green wire (ground), which at the other end is connected to the chassis (or casing) of the instrument (chassis ground). In view of grounding the chassis in this manner, the instrument housing is maintained at zero potential, even in the presence of a fault in the power circuit (e.g., a leakage or a short). The power circuitry of an instrument also has a local ground (signal ground), with reference to which its power signal is measured. This is a sufficiently thick conductor within the instrument, and it provides a common and uniform reference of 0 V. Consider the sensor signal-conditioning example shown in Figure 2.10. The dc power supply can provide both positive (+) and negative (−) outputs. Its zero-voltage reference is denoted by COM (common ground), and it is the signal ground of the device. It should be noted that COM of the dc power supply is not connected to the chassis ground, the latter connected to the earth ground through the round prong of the power plug of the power supply. This is necessary to avoid the danger of an electric shock. Note that COM of the power supply is connected to the signal ground of the signal-conditioning module. In this manner, a common 0 V reference is provided for the dc voltage that is supplied to the signal-conditioning module.

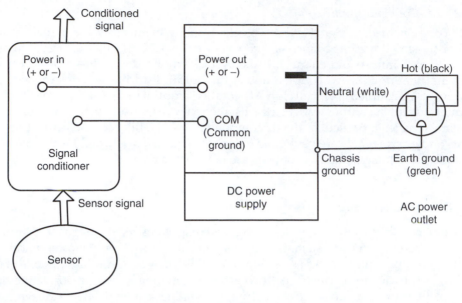

FIGURE 2.10
An example of grounding of instruments.

The main cause of electrical noise is the ground loops, which are created due to improper grounding of instruments. If two interconnected instruments are grounded at two separate locations that are far apart (multiple grounding), ground-loop noise can enter the signal leads because of the possible potential difference between the two ground points. The reason is that ground itself is not generally a uniform-potential medium, and a nonzero (and finite) impedance may exist from point to point within this medium. This is, in fact, the case with a typical ground medium such as a COM wire. An example is shown schematically in Figure 2.11a. In this example, the two leads of a sensor are directly connected to a signal-conditioning device such as an amplifier, with one of its input leads (+) grounded (at point B). The 0 V reference lead of the sensor is grounded through its housing to the earth ground (at point A). Because of nonuniform ground potentials, the two ground points A and B are subjected to a potential difference v_g. This will create a ground loop with the common reference lead, which interconnects the two devices. The solution to this problem is to isolate (i.e., provide an infinite impedance to) either one of the two devices. Figure 2.11b shows internal isolation of the sensor. External isolation, by insulating the housing of the sensor, will also remove the ground loop. Floating off the COM of a power supply (see Figure 2.10) is another approach to eliminating ground loops. Specifically, COM is not connected to earth ground.

2.4 Analog Filters

A filter is a device that allows through only the desirable part of a signal, rejecting the unwanted part. Unwanted signals can seriously degrade the performance of a control system. External disturbances, error components in excitations, and noise generated internally within system components and instrumentation are such spurious signals,

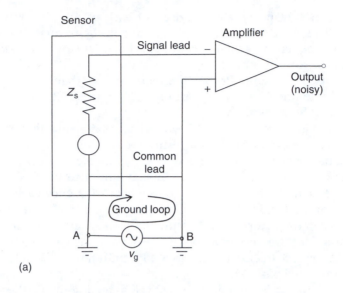

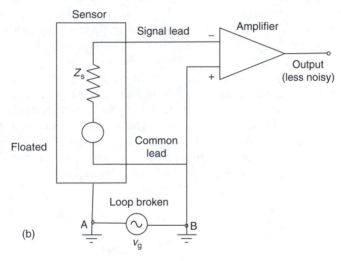

FIGURE 2.11
(a) Illustration of a ground loop. (b) Device isolation to eliminate ground loops (an example of internal isolation).

which may be removed by a filter. As well, a filter is capable of shaping a signal in a desired manner.

In typical applications of acquisition and processing of signals in a control system, the filtering task would involve the removal of signal components in a specific frequency range. In this context, we can identify the following four broad categories of filters:

1. Low-pass filters
2. High-pass filters
3. Band-pass filters
4. Band-reject (or notch) filters

The ideal frequency–response characteristic of each of these four types of filters is shown in Figure 2.12. Note that only the magnitude of the frequency response function (magnitude of the frequency transfer function) is shown. It is understood, however, that the phase distortion of the input signal also should be small within the pass band (the allowed frequency range). Practical filters are less than ideal. Their frequency response functions do not exhibit sharp cutoffs as in Figure 2.12 and, furthermore, some phase distortion will be unavoidable.

A special type of band-pass filter that is widely used in acquisition and monitoring of response signals (e.g., in product dynamic testing) is tracking filter. This is simply a band-pass filter with a narrow pass band that is frequency tunable. The center frequency (mid-value) of the pass band is variable, usually by coupling it to the frequency of a carrier signal (e.g., drive signal). In this manner, signals whose frequency varies with some basic variable in the system (e.g., rotor speed, frequency of a harmonic excitation signal, frequency of a sweep oscillator) can be accurately tracked in the presence of noise. The inputs to a tracking filter are the signal that is tracked and the variable tracking frequency (carrier input). A typical tracking filter that can simultaneously track two signals is schematically shown in Figure 2.13.

Filtering can be achieved by digital filters as well as analog filters. Before digital signal processing became efficient and economical, analog filters were exclusively used for signal filtering, and are still widely used. An analog filter is typically an active filter containing active components such as transistors or op-amps. In an analog filter, the input signal is passed through an analog circuit. Dynamics of the circuit will determine which (desired) signal components would be allowed through and which (unwanted) signal components would be rejected. Earlier versions of analog filters employed discrete circuit elements such as discrete transistors, capacitors, resistors, and even discrete inductors. Since inductors have several shortcomings such as susceptibility to electromagnetic noise, unknown resistance effects, and large size, today they are rarely used in filter circuits. Furthermore, due to well-known advantages of IC devices, today analog filters in the form of monolithic IC chips are extensively used in modem applications and are preferred

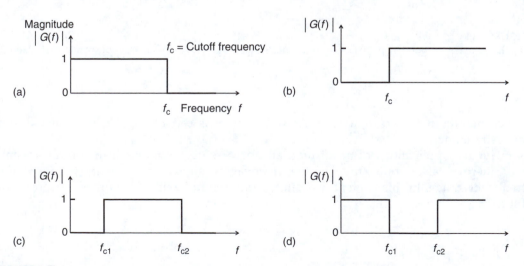

FIGURE 2.12
Ideal filter characteristics: (a) Low-pass filter. (b) High-pass filter. (c) Band-pass filter. (d) Band-reject (notch) filter.

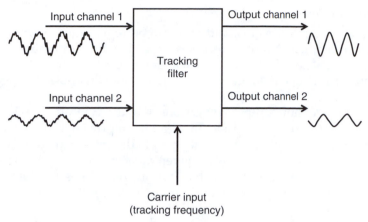

FIGURE 2.13
Schematic representation of a two-channel tracking filter.

over discrete-element filters. Digital filters, which employ digital signal processing to achieve filtering, are also widely used today.

2.4.1 Passive Filters and Active Filters

Passive analog filters employ analog circuits containing passive elements such as resistors and capacitors (and sometimes inductors) only. An external power source is not needed in a passive filter. Active analog filters employ active elements and components such as transistors and op-amps in addition to passive elements. Since external power is needed for the operation of the active elements and components, an active filter is characterized by the need for an external power supply. Active filters are widely available in a monolithic IC package and are usually preferred over passive filters.

Advantages of active filters include the following:

1. Loading effects and interaction with other components are negligible because active filters can provide a very high input impedance and a very low output impedance.
2. They can be used with low signal levels because both signal amplification and filtering can be provided by the same active circuit.
3. They are widely available in a low-cost and compact IC form.
4. They can be easily integrated with digital devices.
5. They are less susceptible to noise from EMI.

Commonly mentioned disadvantages of active filters are the following:

1. They need an external power supply.
2. They are susceptible to saturation-type nonlinearity at high signal levels.
3. They can introduce many types of internal noise and unmodeled signal errors (offset, bias signals, etc.).

Note that advantages and disadvantages of passive filters can be directly inferred from the disadvantages and advantages of active filters, as given earlier.

2.4.1.1 *Number of Poles*

Analog filters are dynamic systems, and they can be represented by transfer functions, assuming linear dynamics. Number of poles of a filter is the number of poles in the associated transfer function. This is also equal to the order of the characteristic polynomial of the filter transfer function (i.e., order of the filter). Note that poles (or eigenvalues) are the roots of the characteristic equation.

In our discussion, we show simplified versions of filters, typically consisting of a single filter stage. Performance of such a basic filter can be improved at the expense of circuit complexity (and increased pole count). Only simple discrete-element circuits are shown for passive filters. Basic op-amp circuits are given for active filters. Even here, much more complex devices are commercially available, but our purpose is to illustrate the underlying principles rather than to provide complete descriptions and data sheets for commercial filters.

2.4.2 Low-Pass Filters

The purpose of a low-pass filter is to allow through all signal components below a certain (cutoff) frequency and block off all signal components above that cutoff. Analog low-pass filters are widely used as antialiasing filters in digital signal processing. An error known as aliasing enters the digitally processed results of a signal if the original signal has frequency components above half the sampling frequency (half the sampling frequency is called the Nyquist frequency). Hence, aliasing distortion can be eliminated if the signal is filtered using a low-pass filter with its cutoff set at Nyquist frequency, before sampling and digital processing. This is one of the numerous applications of analog low-pass filters. Another typical application would be to eliminate high-frequency noise in a measured system response.

A single-pole passive low-pass filter circuit is shown in Figure 2.14a. An active filter corresponding to the same low-pass filter is shown in Figure 2.14b. It can be shown that the two circuits have identical transfer functions. Hence, it might seem that the op-amp in Figure 2.14b is redundant. This is not true, however. If two passive filter stages, each similar to Figure 2.14a, are connected together, the overall transfer function is not equal to the product of the transfer functions of the individual stages. The reason for this apparent ambiguity is the circuit loading (interaction) that arises due to the fact that the input impedance of the second stage is not sufficiently larger than the output impedance of the first stage. But, if two active filter stages similar to Figure 2.14b are connected together, such loading errors will be negligible because the op-amp with feedback (i.e., a voltage follower) introduces a high input impedance and low output impedance, while maintaining the voltage gain at unity. With similar reasoning, it can be concluded that an active filter has the desirable property of very low interaction with any other connected component.

To obtain the filter equation for Figure 2.14a, note that since the output is open circuit (zero load current), the current through capacitor C is equal to the current through resistor R. Hence,

$$C\frac{dv_o}{dt} = \frac{v_i - v_o}{R},$$

or

$$\tau\frac{dv_o}{dt} + v_o = v_i, \tag{2.23}$$

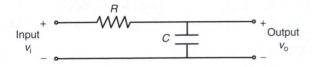

(a)

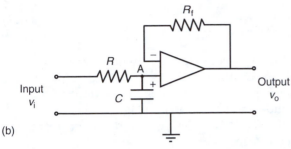

(b)

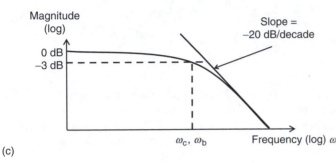

(c)

FIGURE 2.14
A single-pole low-pass filter. (a) A passive filter stage. (b) An active filter stage. (c) The frequency response characteristic.

where the filter time constant is

$$\tau = RC. \tag{2.24}$$

Now, from Equation 2.23, it follows that the filter transfer function is

$$\frac{v_o}{v_i} = G(s) = \frac{1}{(\tau s + 1)}. \tag{2.25}$$

From this transfer function, it is clear that an analog low-pass filter is essentially a lag circuit (i.e., it provides a phase lag).

It can be shown that the active filter stage in Figure 2.14b has the same input/output equation. First, since current through an op-amp lead is almost zero, it follows from the previous analysis of the passive circuit stage that

$$\frac{v_A}{v_i} = \frac{1}{(\tau s + 1)}, \tag{i}$$

where v_A is the voltage at the node point A. Now, since the op-amp with feedback resistor is in fact a voltage follower, we have

$$\frac{v_o}{v_A} = 1. \tag{ii}$$

Next, by combining Equation i and Equation ii, we get Equation 2.25 as required. Repeating, a main advantage of the active filter version is that the resulting loading error is negligible.

The frequency response function corresponding to Equation 2.25 is obtained by setting $s = j\omega$; thus,

$$G(j\omega) = \frac{1}{(\tau j\omega + 1)}. \tag{2.26}$$

This gives the response of the filter when a sinusoidal signal of frequency ω is applied. The magnitude $|G(j\omega)|$ of the frequency transfer function gives the signal amplification, and the phase angle $\angle G(j\omega)$ gives the phase lead of the output signal with respect to the input. The magnitude curve (Bode magnitude curve) is shown in Figure 2.14c. Note from Equation 2.26 that for small frequencies (i.e., $\omega \ll 1/\tau$) the magnitude is approximately unity. Hence, $1/\tau$ can be considered the cutoff frequency ω_c:

$$\omega_c = \frac{1}{\tau}. \tag{2.27}$$

Example 2.6
Show that the cutoff frequency given by Equation 2.27 is also the half-power bandwidth for the low-pass filter. Show that for frequencies much larger than this, the filter transfer function on the Bode magnitude plane (i.e., log magnitude vs. log frequency) can be approximated by a straight line with slope -20 dB/decade. This slope is known as the *roll-off rate*.

Solution
The frequency corresponding to half power (or $1/\sqrt{2}$ magnitude) is given by

$$\frac{1}{|\tau j\omega + 1|} = \frac{1}{\sqrt{2}}.$$

By cross-multiplying, squaring, and simplifying the equation we get

$$\tau^2 \omega^2 = 1.$$

Hence, the half-power bandwidth is

$$\omega_b = \frac{1}{\tau}. \tag{2.28}$$

This is identical to the cutoff frequency given by Equation 2.17.

Now for $\omega \gg 1/\tau$ (i.e., $\tau\omega \gg 1$), Equation 2.26 can be approximated by

$$G(j\omega) = \frac{1}{\tau j\omega}.$$

This has the magnitude

$$|G(j\omega)| = \frac{1}{\tau\omega}.$$

Converting to the log scale,

$$\log_{10} |G(j\omega)| = -\log_{10} \omega - \log_{10} \tau.$$

It follows that the $\log_{10}$ (magnitude) vs. $\log_{10}$ (frequency) curve is a straight line with slope -1. In other words, when frequency increases by a factor of 10 (i.e., a decade), the $\log_{10}$ magnitude decreases by unity (i.e., by 20 dB). Hence, the roll-off rate is -20 dB/decade. These observations are shown in Figure 2.14c. Note that an amplitude change by a factor of $\sqrt{2}$ (or power by a factor of 2) corresponds to 3 dB. Hence, when the dc (zero-frequency magnitude) value is unity (0 dB), the half-power magnitude is -3 dB.

Cutoff frequency and the roll-off rate are the two main design specifications for a low-pass filter. Ideally, we would like a low-pass filter magnitude curve to be flat up to the required pass-band limit (cutoff frequency) and then roll off very rapidly. The low-pass filter shown in Figure 2.14 only approximately meets these requirements. In particular, the roll-off rate is not large enough. We would prefer a roll-off rate of at least -40 dB/decade and even -60 dB/decade in practical filters. This can be realized by using a high-order filter (i.e., a filter with many poles). Low-pass Butterworth filter is of this type and is widely used.

2.4.2.1 Low-Pass Butterworth Filter

A low-pass Butterworth filter with two poles can provide a roll-off rate of -40 dB/decade, and one with three poles can provide a roll-off rate of -60 dB/decade. Furthermore, the steeper the roll-off slope, the flatter the filter magnitude curve within the pass band.

A two-pole, low-pass Butterworth filter is shown in Figure 2.15. We could construct a two-pole filter simply by connecting together two single-pole stages of the type shown in Figure 2.14b. Then, we would require two op-amps, whereas the circuit shown in Figure 2.15 achieves the same objective by using only one op-amp (i.e., at a lower cost).

Example 2.7
Show that the op-amp circuit in Figure 2.15 is a low-pass filter with two poles. What is the transfer function of the filter? Estimate the cutoff frequency under suitable conditions. Show that the roll-off rate is -40 dB/decade.

Solution
To obtain the filter equation, we write the current balance equations first. Specifically, the sum of the currents through R_1 and C_1 passes through R_2. The same current has to pass through C_2 because the current through the op-amp lead is zero (a property of an op-amp). Hence,

$$\frac{v_i - v_A}{R_1} + C_1 \frac{d}{dt}(v_o - v_A) = \frac{v_A - v_B}{R_2} = C_2 \frac{dv_B}{dt}. \tag{i}$$

Also, since the op-amp with a feedback resistor R_f is a voltage follower (with unity gain), we have

$$v_B = v_o. \tag{ii}$$

From Equation i and Equation ii we get

$$\frac{v_i - v_A}{R_1} + C_1 \frac{dv_o}{dt} - C_1 \frac{dv_A}{dt} = C_2 \frac{dv_o}{dt} \tag{iii}$$

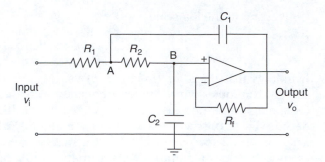

FIGURE 2.15
A two-pole low-pass Butterworth filter.

$$\frac{v_A - v_o}{R_2} = C_2 \frac{dv_o}{dt}.$$ (iv)

Now, defining the constants

$$\tau_1 = R_1 C_1,$$ (2.29)
$$\tau_2 = R_2 C_2,$$ (2.30)
$$\tau_3 = R_1 C_2,$$ (2.31)

and introducing the Laplace variable s, we can eliminate v_A by substituting Equation iv into Equation iii. Thus,

$$\frac{v_o}{v_i} = \frac{1}{[\tau_1 \tau_2 s^2 + (\tau_2 + \tau_3 + 1]} = \frac{\omega_n^2}{[s^2 + 2\zeta\omega_n^2 + \omega_n^2]}.$$ (2.32)

This second-order transfer function becomes oscillatory if $(\tau_2 + \tau_3)^2 < 4\tau_1\tau_2$. Ideally, we would like to have a zero-resonant frequency, which corresponds to a damping ratio value $\zeta = 1/\sqrt{2}$. Since the undamped natural frequency is

$$\omega_n = \frac{1}{\sqrt{\tau_1 \tau_2}},$$ (2.33)

the damping ratio is

$$\zeta = \frac{\tau_2 + \tau_3}{\sqrt{4\tau_1\tau_2}}$$ (2.34)

and the resonant frequency is

$$\omega_r = \sqrt{1 - 2\zeta^2}\,\omega_n.$$ (2.35)

We have, under ideal conditions (i.e., for $\omega_r = 0$),

$$(\tau_2 + \tau_3)^2 = 2\tau_1\tau_2.$$ (2.36)

The frequency response function of the filter is (see Equation 2.32)

$$G(j\omega) = \frac{\omega_n^2}{[\omega_n^2 - \omega^2 + 2j\zeta\omega_n\omega]}.$$ (2.37)

Now, for $\omega \ll \omega_n$, the filter frequency response is flat with a unity gain. For $\omega \gg \omega_n$, the filter frequency response can be approximated by

$$G(j\omega) = -\frac{\omega_n^2}{\omega^2}.$$

In a log (magnitude) vs. log (frequency) scale, this function is a straight line with slope of -2. Hence, when the frequency increases by a factor of 10 (i.e., one decade), the $\log_{10}$ (magnitude) drops by 2 units (i.e., 40 dB). In other words, the roll-off rate is -40 dB/decade. Also, ω_n can be taken as the filter cutoff frequency. Hence,

$$\omega_c = \frac{1}{\sqrt{\tau_1 \tau_2}}. \tag{2.38}$$

It can be easily verified that when $\zeta = 1/\sqrt{2}$, this frequency is identical to the half-power bandwidth (i.e, the frequency at which the transfer function magnitude becomes $1/\sqrt{2}$).

Note that if two single-pole stages (of the type shown in Figure 2.14b) are cascaded, the resulting two-pole filter has an overdamped (nonoscillatory) transfer function, and it is not possible to achieve $\zeta = 1/\sqrt{2}$ as in the present case. Also, note that a three-pole low-pass Butterworth filter can be obtained by cascading the two-pole unit shown in Figure 2.15 with a single-pole unit shown in Figure 2.14b. Higher-order low-pass Butterworth filters can be obtained in a similar manner by cascading an appropriate selection of basic units.

2.4.3 High-Pass Filters

Ideally, a high-pass filter allows through it all signal components above a certain (cutoff) frequency and blocks off all signal components below that frequency. A single-pole high-pass filter is shown in Figure 2.16. As for the low-pass filter that was discussed earlier, the passive filter stage (Figure 2.16a) and the active filter stage (Figure 2.16b) have identical transfer functions. The active filter is desired, however, because of its many advantages, including negligible loading error due to high input impedance and low output impedance of the op-amp voltage follower that is present in this circuit.

Filter equation is obtained by considering current balance in Figure 2.16a, noting that the output is in open circuit (zero-load current). Accordingly,

$$C\frac{d}{dt}(v_1 - v_o) = \frac{v_o}{R},$$

or

$$\left(\tau\frac{dv_o}{dt}\right) + v_o = \left(\tau\frac{dv_i}{dt}\right), \tag{2.39}$$

where the filter time constant

$$\tau = RC. \tag{2.40}$$

Introducing the Laplace variable s, the filter transfer function is obtained as

$$\frac{v_o}{v_i} = G(s) = \frac{\tau s}{(\tau s + 1)}. \tag{2.41}$$

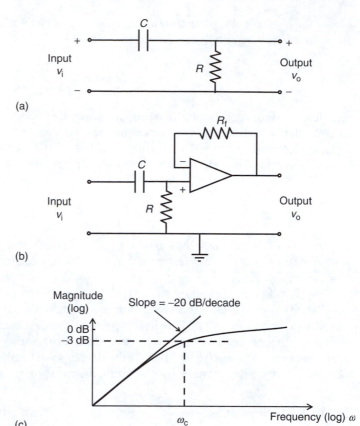

FIGURE 2.16
A single-pole high-pass filter. (a)
A passive filter stage. (b) An active
filter stage. (c) Frequency response
characteristic.

Note that this corresponds to a lead circuit (i.e., an overall phase lead is provided by this transfer function). The frequency response function is

$$G(j\omega) = \frac{\tau j\omega}{(\tau j\omega + 1)}. \tag{2.42}$$

Since its magnitude is zero for $\omega \ll 1/\tau$, and it is unity for $\omega \gg 1/\tau$, we have the cutoff frequency

$$\omega_c = \frac{1}{\tau}. \tag{2.43}$$

Signals above this cutoff frequency should be allowed undistorted, by an ideal high-pass filter, and signals below the cutoff should be completely blocked off. The actual behavior of the basic high-pass filter discussed earlier is not that perfect, as observed from the frequency-response characteristic shown in Figure 2.16c. It can be easily verified that the half-power bandwidth of the basic high-pass filter is equal to the cutoff frequency given by Equation 2.43, as in the case of the basic low-pass filter. The roll-up slope of the single-pole high-pass filter is 20 dB/decade. Steeper slopes are desirable. Multiple-pole, high-pass Butterworth filters can be constructed to give steeper roll-up slopes and reasonably flat pass-band magnitude characteristics.

2.4.4 Band-Pass Filters

An ideal band-pass filter passes all signal components within a finite frequency band and blocks off all signal components outside that band. The lower frequency limit of the pass band is called the *lower cutoff frequency* (ω_{c1}), and the upper frequency limit of the band is called the *upper cutoff frequency* (ω_{c2}). The most straightforward way to form a band-pass filter is to cascade a high-pass filter of cutoff frequency ω_{c1} with a low-pass filter of cutoff frequency ω_{c2}. Such an arrangement is shown in Figure 2.17. The passive circuit shown in Figure 2.17a is obtained by connecting together the circuits shown in Figure 2.14a and Figure 2.16a. The active circuit shown in Figure 2.17b is obtained by connecting a voltage follower op-amp circuit to the original passive circuit. Passive and active filters have the same transfer function, assuming that loading problems (component interaction) are not present in the passive filter. Since loading errors and interactions can be serious in practice, however, the active version is preferred.

To obtain the filter equation, first consider the high-pass portion of the circuit shown in Figure 2.17a. Since the output is in open-circuit (zero current), we have from Equation 2.41

$$\frac{v_o}{v_A} = \frac{\tau_2 s}{(\tau_2 s + 1)},$$ (i)

where

$$\tau_2 = R_2 C_2.$$ (2.44)

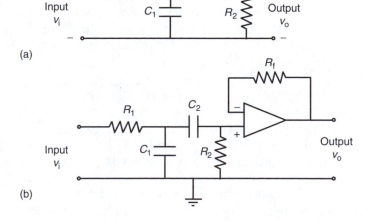

(a)

(b)

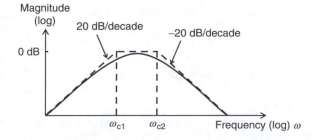

(c)

FIGURE 2.17
Band-pass filter. (a) A basic passive filter stage. (b) A basic active filter stage. (c) Frequency response characteristic.

Next, on writing the current balance at node A of the circuit we have

$$\frac{v_i - v_A}{R_1} = C_1 \frac{dv_A}{dt} + C_2 \frac{d}{dt}(v_A - v_o). \tag{ii}$$

Introducing the Laplace variable s we get

$$v_i = (\tau_1 s + \tau_3 s + 1)v_A - \tau_3 s v_o, \tag{iii}$$

where

$$\tau_1 = R_1 C_1 \tag{2.45}$$

and

$$\tau_3 = R_1 C_2. \tag{2.46}$$

Now on eliminating v_A by substituting Equation i in Equation iii, we get the band-pass filter transfer function

$$\frac{v_o}{v_i} = G(s) = \frac{\tau_2 s}{[\tau_1 \tau_2 s^2 + (\tau_1 + \tau_2 + \tau_3)s + 1]}. \tag{2.47}$$

We can show that the roots of the characteristic equation

$$\tau_1 \tau_2 s^2 + (\tau_1 + \tau_2 + \tau_3)s + 1 = 0 \tag{2.48}$$

are real and negatives. The two roots are denoted by $-\omega_{c1}$ and $-\omega_{c2}$, and they provide the two cutoff frequencies shown in Figure 2.17c. It can be verified that, for this basic band-pass filter, the roll-up slope is $+20$ dB/decade and the roll-down slope is -20 dB/decade. These slopes are not sufficient in many applications. Furthermore, the flatness of the frequency response within the pass band of the basic filter is not adequate as well. More complex (higher-order) band-pass filters with sharper cutoffs and flatter pass bands are commercially available.

2.4.4.1 *Resonance-Type Band-Pass Filters*

There are many applications where a filter with a very narrow pass band is required. The tracking filter mentioned at the beginning of the section on analog filters is one such application. A filter circuit with a sharp resonance can serve as a narrow-band filter. Note that the cascaded RC circuit shown in Figure 2.17 does not provide an oscillatory response (filter poles are all real) and, hence, it does not form a resonance-type filter. A slight modification to this circuit using an additional resistor R_1, as shown in Figure 2.18a, will produce the desired effect.

To obtain the filter equation, note that for the voltage follower unit

$$v_A = v_o. \tag{i}$$

Next, since current through an op-amp lead is zero, for the high-pass circuit unit (see Equation 2.41), we have

$$\frac{v_A}{v_B} = \frac{\tau_2 s}{(\tau_2 s + 1)}, \tag{ii}$$

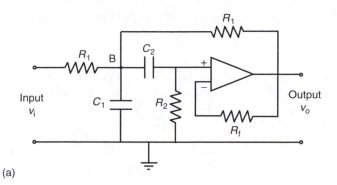

(a)

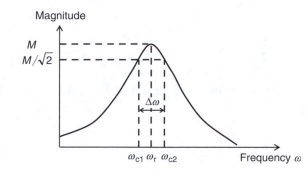

(b)

FIGURE 2.18
A resonance-type narrow band-pass filter. (a) An active filter stage. (b) Frequency response characteristic.

where

$$\tau_2 = R_2 C_2.$$

Finally, the current balance at node B gives

$$\frac{v_i - v_B}{R_1} = C_1 \frac{dv_B}{dt} + C_2 \frac{d}{dt}(v_B - v_A) + \frac{v_B - v_o}{R_1}$$

or, by using the Laplace variable, we get

$$v_i = (\tau_1 s + \tau_3 s + 2)v_B - \tau_3 s v_A - v_o. \tag{iii}$$

Now, by eliminating v_A and v_B in Equation i through Equation iii, we get the filter transfer function

$$\frac{v_o}{v_i} = G(s) = \frac{\tau_2 s}{[\tau_1 \tau_2 s^2 + (\tau_1 + \tau_2 + \tau_3)s + 2]}. \tag{2.49}$$

It can be shown that, unlike Equation 2.47, the present characteristic equation

$$\tau_1 \tau_2 s^2 + (\tau_1 + \tau_2 + \tau_3)s + 2 = 0 \tag{2.50}$$

can possess complex roots.

Example 2.8
Verify that the band-pass filter shown in Figure 2.18a can have a frequency response with a resonant peak, as shown in Figure 2.18b. Verify that the half-power bandwidth $\Delta\omega$ of the

filter is given by $2\zeta \omega_r$ at low damping values (note that ζ is the damping ratio and ω_r is the resonant frequency).

Solution
We may verify that the transfer function given by Equation 2.49 can have a resonant peak by showing that the characteristic Equation 2.50 can have complex roots. For example, if we use parameter values $C_1 = 2$, $C_2 = 1$, $R_1 = 1$, and $R_2 = 2$, we have $\tau_1 = 2$, $\tau_2 = 2$, and $\tau_3 = 1$. The corresponding characteristic equation is

$$4s^2 + 5s + 2 = 0,$$

which has the roots

$$-\frac{5}{8} \pm j\frac{\sqrt{7}}{8}$$

and is obviously complex.

To obtain an expression for the half-power bandwidth of the filter, note that the filter transfer function may be written as

$$G(s) = \frac{ks}{(s^2 + 2\zeta\omega_n s + \omega_n^2)}, \tag{2.51}$$

where ω_n is the undamped natural frequency, ζ is the damping ratio, and k is a gain parameter.

The frequency response function is given by

$$G(j\omega) = \frac{k\,j\omega}{[\omega_n^2 - \omega^2 + 2j\zeta\omega_n\omega]}. \tag{2.52}$$

For low damping, resonant frequency $\omega_r \cong \omega_n$. The corresponding peak magnitude M is obtained by substituting $\omega = \omega_n$ in Equation 2.52 and taking the transfer function magnitude. Thus,

$$M = \frac{k}{2\zeta\omega_n}. \tag{2.53}$$

At half-power frequencies we have

$$|G(j\omega)| = \frac{M}{\sqrt{2}}$$

or

$$\frac{k\omega}{\sqrt{(\omega_n^2 - \omega^2)^2 + 4\zeta^2\omega_n^2\omega^2}} = \frac{k}{2\sqrt{2}\zeta\omega_n}.$$

This gives

$$(\omega_n^2 - \omega^2)^2 = 4\zeta^2\omega_n^2\omega^2, \tag{2.54}$$

the positive roots of which provide the pass band frequencies ω_{c1} and ω_{c2}. Note that the roots are given by

$$\omega_n^2 - \omega^2 = \pm 2\zeta\omega_n\omega.$$

Hence, the two roots ω_{c1} and ω_{c2} satisfy the following two equations

$$\omega_{c1}^2 + 2\zeta\omega_n\omega_{c1} - \omega_n^2 = 0$$

and

$$\omega_{c2}^2 - 2\zeta\omega_n\omega_{c2} - \omega_n^2 = 0.$$

Accordingly, by solving these two quadratic equations and selecting the appropriate sign, we get

$$\omega_{c1} = -\zeta\omega_n + \sqrt{\omega_n^2 + \zeta^2\omega_n^2} \tag{2.55}$$

and

$$\omega_{c2} = \zeta\omega_n + \sqrt{\omega_n^2 + \zeta^2\omega_n^2}. \tag{2.56}$$

The half-power bandwidth is

$$\Delta\omega = \omega_{c2} - \omega_{c1} = 2\zeta\omega_n. \tag{2.57}$$

Now, since $\omega_n \cong \omega_r$ for low ζ we have

$$\Delta\omega = 2\zeta\omega_r. \tag{2.58}$$

A notable shortcoming of a resonance-type filter is that the frequency response within the bandwidth (pass band) is not flat. Hence, quite nonuniform signal attenuation takes place inside the pass band.

2.4.5 Band-Reject Filters

Band-reject filters or notch filters are commonly used to filter out a narrow band of noise components from a signal. For example, 60 Hz line noise in a signal can be eliminated by using a notch filter with a notch frequency of 60 Hz.

An active circuit that could serve as a notch filter is shown in Figure 2.19a. This is known as the Twin T circuit because its geometric configuration resembles two T-shaped circuits connected together. To obtain the filter equation, note that the voltage at point P is v_o because of unity gain of the voltage follower. Now, we write the current balance at nodes A and B. Thus,

$$\frac{v_i - v_B}{R} = 2C\frac{dv_B}{dt} + \frac{v_B - v_o}{R}$$

$$C\frac{d}{dt}(v_i - v_A) = \frac{v_A}{R/2} + C\frac{d}{dt}(v_A - v_o).$$

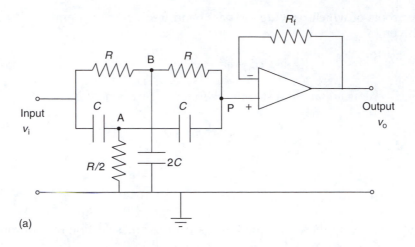

(a)

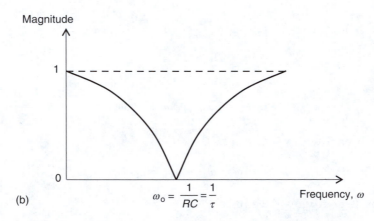

(b)

FIGURE 2.19
A notch filter. (a) An active twin T filter circuit. (b) Frequency response characteristic.

Next, since the current through the positive lead of the op-amp (voltage follower) is zero, we have the current through point P as

$$\frac{v_B - v_o}{R} = C\frac{d}{dt}(v_o - v_A).$$

These three equations are written in the Laplace form

$$v_i = 2(\tau s + 1)v_B - v_o, \tag{i}$$
$$\tau s v_i = 2(\tau s + 1)v_A - \tau s v_o, \tag{ii}$$
$$v_B = (\tau s + 1)v_o - \tau s v_A, \tag{iii}$$

where

$$\tau = RC. \tag{2.59}$$

Finally, eliminating v_A and v_B in Equation i through Equation iii we get

$$\frac{v_o}{v_i} = G(s) = \frac{(\tau^2 s^2 + 1)}{(\tau^2 s^2 + 4\tau s + 1)}. \tag{2.60}$$

The frequency response function of the filter (with $s = j\omega$) is

$$G(j\omega) = \frac{(1 - \tau^2\omega^2)}{(1 - \tau^2\omega^2 + 4j\tau\omega)}. \tag{2.61}$$

Note that the magnitude of this function becomes zero at frequency

$$\omega_o = \frac{1}{\tau}. \tag{2.62}$$

This is known as the notch frequency. The magnitude of the frequency response function of the notch filter is sketched in Figure 2.19b. It is noticed that any signal component at frequency ω_o will be completely eliminated by the notch filter. Sharp roll-down and roll-up are needed to allow the other (desirable) signal components through without too much attenuation.

While the previous three types of filters achieve their frequency response characteristics through the poles of the filter transfer function, a notch filter achieves its frequency response characteristic through its zeros (roots of the numerator polynomial equation). Some useful information about filters is summarized in Box 2.2.

2.5 Modulators and Demodulators

Sometimes signals are deliberately modified to maintain the accuracy during their transmission, conditioning, and processing. In signal modulation, the data signal, known as the *modulating signal*, is used to vary a property (such as amplitude or frequency) of a carrier signal. In this manner, the carrier signal is modulated by the data signal. After transmitting or conditioning the modulated signal, typically the data signal has to be recovered by removing the carrier signal. This is known as demodulation or *discrimination*.

A variety of modulation techniques exist, and several other types of signal modification (e.g., digitizing) could be classified as signal modulation even though they might not be commonly termed as such. Four types of modulation are illustrated in Figure 2.20. In amplitude modulation (AM), the amplitude of a periodic carrier signal is varied according to the amplitude of the data signal (modulating signal), keeping the frequency of the carrier signal (carrier frequency) constant. Suppose that the transient signal shown in Figure 2.20a is the modulating signal and a high-frequency sinusoidal signal is used as the carrier signal. The resulting amplitude-modulated signal is shown in Figure 2.20b. AM is used in telecommunication, transmission of radio and TV signals, instrumentation, and signal conditioning. The underlying principle is particularly useful in applications such as sensing and control instrumentation of control systems, and fault detection and diagnosis in rotating machinery.

In frequency modulation (FM), the frequency of the carrier signal is varied in proportion to the amplitude of the data signal (modulating signal), while keeping the amplitude of the carrier signal constant. Suppose that the data signal shown in Figure 2.20a is used to frequency modulate a sinusoidal carrier signal. The modulated result will appear as in Figure 2.20c. Since information is carried as frequency rather than amplitude, any noise that might alter the signal amplitude will have virtually no effect on the transmitted data. Hence, FM is less susceptible to noise than AM. Furthermore, since in FM the carrier amplitude is kept constant, signal weakening and noise effects that are unavoidable in long-distance data communication will have less effect than in the case of AM, particularly if the data signal

Box 2.2   Filters

Active Filters (Need External Power)
  Advantages

  - Smaller loading errors and interaction (have high input impedance and low output impedance, and hence do not affect the input circuit conditions, output signals, and other components)
  - Lower cost
  - Better accuracy.

Passive Filters (No External Power, Use Passive Elements)
  Advantages

  - Useable at very high frequencies (e.g., radio frequency)
  - No need for power supply.

Filter Types

  - Low pass: Allows frequency components up to cutoff and rejects the higher-frequency components.
  - High pass: Rejects frequency components up to cutoff and allows the higher-frequency components.
  - Band pass: Allows frequency components within an interval and rejects the rest.
  - Notch (or band reject): Rejects frequency components within an interval (usually, a narrow band) and allows the rest.

Definitions

  - Filter order: Number of poles in the filter circuit or transfer function.
  - Antialiasing filter: Low-pass filter with cutoff at less than half the sampling rate (i.e., at less than Nyquist frequency) for digital processing.
  - Butterworth filter: A high-order filter with a flat pass band.
  - Chebyshev filter: An optimal filter with uniform ripples in the pass band.
  - Sallen-key filter: An active filter whose output is in phase with input.

level is low in the beginning. However, more sophisticated techniques and hardware are needed for signal recovery (demodulation) in FM transmission because FM demodulation involves frequency discrimination rather than amplitude detection. FM is also widely used in radio transmission and in data recording and replay.

In pulse-width modulation (PWM), the carrier signal is a pulse sequence. The pulse width is changed in proportion to the amplitude of the data signal, while keeping the pulse spacing constant. This is illustrated in Figure 2.20d. Suppose that the high level of the PWM signal corresponds to the "on" condition of a circuit and the low level corresponds to the "off" condition. Then, as shown in Figure 2.21, the pulse width is equal to the on time ΔT of

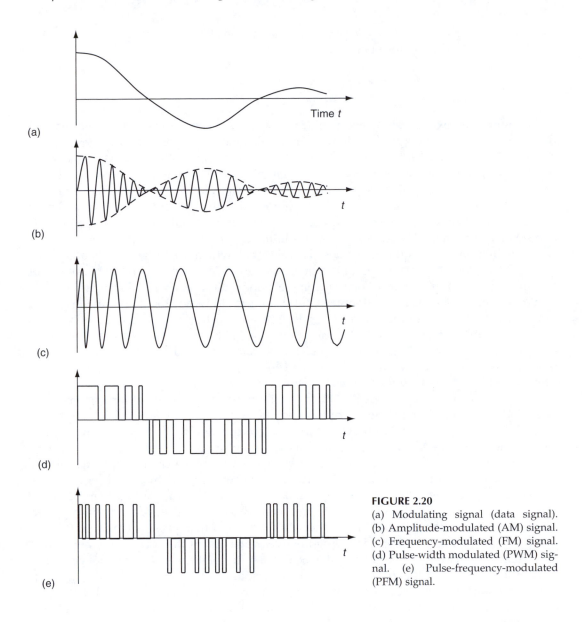

(a)

(b)

(c)

(d)

(e)

FIGURE 2.20
(a) Modulating signal (data signal).
(b) Amplitude-modulated (AM) signal.
(c) Frequency-modulated (FM) signal.
(d) Pulse-width modulated (PWM) signal. (e) Pulse-frequency-modulated (PFM) signal.

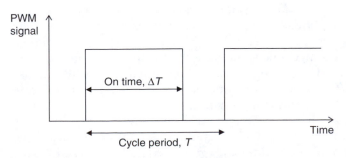

FIGURE 2.21
Duty cycle of a PWM signal.

the circuit within each signal cycle period T. The duty cycle of the PWM is defined as the percentage on time in a pulse period and is given by

$$\text{Duty cycle} = \frac{\Delta T}{T} \times 100\%. \tag{2.63}$$

PWM signals are extensively used in control systems, for controlling electric motors and other mechanical devices such as valves (hydraulic and pneumatic) and machine tools. Note that in a given (short) time interval, the average value of the PWM signal is an estimate of the average value of the data signal in that period. Hence, PWM signals can be used directly in controlling a process, without demodulating it. Advantages of PWM include better energy efficiency (less dissipation) and better performance with nonlinear devices. For example, a device may stick at low speeds due to Coulomb friction. This can be avoided by using a PWM signal with an amplitude that is sufficient to overcome friction, while maintaining the required average control signal, which might be very small.

In PFM, as well, the carrier signal is a pulse sequence. In this method, the frequency of the pulses is changed in proportion to the value of the data signal, while keeping the pulse width constant. PFM has the advantages of ordinary FM. Additional advantages result due to the fact that electronic circuits (digital circuits in particular) can handle pulses very efficiently. Furthermore, pulse detection is not susceptible to noise because it involves distinguishing between the presence and the absence of a pulse, rather than accurate determination of the pulse amplitude (or width). PFM may be used in place of PWM in most applications, with better results.

Another type of modulation is PM. In this method, the phase angle of the carrier signal is varied in proportion to the amplitude of the data signal. Conversion of discrete (sampled) data into the digital (binary) form is also considered a form of modulation. In fact, this is termed PCM. In this case, each discrete data sample is represented by a binary number containing a fixed number of binary digits (bits). Since each digit in the binary number can take only two values, 0 or 1, it can be represented by the absence or the presence of a voltage pulse. Hence, each data sample can be transmitted using a set of pulses. This is known as *encoding*. At the receiver, the pulses have to be interpreted (or decoded) to determine the data value. As with any other pulse technique, PCM is quite immune to noise because decoding involves detection of the presence or absence of a pulse, rather than determination of the exact magnitude of the pulse signal level. Also, since pulse amplitude is constant, long-distance signal transmission (of this digital data) can be accomplished without the danger of signal weakening and associated distortion. Of course, there will be some error introduced by the digitization process itself, which is governed by the finite word size (or dynamic range) of the binary data element. This is known as the *quantization error* and is unavoidable in signal digitization.

In any type of signal modulation, it is essential to preserve the algebraic sign of the modulating signal (data). Different types of modulators handle this in different ways. For example, in PCM, an extra sign bit is added to represent the sign of the transmitted data sample. In AM and FM, a phase-sensitive demodulator is used to extract the original (modulating) signal with the correct algebraic sign. Note that in these two modulation techniques a sign change in the modulating signal can be represented by a 180° phase change in the modulated signal. This is not quite noticeable in Figure 2.20b and Figure 2.20c. In PWM and PFM, a sign change in the modulating signal can be represented by changing the sign of the pulses, as shown in Figure 2.20d and Figure 2.20e. In PM, a positive range of phase angles (say 0 to π) could be assigned for the positive values of the data signal, and a negative range of phase angles (say $-\pi$ to 0) could be assigned for the negative values of the signal.

2.5.1 Amplitude Modulation

AM can naturally enter into many physical phenomena. More important, perhaps, is the deliberate (artificial) use of AM to facilitate data transmission and signal conditioning. Let us first examine the related mathematics.

AM is achieved by multiplying the data signal (modulating signal) $x(t)$ by a high-frequency (periodic) carrier signal $x_c(t)$. Hence, amplitude-modulated signal $x_a(t)$ is given by

$$x_a(t) = x(t)x_c(t). \tag{2.64}$$

Note that the carrier could be any periodic signal such as harmonic (sinusoidal), square wave, or triangular. The main requirement is that the fundamental frequency of the carrier signal (carrier frequency) f_c be significantly large (say, by a factor of 5 or 10) than the highest frequency of interest (bandwidth) of the data signal. Analysis can be simplified by assuming a sinusoidal carrier frequency. Thus,

$$x_c(t) = a_c \cos 2\pi f_c t. \tag{2.65}$$

2.5.1.1 Modulation Theorem

This is also known as the *frequency-shifting theorem* and relates the fact that if a signal is multiplied by a sinusoidal signal, the Fourier spectrum of the product signal is simply the Fourier spectrum of the original signal shifted through the frequency of the sinusoidal signal. In other words, the Fourier spectrum $X_a(f)$ of the amplitude-modulated signal $x_a(t)$ can be obtained from the Fourier spectrum $X(f)$ of the original data signal $x(t)$, simply by shifting it through the carrier frequency f_c.

To mathematically explain the modulation theorem, we use the definition of the Fourier integral transform to get

$$X_a(f) = a_c \int\limits_{-\infty}^{\infty} x(t) \cos 2\pi f_c t \exp(-j2\pi ft)dt.$$

Next, since

$$\cos 2\pi f_c t = \frac{1}{2}[\exp(j2\pi f_c t) + \exp(-j2\pi f_c t)],$$

we have

$$X_a(f) = \frac{1}{2}a_c \int\limits_{-\infty}^{\infty} x(t) \exp[-j2\pi(f - f_c)t]dt + \frac{1}{2}a_c \int\limits_{-\infty}^{\infty} x(t) \exp[-j2\pi(f + f_c)t]dt$$

or,

$$X_a(f) = \frac{1}{2}a_c[X(f - f_c) + X(f + f_c)]. \tag{2.66}$$

Equation 2.66 is the mathematical statement of the modulation theorem. It is illustrated by an example in Figure 2.22. Consider a transient signal $x(t)$ with a (continuous) Fourier spectrum $X(f)$, whose magnitude $|X(f)|$ is as shown in Figure 2.22a. If this signal is used

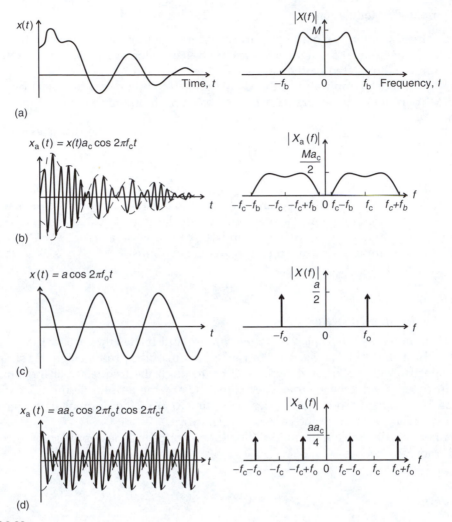

FIGURE 2.22
Illustration of the modulation theorem. (a) A transient data signal and its Fourier spectrum magnitude. (b) Amplitude-modulated signal and its Fourier spectrum magnitude. (c) A sinusoidal data signal. (d) Amplitude modulation by a sinusoidal signal.

to amplitude modulate a high-frequency sinusoidal signal, the resulting modulated signal $x_a(t)$ and the magnitude of its Fourier spectrum are as shown in Figure 2.22b. It should be kept in mind that the magnitude has been multiplied by $a_c/2$. Furthermore, the data signal is assumed to be band limited, with bandwidth f_b. Of course, the theorem is not limited to band-limited signals, but for practical reasons, we need to have some upper limit on the useful frequency of the data signal. Also, for practical reasons (not for the theorem itself), the carrier frequency f_c should be several times larger than f_c so that there is a reasonably wide frequency band from 0 to (f_c-f_b), within which the magnitude of the modulated signal is virtually zero. The significance of this should be clear when we discuss applications of AM.

Figure 2.22 shows only the magnitude of the frequency spectra. It should be remembered, however, that every Fourier spectrum has a phase angle spectrum as well. This is not shown for the sake of conciseness. But, clearly, the phase-angle spectrum is also similarly affected (frequency shifted) by AM.

2.5.1.2 *Side Frequencies and Side Bands*

The modulation theorem, as described earlier, assumed transient data signals with associated continuous Fourier spectra. The same ideas are applicable as well to periodic signals (with discrete spectra). Periodic signals represent merely a special case of what was discussed earlier. This case can be analyzed by directly using the Fourier integral transform. In that case, however, we have to cope with impulsive spectral lines. Alternatively, Fourier series expansion may be employed thereby avoiding the introduction of impulsive discrete spectra into the analysis. As shown in Figure 2.22c and Figure 2.22d, however, no analysis is actually needed for the case of periodic signals because the final answer can be deduced from the results for a transient signal. Specifically, in the Fourier series expansion of the data signal, each frequency component f_o with amplitude $a/2$ will be shifted by $\pm f_c$ to the two new frequency locations $f_c + f_o$ and $-f_c + f_o$ with an associated amplitude $aa_c/4$. The negative frequency component $-f_o$ should also be considered in the same way, as illustrated in Figure 2.22d. Note that the modulated signal does not have a spectral component at the carrier frequency f_c but rather, on each side of it, at $f_c \pm f_o$. Hence, these spectral components are termed side frequencies. When a band of side frequencies is present, it is termed a *side band*. Side frequencies are very useful in fault detection and diagnosis of rotating machinery.

2.5.2 Application of Amplitude Modulation

The main hardware component of an amplitude modulator is an analog multiplier. It is commercially available in the monolithic IC form. Alternatively, one can be assembled using IC op-amps and various discrete circuit elements. Schematic representation of an amplitude modulator is shown in Figure 2.23. In practice, to achieve satisfactory modulation, other components such as signal preamplifiers and filters would be needed.

There are many applications of AM. In some applications, modulation is performed intentionally. In others, modulation occurs naturally as a consequence of the physical process, and the resulting signal is used to meet a practical objective. Typical applications of AM include the following:

1. Conditioning of general signals (including dc, transient, and low frequency) by exploiting the advantages of ac signal-conditioning hardware
2. Improvement of the immunity of low-frequency signals to low-frequency noise
3. Transmission of general signals (dc, low frequency, etc.) by exploiting the advantages of ac signals
4. Transmission of low-level signals under noisy conditions

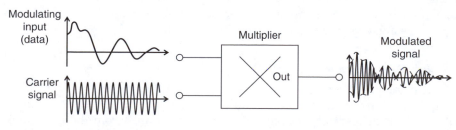

FIGURE 2.23
Representation of an amplitude modulator.

5. Transmission of several signals simultaneously through the same medium (e.g., same telephone line, same transmission antenna, etc.)

6. Fault detection and diagnosis of rotating machinery.

The role of AM in many of these applications should be obvious if one understands the frequency-shifting property of AM. Several other types of applications are also feasible due to the fact that power of the carrier signal can be increased somewhat arbitrarily, irrespective of the power level of the data (modulating) signal. Let us discuss, one by one, the six categories of applications already mentioned.

AC signal-conditioning devices such as ac amplifiers are known to be more stable than their dc counterparts. In particular, drift problems are not as severe and nonlinearity effects are lower in ac signal-conditioning devices. Hence, instead of conditioning a dc signal using dc hardware, we can first use the signal to modulate a high-frequency carrier signal. Then, the resulting high-frequency modulated signal may be conditioned more effectively using ac hardware.

The frequency-shifting property of AM can be exploited in making low-frequency signals immune to low-frequency noise. Note from Figure 2.22 that using AM, low-frequency spectrum of the modulating signal can be shifted out into a very high frequency region, by choosing a carrier frequency f_c that is sufficiently large. Then, any low-frequency noise (within the band 0 to $f_c - f_b$) would not distort the spectrum of the modulated signal. Hence, this noise could be removed by a high-pass filter (with cutoff at $f_c - f_b$) so that it would not affect the data. Finally, the original data signal can be recovered using demodulation. Since the frequency of a noise component can very well be within the bandwidth f_b of the data signal, if AM was not employed, noise could directly distort the data signal.

Transmission of ac signals is more efficient than that of dc signals. Advantages of ac transmission include lower energy dissipation problems. As a result, a modulated signal can be transmitted over long distances more effectively than could the original data signal alone. Furthermore, the transmission of low-frequency (large wave-length) signals requires large antennas. Hence, when AM is employed (with an associated reduction in signal wave length), the size of broadcast antenna can be effectively reduced.

Transmission of weak signals over long distances is not desirable because further signal weakening and corruption by noise could produce disastrous results. By increasing the power of the carrier signal to a sufficiently high level, the strength of the modulated signal can be elevated to an adequate level for long-distance transmission.

It is impossible to transmit two or more signals in the same frequency range simultaneously using a single telephone line. This problem can be resolved by using carrier signals with significantly different carrier frequencies to amplitude modulate the data signals. By picking the carrier frequencies sufficiently farther apart, the spectra of the modulated signals can be made nonoverlapping, thereby making simultaneous transmission possible. Similarly, with AM, simultaneous broadcasting by several radio (AM) broadcast stations in the same broadcast area has become possible.

2.5.2.1 *Fault Detection and Diagnosis*

A use of the principle of AM, particularly important in the practice of electromechanical systems, is in the fault detection and diagnosis of rotating machinery. In this method, modulation is not deliberately introduced, but rather results from the dynamics of the machine. Flaws and faults in a rotating machine are known to produce periodic forcing signals at frequencies higher than, and typically at an integer multiple of, the rotating speed of the machine. For example, backlash in a gear pair will generate forces at the tooth-meshing frequency (equal to the product: number of teeth × gear rotating speed). Flaws in roller bearings can generate forcing signals at frequencies proportional to the rotating speed times the number of rollers in the bearing race. Similarly, blade passing in turbines and compressors, and eccentricity and unbalance in the rotor can generate forcing components at frequencies that are integer multiples of the rotating speed. The resulting system response is clearly an amplitude-modulated signal, where the rotating response of the machine modulates the high-frequency forcing response. This can be confirmed experimentally through Fourier analysis (fast Fourier transform or FFT) of the resulting response signals. For a gearbox, for example, it will be noticed that, instead of getting a spectral peak at the gear tooth-meshing frequency, two side bands are produced around that frequency. Faults can be detected by monitoring the evolution of these side bands. Furthermore, since side bands are the result of modulation of a specific forcing phenomenon (e.g., gear-tooth meshing, bearing-roller hammer, turbine-blade passing, unbalance, eccentricity, misalignment, etc.), one can trace the source of a particular fault (i.e., diagnose the fault) by studying the Fourier spectrum of the measured response.

AM is an integral part of many types of sensors. In these sensors, a high-frequency carrier signal (typically the ac excitation in a primary winding) is modulated by the motion. Actual motion can be detected by demodulation of the output. Examples of sensors that generate modulated outputs are differential transformers [linear variable differential transducer or transformer (LVDT), RVDT], magnetic-induction proximity sensors, eddy-current proximity sensors, ac tachometers, and strain-gage devices that use ac bridge circuits. Signal conditioning and transmission would be facilitated by AM in these cases. The signal has to be demodulated at the end, for most practical purposes such as analysis and recording.

2.5.3 Demodulation

Demodulation or discrimination, or detection is the process of extracting the original data signal from a modulated signal. In general, demodulation has to be phase sensitive in the sense that, algebraic sign of the data signal should be preserved and determined by the demodulation process. In full-wave demodulation, an output is generated continuously. In half-wave demodulation, no output is generated for every alternate half period of the carrier signal.

A simple and straightforward method of demodulation is by detection of the envelope of the modulated signal. For this method to be feasible, the carrier signal must be quite powerful (i.e., signal level has to be high) and the carrier frequency also should be very high. An alternative method of demodulation, which generally provides more reliable results, involves a further step of modulation performed on the already modulated signal, followed by low-pass filtering. This method can be explained by referring to Figure 2.22.

Consider the amplitude-modulated signal $x_a(t)$ shown in Figure 2.22b. If this signal is multiplied by the sinusoidal carrier signal $2/a_c \cos 2\pi f_c t$, we get

$$\tilde{x}(t) = \frac{2}{a_c} x_a(t) \cos 2\pi f_c t. \tag{2.67}$$

Now, by applying the modulation theorem (Equation 2.66) to Equation 2.67, we get the Fourier spectrum of $\tilde{x}(t)$ as

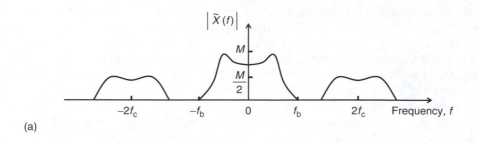

(a)

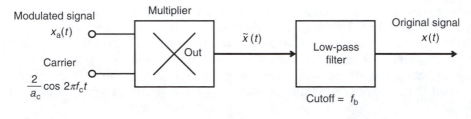

(b)

FIGURE 2.24
Amplitude demodulation. (a) Spectrum of the signal after the second modulation. (b) Demodulation schematic diagram (modulation + filtering).

$$\tilde{X}(f) = \frac{1}{2}\frac{2}{a_c}\left[\frac{1}{2}a_c\{X(f - 2f_c) + X(f)\} + \frac{1}{2}a_c\{X(f) + X(f + 2f_c)\}\right]$$

or

$$\tilde{X}(f) = X(f) + \frac{1}{2}X(f - 2f_c) + \frac{1}{2}X(f + 2f_c). \tag{2.68}$$

The magnitude of this spectrum is shown in Figure 2.24a. Observe that we have recovered the spectrum $X(f)$ of the original data signal, except for the two side bands that are present at locations far removed (centered at $\pm 2f_c$) from the bandwidth of the original signal. We can conveniently low-pass filter the signal $\tilde{x}(t)$ using a filter with cutoff at f_b to recover the original data signal. A schematic representation of this method of amplitude demodulation is shown in Figure 2.24b.

2.6   Analog–Digital Conversion

Control systems use digital data acquisition for a variety of purposes such as process condition monitoring and performance evaluation, fault detection and diagnosis, product quality assessment, dynamic testing, system identification (i.e., experimental modeling), and feedback control. Consider the feedback control system shown in Figure 2.25. Typically, the measured responses (outputs) of a physical system (process, plant) are available in the analog form as a continuous signal (function of continuous time).

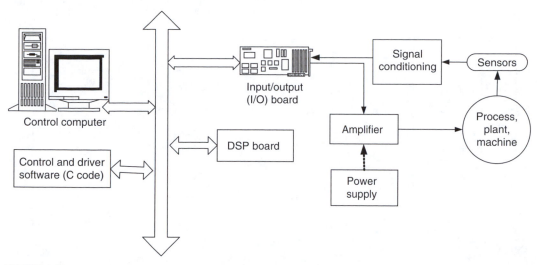

FIGURE 2.25
Components of a data acquisition and control loop.

Furthermore, typically, the excitation signals (or control inputs) for a physical system have to be provided in the analog form. A digital computer is an integral component of a modern control system, and is commonly incorporated in the form of microprocessors and single-board computers, together with such components as digital signal processors (DSP). In a control system, a digital computer will perform tasks such as signal processing, data analysis and reduction, parameter estimation and model identification, decision-making, and control.

Inputs to a digital device (typically, a digital computer) and outputs from a digital device are necessarily present in the digital form. Hence, when a digital device is interfaced with an analog device, the interface hardware and associated driver software have to perform several important functions. Two of the most important interface functions are digital to analog conversion (DAC) and analog to digital conversion (ADC). A digital output from a digital device has to be converted into the analog form for feeding into an analog device, such as actuator or analog recording or display unit. Also, an analog signal has to be converted into the digital form, according to an appropriate code, before it is read by a digital processor or computer.

Both ADC and DAC are elements or components in a typical input/output board (or I/O board or data acquisition and control card or DAC or DAQ). Complete I/O cards for control applications are available from companies such as National Instruments, Servo to Go, Precision MicroDynamics, and Keithly Instruments (Metrabyte). An I/O board can be plugged into slot of a personal computer (PC) and automatically linked with the bus of the PC. The main components of an I/O board are shown in Figure 2.26. The multiplexer (MUX) selects the appropriate input channel. The signal is amplified by a programmable amplifier before ADC. As discussed in a later section, the S/H samples the analog signal and maintains its value at the sampled level until conversion by the ADC. The first-in-first-out element stores the ADC output until it is accessed by the PC for digital processing. The I/O board can provide an analog output through the DAC. Furthermore, a typical I/O board can provide digital outputs as well. An encoder (i.e., a pulse-generating position sensors) can be directly interfaced to I/O boards that are intended for use in motion control applications. Specifications of a typical I/O board are given in Box 2.3. Many of the indicated parameters are discussed in this chapter. Others are either self explanatory or discussed elsewhere in the book. Particular note should be made about the sampling rate.

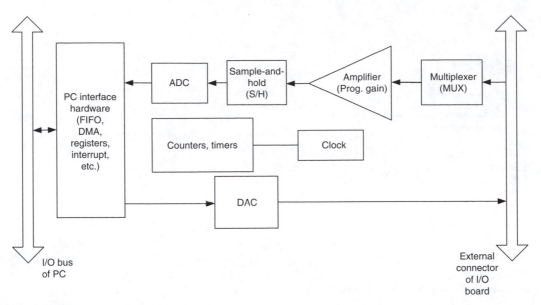

FIGURE 2.26
Main components of an I/O board of a PC.

This is the rate at which an analog input signal is sampled by the ADC. The Nyquist frequency (or the bandwidth limit) of the sampled data would be half this number (50 kHz for the I/O board specified in Box 2.3). When multiplexing is used (i.e., several input channels are read at the same time), the effective sampling rate for each channel will be reduced by a factor equal to the number of channels. For the I/O board specified in Box 2.3, when 16 channels are sampled simultaneously, the effective sampling rate will be 100 kHz/16 = 6.25 kHz, giving a Nyquist frequency of 3.125 kHz.

Box 2.3   Typical Specifications of a Plug-In Input/Output (I/O) Board for a PC

Number of analog input channels = 16 single ended or 8 differential
Analog input ranges = ±5 V; 0 to 10 V; ±10 V; 0 to 20 V
Input gain ranges (programmable) = 1, 2, 5, 10, 20, 50, 100
Sampling rate for A/D conversion = 100, 000 samples/s (100 kHz)
Word size (resolution) of ADC = 12 bits
Number of D/A output channels = 4
Word size (resolution) of DAC = 12 bits
Ranges of analog output = 0 to 10 V (unipolar mode); ±10 V (bipolar mode)
Number of digital input lines = 12
Low voltage of input logic = 0.8 V (maximum)
High voltage of input logic = 2.0 V (minimum)
Number of digital output lines = 12
Low voltage of output logic = 0.45 V (maximum)
High voltage of output logic = 2.4 V (minimum)
Number of counters/timers = 3
Resolution of a counter/timer = 16 bits

Since DAC and ADC play important functions in a control system, they are discussed now. DACs are simpler and cheaper in cost than ADCs. Furthermore, some types of ADCs employ a DAC to perform their function. For these reasons, we will first discuss DAC.

2.6.1 Digital to Analog Conversion

The function of a digital to analog converter (DAC) is to convert a sequence of digital words stored in a data register (called DAC register), typically in the straight binary form, into an analog signal. The data in the DAC register may be arriving from a data bus of a computer. Each binary digit (bit) of information in the register may be present as a state of a bistable (two-stage) logic device, which can generate a voltage pulse or a voltage level to represent that bit. For example, the off state of a bistable logic element or absence of a voltage pulse or low level of a voltage signal or no change in a voltage level can represent binary 0. Conversely, the on state of a bistable device or presence of a voltage pulse or high level of a voltage signal or change in a voltage level will represent binary 1. The combination of these bits forming the digital word in the DAC register will correspond to some numerical value for the analog output signal. Then, the purpose of the DAC is to generate an output voltage (signal level) that has this numerical value and maintain the value until the next digital word is converted into the analog form. Since a voltage output cannot be arbitrarily large or small for practical reasons, some form of scaling would have to be employed in the DAC process. This scale will depend on the reference voltage v_{ref} used in the particular DAC circuit.

A typical DAC unit is an active circuit in the IC form and may consist of a data register (digital circuits), solid-state switching circuits, resistors, and op-amps powered by an external power supply, which can provide the reference voltage for the DAC. The reference voltage will determine the maximum value of the output (full-scale voltage). As noted before, the IC chip that represents the DAC is usually one of the many components mounted on a printed circuit (PC) board, which is the input/output (I/O) board (or I/O card or interface board or data acquisition and control board). This board is plugged into a slot of the data acquisition and control PC (see Figure 2.25 and Figure 2.26).

There are many types and forms of DAC circuits. The form will depend mainly on the manufacturer and requirements of the user or of the particular application. Most types of DAC are variations of two basic types: the weighted type (or summer type or adder type) and the ladder type. The latter type of DAC is more desirable even though the former type could be somewhat simpler and less expensive.

2.6.1.1 Weighted Resistor DAC

A schematic representation of a weighted-resistor DAC (or summer DAC or adder DAC) is shown in Figure 2.27. Note that this is a general n-bit DAC, and n is the number of bits in the output register. The binary word in the register is

$$w = [b_{n-1}b_{n-1}b_{n-3}\ldots b_1b_0], \tag{2.69}$$

where b_i is the bit in the ith position and it can take the value 0 or 1, depending on the value of the digital output. The decimal value of this binary word is given by

$$D = 2^{n-1}b_{n-1} + 2^{n-2}b_{n-2} + \ldots + 2^0b_0. \tag{2.70}$$

Note that the least significant bit (LSB) is b_0 and the most significant bit (MSB) is b_{n-1}. The analog output voltage v of the DAC has to be proportional to D.

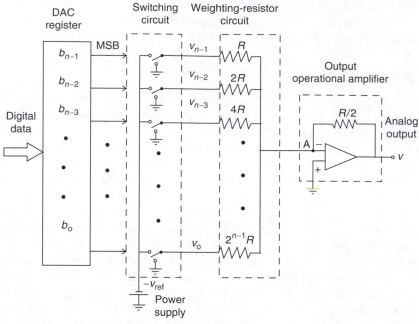

FIGURE 2.27
Weighted-resistor (adder) DAC.

Each bit b_i in the digital word w will activate a solid-state microswitch in the switching circuit, typically by sending a switching voltage pulse. If $b_i = 1$, the circuit lead will be connected to the $-v_{ref}$ supply, providing an input voltage $v_i = -v_{ref}$ to the corresponding weighting resistor $2^{n-i-1} R$. If, on the other hand, $b_i = 0$, then the circuit lead will be connected to ground, thereby providing an input voltage $v_i = 0$ to the same resistor. Note that the MSB is connected to the smallest resistor (R) and the LSB is connected to the largest resistor $(2^{n-1} R)$. By writing the summation of currents at node A of the output op-amp, we get

$$\frac{v_{n-1}}{R} + \frac{v_{n-2}}{2R} + \cdots \frac{v_0}{2^{n-1}R} + \frac{v}{R/2} = 0.$$

In writing this equation, we have used the two principal facts for an op-amp: the voltage is the same at both input leads and the current through each lead is zero. Note that the positive lead is grounded and hence node A should have zero voltage. Now, since $v_i = -b_i v_{ref}$, where $b_i = 0$ or 1 depending on the bit value (state of the corresponding switch), we have

$$v = \left[b_{n-1} + \frac{b_{n-2}}{2} + \cdots + \frac{b_0}{2^{n-1}} \right] \frac{v_{ref}}{2}. \tag{2.71}$$

Clearly, as required, the output voltage v is proportional to the value D of the digital word w.

The full-scale value (FSV) of the analog output occurs when all b_i are equal to 1. Hence,

$$\text{FSV} = \left[1 + \frac{1}{2} + \cdots + \frac{1}{2^{n-1}} \right] \frac{v_{ref}}{2}.$$

Using the commonly known formula for the sum of a geometric series

$$1 + r + r^2 + \ldots + r^{n-1} = \frac{(1 - r^n)}{(1 - r)}, \tag{2.72}$$

we get

$$\text{FSV} = \left(1 - \frac{1}{2^n}\right) v_{\text{ref}}. \tag{2.73}$$

Note that this value is slightly smaller than the reference voltage v_{ref}.

A major drawback of the weighted-resistor DAC is that the range of the resistance value in the weighting circuit is very wide. This presents a practical difficulty, particularly when the size (number of bits n) of the DAC is large. Use of resistors with widely different magnitudes in the same circuit can create accuracy problems. For example, since the MSB corresponds to the smallest weighting resistor, it follows that the resistors must have a very high precision.

2.6.1.2 Ladder DAC

A DAC that uses an R–2R ladder circuit is known as a ladder DAC. This circuit uses only two types of resistors, one with resistance R and the other with $2R$. Hence, the precision of the resistors is not as stringent as what is needed for the weighted-resistor DAC. Schematic representation of an R–2R ladder DAC is shown in Figure 2.28. In this case, the switching circuit can operate just like in the previous case of weighted-resistor DAC. To obtain the input/output equation for the ladder DAC, suppose that, as before, the voltage output from the solid-state switch associated with b_i of the digital word is v_i. Furthermore, suppose that $\tilde{v}_i$ is the voltage at node i of the ladder circuit, as shown in Figure 2.28. Now, writing the current summation at node i we get,

$$\frac{v_i - \tilde{v}_i}{2R} + \frac{\tilde{v}_{i+1} - \tilde{v}_i}{R} + \frac{\tilde{v}_{i-1} - \tilde{v}_i}{R} = 0$$

or

$$\frac{1}{2} v_i = \frac{5}{2} \tilde{v}_i - \tilde{v}_{i-1} - \tilde{v}_{i+1}$$

$$\text{for } i = 1, 2, \ldots, n - 2. \tag{i}$$

Note that Equation i is valid for all nodes, except node 0 and node $n - 1$. It is seen that the current summation for node 0 gives

$$\frac{v_0 - \tilde{v}_0}{2R} + \frac{\tilde{v}_1 - \tilde{v}_0}{R} + \frac{0 - \tilde{v}_0}{2R} = 0$$

or

$$\frac{1}{2} v_0 = 2\tilde{v}_0 - \tilde{v}_1, \tag{ii}$$

and the current summation for node $n-1$ gives

$$\frac{v_{n-1} - \tilde{v}_{n-1}}{2R} + \frac{v - \tilde{v}_{n-1}}{R} + \frac{\tilde{v}_{n-2} - \tilde{v}_{n-1}}{R} = 0.$$

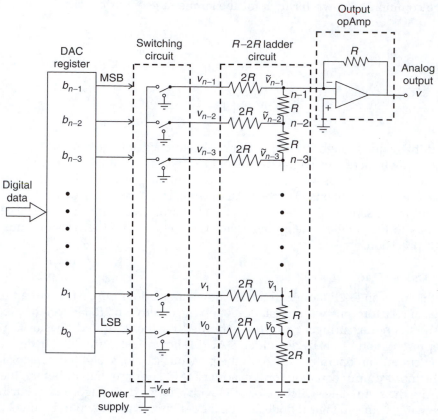

FIGURE 2.28
Ladder DAC.

Now, since the positive lead of the op-amp is grounded, we have $\tilde{v}_{n-1} = 0$. Hence,

$$\frac{1}{2}v_{n-1} = -\tilde{v}_{n-2} - v. \tag{iii}$$

Next, by using Equation i through Equation iii, along with the fact that $\tilde{v}_{n-1} = 0$, we can write the following series of equations:

$$\frac{1}{2}v_{n-1} = -\tilde{v}_{n-2} - v,$$

$$\frac{1}{2^2}v_{n-2} = \frac{1}{2}\frac{5}{2}\tilde{v}_{n-2} - \frac{1}{2}\tilde{v}_{n-3},$$

$$\frac{1}{2^3}v_{n-3} = \frac{1}{2^2}\frac{5}{2}\tilde{v}_{n-3} - \frac{1}{2^2}\tilde{v}_{n-4} - \frac{1}{2^2}\tilde{v}_{n-2},$$

$$\frac{1}{2^{n-1}}v_1 = \frac{1}{2^{n-2}}\frac{5}{2}\tilde{v}_1 - \frac{1}{2^{n-2}}\tilde{v}_0 - \frac{1}{2^{n-2}}\tilde{v}_2,$$

$$\frac{1}{2^2}v_0 = \frac{1}{2^{n-1}}2\tilde{v}_0 - \frac{1}{2^{n-1}}\tilde{v}_1. \tag{iv}$$

If we sum these n equations, first denoting

$$S = \frac{1}{2^2}\tilde{v}_{n-2} + \frac{1}{2^3}\tilde{v}_{n-3} + \ldots + \frac{1}{2^{n-1}}\tilde{v}_1,$$

we get

$$\frac{1}{2}v_{n-1} + \frac{1}{2^2}v_{n-2} + \ldots + \frac{1}{2^n}v_0 = 5S - 4S - S + \frac{1}{2^{n-1}}2\tilde{v}_0 - \frac{1}{2^{n-2}}\tilde{v}_0 - v = -v.$$

Finally, since $v_i = -b_i v_{\text{ref}}$, we have the analog output as

$$v = \left[\frac{1}{2}b_{n-1} + \frac{1}{2^2}b_{n-2} + \ldots + \frac{1}{2^n}b_0\right]v_{\text{ref}}. \tag{2.74}$$

This result is identical to Equation 2.71, which we obtained for the weighted-resistor DAC. Hence, as before, the analog output is proportional to the value D of the digital word and, furthermore, the FSV of the ladder DAC as well is given by the previous Equation 2.73.

2.6.1.3 DAC Error Sources

For a given digital word, the analog output voltage from a DAC would not be exactly equal to what is given by the analytical formulas (e.g., Equation 2.71) that were derived earlier. The difference between the actual output and the ideal output is the error. The DAC error could be normalized with respect to the FSV.

There are many causes of DAC error. Typical error sources include parametric uncertainties and variations, circuit time constants, switching errors, and variations and noise in the reference voltage. Several types of error sources and representations are discussed later.

1. Code ambiguity: In many digital codes (e.g., in the straight binary code), incrementing a number by an LSB will involve more than one bit-switching. If the speed of switching from 0 to 1 is different from that for 1 to 0, and if switching pulses are not applied to the switching circuit simultaneously, the switching of the bits will not take place simultaneously. For example, in a four-bit DAC, incrementing from decimal 2 to decimal 4 will involve changing the digital word from 0011 to 0100. This requires two bit-switchings from 1 to 0 and one bit-switching from 0 to 1. If 1 to 0 switching is faster than the 0 to 1 switching, then an intermediate value given by 0000 (decimal zero) will be generated, with a corresponding analog output. Hence, there will be a moment-ary code ambiguity and associated error in the DAC signal. This problem can be reduced (and eliminated in the case of single-bit increments) if a gray code is used to represent the digital data. Improving the switching circuitry will also help reduce this error.

2. Settling time: The circuit hardware in a DAC unit will have some dynamics, with associated time constants and perhaps oscillations (underdamped response). Hence, the output voltage cannot instantaneously settle to its ideal value upon switching. The time required for the analog output to settle within a certain band (say $\pm 2\%$ of the final value or $\pm 1/2$ resolution), following the application of the digital data, is termed settling time. Naturally, settling time

should be smaller for better (faster and more accurate) performance. As a rule of the thumb, the settling time should be approximately half the data arrival time. Note that the data arrival time is the time interval between the arrival of two successive data values, and is given by the inverse of the data arrival rate.

3. Glitches: Switching of a circuit will involve sudden changes in magnetic flux due to current changes. This will induce the voltages that produce unwanted signal components. In a DAC circuit, these induced voltages due to rapid switching can cause signal spikes, which will appear at the output. The error due to these noise signals is not significant at low conversion rates.

4. Parametric errors: As discussed earlier, resistor elements in a DAC might not be very precise, particularly when resistors within a wide range of magnitudes are employed, as in the case of weighted-resistor DAC. These errors appear at the analog output. Furthermore, aging and environmental changes (primarily, change in temperature) will change the values of circuit parameters, resistance in particular. This will also result in DAC error. These types of errors due to imprecision of circuit parameters and variations of parameter values are termed parametric errors. Effects of such errors can be reduced by several ways including the use of compensation hardware (and perhaps software), and directly by using precise and robust circuit components and employing good manufacturing practices.

5. Reference voltage variations: Since the analog output of a DAC is proportional to the reference voltage v_{ref}, any variations in the voltage supply will directly appear as an error. This problem can be overcome by using stabilized voltage sources with sufficiently low output impedance.

6. Monotonicity: Clearly, the output of a DAC should change by its resolution ($\delta y = v_{\mathrm{ref}}/2^n$) for each step of one LSB increment in the digital value. This ideal behavior might not exist in some practical DACs due to such errors as those mentioned earlier. At least the analog output should not decrease as the value of the digital input increases. This is known as the monotonicity requirement, and it should be met by a practical DAC.

7. Nonlinearity: Suppose that the digital input to a DAC is varied from [0 0 ... 0] to [1 1 ... 1] in steps of one LSB. As mentioned earlier, ideally the analog output should increase in constant jumps of $\delta y = v_{\mathrm{ref}}/2^n$, giving a staircase-shaped analog output. If we draw the best linear fit for this ideally montonic staircase response, it will have a slope equal to the resolution/step. This slope is known as the ideal scale factor. Nonlinearity of a DAC is measured by the largest deviation of the DAC output from this best linear fit. Note that in the ideal case, the nonlinearity is limited to half the resolution ($1/2\ \delta y$).

One cause of nonlinearity is clearly the faulty bit transitions. Another cause is circuit nonlinearity in the conventional sense. Specifically, due to nonlinearities in circuit elements such as op-amps and resistors, the analog output will not be proportional to the value of the digital word dictated by the bit switchings (faulty or not). This latter type of nonlinearity can be accounted for by using calibration.

2.6.2 Analog to Digital Conversion

Analog signals, which are continuously defined with respect to time, have to be sampled at discrete time points, and the sample values have to be represented in the digital form (according to a suitable code) to be read into a digital system such as a microcomputer. An ADC is used to accomplish this. For example, since response measurements of a control system are usually available as analog signals, these signals have to be converted into the digital form before passing on to a digital computer for analysis and possibly generating a control command. Hence, the computer interface for the measurement channels should contain one or more ADCs (see Figure 2.25).

DACs and ADCs are usually situated on the same digital interface board (see Figure 2.26). But, the ADC process is more complex and time consuming than the DAC process. Furthermore, many types of ADCs use DACs to accomplish the ADC. Hence, ADCs are usually more costly, and their conversion rate is usually slower in comparison to DACs. Several types of ADCs are commercially available. The principle of operation may vary depending on the type. A few commonly known types are discussed here.

2.6.2.1 Successive Approximation ADC

This type of ADC is very fast, and is suitable for high-speed applications. The speed of conversion depends on the number of bits in the output register of ADC but is virtually independent of the nature of the analog input signal. A schematic diagram for a successive approximation ADC is shown in Figure 2.29. Note that a DAC is an integral component of this ADC. The sampled analog signal (from an S/H) is applied to a comparator (typically a differential amplifier). Simultaneously, a start conversion (SC) control pulse is sent into the control logic unit by the external device (perhaps a microcomputer) that controls the operation of the ADC. Then, no new data will be accepted by the ADC until a conversion complete (CC) pulse is sent out by the control logic unit. Initially, the registers are cleared so that they contain all zero bits. Now, the ADC is ready for its first conversion approximation.

The first approximation begins with a clock pulse. Then, the control logic unit will set the most significant bit (MSB) of the temporary register (DAC control register) to one, all the remaining bits in that register being zero. This digital word in the temporary register is supplied to the DAC. Note that the analog output of the DAC is now equal to half the FSV. This analog signal is subtracted from the analog input by the comparator. If the output of the comparator is positive, the control logic unit will keep the MSB of the temporary register at binary 1 and will proceed to the next approximation. If the comparator output

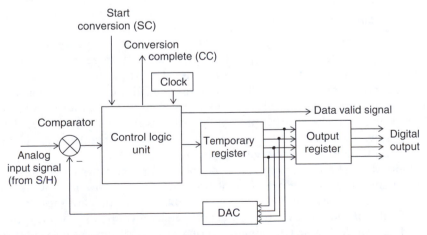

FIGURE 2.29
Successive approximation ADC.

is negative, the control logic unit will change the MSB to binary 0 before proceeding to the next approximation.

The second approximation will start at another clock pulse. This approximation will consider the second MSB of the temporary register. As before, this bit is set to 1 and the comparison is made. If the comparator output is positive, this bit is left at value 1 and the third MSB is considered. If the comparator output is negative, the bit value will be changed to 0 before proceeding to the third MSB.

In this manner, all bits in the temporary register are set successively starting from the MSB and ending with the LSB. The contents of the temporary register are then transferred to the output register, and a data valid signal is sent by the control logic unit, signaling the interfaced device (computer) to read the contents of the output register. The interfaced device will not read the register if a data valid signal is not present. Next, a CC pulse is sent out by the control logic unit, and the temporary register is cleared. The ADC is now ready to accept another data sample for digital conversion. Note that the conversion process is essentially the same for every bit in the temporary register. Hence, the total conversion time is approximately n times the conversion time for one bit. Typically, one bit conversion can be completed within one clock period.

It should be clear that if the maximum value of an analog input signal exceeds the full scale value (FSV) of a DAC, then the excess signal value cannot be converted by the ADC. The excess value will directly contribute to error in the digital output of the ADC. Hence, this situation should be avoided, either by properly scaling the analog input or by properly selecting the reference voltage for the internal DAC unit.

In the foregoing discussion, we have assumed that the value of the analog input signal is always positive. Otherwise, the sign of the signal has to be accounted for by some means. For example, the sign of the signal can be detected from the sign of the comparator output initially, when all bits are zero. If the sign is negative, then the same A/D conversion process, as for a positive signal, is carried out after switching the polarity of the comparator. Finally, the sign is correctly represented in the digital output (e.g., by the two's complement representation for negative quantities). Another approach to account for signed (bipolar) input signals is to offset the signal by a sufficiently large constant voltage, such that the analog input is always positive. After the conversion, the digital number corresponding to this offset is subtracted from the converted data in the output register to obtain the correct digital output. In what follows, we shall assume that the analog input signal is positive.

2.6.2.2 Dual-Slope ADC

This ADC uses an RC integrating circuit. Hence, it is also known as an *integrating ADC*. This ADC is simple and inexpensive. In particular, an internal DAC is not utilized and hence, DAC errors, as mentioned previously, will not enter the ADC output. Furthermore, the parameters R and C in the integrating circuit do not enter the ADC output. As a result, the device is self-compensating in terms of circuit-parameter variations due to temperature, aging, so on. A shortcoming of this ADC is its slow conversion rate because, for accurate results, the signal integration has to proceed for a longer time in comparison to the conversion time for a successive approximation ADC.

A dual-slope ADC is based on timing (i.e., counting the number of clock pulses during) a capacitor-charging process. The principle of operation can be explained with reference to the integrating circuit shown in Figure 2.30a. Note hat v_i is a constant input voltage to the circuit and v is the output voltage. Since the positive lead of the op-amp is grounded, the negative lead (and node A) also will have zero voltage. In addition, the currents through the op-amp leads are negligible. Hence, the current balance at node A gives

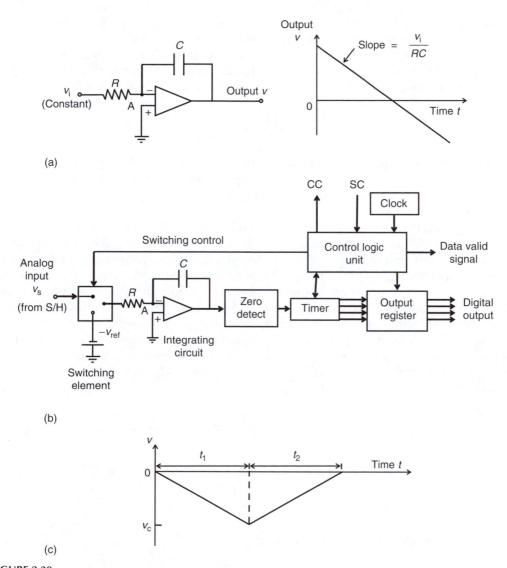

FIGURE 2.30
(a) RC integrating circuit. (b) Dual-slope ADC. (c) Dual-slope charging–discharging curve.

$$\frac{v_i}{R} + C\frac{dv}{dt} = 0.$$

Integrating this equation for constant v_i, we have

$$v(t) = v(0) - \frac{v_i t}{RC}. \tag{2.75}$$

Equation 2.75 will be used in obtaining a principal result for the dual-slope ADC.

A schematic diagram for a dual-slope ADC is shown in Figure 2.30b. Initially, the capacitor C in the integrating circuit is discharged (zero voltage). Then, the analog signal v_s is supplied to the switching element and held constant by the S/H. Simultaneously, a conversion start control signal is sent to the control logic unit. This will clear the timer and the output register (i.e., all bits are set to zero) and will send a pulse to the switching element to connect the input v_s to the integrating circuit. Also, a signal is sent to the timer to initiate timing (counting). The capacitor C will begin to charge. Equation 2.75 is now applicable with input $v_i = v_s$ and the initial state $v(0) = 0$. Suppose that the integrator output v becomes $-v_c$ at time $t = t_1$. Hence, from Equation 2.75, we have

$$v_c = \frac{v_s t_1}{RC}. \tag{i}$$

The timer will keep track of the capacitor charging time (as a clock pulse count n) and will inform the control logic unit when the elapsed time is t_1 (i.e., when the count is n_1). Note that t_1 and n_1 are fixed (and known) parameters, but voltage v_c depends on the value of v_s, and is unknown.

At this point, the control logic unit will send a signal to the switching unit, which will connect the input lead of the integrator to a negative supply voltage $-v_{ref}$. Simultaneously, a signal is sent to the timer to clear its contents and start timing (counting) again. Now the capacitor begins to discharge. The output of the integrating circuit is monitored by the zero-detect unit. When this output becomes zero, the zero-detect unit sends a signal to the timer to stop counting. The zero-detect unit could be a comparator (differential amplifier) with one of the two input leads set at zero potential.

Now suppose that the elapsed time is t_2 (with a corresponding count of n_2). It should be clear that Equation 2.75 is valid for the capacitor discharging process as well. Note that $v_i = -v_{ref}$ and $v(0) = -v_c$ in this case. Also, $v(t) = 0$ at $t = t_2$. Hence, from Equation 2.75, we have

$$0 = -v_c + \frac{v_{ref} t_2}{RC}$$

or

$$v_c = \frac{v_{ref} t_2}{RC}. \tag{ii}$$

On dividing Equation i by Equation ii, we get

$$v_s = v_{ref} \frac{t_2}{t_1}.$$

However, the timer pulse count is proportional to the elapsed time. Hence,

$$\frac{t_2}{t_1} = \frac{n_2}{n_1}.$$

Now we have

$$v_s = \frac{v_{ref}}{n_1} n_2. \tag{2.76}$$

Since v_{ref} and n_1 are fixed quantities, v_{ref}/n_1 can be interpreted as a scaling factor for the analog input. Then, it follows from Equation 2.76 that the second count n_2 is proportional to the analog signal sample v_s. Note that the timer output is available in the digital form. Accordingly, the count n_2 is used as the digital output of the ADC.

At the end of the capacitor discharge period, the count n_2 in the timer is transferred to the output register of the ADC, and the data valid signal is set. The contents of the output register are now ready to be read by the interfaced digital system, and the ADC is ready to convert a new sample.

The charging–discharging curve for the capacitor during the conversion process is shown in Figure 2.30c. The slope of the curve during charging is $-v_s/RC$, and the slope during discharging is $+v_{ref}/RC$. The reason for the use of the term dual slope to denote this ADC is therefore clear.

As mentioned before, any variations in R and C do not affect the accuracy of the output. But, it should be clear from the foregoing discussion that the conversion time depends on the capacitor discharging time t_2 (note that t_1 is fixed), which in turn depends on v_c and hence on the input signal value v_s (see Equation i). It follows that, unlike the successive approximation ADC, the dual-slope ADC has a conversion time that directly depends on the magnitude of the input data sample. This is a disadvantage in a way because in many applications we prefer to have a constant conversion rate.

The earlier discussion assumed that the input signal is positive. For a negative signal, the polarity of the supply voltage v_{ref} has to be changed. Furthermore, the sign has to be properly represented in the contents of the output register as, for example, in the case of successive approximation ADC.

2.6.2.3 Counter ADC

The counter-type ADC has several aspects in common with the successive approximation ADC. Both are comparison-type (or closed loop) ADCs. Both use a DAC unit internally to compare the input signal with the converted signal. The main difference is that in a counter ADC the comparison starts with the LSB and proceeds down. It follows that, in a counter ADC, the conversion time depends on the signal level, because the counting (comparison) stops when a match is made, resulting in shorter conversion times for smaller signal values.

A schematic diagram for a counter ADC is shown in Figure 2.31. Note that this is quite similar to Figure 2.29. Initially, all registers are cleared (i.e., all bits and counts are set to zero). As an analog data signal (from the S/H) arrives at the comparator, an

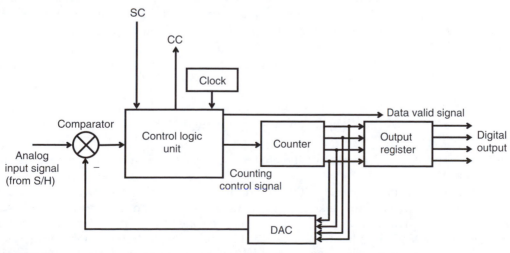

FIGURE 2.31
Counter ADC.

SC pulse is sent to the control logic unit. When the ADC is ready for conversion (i.e., when data valid signal is on), the control logic unit initiates the counter. Now, the counter sets its count to 1, and the LSB of the DAC register is set to 1 as well. The resulting DAC output is subtracted from the analog input, by means of the comparator. If the comparator output is positive, the count is incremented by one, and this causes the binary number in the DAC register to be incremented by one LSB. The new (increased) output of the DAC is now compared with the input signal. This cycle of count incrementing and comparison is repeated until the comparator output becomes less than or equal to zero. At that point, the control logic unit sends out a CC signal and transfers the contents of the counter to the output register. Finally, the data valid signal is turned on, indicating that the ADC is ready for a new conversion cycle, and the contents of the output register (the digital output) is available to be read by the interfaced digital system.

The count of the counter is available in the binary form, which is compatible with the output register as well as the DAC register. Hence, the count can be transferred directly to these registers. The count when the analog signal is equal to (or slightly less than) the output of the DAC, is proportional to the analog signal value. Hence, this count represents the digital output. Again, sign of the input signal has to be properly accounted for in the bipolar operation.

2.6.2.4 ADC Performance Characteristics

For ADCs that use a DAC internally, the same error sources that were discussed previously for DACs will apply. Code ambiguity at the output register will not be a problem because the converted digital quantity is transferred instantaneously to the output register. Code ambiguity in the DAC register can still cause error in ADCs that use a DAC. Conversion time is a major factor, because this is much larger for an ADC. In addition to resolution and dynamic range, quantization error will be applicable to an ADC. These considerations, which govern the performance of an ADC, are discussed next.

1. Resolution and quantization error: The number of bits n in an ADC register determines the resolution and dynamic range of an ADC. For an n-bit ADC, the size of the output register is n bits. Hence, the smallest possible increment of the digital output is one LSB. The change in the analog input that results in a change of one LSB at the output is the resolution of the ADC. For the unipolar (unsigned) case, the available range of the digital outputs is from 0 to $2^n - 1$. This represents the dynamic range. It follows that, as for a DAC, the dynamic range of an n bit ADC is given by the ratio

$$DR = 2^n - 1 \qquad (2.77)$$

or in decibels

$$DR = 20 \log_{10} (2^n - 1) \, dB. \qquad (2.78)$$

The FSV of an ADC is the value of the analog input that corresponds to the maximum digital output.

Suppose that an analog signal within the dynamic range of a particular ADC is converted by that ADC. Since the analog input (sample value) has an infinitesimal resolution and the digital representation has a finite resolution (one LSB), an error is introduced in the process of ADC. This is known as the quantization error. A digital number undergoes successive increments in constant steps of 1 LSB. If an analog value falls at an intermediate point within a step of single LSB, a quantization error is caused as a result. Rounding off the digital output can be accomplished as follows: the magnitude of the error when quantized up, is compared with that when quantized down; say, using two hold elements and a differential amplifier. Then, we retain the digital value corresponding to the lower error magnitude. If the analog value is below the 1/2 LSB mark, then the corresponding digital value is represented by the value at the beginning of the step. If the analog value is above the 1/2 LSB mark, then the corresponding digital value is the value at the end of the step. It follows that with this type of rounding off, the quantization error does not exceed 1/2 LSB.

2. Monotonicity, nonlinearity, and offset error: Considerations of monotonicity and nonlinearity are important for an ADC as well as for a DAC. In the case of an ADC, the input is an analog signal and the output is digital. Disregarding quantization error, the digital output of an ADC will increase in constant steps in the shape of an ideal staircase function, when the analog input is increased from 0 in steps of the device resolution (δy). This is the ideally monotonic case. The best straight-line fit to this curve has a slope equal to $1/\delta y$(LSB/Volts). This is the ideal gain or ideal scale factor. Still there will be an offset error of 1/2 LSB because the best linear fit will not pass through the origin. Adjustments can be made for this offset error.

Incorrect bit transitions can take place in an ADC, due to various errors that might be present and also possibly due to circuit malfunctions. The best linear fit under such faulty conditions will have a slope different from the ideal gain. The difference is the gain error. Nonlinearity is the maximum deviation of the output from the best linear fit. It is clear that with perfect bit transitions, in the ideal case, a nonlinearity of 1/2 LSB would be present. Nonlinearities larger than this would result due to incorrect bit transitions. As in the case of a DAC, another source of nonlinearity in an ADC is circuit nonlinearities, which would deform the analog input signal before converting into the digital form.

3. ADC conversion rate: It is clear that ADC is much more time consuming than DAC. The conversion time is a very important factor because the rate at which the conversion can take place governs many aspects of data acquisition, particularly in real-time applications. For example, the data sampling rate has to synchronize with the ADC conversion rate. This, in turn, will determine the Nyquist frequency (half the sampling rate), which corresponds to the bandwidth of the sampled signal, and is the maximum value of useful frequency that is retained as a result of sampling. Furthermore, the sampling rate will dictate the requirements of storage and memory. Another important consideration related to the conversion rate of an ADC is the fact that a signal sample has to be maintained at the same value during the entire process of conversion into the digital form. This would require a hold circuit, and this circuit should be able to perform accurately at the largest possible conversion time for the particular ADC unit.

The time needed for a sampled analog input to be converted into the digital form will depend on the type of ADC. Usually, in a comparison-type ADC (which uses an internal DAC), each bit transition will take place in one clock period Δt. Also, in an integrating (dual slope) ADC, each clock count will need a time of Δt. On this basis, for the three types of ADC that we have discussed, the following figures can be given for their conversion times:

1. Successive-approximation ADC: In this case, for an n bit ADC, n comparisons are needed. Hence, the conversion time is given by

$$t_c = n \cdot \Delta t, \tag{2.79}$$

where Δt is the clock period. Note that for this ADC, t_c does not depend on the signal level (analog input).

2. Dual-slope (integrating) ADC: In this case, the conversion time is the total time needed to generate the two counts n_1 and n_2 (see Figure 2.30c). Hence,

$$t_c = (n_1 + n_2)\Delta t. \tag{2.80}$$

Note that n_1 is a fixed count. However, n_2 is a variable count, which represents the digital output, and is proportional to the analog input (signal level). Hence, in this type of ADC, conversion time depends on the analog input level. The largest output for an n bit converter is $2n-1$. Hence, the largest conversion time may be given by

$$t_{c\,max} = (n_1 + 2^n - 1)\Delta T. \tag{2.81}$$

3. Counter ADC: For a counter ADC, the conversion time is proportional to the number of bit transitions (1 LSB/step) from zero to the digital output n_o. Hence, the conversion time is given by

$$t_c = n_o \Delta t, \tag{2.82}$$

where n_o is the digital output value (in decimal).

Note that for this ADC as well, t_c depends on the magnitude of the input data sample. For an n bit ADC, since the maximum value of n_o is $2^n - 1$, we have the maximum conversion time

$$t_{c\,max} = (2^n - 1)\Delta t. \tag{2.83}$$

By comparing Equation 2.79, Equation 2.81, and Equation 2.83, it can be concluded that the successive-approximation ADC is the fastest of the three types discussed.

The total time taken to convert an analog signal will depend on other factors besides the time taken for the conversion of sampled data into digital form. For example, in multiple-channel data acquisition (multiplexing), the time taken to select the channels has to be counted in. Furthermore, time needed to sample the data and time needed to transfer the converted digital data into the output register have to be included. In fact, the conversion rate for an ADC is the inverse of this overall time needed for a conversion cycle. Typically, however, the conversion rate depends primarily on the bit conversion time, in the case of a comparison-type ADC, and on the integration time, in the case of an integration-type ADC. A typical time period for a comparison step or counting step in an ADC is $\Delta t = 5$ μs. Hence, for an 8 bit successive approximation ADC, the conversion time is 40 μs. The corresponding sampling rate would be of the order of (less than) $1/40 \times 10^{-6} = 25 \times 10^3$ samples/s (or 25 kHz). The maximum conversion rate for an 8 bit counter ADC would be about $5 \times (2^8 - 1) = 1275$ μs. The corresponding sampling rate would be of the order of 780 samples/s. Note that this is considerably slow. The maximum conversion time for a dual-slope ADC would likely be larger (i.e., slower rate).

2.7 Sample-and-Hold Circuitry

Typical applications of data acquisition use ADC. The analog input to an ADC can be very transient, and furthermore, the process of ADC itself is not instantaneous (ADC time can be much larger than the DAC time). Specifically, the incoming analog signal might be changing at a rate higher than the ADC conversion rate. Then, the input signal value will vary during the conversion period, and there will be an ambiguity as to what analog input value corresponds to a particular digital output value. Hence, it is necessary to sample the analog input signal and maintain the input to the ADC at this sampled value, until the ADC is completed. In other words, since we are typically dealing with analog signals that can vary at a high speed, it would be necessary to S/H the input signal during each ADC cycle. Each data sample must be generated and captured by the S/H circuit on the issue of the SC control signal, and the captured voltage level has to be maintained constant until a CC control signal is issued by the ADC unit.

The main element in an S/H circuit is the holding capacitor. A schematic diagram of an S/H is shown in Figure 2.32. The analog input signal is supplied through a voltage follower to a solid-state switch. The switch typically uses a field-effect transistor (FET), such as the MOSFET. The switch is closed in response to a sample pulse and is opened in response to a hold pulse. Both control pulses are generated by the control logic unit of the ADC. During the time interval between these two pulses, the holding capacitor is charged to the voltage of the sampled input. This capacitor voltage is then supplied to the ADC through a second voltage follower.

The functions of the two voltage followers are explained now. When the FET switch is closed in response to a sample command (pulse), the capacitor has to be charged as quickly as possible. The associated time constant (charging time constant) τ_c is given by

$$\tau_c = R_s C, \qquad (2.84)$$

where R_s is the source resistance and C is the capacitance of the holding capacitor. Since τ_c has to be very small for fast charging, and since C is fixed by the holding requirements (typically C is of the order of 100 pF, where 1 pF $= 1 \times 10^{-12}$ F), we need a very small source resistance. The requirement is met by the input voltage follower (which is known to have a very low output impedance), thereby providing a very small R_s. Furthermore, since a voltage follower has a unity gain, the voltage at the output of this input voltage follower would be equal to the voltage of the analog input signal, as required.

Next, once the FET switch is opened in response to a hold command (pulse), the capacitor should not discharge. This requirement is met due to the presence of the output voltage follower. Since the input impedance of a voltage follower is very high, the current through its leads would be almost zero. Because of this, the holding capacitor will have a

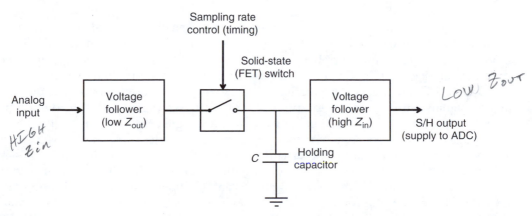

FIGURE 2.32
A sample-and-hold circuit.

virtually zero discharge rate under hold conditions. Furthermore, we like the output of this second voltage follower to be equal to the voltage of the capacitor. This condition is also satisfied due to the fact that a voltage follower has a unity gain. Hence, the sampling would be almost instantaneous, and the output of the S/H circuit would be maintained (almost) constant during the holding period, due to the presence of the two voltage followers. Note that the practical S/H circuits are zero-order hold devices, by definition.

2.8 Multiplexers

A multiplexer (MUX; also known as a scanner) is used to select one channel at a time from a bank of signal channels and connect it to a common hardware unit. In this manner, a costly and complex hardware unit can be time shared among several signal channels. Typically, channel selection is done in a sequential manner at a fixed channel-select rate.

There are two types of MUX: analog MUX and digital MUX. An analog MUX is used to scan a group of analog channels. Alternatively, a digital MUX is used to read one data word at a time sequentially from a set of digital data words.

The process of distributing a single channel of data among several output channels is known as demultiplexing. A demultiplexer (or data distributor) performs the reverse function of a MUX (or scanner). A demultiplexer may be used, for example, when the same (processed) signal from a digital computer is needed for several purposes (e.g., digital display, analog reading, digital plotting, and control).

Multiplexing used in short-distance signal transmission applications (e.g., data logging and process control) is usually time-division multiplexing. In this method, channel selection is made with respect to time. Hence, only one input channel is connected to the output channel of the MUX. This is the method described here. Another method of multiplexing, used particularly in long-distance transmission of several data signals, is known as frequency-division multiplexing. In this method, the input signals are modulated (e.g., by AM, as discussed previously) with carrier signals with different frequencies and are transmitted simultaneously through the same data channel. The signals are separated by demodulation at the receiving end.

2.8.1 Analog Multiplexers

Monitoring of a control system often requires the measurement of several process responses. These signals have to be conditioned (e.g., amplification and filtering) and modified in some manner (e.g., ADC) before supplying to a common-purpose system such as a digital computer or data logger. Usually, data modification devices are costly. In particular, we have noted that ADCs are more expensive than DACs. An expensive option for interfacing several analog signals with a common-purpose system such as a digital computer would be to provide separate data modification hardware for each signal channel. This method has the advantage of high speed. An alternative, low-cost method is to use an analog MUX (analog scanner) to select one signal channel at a time sequentially and connect it to a common signal-modification hardware unit (consisting of amplifiers, filters, S/H, ADC, etc.). In this way, by time-sharing expensive hardware among many data channels, the data acquisition speed is traded off to some extent for significant cost savings. Because very high speeds of channel selection are possible with solid-state switching (e.g., solid-state speeds of the order of 10 MHz), the speed reduction due to multiplexing is not a significant drawback in most applications. On the other hand, since the cost of hardware components such as ADC is declining due to rapid advances in solid-state technologies, cost reductions attainable through the use of multiplexing might not be substantial in some applications. Hence, some economic evaluation and engineering judgment would be needed when deciding on the use of signal multiplexing for a particular data acquisition and control application.

A schematic diagram of an analog MUX is shown in Figure 2.33. The figure represents the general case of N input channels and one output channel. This is called an $N \times 1$ analog MUX. Each input channel is connected to the output through a solid-state switch, typically a FET switch. One switch is closed (turned on) at a time. A switch is selected by a digital word, which contains the corresponding channel address. Note that an n bit address can assume 2^n digital values in the range of 0 to $2^n - 1$. Hence, a MUX with an n bit address can handle $N = 2^n$ channels. Channel selection can be done by an external microprocessor, which places the address of the channel on the address bus and simultaneously sends a control signal to the MUX to enable the MUX. The address decoder decodes the address and activates the corresponding FET switch. In this manner, channel selection can be done in an arbitrary order and with arbitrary timing, controlled by the microprocessor. In simple versions of MUX, the channel selection is made in a fixed order at a fixed speed, however.

Typically, the output of an analog MUX is connected to an S/H circuit and an ADC. Voltage followers can be provided both at the input and the output to reduce loading problems. A differential amplifier (or instrumentation amplifier) could be used at the output to reduce noise problems, particularly to reject common-mode interference, as discussed earlier in this chapter. Note that the channel-select speed has to be synchronized

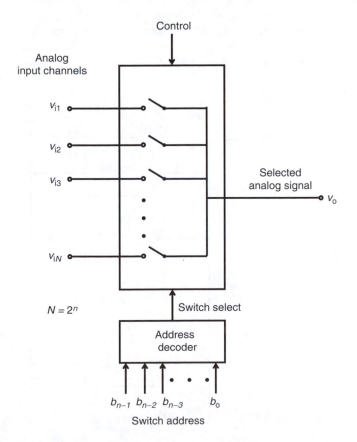

FIGURE 2.33
An N-channel analog multiplexer (MUX) (analog scanner).

with the sampling and ADC speeds for each signal channel. The MUX speed is not a major limitation because very high speeds (solid-state speeds of 10 MHz or more) are available with solid-state switching.

2.8.2 Digital Multiplexers

Sometimes it is required to select one data word at a time from a set of digital data words, to be fed into a common device. For example, the set of data may be the outputs from a bank of digital transducers (e.g., shaft encoders that measure angular motions) or outputs from a set of ADCs that are connected to a series of analog signal channels. Then the selection of the particular digital output (data word) can be made using techniques of addressing and data-bus transfer, which are commonly used in digital systems.

A digital multiplexing (or logic multiplexing) configuration is shown in Figure 2.34. The N registers of the MUX hold a set of N data words. The contents of each register may correspond to a response measurement, and, hence, will change regularly. The registers may represent separate hardware devices (e.g., output registers of a bank of ADCs) or may represent locations in a computer memory to which data are transferred (read in) regularly. Each register has a unique binary address. As in the case of analog MUX, an n bit address can select (address) 2^n registers. Hence, the number of registers will be given by $N = 2^n$, as before. When the address of the register to be selected is placed on the address bus, it enables the particular register. This causes the contents of that register to be placed on the data bus. Now the data bus is read by the device (e.g., computer), which is time-shared among the N data registers. Placing a different address on the address bus will result in selecting another register and reading the contents of that register, as before.

Digital multiplexing is usually faster than analog multiplexing, and has the usual advantages of digital devices; for example, high accuracy, better noise immunity, robustness (no drift and errors due to parameter variations), long-distance data transmission capability without associated errors due to signal weakening, and capability to handle very large numbers of data channels. Furthermore, a digital MUX can be modified using

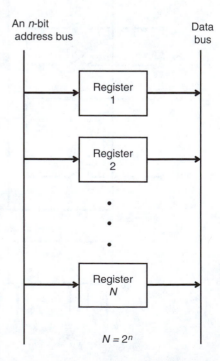

FIGURE 2.34
An $N \times 1$ digital multiplexer (MUX).

software, usually without the need for hardware changes. If, however, instead of using an analog MUX followed by a single ADC, a separate ADC is used for each analog signal channel and then digital multiplexing is used, it would be quite possible for the digital multiplexing approach to be more costly. If, on the other hand, the measurements are already available in the digital form (for instance, as encoder outputs of displacement measurement), then digital multiplexing tends to be very cost effective and most desirable.

Transfer of a digital word from a single data source (e.g., a data bus) into several data registers, which are to be accessed independently, may be interpreted as digital demultiplexing. This is also a straightforward process of digital data transfer and reading.

2.9 Digital Filters

A filter is a device that eliminates undesirable frequency components in a signal and passes only the desirable frequency components through it. In analog filtering, the filter is a physical dynamic system; typically an electric circuit. The signal to be filtered is applied (input) to this dynamic system. The output of the dynamic system is the filtered signal. It follows that any physical dynamic system can be interpreted as an analog filter.

An analog filter can be represented by a differential equation with respect to time. It takes an analog input signal $u(t)$, which is defined continuously in time t and generates an analog output $y(t)$. A digital filter is a device that accepts a sequence of discrete input values (say, sampled from an analog signal at sampling period Δt), represented by

$$\{u_k\} = \{u_0, u_1, u_2, \ldots\}, \tag{2.85}$$

and generates a sequence of discrete output values

$$\{y_k\} = \{y_0, y_1, y_2, \ldots\}. \tag{2.86}$$

It follows that a digital filter is a discrete-time system and it can be represented by a difference equation.

An nth-order linear difference equation can be written in the form

$$a_0 y_k + a_1 y_{k-1} + \ldots + a_n y_{k-n} = b_0 u_k + b_1 u_{k-1} + \ldots + b_m u_{k-m}. \tag{2.87}$$

This is a recursive algorithm, in the sense that it generates one value of the output sequence using previous values of the output sequence, and all values of the input sequence up to the present time point. Digital filters represented in this manner are termed *recursive digital filters*. There are filters that employ digital processing where a block (a collection of samples) of the input sequence is converted by a one-shot computation into a block of the output sequence. They are not recursive filters. Nonrecursive filters usually employ digital Fourier analysis, the FFT algorithm, in particular. We restrict our discussion later to recursive digital filters. Our intention in the present section is to give a brief (and nonexhaustive) introduction to the subject of digital filtering.

2.9.1 Software Implementation and Hardware Implementation

In digital filters, signal filtering is accomplished through digital processing of the input signal. The sequence of input data (usually obtained by sampling and digitizing the corresponding analog signal) is processed according to the recursive algorithm of the particular digital filter. This generates the output sequence. The resulting digital output can be converted into an analog signal using a DAC, if so desired.

A recursive digital filter is an implementation of a recursive algorithm that governs the particular filtering scheme (e.g., low pass, high pass, band pass, and band reject). The filter algorithm can be implemented either by software or by hardware. In software implementation, the filter algorithm is programmed into a digital computer. The processor (e.g., microprocessor or DSP) of the computer can process an input data sequence according to the run-time filter program stored in the memory (in machine code) to generate the filtered output sequence.

Digital processing of data is accomplished by means of logic circuitry that can perform basic arithmetic operations such as addition. In the software approach, the processor of a digital computer makes use of these basic logic circuits to perform digital processing according to the instructions of a software program stored in the computer memory. Alternatively, a hardware digital processor can be built to perform a somewhat complex, yet fixed, processing operation. In this approach, the program of computation is said to be in the hardware. The hardware processor is then available as an IC chip, whose processing operation is fixed and cannot be modified. The logic circuitry in the IC chip is designed to accomplish the required processing function. Digital filters implemented by this hardware approach are termed *hardware digital filters*.

The software implementation of digital filters has the advantage of flexibility; specifically, the filter algorithm can be easily modified by changing the software program that is stored in the computer. If, on the other hand, a large number of filters of a particular (fixed) structure are commercially needed, then it would be economical to design the filter as an IC chip and replicate the chip in mass production. In this manner, very low-cost digital filters can be produced. A hardware filter can operate at a much faster speed in comparison to a software filter because in the former case, processing takes place automatically through logic circuitry in the filter chip without accessing by the processor, a software program, and various data items stored in the memory. The main disadvantage of a hardware filter is that its algorithm and parameter values cannot be modified, and the filter is dedicated to perform a fixed function.

2.10 Bridge Circuits

A full bridge is a circuit with four arms connected in a lattice form. Four nodes are formed in this manner. Two opposite nodes are used for excitation (voltage or current supply) of the bridge, and the remaining two opposite nodes provide the bridge output.

A bridge circuit is used to make some form of measurement. Typical measurements include change in resistance, change in inductance, change in capacitance, oscillating frequency, or some variable (stimulus) that causes these changes. There are two basic methods of making the measurement:

1. Bridge balance method
2. Imbalance output method

A bridge is said to be balanced when its output voltage is zero.

In the bridge-balance method, we start with a balanced bridge. When making a measurement, the balance of the bridge will be upset due to the associated variation. As a result, a nonzero output voltage will be produced. The bridge can be balanced again by varying one of the arms of the bridge (assuming, of course, that some means is provided for fine adjustments that may be required). The change that is required to restore the balance is in fact the "measurement." The bridge can be balanced precisely using a servo device, in this method.

In the imbalance output method as well, we usually start with a balanced bridge. As before, the balance of the bridge will be upset as a result of the change in the variable that is measured. Now, instead of balancing the bridge again, the output voltage of the bridge due to the resulted imbalance is measured and used as the bridge measurement.

There are many types of bridge circuits. If the supply to the bridge is dc, then we have a dc bridge. Similarly, an ac bridge has an ac excitation. A resistance bridge has only resistance elements in its four arms, and it is typically a dc bridge. An impedance bridge has impedance elements consisting of resistors, capacitors, and inductors in one or more of its arms. This is necessarily an ac bridge. If the bridge excitation is a constant voltage supply, we have a constant-voltage bridge. If the bridge supply is a constant current source, we get a constant-current bridge.

2.10.1 Wheatstone Bridge

Wheatstone bridge is a resistance bridge with a constant dc voltage supply (i.e., it is a constant-voltage resistance bridge). A Wheatstone bridge is particularly useful in strain-gage measurements, and consequently in force, torque, and tactile sensors that employ strain-gage techniques. Since a Wheatstone bridge is used primarily in the measurement of small changes in resistance, it could be used in other types of sensing applications as well. For example, in resistance temperature detectors (RTD), the change in resistance in a metallic (e.g., platinum) element, as caused by a change in temperature, is measured using a bridge circuit. Note that the temperature coefficient of resistance is positive for a typical metal (i.e., the resistance increases with temperature). For platinum, this value (change in resistance per unit resistance per unit change in temperature) is about 0.00385 per $°C$.

Consider the Wheatstone bridge circuit shown in Figure 2.35a. Assuming that the bridge output is open circuit (i.e., very high load resistance), the output v_o may be expressed as

$$v_o = v_A - v_B = \frac{R_1 v_{ref}}{(R_1 + R_2)} - \frac{R_3 v_{ref}}{(R_3 + R_4)} = \frac{(R_1 R_4 - R_2 R_3)}{(R_1 + R_2)(R_3 + R_4)} v_{ref}. \tag{2.88}$$

For a balanced bridge, the numerator of the RHS expression of Equation 2.88 must vanish. Hence, the condition for bridge balance is

$$\frac{R_1}{R_2} = \frac{R_3}{R_4}. \tag{2.89}$$

Suppose that at first $R_1 = R_2 = R_3 = R_4 = R$. Then, according to Equation 2.89, the bridge is balanced. Now increase R_1 by δR. For example, R_1 may represent the only active strain gage, while the remaining three elements in the bridge are identical dummy elements. In view of Equation 2.88, the change in the bridge output due to the change δR is given by

$$\delta v_o = \frac{[(R + \delta R)R - R^2]}{(R + \delta R + R)(R + R)} v_{ref} - 0$$

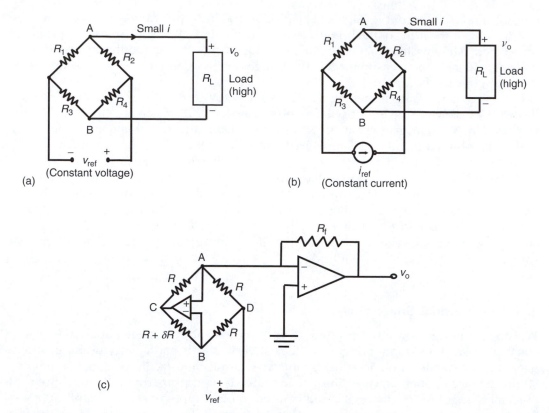

FIGURE 2.35
(a) Wheatstone bridge (constant-voltage resistance bridge). (b) Constant-current resistance bridge. (c) A linearized bridge.

or

$$\frac{\delta v_o}{v_{\text{ref}}} = \frac{\delta R/R}{(4 + 2\delta R/R)}. \tag{2.90a}$$

Note that the output is nonlinear in $\delta R/R$. If, however, $\delta R/R$ is assumed small in comparison to 2, we have the linearized relationship

$$\frac{\delta v_o}{v_{\text{ref}}} = \frac{\delta R}{4R}. \tag{2.91}$$

The factor $1/4$ on the RHS of Equation 2.91 represents the sensitivity is the bridge, as gives the change in the bridge output for a given change in the active resistance, while the other parameters are kept fixed. Strictly speaking, the bridge sensitivity is given by $\delta v_o/\delta R$, which is equal to $v_{\text{ref}}/(4R)$.

The error due to linearization, which is a measure of nonlinearity, may be given as the percentage,

$$N_P = 100\left(1 - \frac{\text{Linearized output}}{\text{Actual output}}\right)\%. \tag{2.92}$$

Hence, from Equation 2.90 and Equation 2.91 we have

$$N_P = 50 \frac{\delta R}{R} \%.$$ (2.93)

Example 2.9

Suppose that in Figure 2.35a, at first $R_1 = R_2 = R_3 = R_4 = R$. Now increase R_1 by δR, decrease R_2 by δR. This will represent two active elements that act in reverse, as in the case of two strain gage elements mounted on the top and the bottom surfaces of a beam in bending. Show that the bridge output is linear in δR in this case.

Solution

From Equation 2.88, we get

$$\delta v_o = \frac{[(R + \delta R)R - R^2]}{(R + \delta R + R - \delta R)(R + R)} v_{ref} - 0.$$

This simplifies to

$$\frac{\delta v_o}{v_{ref}} = \frac{\delta R}{4R},$$

which is linear.

Similarly, it can be shown using Equation 2.88 that the pair of changes: $R_3 \rightarrow R + \delta R$ and $R_4 \rightarrow R - \delta R$ will result in a linear relation for the bridge output.

2.10.2 Constant-Current Bridge

When large resistance variations δR are required for a measurement, the Wheatstone bridge may not be satisfactory due to its nonlinearity, as indicated by Equation 2.90. The constant-current bridge is less nonlinear and is preferred in such applications. However, it needs a current-regulated power supply, which is typically more costly than a voltage-regulated power supply.

As shown in Figure 2.35b, the constant-current bridge uses a constant-current excitation i_{ref}, instead of a constant-voltage supply. The output equation for a constant-current bridge can be determined from Equation 2.88, simply by knowing the voltage at the current source. Suppose that this voltage is v_{ref}, with the polarity shown in Figure 2.35a. Now, since the load current is assumed small (high-impedance load), the current through R_2 is equal to the current through R_1, and is given by $v_{ref}/(R_1 + R_2)$. Similarly, current through R_4 and R_3 is given by $v_{ref}/(R_3 + R_4)$. Accordingly, by current summation we get,

$$i_{ref} = \frac{v_{ref}}{(R_1 + R_2)} + \frac{v_{ref}}{(R_3 + R_4)}$$

or

$$v_{ref} = \frac{(R_1 + R_2)(R_3 + R_4)}{(R_1 + R_2 + R_3 + R_4)} i_{ref}.$$ (2.94)

This result may be directly obtained from the equivalent resistance of the bridge, as seen by the current source. Substituting Equation 2.94 in Equation 2.88, we have the output equation for the constant-current bridge. Thus,

$$v_o = \frac{(R_1 R_4 - R_2 R_3)}{(R_1 + R_2 + R_3 + R_4)} i_{ref}. \tag{2.95}$$

Note from Equation 2.95 that the bridge-balance requirement (i.e., $v_o = 0$) is again given by Equation 2.89.

To estimate the nonlinearity of a constant-current bridge, we start with the balanced condition: $R_1 = R_2 = R_3 = R_4 = R$, and change R_1 by δR while keeping the remaining resistors inactive. Again, R_1 will represent the active element (sensing element) of the bridge, and may correspond to an active strain gage. The change in output δv_o is given by

$$\delta v_o = \frac{[(R + \delta R)R - R^2]}{(R + \delta R + R + R + R)} i_{ref} - 0$$

or

$$\frac{\delta v_o}{R i_{ref}} = \frac{\delta R / R}{(4 + \delta R / R)}. \tag{2.96a}$$

By comparing the denominator on the RHS of this equation with Equation 2.90, we observe that the constant-current bridge is less nonlinear. Specifically, using the definition given by Equation 2.92, the percentage nonlinearity may be expressed as

$$N_p = 25 \frac{\delta R}{R} \%. \tag{2.97}$$

It is noted that the nonlinearity is halved by using a constant-current excitation, instead of a constant-voltage excitation.

Example 2.10

Suppose that in the constant-current bridge circuit shown in Figure 2.35b, at first $R_1 = R_2 = R_3 = R_4 = R$. Assume that R_1 and R_4 represent strain gages mounted on the same side of a rod in tension. Due to the tension, R_1 increases by δR and R_4 also increases by δR. Derive an expression for the bridge output (normalized) in this case, and show that it is linear. What would be the result if R_2 and R_3 represent the active tensile strain gages in this example?

Solution

From Equation 2.95, we get

$$\delta v_o = \frac{[(R + \delta R)(R + \delta R) - R^2]}{(R + \delta R + R + R + R + \delta R)} i_{ref} - 0.$$

By simplifying and canceling the common term in the numerator and the denominator, we get the linear relation

$$\frac{\delta v_o}{R i_{ref}} = \frac{\delta R / R}{2}. \tag{2.96b}$$

If R_2 and R_3 are the active elements, it is clear from Equation 2.95 that we get the same linear result, except for a sign change. Specifically,

$$\frac{\delta v_o}{Ri_{ref}} = -\frac{\delta R/R}{2}.$$ (2.96c)

2.10.3 Hardware Linearization of Bridge Outputs

From the foregoing developments and as illustrated in the examples, it should be clear that the output of a resistance bridge is not linear in general, with respect to the change in resistance of the active elements. Particular arrangements of the active elements can result in a linear output. It is seen from Equation 2.88 and Equation 2.95 that, when there is only one active element, the bridge output is nonlinear. Such a nonlinear bridge can be linearized using hardware; particularly op-amp elements. To illustrate this approach, consider a constant-voltage resistance bridge. We modify it by connecting two op-amp elements, as shown in Figure 2.35c. The output amplifier has a feedback resistor R_f. The output equation for this circuit can be obtained by using the properties of an op-amp, in the usual manner. In particular, the potentials at the two input leads must be equal and the current through these leads must be zero. From the first property, it follows that the potentials at the nodes A and B are both zero. Let the potential at node C be denoted by v. Now use the second property, and write current summations at nodes A and B.

$$\text{Node A: } \frac{v}{R} + \frac{v_{ref}}{R} + \frac{v_o}{R_f} = 0$$ (i)

$$\text{Node B: } \frac{v_{ref}}{R} + \frac{v}{R + \delta R} = 0.$$ (ii)

Substitute Equation ii in Equation i to eliminate v, and simplify to get the linear result

$$\frac{\delta v_o}{v_{ref}} = \frac{R_f}{R}\frac{\delta R}{R}$$ (2.90b)

Compare this result with Equation 2.90a for the original bridge with a single active element. Note that, when $\delta R = 0$, from Equation ii, we get, $v = v_{ref}$, and from Equation i we get $v_o = 0$. Hence, v_o and δv_o are identical, as used in Equation 2.90b.

2.10.4 Bridge Amplifiers

The output signal from a resistance bridge is usually very small in comparison to the reference signal, and it has to be amplified to increase its voltage level to a useful value (e.g., for use in system monitoring, data logging, or control). A bridge amplifier is used for this purpose. This is typically an instrumentation amplifier, which is essentially a sophisticated differential amplifier. The bridge amplifier is modeled as a simple gain K_a, which multiplies the bridge output.

2.10.5 Half-Bridge Circuits

A half bridge may be used in some applications that require a bridge circuit. A half bridge has only two arms, and the output is tapped from the mid-point of these two arms. The ends of the two arms are excited by two voltages, one of which is positive and the other negative. Initially, the two arms have equal resistances so that nominally the bridge output is zero. One of the arms has the active element. Its change in resistance results in a nonzero output voltage. It is noted that the half-bridge circuit is somewhat similar to a potentiometer circuit (a voltage divider).

A half-bridge amplifier consisting of a resistance half bridge and an output amplifier is shown in Figure 2.36. The two bridge arms have resistances R_1 and R_2, and the output amplifier uses a feedback resistance R_f. To get the output equation, we use the two basic facts for an unsaturated op-amp; the voltages at the two input leads are equal (due to high gain), and the current in either lead is zero (due to high input impedance). Hence, voltage at node A is zero and the current balance equation at node A is given by

$$\frac{v_{\text{ref}}}{R_1} + \frac{(-v_{\text{ref}})}{R_2} + \frac{v_{\text{o}}}{R_f} = 0.$$

This gives

$$v_{\text{o}} = R_f \left(\frac{1}{R_2} - \frac{1}{R_1} \right) v_{\text{ref}}. \tag{2.98}$$

Now, suppose that initially $R_1 = R_2 = R$, and the active element R_1 changes by δR. The corresponding change in output is

$$\delta v_{\text{o}} = R_f \left(\frac{1}{R} - \frac{1}{R + \delta R} \right) v_{\text{ref}} - 0$$

or

$$\frac{\delta v_{\text{o}}}{v_{\text{ref}}} = \frac{R_f}{R} \frac{\delta R / R}{(1 + \delta R / R)}. \tag{2.99}$$

Note that R_f / R is the amplifier gain. Now in view of Equation 2.92, the percentage nonlinearity of the half-bridge circuit is

$$N_p = 100 \frac{\delta R}{R} \%. \tag{2.100}$$

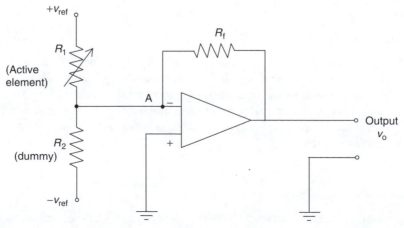

FIGURE 2.36
A half bridge with an output amplifier.

It follows that the nonlinearity of a half-bridge circuit is worse than that for the Wheatstone bridge.

2.10.6 Impedance Bridges

An impedance bridge is an ac bridge. It contains general impedance elements Z_1, Z_2, Z_3, and Z_4 in its four arms, as shown in Figure 2.37a. The bridge is excited by an ac (supply) voltage v_{ref}. Note that v_{ref} would represent a carrier signal, and the output voltage v_o has to be demodulated if a transient signal representative of the variation in one of the bridge elements is needed. Impedance bridges could be used, for example, to measure capacitances in capacitive sensors and changes of inductance in variable-inductance sensors and eddy-current sensors. Also, impedance bridges can be used as oscillator circuits. An oscillator circuit could serve as a constant-frequency source of a signal generator (e.g., in product dynamic testing), or it could be used to determine an unknown circuit parameter by measuring the oscillating frequency.

Analyzing by using frequency-domain concepts, it is seen that the frequency spectrum of the impedance-bridge output is given by

$$v_o(\omega) = \frac{(Z_1 Z_4 - Z_2 Z_3)}{(Z_1 + Z_2)(Z_3 + Z_4)} v_{ref}(\omega). \tag{2.101}$$

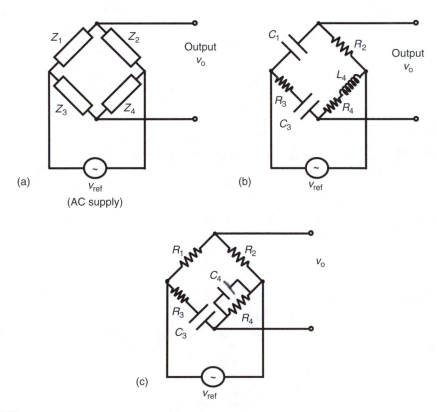

FIGURE 2.37
(a) General impedance bridge. (b) Owen bridge. (c) Wien-bridge oscillator.

This reduces to Equation 2.88 in the dc case of a Wheatstone bridge. The balanced condition is given by

$$\frac{Z_1}{Z_2} = \frac{Z_3}{Z_4}. \tag{2.102}$$

This equation is used to measure an unknown circuit parameter in the bridge. Let us consider two particular impedance bridges.

2.10.6.1 Owen Bridge

The Owen bridge is shown in Figure 2.37b. It may be used, for example, to measure both inductance L_4 and capacitance C_3, by the bridge-balance method. To derive the necessary equation, note that the voltage–current relation for an inductor is

$$v = L\frac{di}{dt} \tag{2.103}$$

and for a capacitor it is

$$i = C\frac{dv}{dt}. \tag{2.104}$$

It follows that the voltage/current transfer function (in the Laplace domain) for an inductor is

$$\frac{v(s)}{i(s)} = Ls, \tag{2.105}$$

and that for a capacitor is

$$\frac{v(s)}{i(s)} = \frac{1}{Cs}. \tag{2.106}$$

Accordingly, the impedance of an inductor element at frequency ω is

$$Z_L = j\omega L, \tag{2.107}$$

and the impedance of a capacitor element at frequency ω is

$$Z_c = \frac{1}{j\omega C}. \tag{2.108}$$

Applying these results for the Owen bridge we have

$$Z_1 = \frac{1}{j\omega C_1},$$
$$Z_2 = R_2,$$
$$Z_3 = R_3 + \frac{1}{j\omega C_3},$$
$$Z_4 = R_4 + j\omega L_4,$$

where ω is the excitation frequency. Now, from Equation 2.102, we have

$$\frac{1}{j\omega C_1}(R_4 + j\omega L_4) = R_2\left(R_3 + \frac{1}{j\omega C_3}\right).$$

By equating the real parts and the imaginary parts of this equation, we get the two equations

$$\frac{L_4}{C_1} = R_2 R_3$$

and

$$\frac{R_4}{C_1} = \frac{R_2}{C_3}.$$

Hence, we have

$$L_4 = C_1 R_2 R_3 \tag{2.109}$$

and

$$C_3 = C_1 \frac{R_2}{R_4}. \tag{2.110}$$

It follows that L_4 and C_3 can be determined with the knowledge of C_1, R_2, R_3, and R_4 under balanced conditions. For example, with fixed C_1 and R_2, an adjustable R_3 could be used to measure the variable L_4, and an adjustable R_4 could be used to measure the variable C_3.

2.10.6.2 Wien-Bridge Oscillator

Now consider the Wien-bridge oscillator shown in Figure 2.37c. For this circuit, we have

$$Z_1 = R_1,$$
$$Z_2 = R_2,$$
$$Z_3 = R_3 + \frac{1}{j\omega C_3},$$
$$\frac{1}{Z_4} = \frac{1}{R_4} + j\omega C_4.$$

Hence, from Equation 2.102, the bridge-balance requirement is

$$\frac{R_1}{R_2} = \left(R_3 + \frac{1}{j\omega C_3}\right)\left(\frac{1}{R_4} + j\omega C_4\right).$$

Equating the real parts, we get

$$\frac{R_1}{R_2} = \frac{R_3}{R_4} + \frac{C_4}{C_3}, \tag{2.111}$$

and by equating the imaginary parts we get

$$0 = \omega C_4 R_3 - \frac{1}{\omega C_3 R_4}.$$

Hence,

$$\omega = \frac{1}{\sqrt{C_3 C_4 R_3 R_4}}. \tag{2.112}$$

Equation 2.112 tells us that the circuit is an oscillator whose natural frequency is given by this equation, under balanced conditions. If the frequency of the supply is equal to the natural frequency of the circuit, large-amplitude oscillations will take place. The circuit can be used to measure an unknown resistance (e.g., in strain gage devices) by first measuring the frequency of the bridge signals at resonance (natural frequency). Alternatively, an oscillator that is excited at its natural frequency can be used as an accurate source of periodic signals (signal generator).

2.11 Linearizing Devices

Nonlinearity is present in any physical device, to varying levels. If the level of nonlinearity in a system (component, device, or equipment) can be neglected without exceeding the error tolerance, then the system can be assumed linear.

In general, a linear system is one that can be expressed as one or more linear differential equations. Note that the principle of superposition holds for linear systems. Specifically, if the system response to an input u_1 is y_1, and the response to another input u_2 is y_2, then the response to $a_1 u_1 + a_2 u_2$ would be $a_1 y_1 + a_2 y_2$.

Nonlinearities in a system can appear in two forms:

1. Dynamic manifestation of nonlinearities
2. Static manifestation of nonlinearities

In many applications, the useful operating region of a system can exceed the frequency range where the frequency response function is flat. The operating response of such a system is said to be dynamic. Examples include a typical control system (e.g., automobile, aircraft, milling machine, robot), actuator (e.g., hydraulic motor), and controller (e.g., proportional-integral-derivative or PID control circuitry). Nonlinearities of such systems can manifest themselves in a dynamic form such as the jump phenomenon (also known as the fold catastrophe), limit cycles, and frequency creation. Design changes, extensive adjustments, or reduction of the operating signal levels and bandwidths would be necessary in general, to reduce or eliminate these dynamic manifestations of nonlinearity. In many instances, such changes would not be practical, and we may have to somehow cope with the presence of these nonlinearities under dynamic conditions. Design changes might involve replacing conventional gear drives by devices such as harmonic drives to reduce backlash, replacing nonlinear actuators by linear actuators, and using components that have negligible Coulomb friction and that make small motion excursions.

A wide majority of sensors, transducers, and signal-modification devices are expected to operate in the flat region of their frequency response function. The input/output relation of these types of devices, in the operating range, is expressed (modeled) as a static curve rather than a differential equation. Nonlinearities in these devices will manifest themselves in the static operating curve in many forms. These manifestations include saturation, hysteresis, and offset.

In the first category of systems (e.g., plants, actuators, and compensators), if a nonlinearity is exhibited in the dynamic form, proper modeling and control practices should be employed to avoid unsatisfactory degradation of the system performance. In the second category of systems (e.g., sensors, transducers, and signal-modification devices), if nonlinearities are exhibited in the static operating curve, again the overall performance of the system will be degraded. Hence it is important to linearize the output of such devices. Note that in dynamic manifestations it is not possible to realistically linearize the output because the response is generated in the dynamic form. The solution in that case is either to minimize nonlinearities within the system by design modifications and adjustments, so that a linear approximation to the system would be valid, or alternatively to take the nonlinearities into account in system modeling and control. In the present section, we are not concerned with this aspect (i.e., dynamic nonlinearities). Instead, we are interested in the linearization of devices in the second category, whose operating characteristics can be expressed by static input/output curves.

Linearization of a static device can be attempted as well by making design changes and adjustments, as in the case of dynamic devices. But, since the response is static, and since we normally deal with an available device (fixed design) whose internal hardware cannot be modified, we should consider ways of linearizing the input/output characteristic by modifying the output itself.

Static linearization of a device can be made in three ways:

1. Linearization using digital software
2. Linearization using digital (logic) hardware
3. Linearization using analog circuitry

In the software approach to linearization, the output of the device is read into a digital processor with software-programmable memory, and the output is modified according to the program instructions. In the hardware approach, the output is read by a device with fixed logic circuitry for processing (modifying) the data. In the analog approach, a linearizing circuit is directly connected at the output of the device, so that the output of the linearizing circuit is proportional to the input to the original device. An example of this type of (analog) linearization was given under Section 2.10.3. We shall discuss these three approaches in the rest of this section, heavily emphasizing the analog-circuit approach.

Hysteresis-type static nonlinearity characteristics have the property that the input/output curve is not one-to-one. In other words, one input value may correspond to more than one (static) output value, and one output value may correspond to more than one input value. If we disregard these types of nonlinearities, our main concern would be with the linearization of a device with a single-valued static response curve that is not a straight line. An example of a typical nonlinear input/output characteristic is shown in Figure 2.38a. Strictly speaking, a straight-line characteristic with a simple offset, as shown in Figure 2.38b, is also a nonlinearity. In particular, note that superposition does not hold for an input/output characteristic of this type, given by

$$y = ku + c. \tag{2.113}$$

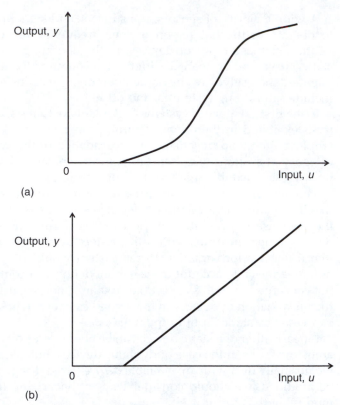

FIGURE 2.38
(a) A general static nonlinear characteristic. (b) An offset nonlinearity.

It is very easy, however, to linearize such a device because a simple addition of a dc component will convert the characteristic into the linear form given by

$$y = ku. \tag{2.114}$$

This method of linearization is known as *offsetting*. Linearization is more difficult in the general case where the characteristic curve could be much more complex.

2.11.1 Linearization by Software

If the nonlinear relationship between the input and the output of a nonlinear device is known, the input can be computed for a known value of the output. In the software approach of linearization, a processor and memory that can be programmed using software (i.e., a digital computer) is used to compute the input using output data. Two approaches can be used. They are

1. equation inversion
2. table lookup.

In the first method, the nonlinear characteristic of the device is known in the analytic (equation) form

$$y = f(u), \tag{2.115}$$

where u is the device input and y is the device output.

Assuming that this is a one-to-one relationship, a unique inverse given by the equation

$$u = f^{-1}(y) \tag{2.116}$$

can be determined. This equation is programmed as a computation algorithm, into the read-and-write memory (RAM) of the computer. When the output values y are supplied to the computer, the processor will compute the corresponding input values u using the instructions (executable program) stored in the RAM.

In the table lookup method, a sufficiently large number of pairs of values (y, u) are stored in the memory of the computer in the form of a table of ordered pairs. These values should cover the entire operating range of the device. Then, when a value for y is entered into the computer, the processor scans the stored data to check whether that value is present. If so, the corresponding value of u will be read, and this is the linearized output. If the value of y is not present in the data table, then the processor will interpolate the data in the vicinity of the value and will compute the corresponding output. In the linear interpolation method, the neighborhood of the data table where the y value falls is fitted with a straight line and the corresponding u value is computed using this straight line. Higher-order interpolations use nonlinear interpolation curves such as quadratic and cubic polynomial equations (splines).

Note that the equation inversion method is usually more accurate than the table lookup method, and the former does not need excessive memory for data storage. But it is relatively slow because data are transferred and processed within the computer using program instructions, which are stored in the memory and which typically have to be accessed in a sequential manner. The table lookup method is fast. Since accuracy depends on the amount of stored data values, this is a memory-intensive method. For better accuracy, more data should be stored. But, since the entire data table has to be scanned to check for a given data value, this increase in accuracy is derived at the expense of speed as well as memory requirements.

2.11.2 Linearization by Hardware Logic

The software approach of linearization is flexible because the linearization algorithm can be modified (e.g., improved, changed) simply by modifying the program stored in the RAM. Furthermore, highly complex nonlinearities can be handled by the software method. As mentioned before, the method is relatively slow, however.

In the hardware logic method of linearization, the linearization algorithm is permanently implemented in the IC form using appropriate digital logic circuitry for data processing and memory elements (e.g., flip-flops). Note that the algorithm and numerical values of parameters (except input values) cannot be modified without redesigning the IC chip, because a hardware device typically does not have programmable memory. Furthermore, it will be difficult to implement very complex linearization algorithms by this method, and unless the chips are mass produced for an extensive commercial market, the initial chip development cost will make the production of linearizing chips economically infeasible. In bulk production, however, the per-unit cost will be very small. Furthermore, since both the access of stored program instructions and extensive data manipulation are not involved, the hardware method of linearization can be substantially faster than the software method.

A digital linearizing unit with a processor and a read-only memory (ROM), whose program cannot be modified, also lacks the flexibility of a programmable software device. Hence, such a ROM-based device also falls into the category of hardware logic devices.

2.11.3   Analog Linearizing Circuitry

Three types of analog linearizing circuitry can be identified:

1. Offsetting circuitry
2. Circuitry that provides a proportional output
3. Curve shapers

We will describe each of these categories now.

An offset is a nonlinearity that can be easily removed using an analog device. This is accomplished by simply adding a dc offset of equal value to the response, in the opposite direction. Deliberate addition of an offset in this manner is known as offsetting. The associated removal of original offset is known as offset compensation. There are many applications of offsetting. Unwanted offsets such as those present in the results of ADC and DAC can be removed by analog offsetting. Constant (dc) error components, such as steady-state errors in dynamic systems due to load changes, gain changes, and other disturbances, can be eliminated by offsetting. Common-mode error signals in amplifiers and other analog devices can also be removed by offsetting. In measurement circuitry such as potentiometer (ballast) circuits, where the actual measurement signal is a small change δv_o of a steady output signal v_o, the measurement can be completely masked by noise. To reduce this problem, first the output should be offset by $-v_o$, so that the net output is δv_o and not $v_o + \delta v_o$. Subsequently, this output can be conditioned through filtering and amplification. Another application of offsetting is the additive change of the scale of a measurement from a relative scale to an absolute scale (e.g., in the case of velocity). In summary, some applications of offsetting are:

1. Removal of unwanted offsets and dc components in signals (e.g., in ADC, DAC, signal integration)
2. Removal of steady-state error components in dynamic system responses (e.g., due to load changes and gain changes in Type 0 systems. Note that Type 0 systems are open-loop systems with no free integrators.)
3. Rejection of common-mode levels (e.g., in amplifiers and filters)
4. Error reduction when a measurement is an increment of a large steady output level (e.g., in ballast circuits for strain-gage and RTD sensors)
5. Scale changes in an additive manner (e.g., conversion from relative to absolute units or from absolute to relative units)

We can remove unwanted offsets in the simple manner as discussed earlier. Let us now consider more complex nonlinear responses that are nonlinear, in the sense that the input/output curve is not a straight line. Analog circuitry can be used to linearize this type of responses as well. The linearizing circuit used will generally depend on the particular device and the nature of its nonlinearity. Hence, often linearizing circuits of this type have to be discussed with respect to a particular application. For example, such linearization circuits are useful in a transverse-displacement capacitive sensor. Several useful circuits are described later.

Consider the type of linearization that is known as *curve shaping*. A curve shaper is a linear device whose gain (output/input) can be adjusted so that response curves with different slopes can be obtained. Suppose that a nonlinear device with an irregular (nonlinear) input/output characteristic is to be linearized. First, we apply the operating input simultaneously to both the device and the curve shaper, and the gain of the curve shaper is adjusted such that it closely matches that of the device in a small range of operation. Now the output of the curve shaper can be utilized for any task that requires the device output. The advantage here is that linear assumptions are valid with the curve shaper, which is not the case for the actual device. When the operating range changes, the curve shaper has to be adjusted to the new range. Comparison (calibration) of the curve shaper and the nonlinear device can be done off line and, once a set of gain values corresponding to a set of operating ranges is determined in this manner for the curve shaper, it is possible to completely replace the nonlinear device by the curve shaper. Then the gain of the curve shaper can be adjusted, depending on the actual operating range during system operation. This is known as *gain scheduling*. Note that we can replace a nonlinear device by a linear device (curve shaper) within a multi-component system in this manner without greatly sacrificing the accuracy of the overall system.

2.11.4 Offsetting Circuitry

Common-mode outputs and offsets in amplifiers and other analog devices can be minimized by including a compensating resistor , which will provide fine adjustments at one of the input leads. Furthermore, the larger the magnitude of the feedback signal in a control system, the smaller the steady-state error. Hence, steady-state offsets can be reduced by reducing the feedback resistance (thereby increasing the feedback signal). Furthermore, since a ballast (potentiometer) circuit provides an output of $v_o + \delta v_o$ and a bridge circuit provides an output of δv_o, the use of a bridge circuit can be interpreted as an offset compensation method.

The most straightforward way of offsetting a nonlinear device is by using a differential amplifier (or a summing amplifier) to subtract (or add) a dc voltage to the output of the device. The dc level has to be variable so that various levels of offset can be provided with the same circuit. This is accomplished by using an adjustable resistance at the dc input lead of the amplifier.

An op-amp circuit that can be used for offsetting is shown in Figure 2.39. Since the input v_i is connected to the negative lead of the op-amp, we have an inverting amplifier, and the input signal will appear in the output v_o with its sign reversed. This is also a

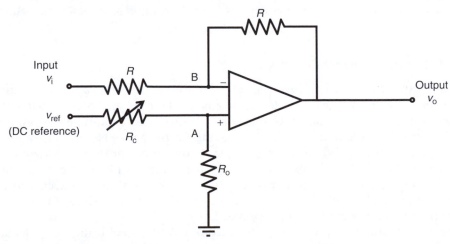

FIGURE 2.39
An inverting amplifier circuit for offset compensation.

summing amplifier because two signals can be added together by this circuit. If the input v_i is connected to the positive lead of the op-amp, we will have a noninverting amplifier.

The dc voltage v_{ref} provides the offsetting voltage. The compensating resistor R_c is variable so that different values of offset can be compensated using the same circuit. To obtain the circuit equation, we write the current balance equation for node A, using the usual assumption that the current through an input lead is zero for an op-amp (because of very high input impedance). Thus,

$$\frac{v_{ref} - v_A}{R_c} = \frac{v_A}{R_o}$$

or

$$v_A = \frac{R_o}{(R_o + R_c)} v_{ref}.$$

Similarly, the current balance at node B gives

$$\frac{v_i - v_B}{R} + \frac{v_o - v_B}{R} = 0$$

or

$$v_o = -v_i + 2v_B. \tag{ii}$$

Since $v_A = v_B$ for the op-amp (because of very high open-loop gain), we can substitute Equation i in Equation ii. Then,

$$v_o = -v_i + \frac{2R_o}{(R_o + R_c)} v_{ref}. \tag{2.117}$$

Note the sign reversal of v_i at the output (because this is an inverting amplifier). This is not a problem because the polarity can be reversed at input or output by connecting this circuit to other circuitry, thereby recovering the original sign. The important result here is the presence of a constant offset term on the RHS of Equation 2.117. This term can be adjusted by picking the proper value for R_c so as to compensate for a given offset in v_i.

2.11.5 Proportional-Output Circuitry

An op-amp circuit may be employed to linearize the output of a capacitive transverse-displacement sensor. We have noted that in constant-voltage and constant-current resistance bridges and in a constant-voltage half bridge, the relation between the bridge output δv_o and the measured (change in resistance in the active element) is nonlinear in general. The nonlinearity is least for the constant-current bridge and it is the highest for the half bridge. As δR is small compared with R the nonlinear relations can be linearized without introducing large errors. However, the linear relations are inexact, and are not suitable if δR cannot be neglected in comparison to R. Under these circumstances, the use of a linearizing circuit would be appropriate.

One way to obtain a proportional output from a Wheatstone bridge is to feedback a suitable factor of the bridge output into the bridge supply v_{ref}. This approach is illustrated previously (see Figure 2.35c). Another way is to use the op-amp circuit shown in

Figure 2.40. This should be compared with the Wheatstone bridge shown in Figure 2.35a. Note that R represents the only active element (e.g., an active strain gage).

First, let us show that the output equation for the circuit in Figure 2.40 is quite similar to Equation 2.88. Using the fact that the current through an input lead of an unsaturated op-amp can be neglected, we have the following current balance equations for nodes A and B

$$\frac{v_{ref} - v_A}{R_4} = \frac{v_A}{R_2}$$

and

$$\frac{v_{ref} - v_B}{R_3} + \frac{v_o - v_B}{R_1} = 0.$$

Hence,

$$v_A = \frac{R_2}{(R_2 + R_4)} v_{ref}$$

and

$$v_B = \frac{R_1 v_{ref} + R_3 v_o}{(R_1 + R_3)}.$$

Now using the fact $v_A = v_B$ for an op-amp, we get

$$\frac{R_1 v_{ref} + R_3 v_o}{(R_1 + R_3)} = \frac{R_2}{(R_2 + R_4)} v_{ref}.$$

Accordingly, we have the circuit output equation

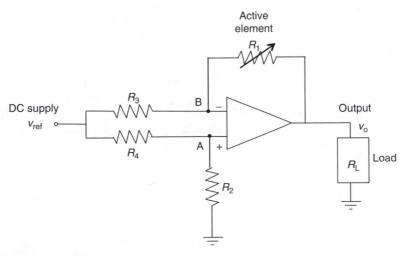

FIGURE 2.40
A proportional-output circuit for an active resistance element (strain gage).

$$v_o = \frac{(R_2 R_3 - R_1 R_4)}{R_3(R_2 + R_4)} v_{ref}. \tag{2.118}$$

Note that this relation is quite similar to the Wheatstone bridge equation (Equation 2.88). The balance condition (i.e., $v_o = 0$) is again given by Equation 2.89.

Suppose that $R_1 = R_2 = R_3 = R_4 = R$ in the beginning (hence, the circuit is balanced), so that $v_o = 0$. Next, suppose that the active resistance R_1 is changed by δR (say, due to a change in strain in the strain gage R_1). Then, using Equation 2.118, we can write an expression for the resulting change in the circuit output as

$$\delta v_o = \frac{[R^2 - R(R + \delta R)]}{R(R + R)} v_{ref} - 0$$

or

$$\frac{\delta v_o}{v_{ref}} = -\frac{1}{2}\frac{\delta R}{R}. \tag{2.119}$$

By comparing this result with Equation 2.90, we observe that the circuit output δv_o is proportional to the measured δR. Furthermore, note that the sensitivity ($1/2$) of the circuit in Figure 2.40 is double that of a Wheatstone bridge ($1/4$) with one active element, which is a further advantage of the proportional-output circuit. The sign reversal is not a drawback because it can be accounted for by reversing the load polarity.

2.11.6   Curve-Shaping Circuitry

A curve shaper can be interpreted as an amplifier whose gain is adjustable. A typical arrangement for a curve-shaping circuit is shown in Figure 2.41. The feedback resistance R_f is adjustable by some means. For example, a switching circuit with a bank of resistors (say, connected in parallel through solid-state switches as in the case of weighted-resistor

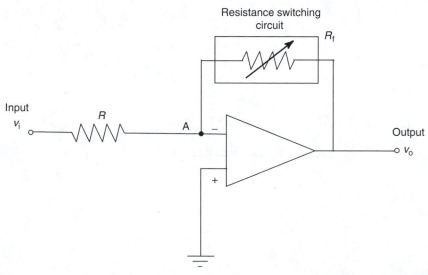

FIGURE 2.41
A curve-shaping circuit.

DAC) can be used to switch the feedback resistance to the required value. Automatic switching can be realized by using Zener diodes, which will start conducting at certain voltage levels. In both cases (external switching by switching pulses and automatic switching using Zener diodes), amplifier gain is variable in discrete steps. Alternatively, a potentiometer may be used as R_f so that the gain can be continuously adjusted (manually or automatically).

The output equation for the curve-shaping circuit shown in Figure 2.41 is obtained by writing the current balance at node A, noting that $v_A = 0$. Thus,

$$\frac{v_i}{R} + \frac{v_o}{R_f} = 0$$

or

$$v_o = -\frac{R_f}{R} v_i. \tag{2.120}$$

It follows that the gain (R_f/R) of the amplifier can be adjusted by changing R_f.

2.12 Miscellaneous Signal-Modification Circuitry

In addition to the signal-modification devices discussed so far in this chapter, there are many other types of circuitry that are used for signal modification and related tasks. Examples are phase shifters, voltage-to-frequency converters (VFC), frequency-to-voltage converters (FVC), voltage-to-current converts, and peak-hold circuits. The objective of the present section is to briefly discuss several of such miscellaneous circuits and components that are useful in the instrumentation of control systems.

2.12.1 Phase Shifters

A sinusoidal signal given by

$$v = v_a \sin(\omega t + \phi) \tag{2.121}$$

has the following three representative parameters:

the amplitude, v_a
the frequency, ω
the phase angle, ϕ

Note that the phase angle represents the time reference (starting point) of the signal. The phase angle is an important consideration only when two or more signal components are compared. In particular, the Fourier spectrum of a signal is presented as its amplitude (magnitude) and the phase angle with respect to frequency.

Phase shifting circuits have many applications. When a signal passes through a system, its phase angle changes due to dynamic characteristics of the system. Consequently, the phase change provides very useful information about the dynamic characteristics of the system. Specifically, for a linear constant-coefficient system, this phase shift is equal to the phase angle of the frequency-response function (i.e., frequency-transfer function) of the system at that particular frequency. This phase shifting behavior is, of course, not limited to electrical systems and is equally exhibited by other types of systems, including mechanical systems and mixed systems. The phase shift between two signals can be determined by converting the signals into the electrical form (using suitable transducers) and shifting the phase angle of one signal through known amounts using a phase-shifting circuit until the two signals are in phase.

Another application of phase shifters is in signal demodulation. For example, as noted earlier in this chapter, one method of amplitude demodulation involves processing the modulated signal together with the carrier signal. This, however, requires the modulated signal and the carrier signal to be in phase. But, usually, since the modulated signal has already transmitted through electrical circuitry with impedance characteristics, its phase angle would have changed. Then, it is necessary to shift the phase angle of the carrier until the two signals are in phase, so that demodulation can be performed accurately. Hence, phase shifters are used in demodulating, for example, the outputs of linear variable transformer (LVDT) displacement sensors.

A phase shifter circuit, ideally, should not change the signal amplitude while changing the phase angle by a required amount. Practical phase shifters could introduce some degree of amplitude distortion (with respect to frequency) as well. A simple phase shifter circuit can be constructed using resistor (R) and capacitor (C) elements. A resistor or a capacitor of such an RC circuit is made fine-adjustable so as to obtain a variable phase shifter.

An op-amp–based phase shifter circuit is shown in Figure 2.42. We can show that this circuit provides a phase shift without distorting the signal amplitude. The circuit equation is obtained by writing the current balance equations at nodes A and B, as usual, noting that the current through the op-amp leads can be neglected due to high input impedance. Thus,

$$\frac{v_i - v_A}{R_C} = C \frac{dv_A}{dt}$$

and

$$\frac{v_i - v_B}{R} + \frac{v_o - v_B}{R} = 0.$$

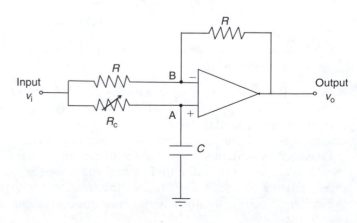

FIGURE 2.42
A phase shifter circuit.

On simplifying and introducing the Laplace variable s, we get

$$v_i = (\tau s + 1)v_A \tag{i}$$

and

$$v_B = \frac{1}{2}(v_i + v_o), \tag{ii}$$

where, the circuit time constant τ is given by

$$\tau = R_c C.$$

Since $v_A = v_B$, as a result of very high gain in the op-amp, we have by substituting Equation ii in Equation i,

$$v_i = \frac{1}{2}(\tau s + 1)(v_i + v_o).$$

It follows that the transfer function $G(s)$ of the circuit is given by

$$\frac{v_o}{v_i} = G(s) = \frac{(1 - \tau s)}{(1 + \tau s)}. \tag{2.122}$$

It is seen that the magnitude of the frequency-response function $G(j\omega)$ is

$$|G(j\omega)| = \frac{\sqrt{1 + \tau^2\omega^2}}{\sqrt{1 + \tau^2\omega^2}}$$

or

$$|G(j\omega)| = 1, \tag{2.123}$$

and the phase angle of $G(j\omega)$ is

$$\angle G(j\omega) = -\tan^{-1}\tau\omega - \tan^{-1}\tau\omega$$

or

$$\angle G(j\omega) = -2\tan^{-1}\tau\omega = -2\tan^{-1}R_c C\omega. \tag{2.124}$$

As needed, the transfer function magnitude is unity, indicating that the circuit does not distort the signal amplitude over the entire bandwidth. Equation 2.124 gives the phase lead of the output v_o with respect to the input v_i. Note that this angle is negative, indicating that actually a phase lag is introduced. The phase shift can be adjusted by varying the resistance R_c.

2.12.2 Voltage-to-Frequency Converters

A VFC generates a periodic output signal whose frequency is proportional to the level of an input voltage. Since such an oscillator generates a periodic output according to the voltage excitation, it is also called a voltage-controlled oscillator (VCO).

A common type of VFC uses a capacitor. The time needed for the capacitor to be charged to a fixed voltage level depends on (inversely proportional to) the charging voltage. Suppose that this voltage is governed by the input voltage. Then if the capacitor is made to periodically charge and discharge, we have an output whose frequency (inverse of the charge–discharge period) is proportional to the charging voltage. The output amplitude will be given by the fixed voltage level to which the capacitor is charged in each cycle. Consequently, we have a signal with a fixed amplitude and a frequency that depends on the charging voltage (input).

A VFC (or VCO) circuit is shown in Figure 2.43a. The voltage-sensitive switch closes when the voltage across it exceeds a reference level v_s and it will open again when the voltage across it falls below a lower limit $v_o(0)$. The programmable unijunction transistor is such a switching device.

Note that the polarity of the input voltage v_i is reversed. Suppose that the switch is open. Then, current balance at node A of the op-amp circuit is given by,

$$\frac{v_i}{R} = C\frac{dv_o}{dt}.$$

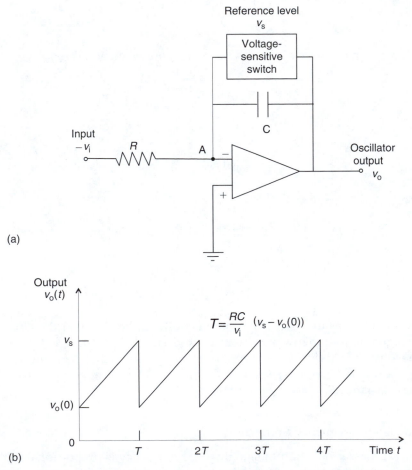

(a)

(b)

FIGURE 2.43

A voltage-to-frequency converter (VFC) or voltage-controlled oscillator (VCO). (a) Circuit. (b) Output signal.

As usual, v_A, voltage at positive lead, is 0 because the op-amp has a very high gain, and current through the op-amp leads is 0 because the op-amp has a very high input impedance. The capacitor-charging equation can be integrated for a given value of v_i. This gives

$$v_o(t) = \frac{1}{RC}v_i t + v_o(0).$$

The switch will be closed when the voltage across the capacitor $v_o(t)$ equals the reference level v_s. Then, the capacitor will be immediately discharged through the closed switch. Hence, the capacitor charging time T is given by

$$v_s = \frac{1}{RC}v_i T + v_o(0).$$

Accordingly,

$$T = \frac{RC}{v_i}(v_s - v_o(0)). \tag{2.125}$$

The switch will be open again when the voltage across the capacitor drops to $v_o(0)$, and the capacitor will again begin to charge from $v_o(0)$ up to v_s. This cycle of charging and instantaneous discharge will repeat periodically. The corresponding output signal will be as shown in Figure 2.43b. This is a periodic (saw tooth) wave with period T. The frequency of oscillation ($1/T$) of the output is given by

$$f = \frac{v_i}{RC(v_s - v_o(0))}. \tag{2.126}$$

It is seen that the oscillator frequency is proportional to the input voltage v_i. The oscillator amplitude is v_s, which is fixed.

VCOs have many applications. One application is in ADC. In the VCO-type ADCs, the analog signal is converted into an oscillating signal using a VCO. Then the oscillator frequency is measured using a digital counter. This count, which is available in the digital form, is representative of the input analog signal level. Another application is in digital voltmeters. Here, the same method as for ADC is used. Specifically, the voltage is converted into an oscillator signal, and its frequency is measured using a digital counter. The count can be scaled and displayed to provide the voltage measurement. A direct application of VCO is apparent from the fact that VCO is actually a FM, providing a signal whose frequency is proportional to the input (modulating) signal. Hence, VCO is useful in applications that require FM. Also, a VCO can be used as a signal (wave) generator for variable-frequency applications; for example, excitation inputs for shakers in product dynamic testing, excitations for frequency-controlled dc motors, and pulse signals for translator circuits of stepping motors.

2.12.3 Frequency-to-Voltage Converter

A FVC generates an output voltage whose level is proportional to the frequency of its input signal. One way to obtain an FVC is to use a digital counter to count the signal frequency and then use a DAC to obtain a voltage proportional to the frequency. A schematic representation of this type of FVC is shown in Figure 2.44a.

An alternative FVC circuit is schematically shown in Figure 2.44b. In this method, the frequency signal is supplied to a comparator along with a threshold voltage level. The

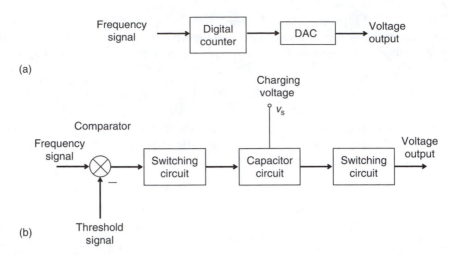

FIGURE 2.44
Frequency-to-voltage converter (FVCs). (a) Digital counter method. (b) Capacitor charging method.

sign of the comparator output will depend on whether the input signal level is larger or smaller than the threshold level. The first sign change (negative to positive) in the comparator output is used to trigger a switching circuit that will respond by connecting a capacitor to a fixed charging voltage. This will charge the capacitor. The next sign change (positive to negative) of the comparator output will cause the switching circuit to short the capacitor, thereby instantaneously discharging it. This charging–discharging process will be repeated in response to the oscillator input. Note that the voltage level to which the capacitor is charged each time will depend on the switching period (charging voltage is fixed), which is in turn governed by the frequency of the input signal. Hence, the output voltage of the capacitor circuit will be representative of the frequency of the input signal. Since the output is not steady due to the ramp-like charging curve and instantaneous discharge, a smoothing circuit is provided at the output to remove the resulting noise ripples.

Applications of FVC include demodulation of frequency-modulated signals, frequency measurement in control applications, and conversion of pulse outputs in some types of sensors and transducers into analog voltage signals.

2.12.4 Voltage-to-Current Converter

Measurement and feedback signals are usually transmitted as current levels in the range of 4 to 20 mA, rather than as voltage levels. This is particularly useful when the measurement site is not close to the monitoring room. Since the measurement itself is usually available as a voltage, it has to be converted into current by using a VCC. For example, pressure transmitters and temperature transmitters in operability testing systems provide current outputs that are proportional to the measured values of pressure and temperature.

There are many advantages to transmitting current rather than voltage. In particular, the voltage level will drop due to resistance in the transmitting path, but the current through a conductor will remain uncharged unless the conductor is branched. Hence, current signals are less likely to acquire errors due to signal weakening. Another advantage of using current, instead of voltage as the measurement signal, is that the same signal can be used to operate several devices in series (e.g., a display, a plotter, and a signal

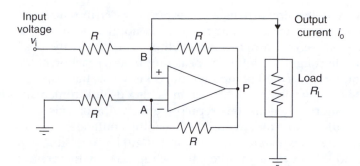

FIGURE 2.45
A voltage-to-current converter.

processor simultaneously), again without causing errors due to signal weakening by the power lost at each device, because the same current is applied to all devices. A VCC should provide a current proportional to an input voltage, without being affected by the load resistance to which the current is supplied.

An op-amp–based voltage-to-current converter circuit is shown in Figure 2.45. Using the fact that the currents through the input leads of an unsaturated op-amp can be neglected (due to very high input impedance), we write the current summation equations for the two nodes A and B. Thus,

$$\frac{v_A}{R} = \frac{v_P - v_A}{R}$$

and

$$\frac{v_i - v_B}{R} + \frac{v_P - v_B}{R} = i_o.$$

Accordingly, we have

$$2v_A = v_P \qquad (i)$$

and

$$v_i - 2v_B + v_P = Ri_o. \qquad (ii)$$

Now using the fact that $v_A = v_B$ for the op-amp (due to very high gain), we substitute Equation i in Equation ii. This gives

$$i_o = \frac{v_i}{R}, \qquad (2.127)$$

where i_o is the output current and v_i is the input voltage.

It follows that the output current is proportional to the input voltage, irrespective of the value of the load resistance R_L, as required for a VCC.

2.12.5 Peak-Hold Circuits

Unlike an simple-and-hold circuit (S/H), which holds every sampled value of a signal, a peak-hold circuit holds only the largest value reached by the signal during the monitored period. Peak holding is useful in a variety of applications. In signal processing for shock and vibration studies of dynamic systems, what is known as *response spectra* (e.g., shock response spectrum) are determined by using a response spectrum analyzer, which exploits a peak-holding scheme. Suppose that a signal is applied to a simple oscillator (a single-degree-of-freedom second-order system with no zeros) and the peak value of the response (output) is determined. A plot of the peak output as a function of the natural frequency of the oscillator, for a specified damping ratio, is known as the response spectrum of the signal for that damping ratio. Peak detection is also useful in machine monitoring and alarm systems. In short, when just one representative value of a signal is needed in a particular application, the peak value would be a leading contender.

Peak detection of a signal can be conveniently done using digital processing. For example, the signal is sampled and the previous sample value is replaced by the present sample value, if and only if the latter is larger than the former. In this manner, the peak value of the signal is retained by sampling and then holding one value. Note that, usually the time instant at which the peak occurs is not retained.

Peak detection can be done using analog circuitry as well. This is in fact the basis of analog spectrum analyzers. A peak-holding circuit is shown in Figure 2.46. The circuit consists of two voltage followers. The first voltage follower has a diode at its output that is forward biased by the positive output of the voltage follower and reverse biased by a low-leakage capacitor, as shown. The second voltage follower presents the peak voltage that is held by the capacitor to the circuit output at a low output impedance, without loading the previous circuit stage (capacitor and first voltage follower). To explain the operation of the circuit, suppose that the input voltage v_i is larger than the voltage to which capacitor is charged (v). Since the voltage at the positive lead of the op-amp is v_i and the voltage at the negative lead is v, the first op-amp will be saturated. Since the differential input to the op-amp is positive under these conditions, the op-amp output will be positive. The output will charge the capacitor until the capacitor voltage v equals the input voltage v_i. This voltage (call it v_o) is in turn supplied to the second voltage follower, which presents the same value to its output (note that the gain is 1 for a voltage follower), but at a very low impedance level. The op-amp output remains at the saturated value only for a very short time (the time taken by the capacitor to charge). Now, suppose that v_i is smaller than v. Then the differential input of the op-amp will be negative, and the op-amp output will be saturated at the negative saturation level. This will reverse bias the diode. Hence, the output of the first op-amp will be in open circuit, and as a result, the voltage supplied to the output voltage follower will still be the capacitor voltage and not the output of the first op-amp. It follows that the voltage level of the capacitor (and hence the output of the second voltage follower) will always be the peak value of the input signal. The circuit can be reset by discharging the capacitor through a solid-state switch that is activated by an external pulse.

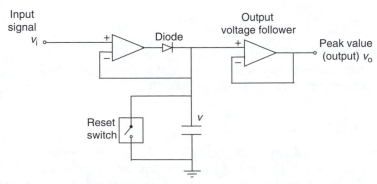

FIGURE 2.46
A peak-holding circuit.

2.13 Signal Analyzers and Display Devices

Since signal analysis involves processing of a signal to generate useful information, it is appropriate to consider the topic within the present context of signal modification as well. Signal analysis may employ both analog and digital procedures. In the present section, we will introduce digital signal analyzers. Signal display devices also make use of at least some basic types of signal processing. This may involve filtering and change of signal level and format. More sophisticated signal display devices, particularly digital oscilloscopes, can carry out more complex signal analysis functions, such as those normally available with digital signal analyzers. Oscilloscopes, which are primarily instruments for signal display and monitoring, are introduced as well in this section. They typically employ basic types of signal analysis, and may be treated under signal analysis and instrumentation.

Signal-recording equipment commonly employed in the control practice includes digital storage devices such as hard drives, floppy disks, and CD-ROMs, analog devices like tape recorders, strip-chart recorders, and X–Y plotters, and digital printers. Tape recorders are used to record system response data (transducer outputs) that are subsequently reproduced for processing or examination. Often, tape-recorded waveforms are also used to generate (by replay) signals that drive dynamic test exciters (shakers). Tape recorders use tapes made of a plastic material that has a thin coating of a specially treated ferromagnetic substance. During the recording process, magnetic flux proportional to the recorded signal is produced by the recording head (essentially an electromagnet), which magnetizes the tape surface in proportion to the signal variation. Reproduction is the reverse process, whereby an electrical signal is generated at the reproduction head by electromagnetic induction, in accordance with the magnetic flux of the magnetized (recorded) tape. Several signal-conditioning circuitries are involved in the recording and reproducing stages. Recording by FM is very common in dynamic testing.

Strip-chart recorders are usually employed to plot time histories (i.e., quantities that vary with time), although they also may be used to plot such data as frequency-response functions and response spectra. In these recorders, a paper roll unwinds at a constant linear speed, and the writing head moves across the paper (perpendicular to the paper motion) proportional to the signal level. There are many kinds of strip-chart recorders, which are grouped according to the type of writing head employed. Graphic-level recorders, which use ordinary paper, employ such heads as ink pens or brushes, fiber pens, and sapphire styli. Visicoders are simple oscilloscopes that are capable of producing permanent records; they employ light-sensitive paper for this. Several channels of input data can be incorporated with a visicoder. Obviously, graphic-level recorders are generally limited by the number of writing heads available (typically, one or two), but visicoders can have many more input channels (typically, 24). Performance specifications of these devices include paper speed, frequency range of operation, dynamic range, and power requirements.

In electro-mechanical experimentation, X–Y plotters are generally employed to plot frequency data [e.g., power spectral densities (psd), frequency-response functions, response spectra, and transmissibility curves], although they can also be used to plot time–history data. Many types of X–Y plotters are available, most of them using ink pens on ordinary paper. There are also hardcopy units that use heat-sensitive paper in conjunction with a heating element as the writing head. The writing head of an X–Y plotter is moved in the X and Y directions on the paper by two input signals, which form the coordinates for the plot. In this manner, a trace is made on stationary plotting paper. Performance specifications of X–Y plotters are governed by such factors as paper size; writing speed (in/sec, cm/sec); dead band (expressed as a percentage of the full scale) which measures the resolution of the plotter head; linearity (expressed as a percentage of the full scale), which measures the accuracy of the plot or deviation from a reference straight line; minimum trace separation (cm) for multiple plots on the same axes, dynamic range; input impedance; and maximum input (mV/in, mV/cm).

Today, the most widespread signal-recording device is in fact the digital computer (memory, storage) and printer combination. Digital computer and other (analog) devices used in signal recording and display, generally make use of some form of signal modification to accomplish their functions. But, we will not discuss these devices in this section.

2.13.1   Signal Analyzers

Modern signal analyzers employ digital techniques of signal analysis to extract useful information that is carried by the signal. Digital Fourier analysis using fast fourier transform (FFT) is perhaps the single common procedure that is used in the vast majority of signal analyzers. Fourier analysis produces the frequency spectrum of a time signal. It should be clear, therefore, why the terms digital signal analyzer, FFT analyzer, frequency analyzer, spectrum analyzer, and digital Fourier analyzer are synonymous to some extent, as used in the commercial instrumentation literature.

A signal analyzer typically has two (dual) or more (multiple) input signal channels. To generate results such as frequency response (transfer) functions, cross-spectra, coherence functions, and cross-correlation functions, we need at least two data signals and hence a dual-channel analyzer.

In hardware analyzers, digital circuitry, rather than software, is used to carry out the mathematical operations. Clearly, they are very fast but less flexible (in terms of programmability and functional capability) for this reason. Digital signal analyzers, regardless of whether they use the hardware approach or the software approach, employ some basic operations. These operations, carried out in sequence, are

1. Antialias filtering (analog)
2. ADC (i.e., single sampling and digitization)
3. Truncation of a block of data and multiplication by a window function
4. FFT analysis of the block of data.

The following facts are important in the present context of digital signal analysis. If the sampling period of the ADC is ΔT (i.e., the sampling frequency is $1/\Delta T$), then the Nyquist frequency $f_c = 1/2\Delta T$. This Nyquist frequency is the upper limit of the useful frequency content of the sampled signal. The cutoff frequency of the antialiasing filter should be set at f_c or less. If there are N data samples in the block of data that is used in the FFT analysis, the corresponding record length is $T = N \cdot \Delta T$. Then, the spectral lines in the FFT results are separated at a frequency spacing of $\Delta F = 1/T$. In view of the Nyquist frequency limit, however, there will be only $N/2$ useful spectral lines in the FFT result.

Strictly speaking, a real-time signal analyzer should analyze a signal instantaneously and continuously, as the signal is received by the analyzer. This is usually the case with an analog signal analyzer. But, in digital signal analyzers, which are usually based on digital Fourier analysis, a block of data (i.e., N samples of record length T) is analyzed together to produce $N/2$ useful spectral lines (at frequency spacing $1/T$). This then is not a truly real-time analysis. But for practical purposes, if the speed of analysis is sufficiently fast, the analyzer may be considered real time, which is usually the case with hardware analyzers and also modern, high-speed, software analyzers.

The bandwidth B of a digital signal analyzer is a measure of its speed of signal processing. Specifically, for an analyzer that uses N data samples in each block of signal analysis, the associated processing time may be given by

$$T_c = \frac{N}{B}. \tag{2.128}$$

Note that the larger the B, the smaller the T_c. Then, the analyzer is considered a real-time one if the analysis time (T_c) of the data record is less than the generation time ($T = N \cdot \Delta T$) of the data record. Hence, we need $T_c < T$ or $N/B < T$ or $N/B < N \cdot \Delta T$, which gives

$$\frac{1}{\Delta T} < B. \tag{2.129}$$

In other words, a real-time analyzer should have a bandwidth greater than its sampling rate.

A multi-channel digital signal analyzer can analyze one or more signals simultaneously and generate (and display) results such as Fourier spectra, psd, cross-spectral densities, frequency response functions, coherence functions, autocorrelations, and cross-correlations. They are able to perform high-resolution analysis on a small segment of the frequency spectrum of a signal. This is termed *zoom analysis*. Essentially, in this case, the spectral line spacing ΔF is decreased while keeping unchanged the number of lines (N) and hence, the number of time data samples. That means the record length ($T = 1/\Delta F$) has to be increased in proportion, for zoom analysis.

2.13.2 Oscilloscopes

An oscilloscope is used to display and observe one or two signals separately or simultaneously. Amplitude, frequency, and phase information of the signals can be obtained using an oscilloscope. In this sense, it is a signal analysis/modification device as well as a measurement (monitoring) and display device. While both analog and digital oscilloscopes are commercially available, the latter is far more common. A typical application of an oscilloscope is to observe (monitor) experimental data such as response signals of machinery and processes, as obtained from sensors and transducers. They are also useful in observing and examining dynamic test results, such as frequency-response plots, psd curves, and response spectra. Typically, only temporary records are available on an analog oscilloscope screen. The main component of an analog oscilloscope is the cathode-ray tube (CRT) consisting of an electron gun (cathode), which deflects an electron ray according to the input-signal level. The oscilloscope screen has a coating of electron-sensitive material, so that the electron ray that impinges on the screen leaves a temporary trace on it. The electron ray sweeps across the screen horizontally, so that waveform traces can be recorded and observed. Typically, two input channels are available. Each input may be observed separately, or the variations in one input may be observed against those of the other. In this manner, signal phasing can be examined. Several sensitivity settings for the input-signal-amplitude scale (in the vertical direction) and sweep-speed selections are available on the oscilloscope panel.

2.13.2.1 *Triggering*

The voltage level of the input signal deflects the electron gun in proportion in the vertical (y-axis) direction on the CRT screen of an oscilloscope. This alone will not show the time evolution of the signal. The true time variation of the signal is achieved by means of a saw-tooth signal, which is generated internally in the oscilloscope and used to move the electron gun in the horizontal (*x*-axis) direction. As the name implies, the saw-tooth signal increases linearly in amplitude up to a threshold value and then suddenly drops to zero, and repeats this cycle over and over again. In this manner, the observed signal is repetitively swept across the screen, and a trace of it can be observed as a result of the temporary retention of the illumination of the electron gun on the fluorescent screen. The saw-tooth signal may be controlled (triggered) in several ways. For example, the external trigger mode uses an external signal from another channel (not the observed channel) to generate and synchronize the saw-tooth signal. In the line trigger mode, the saw-tooth signal is synchronized with the ac line supply (60 Hz or 50 Hz). In the internal trigger mode, the observed signal (which is used to deflect the electron beam in the *y*-direction) itself is used to generate (synchronize) the saw-tooth signal. Since the frequency and the phase of the observed signal and the trigger signal are perfectly synchronized in this case, the trace on the oscilloscope screen will appear stationary. Careful observation of a signal can be made in this manner.

2.13.2.2 *Lissajous Patterns*

Suppose that two signals *x* and *y* are provided to the two channels of an oscilloscope. If they are used to deflect the electron beam in the horizontal and the vertical directions, a pattern known as the Lissajous pattern will be observed on the oscilloscope screen. Useful information about the amplitude and phase of the two signals may be observed by means of these patterns. Consider two sine waves *x* and *y*. Several special cases of Lissajous patterns are given.

1. Same frequency, same phase:

    Here,

    $$x = x_o \sin(\omega t + \phi),$$
    $$y = y_o \sin(\omega t + \phi).$$

    Then we have,

    $$\frac{x}{x_o} = \frac{y}{y_o},$$

    which gives a straight-line trace with a positive slope, as shown in Figure 2.47a.

2. Same frequency, 90° out-of-phase

    Here,

    $$x = x_o \sin(\omega t + \phi),$$
    $$y = y_o \sin(\omega t + \phi + \pi/2),$$
    $$= y_o \cos(\omega t + \phi).$$

    Then we have,

    $$\left(\frac{x}{x_o}\right)^2 + \left(\frac{y}{y_o}\right)^2 = 1,$$

    which gives an ellipse, as shown in Figure 2.47b.

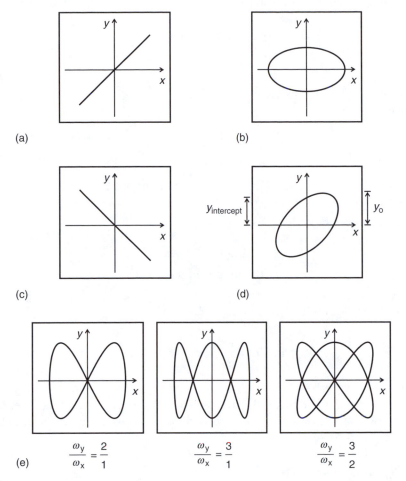

FIGURE 2.47
Some Lissajous patterns for (a) equal frequency and in phase, (b) equal frequency and 90° out of phase, (c) equal frequency and 180° out of phase, (d) equal frequency and θ out of phase, (e) integral frequency ratio.

3. Same frequency, 180° out-of-phase

   Here,

$$x = x_0 \sin(\omega t + \phi),$$
$$y = y_0 \sin(\omega t + \phi + \pi),$$
$$= -y_0 \sin(\omega t + \phi).$$

   Hence,

$$\frac{x}{x_0} + \frac{y}{y_0} = 0,$$

   which corresponds to a straight line with a negative slope, as shown in Figure 2.47c.

4. Same frequency, θ out-of-phase

$$x = x_o \sin(\omega t + \phi),$$
$$y = y_o \sin(\omega t + \phi + \theta).$$

When $\omega t + \phi = 0$, $y = y_{intercept} = y_o \sin \theta$.
Hence,

$$\sin \theta = \frac{y_{intercept}}{y_o}.$$

In this case, we get a tilted ellipse as shown in Figure 2.47d. The phase difference θ is obtained from the Lissajous pattern.

5. Integral frequency ratio

$$\frac{\omega_y}{\omega_x} = \frac{\text{Number of } y - \text{peaks}}{\text{Number of } x - \text{peaks}}.$$

Three examples are shown in Figure 2.47e.

$$\frac{\omega_y}{\omega_x} = \frac{2}{1} \quad \frac{\omega_y}{\omega_x} = \frac{3}{1} \quad \frac{\omega_y}{\omega_x} = \frac{3}{2}.$$

Note that these observations hold true as well for narrowband signals, which can be approximated as sinusoidal signals. Broadband random signals produce scattered (irregular) Lissajous patterns.

2.13.2.3 Digital Oscilloscopes

The basic uses of a digital oscilloscope are quite similar to those of a traditional analog oscilloscope. The main differences stem from the manner in which information is represented and processed internally within the oscilloscope. Specifically, a digital oscilloscope first samples a signal that arrives at one of its input channels and stores the resulting digital data within a memory segment. This is essentially a typical ADC operation. This digital data may be processed to extract and display the necessary information. The sampled data and the processed information may be stored on a floppy disk, if needed, for further processing using a digital computer. Also, some digital oscilloscopes have the communication capability so that the information may be displayed on a video monitor or printed to provide a hard copy.

A typical digital oscilloscope has four channels so that four different signals may be acquired (sampled) into the oscilloscope and displayed. Also, it has various triggering options so that the acquisition of a signal may be initiated and synchronized by means of either an internal trigger or an external trigger. Apart from the typical capabilities that were listed in the context of an analog oscilloscope, a digital oscilloscope can automatically provide other useful features such as the following:

1. Automatic scaling of the acquired signal
2. Computation of signal features such as frequency, period, amplitude, mean, root-mean-square (rms) value, and rise time

3. Zooming into regions of interest of a signal record
4. Averaging of multiple signal records
5. Enveloping of multiple signal records
6. FFT capability, with various window options and antialiasing

These various functions are menu selectable. Typically, first a channel of the incoming data (signal) is selected and then an appropriate operation on the data is chosen from the menu (through menu buttons).

Problems

2.1 Define electrical impedance and mechanical impedance. Identify a defect in these definitions in relation to the force–current analogy. What improvements would you suggest? What roles do input impedance and output impedance play in relation to the accuracy of a measuring device?

2.2 What is meant by loading error in a signal measurement? Also, suppose that a piezoelectric sensor of output impedance Z_s is connected to a voltage-follower amplifier of input impedance Z_i, as shown in Figure P2.2. The sensor signal is v_i volts and the amplifier output is v_o volts. The amplifier output is connected to a device with very high input impedance. Plot to scale the signal ratio v_o/v_i against the impedance ratio Z_i/Z_s for values of the impedance ratio in the range 0.1 to 10.

2.3 Thevenin's theorem states that with respect to the characteristics at an output port, an unknown subsystem consisting of linear passive elements and ideal source elements may be represented by a single across-variable (voltage) source v_{eq} connected in series with a single impedance Z_{eq}. This is illustrated in Figure P2.3a and P2.3b. Note that v_{eq} is equal to the open-circuit across variable v_{oc} at the output port because the current through Z_{eq} is zero. Consider the network shown in Figure P2.3c. Determine the equivalent voltage source v_{eq} and the equivalent series impedance Z_{eq}, in the frequency domain, for this circuit.

2.4 Explain why a voltmeter should have a high resistance and an ammeter should have a very low resistance. What are some of the design implications of these general

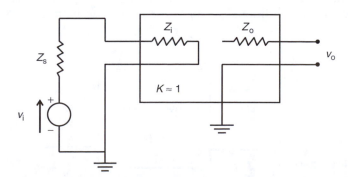

FIGURE P2.2
System with a piezoelectric sensor.

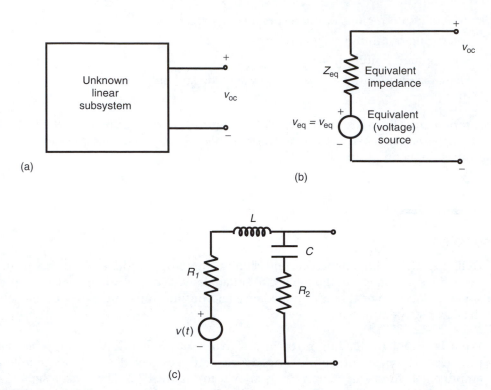

(a)

(b)

(c)

FIGURE P2.3
Illustration of Thevenin's theorem. (a) Unknown linear subsystem. (b) Equivalent representation. (c) Example.

requirements for the two types of measuring instruments, particularly with respect to instrument sensitivity, speed of response, and robustness? Use a classical moving-coil meter as the model for your discussion.

2.5 A two-port nonlinear device is shown schematically in Figure P2.5. The transfer relations under static equilibrium conditions are given by

$$v_o = F_1(f_o, f_i),$$
$$v_i = F_2(f_o, f_i),$$

where v denotes an across variable, f denotes a through variable, and the subscripts o and i represent the output port and the input port, respectively. Obtain expressions for input impedance and output impedance of the system in the neighborhood of an operating point, under static conditions, in terms of partial derivatives of

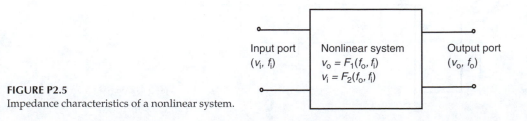

FIGURE P2.5
Impedance characteristics of a nonlinear system.

the functions F_1 and F_2. Explain how these impedances could be determined experimentally.

2.6 Define the terms

a. mechanical loading

b. electrical loading

in the context of motion sensing, and explain how these loading effects can be reduced. The following table gives ideal values for some parameters of an op-amp. Give typical, practical values for these parameters (e.g., output impedance of 50 Ω).

Parameter	Ideal Value	Typical Value
Input impedance	Infinity	?
Output impedance	Zero	50 Ω
Gain	Infinity	?
Bandwidth	Infinity	?

In addition, note that, under ideal conditions, inverting-lead voltage is equal to the noninverting-lead voltage (i.e., offset voltage is zero).

2.7 LVDT is a displacement sensor, which is commonly used in control systems. Consider a digital control loop that uses an LVDT measurement for position control of a machine. Typically, the LVDT is energized by a dc power supply. An oscillator provides an excitation signal in the kilohertz range to the primary windings of the LVDT. The secondary winding segments are connected in series opposition. An ac amplifier, demodulator, low-pass filter, amplifier, and ADC are used in the monitoring path. Figure P2.7 shows the various hardware components in the control loop. Indicate the functions of these components.

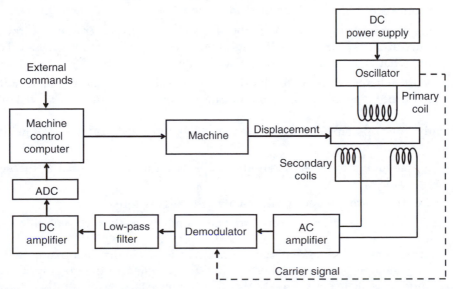

FIGURE P2.7
Components of an LVDT-based machine control loop.

At null position of the LVDT stroke, there was a residual voltage. A compensating resistor is used to eliminate this voltage. Indicate the connections for this compensating resistor.

2.8 Today, machine vision is used in many industrial tasks including process control and monitoring. In an industrial system based on machine vision, an imaging device such as a charge-coupled-device (CCD) camera is used as the sensing element. The camera provides to an image processor an image (picture) of a scene related to the industrial process (the measurement). The computed results from the image processor are used to determine the necessary information about the process (plant).

A CCD camera has an image plate consisting of a matrix of MOSFET elements. The electrical charge that is held by each MOSFET element is proportional to the intensity of light falling on the element. The output circuit of the camera has a charge-amplifier–like device (capacitor coupled), which is supplied by each MOSFET element. The MOSFET element that is to be connected to the output circuit at a given instant is determined by the control logic, which systematically scans the matrix of MOSFET elements. The capacitor circuit provides a voltage that is proportional to the charge in each MOSFET element.

a. Draw a schematic diagram for a process monitoring system based on machine vision, which uses a CCD camera. Indicate the necessary signal modification operations at various stages in the monitoring loop, showing whether analog filters, amplifiers, ADC, and DAC are needed and if so, at which locations.

An image may be divided into pixels (or picture elements) for representation and subsequent processing. A pixel has a well-defined coordinate location in the picture frame, relative to some reference coordinate frame. In a CCD camera, the number of pixels per image frame is equal to the number of CCD elements in the image plate. The information carried by a pixel (in addition to its location) is the photointensity (or gray level) at the image location. This number has to be expressed in the digital form (using a certain number of bits) for digital image processing. The need for very large data-handling rates is a serious limitation on a real-time controller that uses machine vision.

b. Consider a CCD image frame of the size 488×380 pixels. The refresh rate of the picture frame is 30 frames/s. If 8 bits are needed to represent the gray level of each pixel, what is the associated data (baud) rate?

c. Discuss whether you prefer hardware processing or programmable-software–based processing in a process monitoring system based on machine vision.

2.9 Usually, an op-amp circuit is analyzed making use of the following two assumptions:

1. The potential at the positive input lead is equal to the potential at the negative input lead.

2. The current through each of the two input leads is zero.

Explain why these assumptions are valid under unsaturated conditions of an op-amp.

An amateur electronics enthusiast connects to a circuit an op-amp without a feedback element. Even when there is no signal applied to the op-amp, the output was found to oscillate between $+12$ V and -12 V once the power supply is turned on. Give a reason for this behavior.

An op-amp has an open-loop gain of 5×10^5 and a saturated output of ± 14 V. If the noninverting input is $-1\ \mu V$ and the inverting input is $+0.5\ \mu V$, what is the output? If the inverting input is $5\ \mu V$ and the noninverting input is grounded, what is the output?

2.10 Define the following terms in connection with an op-amp:

a. offset current

b. offset voltage (at input and output)

c. unequal gains

d. slew rate

Give typical values for these parameters. The open-loop gain and the input impedance of an op-amp are known to vary with frequency and are known to drift (with time) as well. Still, the op-amp circuits are known to behave very accurately. What is the main reason for this?

2.11 What is a voltage follower? Discuss the validity of the following statements:

a. Voltage follower is a current amplifier.

b. Voltage follower is a power amplifier.

c. Voltage follower is an impedance transformer.

Consider the amplifier circuit shown in Figure P2.11. Determine an expression for the voltage gain K_v of the amplifier in terms of the resistances R and R_f. Is this an inverting amplifier or a noninverting amplifier?

2.12 The speed of response of an amplifier may be represented using the three parameters: bandwidth, rise time, and slew rate. For an idealized linear model (transfer function), it can be verified that the rise time and the bandwidth are independent of the size of the input and the dc gain of the system. Since the size of the output (under steady conditions) may be expressed as the product of the input size and the dc gain, it is seen that rise time and the bandwidth are independent of the amplitude of the output, for a linear model.

Discuss how slew rate is related to bandwidth and rise time of a practical amplifier. Usually, amplifiers have a limiting slew rate value. Show that bandwidth decreases with the output amplitude in this case.

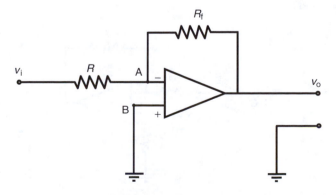

FIGURE P2.11
An amplifier circuit.

A voltage follower has a slew rate of 0.5 V/μs. If a sinusoidal voltage of amplitude 2.5 V is applied to this amplifier, estimate the operating bandwidth. If, instead, a step input of magnitude 5 V is applied, estimate the time required for the output to reach 5 V.

2.13 Define the terms

a. common-mode voltage

b. common-mode gain

c. CMRR

What is a typical value for the CMRR of an op-amp? Figure P2.13 shows a differential amplifier circuit with a flying capacitor. The switch pairs A and B are turned on and off alternately during operation. For example, first the switches denoted by A are turned on (closed) with the switches B off (open). Next, the switches A are opened and the switches B are closed. Explain why this arrangement provides good common-mode rejection characteristics.

2.14 Compare the conventional (textbook) meaning of system stability and the practical interpretation of instrument stability.

An amplifier is known to have a temperature drift of 1 mV/°C and a long-term drift of 25 μV/month. Define the terms temperature drift and long-term drift. Suggest ways to reduce drift in an instrument.

2.15 Electrical isolation of a device (or circuit) from another device (or circuit) is very useful in the control practice. An isolation amplifier may be used to achieve this. It provides a transmission link, which is almost one way and avoids loading problems. In this manner, damage in one component due to increase in signal levels in the other components (perhaps due to short circuits, malfunctions, noise, high common-mode signals, etc.) could be reduced. An isolation amplifier can be constructed from a transformer and a demodulator with other auxiliary components such as filters and amplifiers. Draw a suitable schematic diagram for an isolation amplifier and explain the operation of this device.

2.16 What are passive filters? List several advantages and disadvantages of passive (analog) filters in comparison to active filters.

A simple way to construct an active filter is to start with a passive filter of the same type and add a voltage follower to the output. What is the purpose of such a voltage follower?

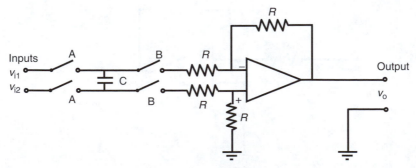

FIGURE P2.13
A differential amplifier with a flying capacitor for common-mode rejection.

2.17 Give one application each for the following types of analog filters:

    a. Low-pass filter

    b. High-pass filter

    c. Band-pass filter

    d. Notch filter

    Suppose that several single-pole active filter stages are cascaded. Is it possible for the overall (cascaded) filter to possess a resonant peak? Explain.

2.18 Butterworth filter is said to have a maximally flat magnitude. Explain what is meant by this. Give another characteristic that is desired from a practical filter.

2.19 An active filter circuit is given in Figure P2.19. *P.61*

    a. Obtain the input/output differential equation for the circuit.

    b. What is the filter transfer function?

    c. What is the order of the filter?

    d. Sketch the magnitude of the frequency transfer function and state what type of filter it represents. *→ MATLAB*

    e. Estimate the cutoff frequency and the roll-off slope. *→ FROM BODE*

2.20 What is meant by each of the following terms: modulation, modulating signal, carrier signal, modulated signal, and demodulation? Explain the following types of signal modulation giving an application for each case:

    a. AM

    b. FM

    c. PM

    d. PWM

    e. PFM

    f. PCM

    How could the sign of the modulating signal be accounted for during demodulation in each of these types of modulation?

2.16 (b)
P. 62

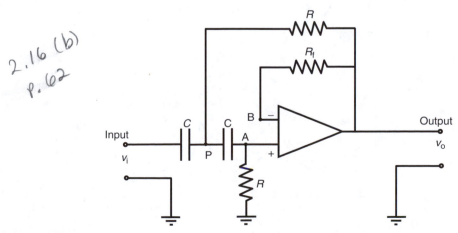

FIGURE P2.19
An active filter circuit.

2.21 Give two situations where AM is intentionally introduced, and in each situation explain how AM is beneficial. Also, describe two devices where AM might be naturally present. Could the fact that AM is present be exploited to our advantage in these two natural situations as well? Explain.

2.22 A monitoring system for a ball bearing of a rotating machine is schematically shown in Figure P2.22a. It consists of an accelerometer to measure the bearing vibration and an FFT analyzer to compute the Fourier spectrum of the response signal. This spectrum is examined over a period of one month after installation of the rotating machine to detect any degradation in the bearing performance. An interested segment of the Fourier spectrum can be examined with high resolution by using the zoom analysis capability of the FFT analyzer. The magnitude of the original spectrum and that of the spectrum determined one month later, in the same zoom region, are shown in Figure P2.22b.

   a. Estimate the operating speed of the rotating machine and the number of balls in the bearing.

   b. Do you suspect any bearing problems?

2.23 Explain the following terms:

   a. phase-sensitive demodulation

   b. half-wave demodulation

   c. full-wave demodulation

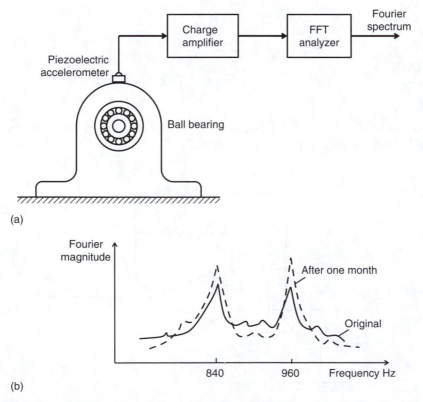

FIGURE P2.22
(a) A monitoring system for a ball bearing. (b) A zoomed Fourier spectrum.

When vibrations in rotating machinery such as gearboxes, bearings, turbines, and compressors are monitored, it is observed that a peak of the spectral magnitude curve does not usually occur at the frequency corresponding to the forcing function (e.g., tooth meshing, ball or roller hammer, blade passing). But, instead, two peaks occur on the two sides of this frequency. Explain the reason for this fact.

2.24 Define the following terms in relation to an ADC:

a. resolution

b. dynamic range

c. FSV

d. quantization error

2.25 Single-chip amplifiers with built-in compensation and filtering circuits are becoming popular for signal-conditioning tasks in control systems, particularly those associated with data acquisition, machine monitoring, and control. Signal processing such as integration that would be needed to convert, say, an accelerometer into a velocity sensor, can also be accomplished in the analog form using an IC chip. What are the advantages of such signal-modification chips in comparison with the conventional analog signal-conditioning hardware that employ discrete circuit elements and separate components to accomplish various signal-conditioning tasks?

2.26 Compare the three types of bridge circuits: constant-voltage bridge, constant-current bridge, and half bridge, in terms of nonlinearity, effect of change in temperature, and cost.

Obtain an expression for the percentage error in a half-bridge circuit output due to an error δv_{ref} in the voltage supply v_{ref}. Compute the percentage error in the output if voltage supply has a 1% error.

2.27 Suppose that in the constant-voltage bridge circuit shown in Figure 2.35a, at first, $R_1 = R_2 = R_3 = R_4 = R$. Assume that R_1 represents a strain gage mounted on the tensile side of a bending beam element and that R_3 represents another strain gage mounted on the compressive side of the bending beam. Due to bending, R_1 increases by δR and R_3 decreases by δR. Derive an expression for the bridge output in this case, and show that it is nonlinear. What would be the result if R_2 represents the tensile strain gage and R_4 represents the compressive strain gage, instead?

2.28 Suppose that in the constant-current bridge circuit shown in Figure 2.35b, at first, $R_1 = R_2 = R_3 = R_4 = R$. Assume that R_1 and R_2 represent strain gages mounted on a rotating shaft, at right angles and symmetrically about the axis of rotation. Also, in this configuration and in a particular direction of rotation of the shaft, suppose that R_1 increases by δR and R_2 decreases by δR. Derive an expression for the bridge output (normalized) in this case, and show that it is linear. What would be the result if R_4 and R_3 were to represent the active strain gages in this example, the former element in tension and the latter in compression?

2.29 Consider the constant-voltage bridge shown in Figure 2.35a. The output Equation 2.88 can be expressed as

$$v_o = \frac{(R_1/R_2 - R_3/R_4)}{(R_1/R_2 + 1)(R_3/R_4 + 1)} v_{\text{ref}}.$$

Now suppose that the bridge is balanced, with the resistors set according to

$$\frac{R_1}{R_2} = \frac{R_3}{R_4} = p.$$

Then, if the active element R_1 increases by δR_1, show that the resulting output of the bridge is given by

$$\delta v_o = \frac{p\delta r}{[p(1 + \delta r) + 1](p + 1)} v_{\text{ref}},$$

where, $\delta r = \delta R_1/R_1$, which is the fractional change in resistance in the active element.

For a given δr, it should be clear that δv_o represents the sensitivity of the bridge. For what value of the resistance ratio p, would the bridge sensitivity be a maximum? Show that this ratio is almost equal to 1.

2.30 The Maxwell bridge circuit is shown in Figure P2.30. Obtain the conditions for a balanced Maxwell bridge in terms of the circuit parameters R_1, R_2, R_3, R_4, C_1, and L_4. Explain how this circuit could be used to measure variations in both C_1 or L_4.

2.31 The standard LVDT arrangement has a primary coil and two secondary coil segments connected in series opposition. Alternatively, some LVDTs use a bridge circuit to produce their output. An example of a half-bridge circuit for an LVDT is shown in Figure P2.31. Explain the operation of this arrangement. Extend this idea to a full impedance bridge, for LVDT measurement.

2.32 The output of a Wheatstone bridge is nonlinear with respect to the variations in a bridge resistance. This nonlinearity is negligible for small changes in resistance. For

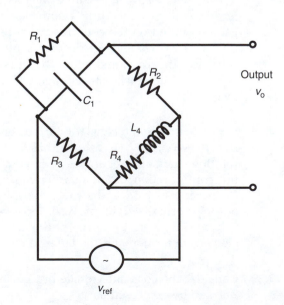

FIGURE P2.30
The Maxwell bridge.

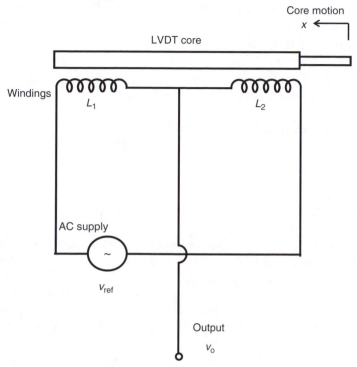

FIGURE P2.31
A half-bridge circuit for an LVDT.

large variations in resistance, however, some method of calibration or linearization should be employed. One way to linearize the bridge output is to positive feedback the output voltage signal into the bridge supply using a feedback op-amp. Consider the Wheatstone bridge circuit shown in Figure 2.35a. Initially, the bridge is balanced with $R_1 = R_2 = R_3 = R_4 = R$. Then, the resistor R_1 is varied to $R + \delta R$. Suppose that the bridge output δv_o is fed back (positive) with a gain of 2 into the bridge supply v_{ref}. Show that this will linearize the bridge equation.

2.33 A furnace used in a chemical process is controlled in the following manner. The furnace is turned on in the beginning of the process. When the temperature within the furnace reaches a certain threshold value T_o, the (temperature) × (time) product is measured in the units of Celsius minutes. When this product reaches a specified value, the furnace is turned off. The available hardware includes a RTD, a differential amplifier, a diode circuit, which does not conduct when the input voltage is negative and conducts with a current proportional to the input voltage when the input is positive, a current-to-voltage converter circuit, a VFC, a counter, and an on/off control unit. Draw a block diagram for this control system and explain its operation. Clearly identify the signal-modification operations in this control system, indicating the purpose of each operation.

2.34 Typically, when a digital transducer is employed to generate the feedback signal for an analog controller, a DAC would be needed to convert the digital output from the transducer into a continuous (analog) signal. Similarly, when a digital controller is used to drive an analog process, a DAC has to be used to convert the digital output from the controller into the analog drive signal. There exist ways, however, to eliminate the need for a DAC in these two types of situations.

1. Show how a shaft encoder and an FVC can replace an analog tachometer in an analog speed-control loop.

2. Show how a digital controller with PWM can be employed to drive a DC motor without the use of a DAC.

2.35 The noise in an electrical circuit can depend on the nature of the coupling mechanism. In particular, the following types of coupling are available:

a. Conductive coupling

b. Inductive coupling

c. Capacitive coupling

d. Optical coupling

Compare these four types of coupling with respect to the nature and level of noise that is fed through or eliminated in each case. Discuss ways to reduce noise that is fed through in each type of coupling.

The noise due to variations in ambient light can be a major problem in optically coupled systems. Briefly discuss a method that could be used in an optically coupled device to make the device immune to variations in the ambient light level.

2.36 What are the advantages of using optical coupling in electrical circuits? For optical coupling, diodes that emit infrared radiation are often preferred over light-emitting diodes that emit visible light. What are the reasons behind this? Discuss why pulse-modulated light (or pulse-modulated radiation) is used in many types of optical systems. List several advantages and disadvantages of laser-based optical systems.

The Young's modulus of a material with known density can be determined by measuring the frequency of the fundamental mode of transverse vibration of a uniform cantilever beam specimen of the material. A photosensor and a timer can be used for this measurement. Describe an experimental setup for this method of determining the modulus of elasticity.

3

Performance Specification and Analysis

A control system consists of an integration of several components such as sensors, transducers, signal-conditioning and modification devices, controllers, and a variety of other electronic and digital hardware. In the design, selection, and prescription of these components their performance requirements have to be specified or established within the functional needs of the overall control system. Engineering parameters for performance specification, particularly for control-system components, may be defined either in the time domain or in the frequency domain. Instrument ratings of commercial products are often developed on the basis of these engineering parameters. The present chapter addresses these and related issues of performance specification.

A sensor detects (feels) the quantity that is measured (measurand). The transducer converts the detected measurand into a convenient form for subsequent use (recording, control, actuation, etc.). The transducer input signal may be filtered, amplified, and suitably modified. Transfer-function models, in the frequency domain, are quite useful in representing, analyzing, designing, and evaluating sensors, transducers, controllers, actuators, and interface devices (including signal-conditioning and modification devices). Bandwidth plays an important role in specifying and characterizing any component of a control system. In particular, useful frequency range, operating bandwidth, and control bandwidth are important considerations in control systems. In this chapter, we study several important issues related to system bandwidth as well.

In any multicomponent system, the overall error depends on the component error. Component error degrades the performance of a control system. This is particularly true for sensors and transducers as their error is directly manifested within the system as incorrectly known system variables and parameters. As error may be separated into a systematic (or deterministic) part and a random (or stochastic) part, statistical considerations are important in error analysis. This chapter also deals with such considerations of error analysis.

3.1 Parameters for Performance Specification

All devices that assist in the functions of a control system can be interpreted as components of the system. Selection of available components for a particular application, or design of new components, should rely heavily on performance specifications for these components. A great majority of instrument ratings provided by manufacturers are in the form of static parameters. In control applications, however, dynamic performance specifications are also very important. In the present section, we study instrument ratings and parameters for performance specification, pertaining to both static and dynamic characteristics of instruments.

3.1.1 Perfect Measurement Device

Consider a measuring device of a control system, for example. A *perfect measuring device* can be defined as one that possesses the following characteristics:

1. Output of the measuring device instantly reaches the measured value (fast response).

2. Transducer output is sufficiently large (high gain, low output impedance, high sensitivity).

3. Device output remains at the measured value (without drifting or getting affected by environmental effects and other undesirable disturbances and noise) unless the measurand (i.e., what is measured) itself changes (stability and robustness).

4. The output signal level of the transducer varies in proportion to the signal level of the measurand (static linearity).

5. Connection of a measuring device does not distort the measurand itself (loading effects are absent and impedances are matched; see Chapter 2).

6. Power consumption is small (high input impedance; see Chapter 2).

All these properties are based on dynamic characteristics and, therefore, can be explained in terms of dynamic behavior of the measuring device. In particular, items 1 through 4 can be specified in terms of the device response, either in the *time domain* or in the *frequency domain*. Items 2, 5, and 6 can be specified using the impedance characteristics of the device. First, we shall discuss response characteristics that are important in performance specification of a component of a control system.

3.2 Time-Domain Specifications

Figure 3.1 shows a typical step response in the dominant mode of a device. Note that the curve is normalized with respect to the steady-state value. We have identified several parameters that are useful for the time-domain performance specification of the device. Definitions of these parameters are given now.

3.2.1 Rise Time

This is the time taken to pass the steady-state value of the response for the first time. In overdamped systems, the response is nonoscillatory; consequently, there is no overshoot. This definition is valid for all systems; rise time is often defined as the time taken to pass 90% of the steady-state value. Rise time is often measured from 10% of the steady-state value in order to leave out start-up irregularities and time lags that might be present in a system. A modified rise time (T_{rd}) may be defined in this manner (see Figure 3.1). An alternative definition of rise time, particularly suitable for nonoscillatory responses, is the reciprocal slope of the step response curve at 50% of the steady-state value, multiplied by the steady-state value. In process control terminology, this is called the *cycle time*. No matter what definition is used, rise time represents the speed of response of a device— a small rise time indicates a fast response.

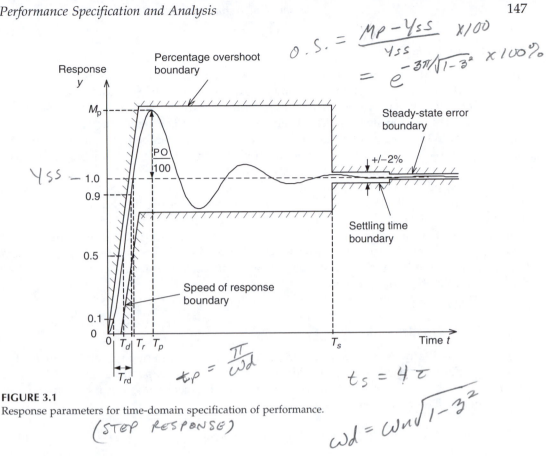

$$O.S. = \frac{M_P - Y_{SS}}{Y_{SS}} \times 100$$

$$= e^{-3\pi/\sqrt{1-3^2}} \times 100\%$$

$$t_p = \frac{\pi}{\omega_d}$$

$$t_s = 4\tau$$

$$\omega_d = \omega_n\sqrt{1-3^2}$$

FIGURE 3.1

Response parameters for time-domain specification of performance.

(STEP RESPONSE)

3.2.2 Delay Time

This is usually defined as the time taken to reach 50% of the steady-state value for the first time. This parameter is also a measure of speed of response.

3.2.3 Peak Time

The time at the first peak of the device response is the peak time. This parameter also represents the speed of response of the device.

3.2.4 Settling Time

This is the time taken for the device response to settle down within a certain percentage (typically $\pm 2\%$) of the steady-state value. This parameter is related to the degree of damping present in the device as well as the degree of stability.

3.2.5 Percentage Overshoot

This is defined as

$$PO = 100(M_p - 1)\%, \tag{3.1}$$

using the normalized-to-unity step response curve, where M_p is the peak value. Percentage overshoot (PO) is a measure of damping or relative stability in the device.

3.2.6   Steady-State Error

This is the deviation of the actual steady-state value of the device response from the desired final value. Steady-state error may be expressed as a percentage with respect to the (desired) steady-state value. In a device output, the steady-state error manifests itself as an offset. This is a systematic (deterministic) error that can be normally corrected by recalibration. In servo-controlled devices, steady-state error can be reduced by increasing loop gain or by introducing lag compensation. Steady-state error can be completely eliminated using the integral control (*reset*) action.

   For the best performance of an output device (e.g., sensor–transducer unit), we wish to have the values of all the foregoing parameters as small as possible. In actual practice, however, it might be difficult to meet all the specifications, particularly for conflicting requirements. For instance, T_r can be decreased by increasing the dominant natural frequency ω_n of the device. This, however, increases the PO and sometimes the T_s. On the other hand, the PO and T_s can be decreased by increasing device damping, but it has the undesirable effect of increasing T_r.

3.2.7   Simple Oscillator Model

The simple oscillator is a versatile model, which can represent the performance of a variety of devices, particularly the desired performance. Depending on the level of damping that is present, both oscillatory and nonoscillatory behavior can be represented by this model. The model can be expressed as

$$\ddot{y} + 2\zeta\omega_n \, \dot{y} + \omega_n^2 \, y = \omega_n^2 \, u(t), \tag{3.2}$$

where u is the excitation or input (normalized), y is the response or output, ω_n = undamped natural frequency, and ζ = damping ratio. The damped natural frequency is given by

$$\omega_d = \sqrt{1 - \zeta^2} \, \omega_n. \tag{3.3}$$

The actual (damped) system executes free (natural) oscillations at this frequency. The response of the system to a unit step excitation, with zero initial conditions, is known to be

$$y = 1 - \frac{1}{\sqrt{1 - \zeta^2}} \, e^{-\zeta\omega_n t} \sin(\omega_d t + \phi) \tag{3.4}$$

where

$$\cos\phi = \zeta. \tag{3.5}$$

As derived in Chapter 2, some important parameters for performance specification in the time domain, using the simple oscillator model, are given in Table 3.1.

   With respect to time-domain specifications of a control-system component such as a transducer, it is desirable to have a very small rise time, and very small settling time in comparison with the time constants of the system whose response is measured, and low percentage overshoot. These conflicting requirements guarantee fast, stable, and steady response.

TABLE 3.1

Time-Domain Performance Parameters Using the Simple Oscillator Model

Performance Parameter	Expression
Rise Time	$T_r = \dfrac{\pi - \phi}{\omega_d}$ with $\cos \phi = \zeta$
Peak Time	$T_p = \dfrac{\pi}{\omega_d}$
Peak Value	$M_p = 1 - e^{-\pi \zeta / \sqrt{1 - \zeta^2}}$
Percentage Overshoot (PO)	$PO = 100\, e^{-\pi \zeta / \sqrt{1 - \zeta^2}}$
Time Constant	$\tau = \dfrac{1}{\zeta \omega_n}$
Settling Time (2%)	$T_s = -\dfrac{\ln\left[0.02\sqrt{1 - \zeta^2}\right]}{\zeta \omega_n} \approx 4\tau = \dfrac{4}{\zeta \omega_n}$

$$T_s\,(5\%) = 3\tau$$

Example 3.1

An automobile weighs 1000 kg. The equivalent stiffness at each wheel, including the suspension system, is approximately 60.0×10^3 N/m. If the suspension is designed for a percentage overshoot of 1%, estimate the damping constant that is needed at each wheel.

Solution

For a quick estimate use a simple oscillator model, which is of the form

$$m\ddot{y} + b\dot{y} + ky = ku(t), \tag{i}$$

where m = equivalent mass = 250 kg; b = equivalent damping constant (to be determined); k = equivalent stiffness = 60.0×10^3 N/m; u = displacement excitation at the wheel.

By comparing Equation (i) with Equation 3.2 we get

$$\zeta = \frac{b}{2\sqrt{km}}. \tag{ii}$$

Note: the equivalent mass at each wheel is taken as one-fourth of the total mass.

For a PO of 1%, from Table 3.1, we have

$$1 = 100 \, \exp\left(-\frac{\pi \zeta}{\sqrt{1 - \zeta^2}}\right),$$

which gives $\zeta = 0.83$. Substitute values in Equation (ii). We get

$$0.83 = \frac{b}{2\sqrt{60 \times 10^3 \times 250.0}}$$

or $b = 6.43 \times 10^3$ N/m/s.

3.2.8 Stability and Speed of Response

The free response of a control device can provide valuable information concerning the natural characteristics of the device. The free (unforced) excitation may be obtained, for example, by giving an initial-condition excitation to the device and then allowing it to respond freely. Two important characteristics that can be determined in this manner are:

1. Stability
2. Speed of response

The stability of a dynamic system implies that the response will not grow without bounds when the excitation force itself is finite. Speed of response of a system indicates how fast the system responds to an excitation force. It is also a measure of how fast the free response (1) rises or falls if the system is oscillatory (i.e., underdamped); or (2) decays, if the system is nonoscillatory (i.e., overdamped). It follows that the two characteristics, stability and speed of response, are not completely independent. In particular, for non-oscillatory systems these two properties are very closely related.

The level of stability of a linear dynamic system depends on the real parts of the eigenvalues (or poles), which are the roots of the characteristic equation. Specifically, if all the roots have real parts that are negative, then the system is stable. Additionally, the more negative the real part of a pole the faster the decay of the free response component corresponding to that pole. The inverse of the negative real part is the *time constant*. Hence, the smaller the time constant, the faster the decay of the corresponding free response, and hence, the higher the level of stability associated with that pole. We can summarize these observations as follows:

Level of stability	Depends on decay rate of free response (and hence on time constants or real parts of poles).
Speed of response	Depends on natural frequency and damping for oscillatory systems and decay rate for nonoscillatory systems.
Time constant	Determines system stability and decay rate of free response (and speed of response in nonoscillatory systems).

Example 3.2

Consider an underdamped system and an overdamped system with the same undamped natural frequency but with damping ratios ζ_u and ζ_o, respectively. Show that the under-damped system is more stable and faster than the overdamped system if and only if:

$$\zeta_o > \frac{\zeta_u^2 + 1}{2\zeta_u},$$

where $\zeta_o > 1 > \zeta_u > 0$ by definition.

Solution

Use the simple oscillator model (Equation 3.2). The characteristic equation is

$$\lambda^2 + 2\zeta\omega_n\lambda + \omega_n^2 = 0. \tag{3.6}$$

The eigenvalues (poles) are

$$\lambda = -\zeta\omega_n \pm \sqrt{\zeta^2 - 1}\,\omega_n. \tag{3.7}$$

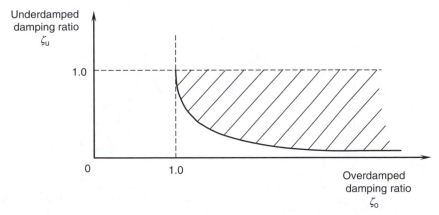

FIGURE 3.2

Region (shaded) where underdamped system is faster and more stable than the corresponding overdamped system.

To be more stable, we should have the underdamped pole located farther away from the origin than the dominant overdamped pole; thus

$$\zeta_u \omega_n > \zeta_o \omega_n - \sqrt{\zeta_o^2 - 1}\,\omega_n.$$

This gives

$$\zeta_o > \frac{\zeta_u^2 + 1}{2\zeta_u}. \tag{3.8}$$

The corresponding region is shown as the shaded area in Figure 3.2.

To explain this result further, consider an undamped ($\zeta = 0$) simple oscillator of natural frequency ω_n. Now, let us add damping and increase ζ gradually from 0 to 1. Then, the complex conjugate poles $-\zeta\omega_n \pm j\omega_d$ will move away from the imaginary axis as ζ increases (because $\zeta\omega_n$ increases) and hence, the level of stability will increase. When ζ reaches the value 1, (critical damping) we get two identical and real poles at $-\omega_n$. When ζ is increased beyond 1, the poles will be real and unequal, with one pole having a magnitude smaller than ω_n and the other having a magnitude larger than ω_n. The former (which is closer to the "origin" of zero value) is the dominant pole that will determine both stability and the speed of response of the resulting overdamped system. It follows that as ζ increases beyond 1, the two poles will branch out from the location $-\omega_n$, one moving toward the origin (becoming less stable) and the other moving away from the origin. It is now clear that as ζ is increased beyond the point of critical damping, the system becomes less stable. Specifically, for a given value of $\zeta_u < 1.0$, there is a value of $\zeta_o > 1$, governed by Equation 3.8, above which the overdamped system is less stable and slower than the underdamped system.

3.3 Frequency-Domain Specifications

Figure 3.3 shows a representative frequency transfer function or FTF (often termed frequency response function or FRF) of a device. This constitutes the plots of *gain* and

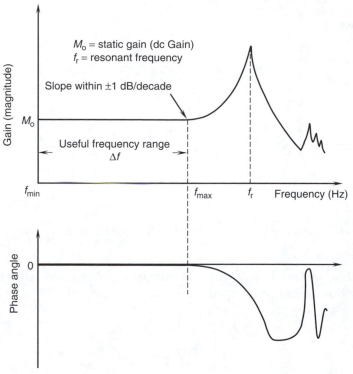

FIGURE 3.3
Response parameters for frequency-domain specification of performance.

phase angle, using frequency as the independent variable. This pair of plots is commonly known as the 2 *Bode diagram*, particularly when the magnitude axis is calibrated in *decibels* (dB) and the frequency axis in a log scale such as *octaves* or *decades*. Experimental determination of these curves can be accomplished either by applying a harmonic excitation and noting the amplitude gain and the phase lead in the response signal at steady state or by Fourier analysis of the excitation and response signals for either transient or random excitations. Experimental determination of transfer functions is known as system identification in the frequency domain. Note that transfer functions provide complete information regarding the system response to a sinusoidal excitation. Since any time signal can be decomposed into sinusoidal components through Fourier transformation, it is clear that the response of a system to an arbitrary input excitation can also be determined using the transfer-function information for that system. In this sense, transfer functions are frequency domain models, which can completely describe a linear system. For this reason, one could argue that it is redundant to use both time-domain specifications and frequency-domain specifications, as they carry the same information. Often, however, both specifications are used simultaneously, because this can provide a better picture of the system performance. Frequency-domain parameters are more suitable in representing some characteristics of a system under some types of excitation.

Some useful parameters for performance specification of a device, in the frequency domain, are:

- Useful frequency range (operating interval)
- Bandwidth (speed of response)
- Static gain (steady-state performance)

- Resonant frequency (speed and critical frequency region)
- Magnitude at resonance (stability)
- Input impedance (loading, efficiency, interconnectability)
- Output impedance (loading, efficiency, interconnectability)
- Gain margin (stability)
- Phase margin (stability)

The first three items are discussed in detail in this chapter, and is also indicated in Figure 3.3. Resonant frequency corresponds to a frequency where the response magnitude peaks. The dominant resonant frequency typically is the lowest resonant frequency, which usually also has the largest peak magnitude. It is shown as f_r in Figure 3.3. The term "Magnitude at Resonance" is self explanatory, and is the peak magnitude mentioned above and shown in Figure 3.3. Resonant frequency is a measure of speed of response and bandwidth, and is also a frequency region that should be avoided during normal operation and whenever possible. This is particularly true for devices that have poor stability (e.g., low damping). Specifically, a high magnitude at resonance is an indication of poor stability. Input impedance and output impedance are discussed in Chapter 2.

3.3.1   Gain Margin and Phase Margin

Gain and phase margins are measures of stability of a device. To define these two parameters consider the feedback system of Figure 3.4a The forward transfer function of the system is $G(s)$ and the feedback transfer function is $H(s)$. These transfer functions are frequency-domain representations of the overall system, which may include a variety

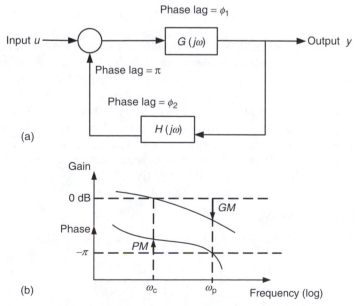

FIGURE 3.4
Illustration of gain and phase margins. (a) A feedback system. (b) Bode diagram.

of components such as the plant, sensors, transducers, actuators, controllers and inter-facing, and signal-modification devices.

The Bode diagram of the system constitutes the magnitude and phase lead plots of the loop transfer function $G(j\omega)H(j\omega)$ as a function of frequency. This is sketched in Figure 3.4b.

Suppose that, at a particular frequency ω the forward transfer function $G(j\omega)$ provides a phase lag of ϕ_1, and the feedback transfer function $H(j\omega)$ provides a phase lag of ϕ_2. Now, in view of the negative feedback, the feedback signal undergoes a phase lag of π,

$$\text{Total phase lag in the loop} = \phi + \pi$$

where,

$$\text{Phase lag of } GH = \phi_1 + \phi_2 = \phi.$$

It follows that, when the overall phase lag of the *loop transfer function* $GH(j\omega)$ is equal to π, the loop phase lag becomes 2π, which means that if a signal of frequency ω travels through the system loop, it will not experience a net phase lag. Additionally, if at this particular frequency, the loop gain $|GH(j\omega)|$ is unity, a sinusoidal signal with this frequency will be able to repeatedly travel through the loop without ever changing its phase or altering its magnitude, even in the absence of any external excitation input. This corresponds to a *marginally stable* condition.

If, on the other hand, the loop gain $|GH(j\omega)| > 1$ at this frequency while the loop phase lag is π, the signal magnitude will monotonically grow as the signal travels through the loop. This is an unstable situation. Furthermore, if the loop gain is <1 at this frequency while the loop phase lag is π, the signal magnitude will monotonically decay as the signal cycles through the loop. This is a stable situation.

In summary,

1. If $|GH(j\omega)| = 1$ when $\angle GH(j\omega) = -\pi$, the system is marginally stable.
2. If $|GH(j\omega)| > 1$ when $\angle GH(j\omega) = -\pi$, the system is unstable.
3. If $|GH(j\omega)| < 1$ when $\angle GH(j\omega) = -\pi$, the system is stable.

It follows that, the margin of smallness of $|GH(j\omega)|$ when compared to 1 at the frequency ω, where $\angle GH(j\omega) = -\pi$, provides a measure of stability, and is termed *gain margin* (see Figure 3.4b). Similarly, at the frequency ω, where $|GH(j\omega)| = 1$, the amount (margin) of phase lag that can be added to the system so as to make the loop phase lag equal to π, is a measure of stability. This amount is termed *phase margin* (see Figure 3.4b).

In terms of frequency-domain specifications, a control-system device such as a transducer should have a wide useful frequency range. For this it must have a high fundamental natural frequency (about 5–10 times the maximum frequency of the operating range) and a somewhat low damping ratio (slightly <1).

3.3.2 Simple Oscillator Model

As discussed in Chapter 2, the transfer function for a simple oscillator is given by

$$\frac{Y(s)}{U(s)} = H(s) = \left[\frac{\omega_n^2}{s^2 + 2\zeta\omega_n s + \omega_n^2}\right]. \tag{3.9}$$

The frequency-transfer function $H(j\omega)$ is defined as $H(s)|_{s=j\omega}$, where ω is the excitation frequency. Note that $H(j\omega)$ is a complex function in ω.

$$\text{Gain} = |H(j\omega)| = \text{magnitude of } H(j\omega)$$
$$\text{Phase Lead} = \angle H(j\omega) = \text{phase angle of } H(j\omega)$$

These represent amplitude gain and phase lead of the output (response) when a sine input signal (excitation) of frequency ω is applied to the system.

Resonant frequency ω_r corresponds to the excitation frequency when the amplitude gain is maximum and is given by

$$\omega_r = \sqrt{1 - 2\zeta^2}\, \omega_n. \tag{3.10}$$

This expression is valid for $\zeta \leq 1/\sqrt{2}$. It can be shown that

$$\text{Gain} = \frac{1}{2\zeta} \text{ and Phase lead} = -\frac{\pi}{2} \text{ when } \omega = \omega_n. \tag{3.11}$$

This concept is used to measure damping in simple systems, in addition to specifying the performance in the frequency domain. Frequency-domain concepts are discussed further under bandwidth considerations.

3.4 Linearity

A device is considered linear if it can be modeled by linear differential equations, with time t as the independent variable. Nonlinear devices are often analyzed using linear techniques by considering small excursions about an operating point. This linearization is accomplished by introducing incremental variables for inputs and outputs. If one increment can cover the entire operating range of a device with sufficient accuracy, it is an indication that the device is linear. If the input–output relations are nonlinear algebraic equations, it represents a *static nonlinearity*. Such a situation can be handled simply by using nonlinear calibration curves, which linearize the device without introducing nonlinearity errors. If, on the other hand, the input–output relations are nonlinear differential equations, analysis usually becomes quite complex. This situation represents a *dynamic nonlinearity*.

Transfer-function representation of an instrument implicitly assumes linearity. According to industrial terminology, a linear measuring instrument provides a measured value that varies linearly with the value of the measurand—the variable that is measured. This is consistent with the definition of static linearity. All physical devices are nonlinear to some degree. This stems due to deviation from the ideal behavior, because of causes such as saturation, deviation from Hooke's law in elastic elements, Coulomb friction, creep at joints, aerodynamic damping, backlash in gears and other loose components, and component wear out.

Nonlinearities in devices are often manifested as some peculiar characteristics. In particular, the following properties are important in detecting nonlinear behavior in dynamic systems.

3.4.1 Saturation

Nonlinear devices may exhibit saturation (see Figure 3.5a). This may be the result of causes such as magnetic saturation, which is common in magnetic-induction devices and

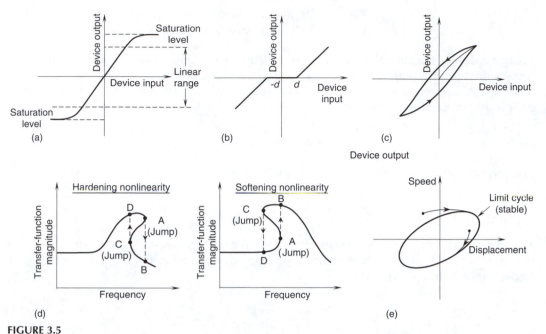

FIGURE 3.5
Common manifestations of nonlinearity in control-system components: (a) Saturation. (b) Dead zone. (c) Hysteresis. (d) The jump phenomenon. (e) Limit cycle response.

transformer-like devices (e.g., differential transformers), plasticity in mechanical components, and nonlinear springs.

3.4.2   Dead Zone

A dead zone is a region in which a device would not respond to an excitation. Stiction in mechanical devices with Coulomb friction is a good example. Because of stiction, a component would not move until the applied force reaches a certain minimum value. Once the motion is initiated, subsequent behavior can be either linear or nonlinear. A dead zone with subsequent linear behavior is shown in Figure 3.5b.

3.4.3   Hysteresis

Nonlinear devices may produce hysteresis. In hysteresis, the input–output curve changes depending on the direction of motion (as indicated in Figure 3.5c), resulting in a hysteresis loop. This behavior is common in loose components such as gears, which have backlash; in components with nonlinear damping, such as Coulomb friction; and in magnetic devices with ferromagnetic media and various dissipative mechanisms (e.g., eddy current dissipation). For example, consider a coil wrapped around a ferromagnetic core. If a dc current is passed through the coil, a magnetic field is generated. As the current is increased from zero, the field strength will also increase. Now, if the current is decreased back to zero, the field strength will not return to zero because of residual magnetism in the ferromagnetic core. A negative current has to be applied to demagnetize the core. It follows that the field strength vs. current curve looks somewhat like Figure 3.5c. This is magnetic hysteresis. Linear viscous damping also exhibits a hysteresis loop in its force–displacement curve. This is a property of any mechanical component that dissipates energy. (Area within the hysteresis loop gives the energy dissipated in one cycle of motion.) In general, if force depends on displacement (as in the case of a spring) and velocity (as in the case of a damping element), the value of force at a given value of displacement will change with velocity. In particular, the force when the component is moving in one direction (say positive velocity) will be different from the force at the same location when the component is moving in the opposite direction (negative velocity), thereby giving a hysteresis loop in the force–displacement plane. If the relationship of displacement and velocity to force is linear (as in viscous damping), the hysteresis effect is linear. If on the other hand the relationship is nonlinear (as in Coulomb damping and aerodynamic damping), the resulting hysteresis is nonlinear.

3.4.4 The Jump Phenomenon

Some nonlinear devices exhibit an instability known as the jump phenomenon (or *fold catastrophe*) in the frequency response (transfer) function curve. This is shown in Figure 3.5d for both *hardening* devices and *softening* devices. With increasing frequency, jump occurs from A to B; and with decreasing frequency, it occurs from C to D. Furthermore, the transfer function itself may change with the level of input excitation in the case of nonlinear devices.

3.4.5 Limit Cycles

Nonlinear devices may produce limit cycles. An example is given in Figure 3.5e on the phase plane of velocity vs. displacement. A limit cycle is a closed trajectory in the state space that corresponds to sustained oscillations at a specific frequency and amplitude, without decay or growth. Amplitude of these oscillations is independent of the initial location from which the response started. In addition, an external input is not needed to sustain a limit-cycle oscillation. In the case of a stable limit cycle, the response will move onto the limit cycle irrespective of the location in the neighborhood of the limit cycle from which the response was initiated (see Figure 3.5e). In the case of an unstable limit cycle, the response will move away from it with the slightest disturbance.

3.4.6 Frequency Creation

At steady state, nonlinear devices can create frequencies that are not present in the excitation signals. These frequencies might be harmonics (integer multiples of the excitation frequency), subharmonics (integer fractions of the excitation frequency), or nonharmonics (usually rational fractions of the excitation frequency).

Example 3.3
Consider a nonlinear device modeled by the differential equation

$$\left\{ \frac{dy}{dt} \right\}^{1/2} = u(t),$$

where $u(t)$ is the input and y is the output. Show that this device creates frequency components that are different from the excitation frequencies.

Solution

First, note that the response of the system is given by

$$y = \int_0^t u^2(t)\mathrm{d}t + y(0).$$

Now, for an input given by

$$u(t) = a_1 \sin \omega_1 t + a_2 \sin \omega_2 t$$

straight forward integration using properties of trigonometric functions gives the following response:

$$y = (a_1^2 + a_2^2)\frac{t}{2} - \frac{a_1^2}{4\omega_1} \sin 2\omega_1 t - \frac{a_2^2}{4\omega_2} \sin 2\omega_2 t$$

$$+ \frac{a_1 a_2}{2(\omega_1 - \omega_2)} \sin(\omega_1 - \omega_2)t - \frac{a_1 a_2}{2(\omega_1 + \omega_2)} \sin(\omega_1 + \omega_2)t - y(0).$$

Note that the discrete frequency components $2\omega_1$, $2\omega_2$, $(\omega_1 - \omega_2)$ and $(\omega_1 + \omega_2)$ are created. Additionally, there is a continuous spectrum that is contributed by the linear function of t that is present in the response.

∎

Nonlinear systems can be analyzed using the *describing function* approach. When a harmonic input (at a specific frequency) is applied to a nonlinear device, the resulting output at steady state will have a component at this fundamental frequency and also components at other frequencies (as a result of frequency creation by the nonlinear device), typically harmonics. The response may be represented by a Fourier series, which has frequency components that are multiples of the input frequency. The describing function approach neglects all the higher harmonics in the response and retains only the fundamental component. This output component, when divided by the input, produces the describing function of the device. This is similar to the transfer function of a linear device, but unlike for a linear device, the gain and the phase shift will be dependent on the input amplitude. Details of the describing function approach can be found in textbooks on nonlinear control theory.

Several methods are available to reduce or eliminate nonlinear behavior in devices. They include calibration (in the static case), use of linearizing elements, such as resistors and amplifiers to neutralize the nonlinear effects, and the use of nonlinear feedback. It is also a good practice to take the following precautions:

1. Avoid operating the device over a wide range of signal levels.
2. Avoid operation over a wide frequency band.
3. Use devices that do not generate large mechanical motions.
4. Minimize Coulomb friction and stiction (e.g., using proper lubrication).
5. Avoid loose joints and gear coupling (i.e., use *direct-drive* mechanisms).

3.5 Instrument Ratings

Instrument manufacturers do not usually provide complete dynamic information for their products. In most cases, it is unrealistic to expect complete dynamic models (in the time domain or the frequency domain) and associated parameter values for complex instruments in a control system. Performance characteristics provided by manufacturers and vendors are primarily static parameters. Known as instrument ratings, these are available as parameter values, tables, charts, calibration curves, and empirical equations. Dynamic characteristics such as transfer functions (e.g., transmissibility curves expressed with respect to excitation frequency) might also be provided for more sophisticated instruments, but the available dynamic information is never complete. Furthermore, definitions of rating parameters used by manufacturers and vendors of instruments are in some cases not the same as analytical definitions used in textbooks. This is particularly true in relation to the terms *linearity* and *stability*. Nevertheless, instrument ratings provided by manufacturers and vendors are very useful in the selection, installation, operation, and maintenance of components in a control system. Let us examine some of these performance parameters.

3.5.1 Rating Parameters

Typical rating parameters supplied by instrument manufacturers are

1. Sensitivity
2. Dynamic range
3. Resolution
4. Linearity
5. Zero drift and full scale drift (Stability)
6. Useful frequency range
7. Bandwidth
8. Input and output impedances

We have already discussed the meaning and significance of some of these terms. In this section, we look at the conventional definitions given by instrument manufacturers and vendors.

Sensitivity of a device (e.g., transducer) is measured by the magnitude (peak, rms value, etc.) of the output signal corresponding to unit input (e.g., measurand). This may be expressed as the ratio of incremental output and incremental input (e.g., slope of a data curve) or, analytically, as the corresponding partial derivative. In the case of vectorial or tensorial signals (e.g., displacement, velocity, acceleration, strain, force), the direction of sensitivity should be specified.

Cross-sensitivity is the sensitivity along directions that are orthogonal to the primary direction of sensitivity. It is normally expressed as a percentage of direct sensitivity. High sensitivity and low cross-sensitivity are desirable for any input–output device (e.g., measuring instrument). Sensitivity to parameter changes and noise has to be small in any device, however, and this is an indication of its robustness. On the other hand, in *adaptive control* and *self-tuning control*, the sensitivity of the system to control parameters has to be sufficiently high. Often, sensitivity and robustness are conflicting requirements.

Dynamic range of an instrument is determined by the allowed lower and upper limits of its input or output (response) so as to maintain a required level of output accuracy. This range is usually expressed as a ratio (e.g., a log value in decibels). In many situations, the lower limit of dynamic range is equal to the resolution of the device. Hence, the dynamic range (ratio) is usually expressed as (range of operation)/(resolution) in dB.

Resolution of an input–output instrument is the smallest change in a signal (input) that can be detected and accurately indicated (output) by a transducer, a display unit, or any pertinent instrument. It is usually expressed as a percentage of the maximum range of the instrument or as the inverse of the dynamic range ratio. It follows that dynamic range and resolution are very closely related.

Example 3.4
The meaning of dynamic range (and resolution) can easily be extended to cover digital instruments. For example, consider an instrument that has a 12 bit analog-to-digital converter (ADC). Estimate the dynamic range of the instrument.

Solution
In this example, dynamic range is determined (primarily) by the word size of the ADC. Each bit can take the binary value 0 or 1. Since the resolution is given by the smallest possible increment, that is, a change by the least significant bit (LSB), it is clear that digital resolution = 1. The largest value represented by a 12 bit word corresponds to the case when all 12 bits are unity. This value is decimal $2^{12} - 1$. The smallest value (when all 12 bits are zero) is zero. Now, use the definition

$$\text{Dynamic range} = 20 \log_{10} \left[\frac{\text{Range of operation}}{\text{Resolution}} \right]. \tag{3.12}$$

The dynamic range of the instrument is given by

$$20 \log_{10} \left[\frac{2^{12} - 1}{1} \right] = 72 \text{ dB}.$$

Another (perhaps more correct) way of looking at this problem is to consider the resolution to be some value δy, rather than unity, depending on the particular application. For example, δy may represent an output signal increment of 0.0025 V. Since a 12 bit word can represent a combination of 2^{12} values (i.e., 4096 values), if the smallest value is denoted by $y_{\min}$, the largest value is $y_{\max} = y_{\min} + (2^{12} - 1) \delta y$.

Note: $y_{\min}$ can be zero, positive, or negative. The smallest increment between values is δy, which is by definition, the resolution. Then,

$$\text{Dynamic range} = \frac{y_{\max} - y_{\min}}{\delta y} = \frac{(2^{12} - 1)\delta y}{\delta y} = 2^2 - 1 = 4095 = 72 \text{ dB}.$$

So we end up with the same result as before for dynamic range, but the interpretation of resolution is somewhat different; the first one representing the resolution of the digital representation and the second on representing the resolution of the engineering quantity that is of interest.

■

Linearity is determined by the calibration curve of an instrument. The curve of output value (e.g., peak or rms value) vs. input value under static (or steady-state) conditions within the dynamic range of an instrument is known as the *static calibration curve*. Its closeness to a straight line measures the degree of linearity of the instrument. Manufacturers provide this information either as the maximum deviation of the calibration curve from the least squares straight-line fit of the calibration curve or from some other reference straight line. If the least-squares fit is used as the reference straight line, the maximum deviation is called *independent linearity* (more correctly, independent nonlinearity, because the larger the deviation, the greater the nonlinearity). Nonlinearity may be expressed as a percentage of either the actual reading at an operating point or the full scale reading.

Zero drift is defined as the drift from the null reading of the instrument when the input is maintained steady for a long period. Note that in this context, the input is kept at zero or any other level that corresponds to the null reading of the instrument. Similarly, *full scale drift* is defined with respect to the full scale reading (i.e., the input is maintained at the full scale value). In the instrumentation practice, drift is a consideration of stability. This interpretation, however, is not identical to the standard textbook definitions of stability. Usual causes of drift include instrument instability (e.g., instability in amplifiers), ambient changes (e.g., changes in temperature, pressure, humidity, and vibration level), changes in power supply (e.g., changes in reference dc voltage or ac line voltage), and parameter changes in an instrument (because of aging, wear and tear, nonlinearities, etc.). Drift due to linear parameter changes that are caused by instrument nonlinearities is known as *parametric drift, sensitivity drift*, or *scale-factor drift*. For example, a change in spring stiffness or electrical resistance because of changes in ambient temperature results in a parametric drift. Note that parametric drift depends on the input level. Zero drift, however, is assumed to be the same at any input level if the other conditions are kept constant. For example, a change in reading caused by thermal expansion of the readout mechanism because of changes in ambient temperature is considered a zero drift. Drift in electronic devices can be reduced by using alternating current (ac) circuitry rather than direct current (dc) circuitry. For example, ac-coupled amplifiers have fewer drift problems than ac amplifiers. Intermittent checking for instrument response level with zero input is a popular way to calibrate for zero drift. In digital devices, for example, this can be done automatically from time to time between sample points, when the input signal can be bypassed without affecting the system operation.

Useful frequency range corresponds to a flat gain curve and a zero phase curve in the frequency response characteristics of an instrument. The maximum frequency in this band is typically less than half (say, one-fifth) of the dominant resonant frequency of the instrument. This is a measure of the instrument bandwidth.

Bandwidth of an instrument determines the maximum speed or frequency at which the instrument is capable of operating. High bandwidth implies faster speed of response (the speed at which a instrument reacts to an input signal). Bandwidth is determined by the dominant natural frequency ω_n or the dominant resonant frequency ω_r of the device. (*Note*: For low damping, ω_r is approximately equal to ω_n.) It is inversely proportional to rise time and the dominant time constant. Half-power bandwidth is also a useful parameter (see the next section). Instrument bandwidth has to be several times greater than the maximum frequency of interest in the input signals. For example, bandwidth of a measuring device is important particularly when measuring transient signals. Note further that bandwidth is directly related to the useful frequency range.

3.6 Bandwidth Design

Bandwidth plays an important role in specifying and characterizing the components of a control system. In particular, useful frequency range, operating bandwidth, and control bandwidth are important considerations. In this section, we study several important issues related to these topics.

3.6.1 Bandwidth

Bandwidth has different meanings depending on the particular context and application. For example, when studying the response of a dynamic system, the bandwidth relates to the fundamental resonant frequency and correspondingly to the speed of response for a given excitation. In band-pass filters, the bandwidth refers to the frequency band within which the frequency components of the signal are allowed through the filter, while the frequency components outside the band are rejected by it. With respect to measuring instruments, bandwidth refers to the range frequencies within which the instrument measures a signal accurately. In digital communication networks (e.g., the Internet), the bandwidth denotes the capacity of the network in terms of information rate (bits/s). These various interpretations of bandwidth are somewhat related even though they are not identical. As a particular note, if a signal passes through a band-pass filter we know that its frequency content is within the bandwidth of the filter, but we cannot determine the actual frequency content of the signal on the basis of that observation. In this context, the bandwidth appears to represent a frequency uncertainty in the observation (i.e., the larger the bandwidth of the filter, less certain is our knowledge about the actual frequency content of a signal that passes through the filter).

3.6.1.1 *Transmission Level of a Band-Pass Filter*

Practical filters can be interpreted as dynamic systems. In fact, all physical dynamic systems (e.g., electro-mechanical systems) are analog filters. It follows that the filter characteristic can be represented by the frequency transfer function $G(f)$ of the filter. A magnitude squared plot of such a filter transfer function is shown in Figure 3.6. In a logarithmic plot the magnitude-squared curve is obtained by simply doubling the corresponding magnitude curve (in the Bode plot). Note that the actual filter transfer function (Figure 3.6b) is not quite flat like the ideal filter shown in Figure 3.6a. The reference level G_r is the average value of the transfer function magnitude in the neighborhood of its peak.

3.6.1.2 *Effective Noise Bandwidth*

Effective noise bandwidth of a filter is equal to the bandwidth of an ideal filter that has the same reference level and that transmits the same amount of power from a white noise source. Note that white noise has a constant (flat) power spectral density (psd). Hence, for a noise source of unity psd, the power transmitted by the practical filter is given by

$$\int_0^\infty |G(f)|^2 df,$$

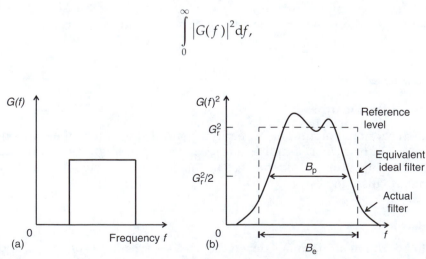

FIGURE 3.6
Characteristics of. (a) An ideal band-pass filter. (b) A practical band-pass filter.

which, by definition, is equal to the power $G_r^2 B_e$ that is transmitted by the equivalent ideal filter. Hence, the effective noise bandwidth B_e is given by,

$$B_e = \int_0^\infty |G(f)|^2 df / G_r^2. \qquad (3.13)$$

3.6.1.3 Half-Power (or 3dB) Bandwidth

Half of the power from a unity-psd noise source as transmitted by an ideal filter is $G_r^2 B_e / 2$. Hence, $G_r / \sqrt{2}$ is referred to as the *half-power level*. This is also known as a 3 db level because $20 \log_{10} \sqrt{2} = 10 \log_{10} 2 = 3$ dB. (*Note*: 3 dB refers to a power ratio of 2 or an amplitude ratio of $\sqrt{2}$. Hence, a 3 dB drop corresponds to a drop of power to half the original value. Furthermore, 20 dB corresponds to an amplitude ratio of 10 or a power ratio of 100.) The 3 dB (or half-power) bandwidth corresponds to the width of the filter transfer function at the half-power level. This is denoted by B_p in Figure 3.6b. Note that B_e and B_p are different in general. In an ideal case where the magnitude-squared filter characteristic has linear rising and fall-off segments, however, these two bandwidths are equal (see Figure 3.7).

3.6.1.4 Fourier Analysis Bandwidth

In Fourier analysis, bandwidth is interpreted as the *frequency uncertainty* in the spectral results. In analytical Fourier integral transform (FIT) results, which assume that the entire signal is available for analysis, the spectrum is continuously defined over the entire frequency range $[-\infty, \infty]$ and the frequency increment df is infinitesimally small ($df \to 0$). There is no frequency uncertainty in this case, and the analysis bandwidth is infinitesimally narrow. In digital Fourier analysis, the discrete spectral lines are generated at frequency intervals of ΔF. This finite frequency increment ΔF, which is the frequency uncertainty, is therefore, the analysis bandwidth B for this analysis (digital computation). It is known that $\Delta F = 1/T$, where T is the record length of the signal (or window length when a rectangular window is used to select the signal segment for analysis). It follows also that the minimum frequency that has a meaningful accuracy is the analysis bandwidth. This interpretation for analysis bandwidth is confirmed by noting the fact that

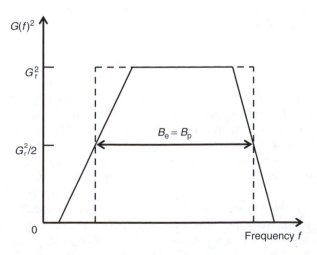

FIGURE 3.7 An idealized filter with linear segments.

harmonic components of frequency less than ΔF (or period greater than T) cannot be studied by observing a signal record of length less than T. Analysis bandwidth carries information regarding distinguishable minimum frequency separation in computed results. In this sense, bandwidth is directly related to the frequency resolution of analyzed (computed) results. The accuracy of analysis (computation) increases by increasing the record length T (i.e., by decreasing the analysis bandwidth B).

When a time window other than the rectangular window is used to truncate a signal, then reshaping of the signal segment (data) occurs according to the shape of the window. This reshaping suppresses the side lobes of the Fourier spectrum of the original rectangular window and hence, reduces the frequency leakage that arises from truncation of the signal. At the same time, however, an error is introduced as a result of the information lost through data reshaping. This error is proportional to the bandwidth of the window itself. The effective noise bandwidth of a rectangular window is only slightly less than $1/T$, because the main lobe of its Fourier spectrum is nearly rectangular, and a lobe has a width of $1/T$. Hence, for all practical purposes, the effective noise bandwidth can be taken as the analysis bandwidth. Data truncation (i.e., multiplication by a window in the time domain) is equivalent to convolution of the Fourier spectrum of the signal with the Fourier spectrum of the window (in the frequency domain). Hence the main lobe of the window spectrum uniformly affects all spectral lines in the discrete spectrum of the data signal. It follows that a window main lobe with a broader effective-noise bandwidth introduces a larger error into the spectral results. Hence, in digital Fourier analysis, bandwidth is taken as the effective-noise bandwidth of the time window that is employed.

3.6.1.5 Useful Frequency Range

This corresponds to the flat region (static region) in the gain curve and the zero-phase-lead region in the phase curve of a device (with respect to frequency). It is determined by the dominant (i.e., the lowest) resonant frequency f_r of the device. The upper frequency limit f_{max} in the useful frequency range is several times smaller than f_r for a typical input–output device (e.g., $f_{max} = 0.25\, f_r$). Useful frequency range may also be determined by specifying the flatness of the static portion of the frequency response curve. For example, since a single pole or a single zero introduces a slope in the order of ± 20 dB/decade to the Bode magnitude curve of the device, a slope within 5% of this value (i.e., ± 1 dB/decade) may be considered flat for most practical purposes. For a measuring instrument, for example, operation in the useful frequency range implies that the significant frequency content of the measured signal is limited to this band. Then, faithful measurement and fast response are guaranteed, because dynamics of the measuring device will not corrupt the measurement.

3.6.1.6 Instrument Bandwidth

This is a measure of the useful frequency range of an instrument. Furthermore, the larger the bandwidth of the device, the faster will be the speed of response. Unfortunately, the larger the bandwidth, the more susceptible the instrument will be to high-frequency noise as well as stability problems. Filtering will be needed to eliminate unwanted noise. Stability can be improved by dynamic compensation. Common definitions of instrument bandwidth include the frequency range over which the transfer-function magnitude is flat; the resonant frequency; and the frequency at which the transfer-function magnitude drops to $1/\sqrt{2}$ (or 70.7 percent) of the zero-frequency (or static) level. As noted before, the last definition corresponds to the *half-power bandwidth*, because a reduction of amplitude level by a factor of $\sqrt{2}$ corresponds to a power drop by a factor of 2.

3.6.1.7 Control Bandwidth

This is used to specify the maximum possible speed of control. It is an important specification in both analog control and digital control. In digital control, the data sampling rate (in samples per second) has to be several times higher than the control bandwidth (in hertz or Hz) so that sufficient data would be available to compute the control action. Moreover, from *Shannon's sampling theorem*, control bandwidth is given by half the rate at which the control action is computed (see later under the topic of aliasing distortion). The control bandwidth provides the frequency range within which a system can be controlled (assuming that all the devices in the system can operate within this bandwidth).

3.6.2 Static Gain

This is the gain (i.e., transfer function magnitude) of a measuring instrument within the useful (flat) range (or at very low frequencies) of the instrument. It is also termed *dc gain*. A high value for static gain results in a high-sensitivity measuring device, which is a desirable characteristic.

Example 3.5

A mechanical device for measuring angular velocity is shown in Figure 3.8. The main element of this tachometer is a rotary viscous damper (damping constant b) consisting of two cylinders. The outer cylinder carries a viscous fluid within which the inner cylinder rotates. The inner cylinder is connected to the shaft whose speed ω_i is to be measured. The outer cylinder is resisted by a linear torsional spring of stiffness k. The rotation θ_o of the outer cylinder is indicated by a pointer on a suitably calibrated scale. Neglecting the inertia of moving parts, perform a bandwidth analysis of this device.

Solution

The damping torque is proportional to the relative velocity of the two cylinders and is resisted by the spring torque. The equation of motion is given by

$$b\left(\omega_i - \dot{\theta}_o\right) = k\theta_o$$

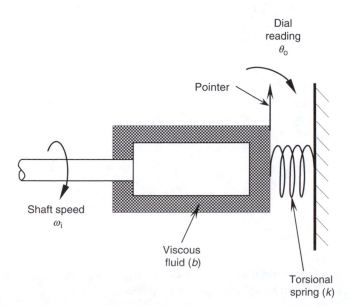

Dial
reading
θ_o

Pointer

Shaft speed
ω_i

Viscous
fluid (b)

Torsional
spring (k)

FIGURE 3.8
A mechanical tachometer.

or

$$b\dot{\theta}_o + k\theta_o = b\omega_i. \tag{i}$$

The transfer function is determined by first replacing the time derivative by the Laplace operator s and taking the ratio: output/input; thus,

$$\frac{\theta_o}{\omega_i} = \frac{b}{[bs+k]} = \frac{b/k}{[(b/k)s+1]} = \frac{k_g}{[\tau s+1]}. \tag{ii}$$

Note that the static gain or dc gain (transfer-function magnitude at $s = 0$) is

$$k_g = \frac{b}{k} \tag{iii}$$

and the time constant is

$$\tau = \frac{b}{k}. \tag{iv}$$

We face conflicting design requirements in this case. On the one hand, we want to have a large static gain so that a sufficiently large reading is available. On the other hand, the time constant must be small to obtain a quick reading that faithfully follows the measured speed. A compromise must be reached here, depending on the specific design requirements. Alternatively, a signal-conditioning device could be employed to amplify the sensor output.

Now, let us examine the half-power bandwidth of the device. The frequency transfer function is

$$G(j\omega) = \frac{k_g}{\tau j\omega + 1}. \tag{v}$$

By definition, the half-power bandwidth ω_b is given by

$$\frac{k_g}{|\tau j\omega_b + 1|} = \frac{k_g}{\sqrt{2}}.$$

Hence

$$(\tau\omega_b)^2 + 1 = 2.$$

As both τ and ω_b are positive we have

$$\tau\omega_b = 1$$

or

$$\omega_b = \frac{1}{\tau}. \tag{vi}$$

Note that the bandwidth is inversely proportional to the time constant. This confirms our earlier statement that bandwidth is a measure of the speed of response.

Example 3.6
Part 1

   i. Briefly discuss any conflicts that can arise in specifying parameters that can be used to predominantly represent the speed of response and the degree of stability of a process (plant).

  ii. Consider a measuring device that is connected to a plant for feedback control. Explain the significance of

     a. Bandwidth

     b. Resolution

     c. Linearity

     d. Input impedance, and

     e. Output impedance
       of the measuring device in the performance of the feedback control system.

Part 2

Consider the speed control system schematically shown in Figure 3.9. Suppose that the plant and the controller together are approximated by the transfer function

$$G_p(s) = \frac{k}{(\tau_p s + 1)},$$

where τ_p is the plant time constant.

    a. Give an expression for the bandwidth ω_p of the plant, without feedback.

    b. If the feedback tachometer is ideal and is represented by a unity (negative) feedback, what is the bandwidth ω_c of the feedback control system?

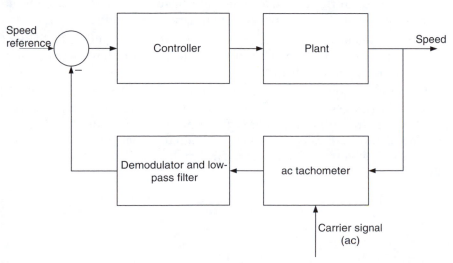

FIGURE 3.9
A speed control system.

c. If the feedback tachometer can be represented by the transfer function

$$G_s(s) = \frac{1}{(\tau_s s + 1)},$$

where τ_s is the sensor time constant, explain why the bandwidth ω_{cs} of the feedback control system is given by the smaller quantity of $1/\tau_s$ and $(k+1)/(\tau_p + \tau_s)$. Assume that both τ_p and τ_s are sufficiently small.

Next suppose that approximately $\tau_p = 0.016$ s. Estimate a sufficient bandwidth in hertz for the tachometer. Additionally, if $k = 1$, estimate the overall bandwidth of the feedback control system. If $k = 49$, what would be the representative bandwidth of the feedback control system?

For the particular ac tachometer (with the bandwidth value as chosen in the present numerical example), what should be the frequency of the carrier signal? In addition, what should be the cutoff frequency of the low-pass filter that is used with its demodulator circuit?

Solution
Part 1

i. Usually speeding up a system has a destabilizing effect. For example, if gain is increased to speed up a system, the % overshoot (PO) can increase as a result.

ii. a. Measuring device bandwidth should cover the entire bandwidth of possible operation of the system. (Typically make it several times larger than the required bandwidth.) Otherwise useful frequency components in the measured could be distorted.

b. Resolution of the measuring device should be less than half the error tolerance of the control system. Otherwise the sensor tolerance alone can provide an unacceptable error level in the control system (even when control itself is satisfactory).

c. If the measuring device has a static nonlinearity, an accurate calibration curve will be needed. Otherwise the operating range has to be limited. Dynamic nonlinearity can cause undesirable effects such as limit cycles, hysteresis, frequency creation, jump phenomenon, saturation, and related errors.

d. Input impedance of the measuring device has to be significantly higher than the output impedance of the process. Otherwise the signal will be subjected to loading error and distortion.

e. The output impedance of the measuring device has to be small. Otherwise, the devices connected to that end should have a very high impedance. Moreover, the output level of a high-output-impedance device will be low in general (not satisfactory). Then, additional, expensive hardware will be necessary to condition the measured signal.

Part 2

a.

$$G_p(s) = \frac{k}{(\tau_p s + 1)},$$

$$\omega_p = 1/\tau_p.$$

b. With unity feedback, closed-loop transfer function is

$$G_c(s) = \frac{k/(\tau_p s + 1)}{1 + k/(\tau_p s + 1)}, \text{ which simplifies to}$$

$$G_c(s) = \frac{k}{(\tau_p s + 1 + k)}.$$

Hence,

$$\omega_c = \frac{1 + k}{\tau_p}.$$

Note that the bandwidth has increased.

c. With feedback sensor of transfer function

$$G_s(s) = \frac{1}{(\tau_s s + 1)}$$

the closed-loop transfer function is

$$\begin{aligned}
G_{cs}(s) &= \frac{k/(\tau_p s + 1)}{1 + k/\{(\tau_p s + 1)(\tau_s s + 1)\}} \\
&= \frac{k(\tau_s s + 1)}{\tau_p \tau_s s^2 + (\tau_p + \tau_s)s + 1 + k} \\
&\cong \frac{k(\tau_s s + 1)}{(\tau_p + \tau_s)s + 1 + k} \quad \{\text{Neglecting } \tau_p \tau_s.\}
\end{aligned}$$

Hence, to avoid the dynamic effect of the sensor (which has introduced a zero in $G_{cs}(s)$) we should limit the bandwidth to $1/\tau_s$.

Additionally, from the denominator of G_{cs}, it is seen that the closed-loop bandwidth is given by $\dfrac{1 + k}{(\tau_p + \tau_s)}$.

Hence, for satisfactory performance, the bandwidth has to be limited to

$$\min\left[\frac{1}{\tau_s}, \frac{1 + k}{(\tau_p + \tau_s)}\right].$$

With $\tau_p = 0.016$ s we have

$$\omega_p = \frac{1}{0.016} = 62.5 \text{ rad/s} = 10.0 \text{ Hz}$$

Hence, pick a sensor bandwidth of 10 times this value.

$$\Rightarrow \quad \omega_s = 100.0 \text{ Hz} = 625.0 \text{ rad/s}.$$

Then $\tau_s = 1/\omega_s = 0.0016$ s. With $k = 1$

$$\frac{1+k}{(\tau_p + \tau_s)} = \frac{(1+1)}{(0.016 + 0.0016)} \text{ rad/s} = 18.0 \text{ Hz}.$$

Also,

$$\frac{1}{\tau_s} = 100.0 \text{ Hz}.$$

Hence,

$$\omega_{cs} \cong \min [100, 18.0] \text{ Hz} = 18.0 \text{ Hz}$$

With $k = 49$

$$\frac{1+k}{(\tau_p + \tau_s)} = \frac{1+49}{(0.016 + 0.0016)} \text{ rad/s} = 450.0 \text{ Hz}$$

and as before

$$\frac{1}{\tau_s} = 100.0 \text{ Hz}$$

Then $\omega_{cs} = \min [100, 450.0] \text{ Hz} = 100.0 \text{ Hz}$. It follows that now the control system bandwidth has increased to about 100 Hz (possibly somewhat lower than 100 Hz).

For a sensor with 100 Hz bandwidth (See Chapter 2 under demodulation and Chapter 4 under LVDT, for the related theory)

Carrier frequency $\cong 10 \times 100 \text{ Hz} = 1000.0 \text{ Hz}$
$2 \times$ carrier frequency $= 2000 \text{ Hz}$
Low-pass filter cutoff $= (1/10) \times 2000 \text{ Hz} = 200.0 \text{ Hz}.$

3.7 Aliasing Distortion due to Signal Sampling

Aliasing distortion is an important consideration when dealing with sampled data from a continuous signal. This error may enter into computation in both the time domain and the frequency domain, depending on the domain in which the data are sampled.

3.7.1 Sampling Theorem

If a time signal $x(t)$ is sampled at equal steps of ΔT, no information regarding its frequency spectrum $X(f)$ is obtained for frequencies higher than $f_c = 1/(2\Delta T)$. This fact is known as Shannon's sampling theorem, and the limiting (cutoff) frequency is called the *Nyquist frequency*.

It can be shown that the aliasing error is caused by folding of the high-frequency segment of the frequency spectrum beyond the Nyquist frequency into the low-frequency segment. This is illustrated in Figure 3.10. The aliasing error becomes more and more prominent for frequencies of the spectrum closer to the Nyquist frequency. In signal analysis, a sufficiently small sample step ΔT should be chosen in order to reduce

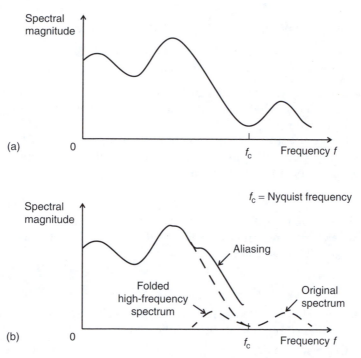

FIGURE 3.10
Aliasing distortion of a frequency spectrum. (a) Original spectrum. (b) Distorted spectrum due to aliasing.

$$\text{CUTOFF FREQUENCY} = 0.8 f_c$$
$$f_s = \text{SAMPLING FREQUENCY}$$
$$f_s = \frac{1}{\Delta T} = 2 f_c$$

aliasing distortion in the frequency domain, depending on the highest frequency of interest in the analyzed signal. This however, increases the signal processing time and the computer storage requirements, which is undesirable particularly in real-time analysis. It can also result in stability problems in numerical computations. The Nyquist sampling criterion requires that the sampling rate $(1/\Delta T)$ for a signal should be at least twice the highest frequency of interest. Instead of making the sampling rate very high, a moderate value that satisfies the Nyquist sampling criterion is used in practice, together with an *antialiasing filter* to remove the frequency components in the original signal that would distort the spectrum of the computed signal.

3.7.2 Antialiasing Filter

It should be clear from Figure 3.10 that, if the original signal is low-pass filtered at a cutoff frequency equal to the Nyquist frequency, then the aliasing distortion because of sampling would not occur. A filter of this type is called an antialiasing filter. Analog hardware filters may be used for this purpose. In practice, it is not possible to achieve perfect filtering. Hence, some aliasing could remain even after using an antialiasing filter, further reducing the valid frequency range of the computed signal. Typically, the useful frequency limit is $f_c/1.28$ so that the last 20% of the spectral points near the Nyquist frequency should be neglected. Sometimes the filter cutoff frequency is chosen to be somewhat lower than the Nyquist frequency. For example, $f_c/1.28$ ($\cong 0.8 f_c$) is used as the filter cutoff frequency. In this case the computed spectrum is accurate up to the filter cutoff frequency $0.8 f_c$ and not up to the Nyquist frequency f_c.

Example 3.7

Consider 1024 data points from a signal, sampled at 1 ms intervals.

Sample rate $f_s = 1/0.001$ samples/s $= 1000$ Hz $= 1$ kHz
Nyquist frequency $= 1000/2$ Hz $= 500$ Hz

Because of aliasing, approximately 20% of the spectrum even in the theoretically useful range (i.e., spectrum beyond 400 Hz) will be distorted. Here we may use an antialiasing filter with a cutoff at 400 Hz.

Suppose that a digital Fourier transform computation provides 1024 frequency points of data up to 1000 Hz. Half of this number is beyond the Nyquist frequency, and will not give any new information about the signal.

Spectral line separation $= 1000/1024$ Hz $= 1$ Hz (approx)

We keep only the first 400 spectral lines as the useful spectrum.
Note: Almost 500 spectral lines may be retained if an accurate antialiasing filter with its cutoff frequency at 500 Hz, is used.

Example 3.8

a. If a sensor signal is sampled at f_s Hz, suggest a suitable cutoff frequency for an antialiasing filter to be used in this application.

b. Suppose that a sinusoidal signal of frequency f_1 Hz was sampled at the rate of f_s samples per second. Another sinusoidal signal of the same amplitude, but of a higher frequency f_2 Hz was found to yield the same data when sampled at f_s. What is the likely analytical relationship between f_1, f_2, and f_s?

c. Consider a plant of transfer function

$$G(s) = \frac{k}{(1 + \tau s)}.$$

What is the static gain of this plant? Show that the magnitude of the transfer function reaches $1/\sqrt{2}$ of the static gain when the excitation frequency is $1/\tau$ rad/s. Note that the frequency, $\omega_b = 1/\tau$ rad/s, may be taken as the operating bandwidth of the plant.

d. Consider a chip refiner that is used in the pulp and paper industry. The machine is used for mechanical pulping of wood chips. It has a fixed plate and a rotating plate, driven by an induction motor. The gap between the plates is sensed and is adjustable as well. As the plate rotates, the chips are ground into a pulp within the gap. A block diagram of the plate-positioning control system is shown in Figure 3.11.

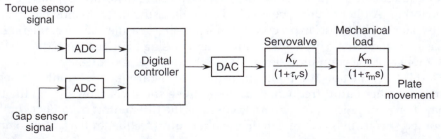

FIGURE 3.11
Block diagram of the plate positioning control system for a chip refiner.

Suppose that the torque sensor signal and the gap sensor signal are sampled at 100 Hz and 200 Hz, respectively, into the digital controller, which takes 0.05 s to compute each positioning command for the servovalve. The time constant of the servovalve is $0.05/2\pi$ s and that of the mechanical load (plant) is $0.2/2\pi$ s. Estimate the control bandwidth and the operating bandwidth of the positioning system.

Solution

a. In theory, the cutoff frequency of the antialiasing filter has to be $1/2\,f_s$, which is the Nyquist frequency. In practice, however, $0.4\,f_s$ would be desirable, providing a useful spectrum of only up to $0.4\,f_s$.

b. It is likely that f_1 and f_2 are symmetrically located on either side of the Nyquist frequency f_c. Then, $f_2 - f_c = f_c - f_1$
This gives

$$f_2 = f_c + (f_c - f_1) = 2f_c - f_1$$

or

$$f_2 = f_s - f_1 \tag{3.14}$$

c.

$$G(j\omega) = \frac{k}{(1 + \tau j\omega)} = \text{frequency transfer function where, } \omega \text{ is in rad/s.}$$

Static gain is the transfer function magnitude at steady state (i.e., at zero frequency).
Hence,
Static gain $= G(0) = k$.
When $\omega = 1/\tau$

$$G(j\omega) = \frac{k}{(1 + j)}.$$

Hence,

$$|G(j\omega)| = k/\sqrt{2} \text{ at this frequency.}$$

This corresponds to the half-power bandwidth.

d. Because of sampling, the torque signal has a bandwidth of $\frac{1}{2} \times 100$ Hz $= 50$ Hz, and the gap sensor signal has a bandwidth of $\frac{1}{2} \times 200 = 100$ Hz. Control cycle time $= 0.05$ s, which generates control signals at a rate of $1/0.05 = 20$ Hz.
Since

$$20 \text{ Hz} < \min (50 \text{ Hz, } 100 \text{ Hz}),$$

we have adequate bandwidth from the sampled sensor signals to compute the control signal. The control bandwidth from the digital controller

$$= \frac{1}{2} \times 20 \text{ Hz (From Shannons sampling theoreom)}$$

$$= 10 \text{ Hz}$$

But, the servovalve is also part of the controller. Its bandwidth

$$= \frac{1}{\tau_v}\,\text{rad/s} = \frac{1}{2\pi\tau_v}\,\text{Hz}$$

$$= \frac{2\pi}{2\pi \times 0.05}\,\text{Hz} = 20\,\text{Hz}$$

Operating bandwidth is limited by both control bandwidth (10 Hz) are Hence,

Control bandwidth = min (10 Hz, 20 Hz) = 10 Hz.

Bandwidth of the mechanical load (plant)

$$= \frac{1}{\tau_m}\,\text{rad/s} = \frac{1}{2\pi\tau_m}\,\text{Hz} = \frac{2\pi}{2\pi \times 0.2}\,\text{Hz} = 5\,\text{Hz}.$$

Operating bandwidth is limited by both control bandwidth (10 Hz) and the plant bandwidth (5 Hz). Hence,

Operating bandwidth of the system = min (10 Hz, 5 Hz) = 5 Hz.

3.7.3   Another Illustration of Aliasing

A simple illustration of aliasing is given in Figure 3.12. Here, two sinusoidal signals of frequency $f_1 = 0.2$ Hz and $f_2 = 0.8$ Hz are shown (Figure 3.12a). Suppose that the two

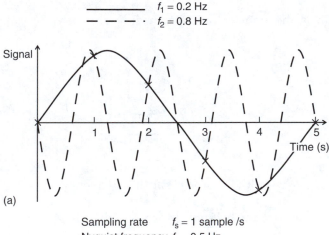

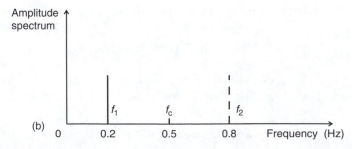

FIGURE 3.12
A simple illustration of aliasing: (a) two harmonic signals with identical sampled data. (b) frequency spectra of the two harmonic signals.

signals are sampled at the rate of $f_s = 1$ sample/s. The corresponding Nyquist frequency is $f_c = 0.5$ Hz. It is seen that, at this sampling rate, the data samples from the two signals are identical. In other words, from the sampled data the high-frequency signal cannot be distinguished from the low-frequency signal. Hence, a high-frequency signal component of frequency 0.8 Hz will appear as a low-frequency signal component of frequency 0.2 Hz. This is aliasing, as clear from the signal spectrum shown in Figure 3.12b. Specifically, the spectral segment of the signal beyond the Nyquist frequency (f_c) folds on to the low frequency side due to data sampling, and cannot be recovered.

Example 3.9

Suppose that the frequency range of interest in a particular signal is 0 to 200 Hz. We are interested in determining the sampling rate (digitization speed) for the data and the cutoff frequency for the antialiasing (low-pass) filter.

The Nyquist frequency f_c is given by $f_c/1.28 = 200$.

Hence, $f_c = 256$ Hz.

The sampling rate (or digitization speed) for the time signal that is needed to achieve this range of analysis is $f_s = 2f_c = 512$ Hz. With this sampling frequency, the cutoff frequency for the antialiasing filter could be set at a value between 200 and 256 Hz.

Example 3.10

Consider the digital control system for a mechanical position application, as schematically shown in Figure 3.13. The control computer generates a control signal according to an algorithm, on the basis of the desired position and actual position, as measured by an optical encoder (see Chapter 5). This digital signal is converted into the analog form using a digital to analog converter (DAC) and is supplied to the drive amplifier. Accordingly, the current signals needed to energize the motor windings are generated by the amplifier. The inertial element, which has to be positioned is directly (and rigidly) linked to the motor rotor and is resisted by a spring and a damper, as shown.

Suppose that the combined transfer function of the drive amplifier and the electromagnetic circuit (torque generator) of the motor is given by

$$\frac{k_e}{(s^2 + 2\zeta_e\omega_e s + \omega_e^2)}$$

and the transfer function of the mechanical system including the inertia of the motor rotor is given by

$$\frac{k_m}{(s^2 + 2\zeta_m\omega_m s + \omega_m^2)}.$$

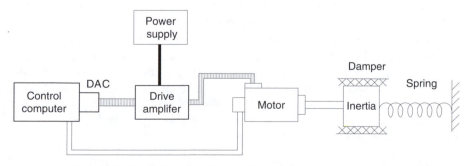

FIGURE 3.13
Digital control system for a mechanical positioning application.

Here, k is equivalent gain, ζ is damping ratio, and ω is natural frequency with the subscripts ( )$_e$ and ( )$_m$ denoting the electrical and mechanical components respectively. Moreover,

ΔT_c = time taken to compute each control action,
ΔT_p = pulse period of the position sensing encoder.

The following numerical values are given:

$$\omega_e = 1000\pi \text{ rad/s, } \zeta_e = 0.5, \omega_m = 100\pi \text{ rad/s, and } \zeta_m = 0.3.$$

For the purpose of this example, you may neglect loading effects and coupling effects that arise from component cascading and signal feedback.

   i. Explain why the control bandwidth of this system cannot be much larger than 50 Hz.

   ii. If $\Delta T_c = 0.02$ s, estimate the control bandwidth of the system.

   iii. Explain the significance of ΔT_p in this application. Why, typically, ΔT_p should not be greater than $0.5 \, \Delta T_c$?

   iv. Estimate the operating bandwidth of the positioning system, assuming that significant plant dynamics are to be avoided.

   v. If $\omega_m = 500\pi$ rad/s and $\Delta T_c = 0.02$ s, with the remaining parameters kept as specified above, estimate the operating bandwidth of the system, again so as not to excite significant plant dynamics.

Solution

   i. The drive system has a resonant frequency less than 500 Hz. Hence the flat region of the spectrum (i.e., operating region) of the drive system would be about one-tenth of this; i.e., 50 Hz. This would limit the maximum spectral component of the drive signal to about 50 Hz. Hence, the control bandwidth would be limited by this value.

   ii. Rate at which the digital control signal is generated $= 1/0.02$ Hz $= 50$ Hz. By Shannon's sampling theorem, the effective (useful) spectrum of the control signal is limited to $\frac{1}{2} \times 50$ Hz $= 25$ Hz. Even though the drive system can accommodate a bandwidth of about 50 Hz, the control bandwidth would be limited to 25 Hz, because of digital control, in this case.

   iii. Note that ΔT_p corresponds to the sampling period of the measurement signal (for feedback). Hence, its useful spectrum would be limited to $1/2 \, \Delta T_p$, by Shannon's sampling theorem. Consequently, the feedback signal will not be able to provide any useful information of the process beyond the frequency $1/2\Delta T_p$. To generate a control signal at the rate of $1/\Delta T_c$ samples per second, the process information has to be provided at least up to $1/\Delta T_c$ Hz. To provide this information we must have:

$$\frac{1}{2\Delta T_p} \geq \frac{1}{\Delta T_c} \text{ or } \Delta T_p \leq 0.5 \, \Delta T_c. \qquad (3.15)$$

This guarantees that at least two points of sampled data from the sensor are used for computing each control action.

iv. The resonant frequency of the plant (positioning system) is approximately (less than)

$$\frac{100\pi}{2\pi}\,\text{Hz} \simeq 50\ \text{Hz}.$$

At frequencies near this, the resonance will interfere with control, and should be avoided if possible, unless the resonances (or modes) of the plant themselves need to be modified through control. At frequencies much larger than this, the process will not significantly respond to the control action, and will not be of much use (the plant will be felt like a rigid wall). Hence, the operating bandwidth has to be sufficiently smaller than 50 Hz, say 25 Hz, in order to avoid plant dynamics.

Note: This is a matter of design judgment, based on the nature of the application (e.g., excavator, disk drive). Typically, however, one needs to control the plant dynamics. In that case, it is necessary to use the entire control bandwidth (i.e., maximum possible control speed) as the operating bandwidth. In the present case, even if the entire control BW (i.e., 25 Hz) is used as the operating BW, it still avoids the plant resonance.

v. The plant resonance in this case is about $500\ \pi/2\ \pi$ Hz $\simeq 250$ Hz. This limits the operating bandwidth to about $250\ \pi/2$ Hz $\simeq 125$ Hz, so as to avoid plant dynamics. But, the control bandwidth is about 25 Hz because $\Delta T_c = 0.02$ s. Hence, the operating bandwidth cannot be greater than this value, and would be $\simeq 25$ Hz.

3.8 Bandwidth Design of a Control System

Based on the foregoing concepts, it is now possible to give a set of simple steps for designing a control system on the basis of bandwidth considerations.

- Step 1: Decide on the maximum frequency of operation (BW_o) of the system based on the requirements of the particular application.
- Step 2: Select the process components (electro-mechanical) that have the capacity to operate at BW_o and perform the required tasks.
- Step 3: Select feedback sensors with a flat frequency spectrum (operating frequency range) greater than $4 \times BW_o$.
- Step 4: Develop a digital controller with a sampling rate greater than $4 \times BW_o$ for the sensor feedback signals (keeping within the flat spectrum of the sensors) and a direct-digital control cycle time (period) of $1/(2 \times BW_o)$. Note that the digital control actions are generated at a rate of $2 \times BW_o$.
- Step 5: Select the control drive system (interface analog hardware, filters, amplifiers, actuators, etc.) that have a flat frequency spectrum of at least BW_o.
- Step 6: Integrate the system and test the performance. If the performance specifications are not satisfied, make necessary adjustments and test again.

$$BW_{SENSOR} = 4\,BW_o \qquad \Delta T_c = \frac{1}{4\,BW_o}$$
$$BW_{FILTER} < BW_o$$

3.8.1   Comment about Control Cycle Time

In the engineering literature it is often used that $\Delta T_c = \Delta T_p$, where ΔT_c = control cycle time (period at which the digital control actions are generated) and ΔT_p = period at which the feedback sensor signals are sampled (see Figure 3.14a). This is acceptable in systems where the useful frequency range of the plant is sufficiently smaller than $1/\Delta T_p$ (and $1/\Delta T_c$). Then, the sampling rate $1/\Delta T_p$ of the feedback measurements (and the Nyquist frequency $0.5/\Delta T_p$) will still be sufficiently larger than the useful frequency range of the plant (see Figure 3.14b), and hence the sampled data will accurately represent the plant response. But, the bandwidth criterion presented earlier in this section satisfies $\Delta T_p = 0.5 < \Delta T_c$. This is a more reasonable option. For example, in Figure 3.14c, the two previous measurement samples are used in computing each control action, and the control computation takes twice the data sampling period. Here, the data sampling period is half the control cycle period, and for a specified control action frequency the Nyquist frequency of the sampled feedback signals is double that of the previous case. As a result the sampled data will cover a larger (double) frequency range of the plant. Of course, a third option would be to still use the two previous data samples

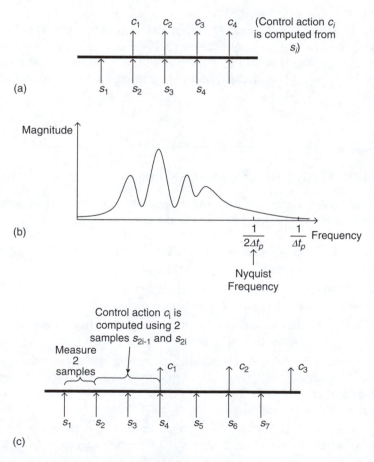

FIGURE 3.14
(a) Conventional sampling of feedback sensor signals for direct digital control. (b) Acceptable frequency characteristic of a plant for case (a). (c) Improved sampling criterion for feedback signals in direct digital control.

to compute a control action, but do the computation faster, in one data sample period (rather than two). This option will require increased processing power as well as larger buffer for storing sampled data for control action computation.

3.9 Instrument Error Analysis

Analysis of error in an instrument or a multicomponent control system is a very challenging task. Difficulties arise for many reasons, particularly the following:

1. True value is usually unknown.
2. The instrument reading may contain random error that cannot be determined exactly.
3. The error may be a complex (i.e., not simple) function of many variables (input variables and state variables or response variables).
4. The control system and instrument may be made up of many components that have complex interrelations (dynamic coupling, multiple degree-of-freedom responses, nonlinearities, etc.), and each component may contribute to the overall error.

The first item is a philosophical issue that would lead to an argument similar to the chicken-and-egg controversy. For instance, if the true value is known, there is no need to measure it; and if the true value is unknown, it is impossible to determine exactly how inaccurate a particular reading is. In fact, this situation can be addressed to some extent by using statistical representation of error, which takes us to the second item listed. The third and fourth items may be addressed by error combination in multivariable systems and by error propagation in complex multicomponent systems. It is not feasible here to provide a full treatment of all these topics. Only an introduction to a useful analytical technique is given, using illustrative examples. The concepts discussed here are applicable not only in statistical error analysis but also in the field of *statistical process control* (SPC) – the use of statistical signals to improve performance of a process. Performing statistical analysis of a response signal and drawing its *control chart*, along with an *upper control line* and a *lower control line*, are key steps in SPC.

3.9.1 Statistical Representation

In general, error in an instrument reading is a random variable. It is defined as:

$$\text{Error} = (\text{instrument reading}) - (\text{true value}).$$

Randomness associated with a measurand (the quantity to be measured) can be interpreted in two ways. First, since the true value of the measurand is a fixed quantity, randomness can be interpreted as the randomness in error that is usually originating from the random factors in instrument response. Second, looking at the issue in a more practical manner, error analysis can be interpreted as an estimation problem in which the objective is to estimate the true value of a measurand from a known set of readings. In this latter point of view, estimated true value itself becomes a random variable. No matter what approach is used, however, the same statistical concepts may be used in representing error.

3.9.2 Accuracy and Precision

The instrument ratings as mentioned before, affect the overall *accuracy* of an instrument. Accuracy can be assigned either to a particular reading or to an instrument. Note that instrument accuracy depends not only on the physical hardware of the instrument but also on the operating conditions (e.g., design conditions, which are the normal, steady operating conditions; or extreme transient conditions, such as emergency start-up and shutdown conditions). *Measurement accuracy* determines the closeness of the measured value (measurement) to true value conditions (measurand). *Instrument accuracy* is related to the worst accuracy obtainable within the dynamic range of the instrument in a specific operating environment. *Measurement error* is defined as

$$\text{error} = (\text{measured value}) - (\text{true value}) \tag{3.16}$$

Correction, which is the negative of error, is defined as

$$\text{correction} = (\text{true value}) - (\text{measured value}). \tag{3.17}$$

Each of these can also be expressed as a percentage of the true values. Accuracy of an instrument may be determined by measuring a parameter whose true value is known, near the extremes of the dynamic range of instrument, under certain operating conditions. For this purpose, standard parameters or signals than can be generated at very high levels of accuracy would be needed. The National Institute for Standards and Testing (NIST) or National Research Council (NRC) is usually responsible for generation of these standards. Nevertheless, accuracy and error values cannot be determined to 100% exactness in typical applications, because the true value is not known to begin with. In a given situation, we can only make estimates for accuracy, by using ratings provided by the instrument manufacturer or by analyzing data from previous measurements and models.

Causes of error include instrument instability, external noise (disturbances), poor calibration, inaccurate information (e.g., poor analytical models, nonlinearity effects on linear models, inaccurate control laws and digital control algorithms), parameter changes (e.g., as a result of environmental changes, aging, and wear out), unknown nonlinearities, and improper use of instrument.

Errors can be classified as *deterministic* (or *systematic*) and *random* (or *stochastic*). Deterministic errors are those caused by well-defined factors, including known nonlinearities and offsets in readings. These usually can be accounted for by proper calibration and analysis practices. Error ratings and calibration charts are used to remove systematic errors from instrument readings. Random errors are caused by uncertain factors entering into instrument response. These include device noise, line noise, and effects of unknown random variations in the operating environment. A statistical analysis using sufficiently large amounts of data is necessary to estimate random errors. The results are usually expressed as a mean error, which is the systematic part of random error, and a standard deviation or confidence interval for instrument response. *Precision* is not synonymous with accuracy. Reproducibility (or repeatability) of an instrument reading determines the precision of an instrument. An instrument that has a high offset error might be able to generate a response at high precision, even though this output is clearly inaccurate. For example, consider a timing device (clock) that very accurately indicates time increments (say, up to the nearest nanosecond). If the reference time (starting time) is set incorrectly, the time readings will be in error, even though the clock has a very high precision.

Instrument error may be represented by a random variable that has a mean value μ_e and a standard deviation σ_e. If the standard deviation is zero, the variable is considered deterministic, for most practical purposes. In that case, the error is said to be deterministic or repeatable. Otherwise, the error is said to be random. The precision of an instrument is determined by the standard deviation of error in the instrument response. Readings of an instrument may have a large mean value of error (e.g., large offset), but if the standard deviation is small, the instrument has high precision. Hence, a quantitative definition for precision would be:

$$\text{Precision} = (\text{measurement range})/\sigma_e. \tag{3.18}$$

Lack of precision originates from random causes and poor construction practices. It cannot be compensated for by recalibration, just as precision of a clock cannot be improved by resetting the time. On the other hand, accuracy can be improved by recalibration. Repeatable (deterministic) accuracy is inversely proportional to the magnitude of the mean error μ_e.

Matching instrument ratings with specifications is very important in selecting instruments for a control application. Several additional considerations should be looked into as well. These include geometric limitations (size, shape, etc.), environmental conditions (e.g., chemical reactions including corrosion, extreme temperatures, light, dirt accumulation, humidity, electromagnetic fields, radioactive environments, shock and vibration), power requirements, operational simplicity, availability, past record and reputation of the manufacturer and of the particular instrument, and cost-related economic aspects (initial cost, maintenance cost, cost of supplementary components such as signal-conditioning and processing devices, design life and associated frequency of replacement, and cost of disposal and replacement). Often, these considerations become the ultimate deciding factors in the selection process.

3.9.3 Error Combination

Error in a response variable of a device or in an estimated parameter of a system would depend on errors present in measured variables and parameter values that are used to determine the unknown variable or parameter. Knowing how component errors are propagated within a multicomponent system and how individual errors in system variables and parameters contribute toward the overall error in a particular response variable or parameter would be important in estimating error limits in complex control systems. For example, if the output power of a rotational manipulator is computed by measuring torque and speed at the output shaft, error margins in the two measured response variables (torque and speed) would be directly combined into the error in the power computation. Similarly, if the natural frequency of a simple suspension system is determined by measuring mass and spring stiffness parameters of the suspension, the natural frequency estimate would be directly affected by possible errors in mass and stiffness measurements. Extending this idea further, the overall error in a control system depends on individual error levels in various components (sensors, actuators, controller hardware, filters, amplifiers, data acquisition devices, etc.) of the system and in the manner in which these components are physically interconnected and physically interrelated. For example, in a robotic manipulator, the accuracy of the actual trajectory of the end effector will depend on the accuracy of sensors and actuators at the manipulator joints and on the accuracy of the robot controller. Note that we are dealing with a generalized idea of error propagation that considers errors in system variables (e.g., input and output signals, such as velocities, forces, voltages, currents, temperatures, heat transfer rates, pressures and fluid flow rates), system parameters (e.g., mass, stiffness, damping, capacitance, inductance, resistance, thermal conductivity, and viscosity), and system components (e.g., sensors, actuators, filters, amplifiers, interface hardware, control circuits, thermal conductors, and valves).

For the analytical development of a basic result in error combination, we will start with a functional relationship of the form

$$y = f(x_1, x_2, \ldots, x_r). \tag{3.19}$$

Here, x_i are the independent system variables or parameter values whose error is propagated into a dependent variable (or parameter value) y. Determination of this functional relationship is not always simple, and the relationship itself may be in error. Since our intention is to make a reasonable estimate for possible error in y because of the combined effect of errors from x_i, an approximate functional relationship would be adequate in most cases. Let us denote error in a variable by the differential of that variable. Taking the differential of Equation 3.19, we get

$$\delta y = \frac{\partial f}{\partial x_1} \delta x_1 + \frac{\partial f}{\partial x_2} \delta x_2 + \ldots + \frac{\partial f}{\partial x_r} \delta x_r \tag{3.20}$$

for small errors. For those who are not familiar with differential calculus, Equation 3.20 should be interpreted as the first-order terms in a *Taylor series expansion* of Equation 3.19. The partial derivatives are evaluated at the operating conditions under which the error assessment is carried out. Now, rewriting Equation 3.20 in the fractional form, we get

$$\frac{\delta y}{y} = \sum_{i=1}^{r} \left[\frac{x_i}{y} \frac{\partial f}{\partial x_i} \frac{\delta x_i}{x_i} \right]. \tag{3.21}$$

Here, $\delta y / y$ represents the overall error and $\delta x_i / x_i$ represents the component error, expressed as fractions. We shall consider two types of estimates for overall error.

3.9.3.1 Absolute Error

Since error δx_i could be either positive or negative, an upper bound for the overall error is obtained by summing the absolute value of each RHS term in Equation 3.21. This estimate e_{ABS}, which is termed *absolute error*, is given by

$$e_{ABS} = \sum_{i=1}^{r} \left| \frac{x_i}{y} \frac{\partial f}{\partial x_i} \right| e_i. \tag{3.22}$$

Note that component error e_i and absolute error e_{ABS} in Equation 3.22 are always positive quantities; when specifying error, however, both positive and negative limits should be indicated or implied (e.g., $\pm e_{ABS}$, $\pm e_i$).

3.9.3.2 SRSS Error

Equation 3.22 provides a conservative (upper bound) estimate for overall error. Since the estimate itself is not precise, it is often wasteful to introduce such a high conservatism. A nonconservative error estimate that is frequently used in practice is the square root of sum of squares (SRSS) error. As the name implies, this is given by

$$e_{SRSS} = \left[\sum_{i=1}^{r} \left(\frac{x_i}{y} \frac{\partial f}{\partial x_i} e_i \right)^2 \right]^{1/2}. \tag{3.23}$$

This is not an upper bound estimate for error. In particular, $e_{SRSS} < e_{ABS}$ when more than one nonzero error contribution is present. The SRSS error relation is particularly suitable when component error is represented by the standard deviation of the associated variable or parameter value and when the corresponding error sources are independent. Now we present several examples of error combination.

Example 3.11
Using the absolute value method for error combination, determine the fractional error in each item x_i so that the contribution from each item to the overall error e_{ABS} is the same.

Solution
For equal contribution, we must have

$$\left|\frac{x_1}{y}\frac{\partial f}{\partial x_1}\right|e_1 = \left|\frac{x_2}{y}\frac{\partial f}{\partial x_2}\right|e_2 = \ldots = \left|\frac{x_r}{y}\frac{\partial f}{\partial x_r}\right|e_r.$$

Hence,

$$r\left|\frac{x_i}{y}\frac{\partial f}{\partial x_i}\right|e_i = e_{ABS}.$$

Thus,

$$e_i = e_{ABS}\bigg/\left(r\left|\frac{x_i}{y}\frac{\partial f}{\partial x_i}\right|\right). \tag{3.24}$$

Example 3.12
The result obtained in the previous example is useful in the design of multicomponent systems and in the cost-effective selection of instrumentation for a particular application. Using Equation 3.24, arrange the items x_i in their order of significance.

Solution
Note that Equation 3.24 may be written as

$$e_i = K\bigg/\left|x_i\frac{\partial f}{\partial x_i}\right|, \tag{3.25}$$

where K is a quantity that does not vary with x_i. It follows that for equal error contribution from all items, error in x_i should be inversely proportional to $|x_i(\partial f/\partial x_i)|$. In particular, the item with the largest $|x(\partial f/\partial x)|$ should be made most accurate. In this manner, allowable relative accuracy for various components can be estimated. Since, in general, the most accurate device is also the costliest, instrumentation cost can be optimized if components are selected according to the required overall accuracy, using a criterion such as that implied by Equation 3.25.

Example 3.13
Figure 3.15 schematically shows an optical device for measuring displacement. This sensor is essentially an optical potentiometer (see Chapter 4). The potentiometer element is uniform and has a resistance R_c. A photoresistive layer is sandwiched between this element and a perfect conductor of electricity. A light source that moves with the object whose displacement is measured, directs a beam of light whose intensity is I, on to a

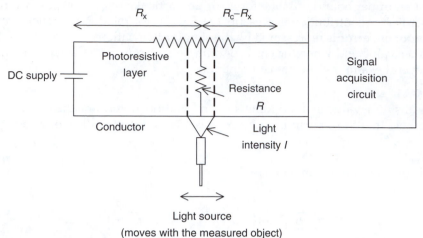

FIGURE 3.15
An optical displacement sensor.

narrow rectangular region of the photoresistive layer. As a result, this region becomes resistive with resistance R, which bridges the potentiometer element and the conductor element, as shown. An empirical relation between R and I was found to be

$$\ln\left(\frac{R}{R_o}\right) = \left(\frac{I_o}{I}\right)^{1/4},$$

where the resistance R is in kΩ and the light intensity I is expressed in watts per square meter (W/m^2). The parameters R_o and I_o are empirical constants having the same units as R and I, respectively. These two parameters generally have some experimental error.

a. Sketch the curve of R vs. I and explain the significance of the parameters R_o and I_o.

b. Using the absolute error method, show that the combined fractional error e_R in the bridging resistance R can be expressed as

$$e_R = e_{R_o} + \frac{1}{4}\left(\frac{I_o}{I}\right)^{1/4}[e_I + e_{I_o}],$$

where e_{R_o}, e_I, and e_{I_o} are the fractional errors in R_o, I, and I_o, respectively.

c. Suppose that the empirical error in the sensor model can be expressed as $e_{R_o} = \pm0.01$ and $e_{I_o} = \pm0.01$, and due to variations in the power supply to the light source and in ambient lighting conditions, the fractional error in I is also ±0.01. If the error E_R is to be maintained within ±0.02, at what light intensity level (I) should the light source operate? Assume that the empirical value of I_o is 2.0 W/m^2.

d. Discuss the advantages and disadvantages of this device as a dynamic displacement sensor.

Solution

a.
$$\ln\frac{R}{R_o} = \left(\frac{I_o}{I}\right)^{1/4}$$

R_o represents the minimum resistance provided by the photoresistive bridge (i.e., at very high light intensity levels). When $I = I_o$, the bridge resistance R is calculated to be about 2.7 R_o, and hence I_o represents a lower bound for the intensity for proper operation of the sensor. A suitable upper bound for the intensity would be 10 I_o, for satisfactory operation. At this value, it can be computed that $R \simeq 1.75\ R_o$, and is shown in Figure 3.16.

b.
$$\ln R - \ln R_o = \left(\frac{I_o}{I}\right)^{1/4}.$$

Differentiate,

$$\frac{\delta R}{R} - \frac{\delta R_o}{R_o} = \frac{1}{4}\left(\frac{I_o}{I}\right)^{-3/4}\left[\frac{\delta I_o}{I} - \frac{I_o}{I^2}\delta I\right]$$

$$= \frac{1}{4}\left(\frac{I_o}{I}\right)^{1/4}\left[\frac{\delta I_o}{I_o} - \frac{\delta I}{I}\right].$$

Hence, with the absolute method of error combination,

$$e_R = e_{R_o} + \frac{1}{4}\left(\frac{I_o}{I}\right)^{1/4}[e_{I_o} + e_I].$$

Note the use of the "+" sign instead of "−" because of the "absolute" method of error combination (i.e., positive magnitudes are used regardless of the actual algebraic sign)

c. With the given numerical values, we have

$$0.02 = 0.01 + \frac{1}{4}\left(\frac{I_o}{I}\right)^{1/4}[0.01 + 0.01]$$

$$\Rightarrow \left(\frac{I_o}{I}\right)^{1/4} = 2$$

or, $I = \dfrac{1}{16}I_o = \dfrac{2.0}{16}\,\mathrm{W/m^2} = 0.125\,\mathrm{W/m^2}$

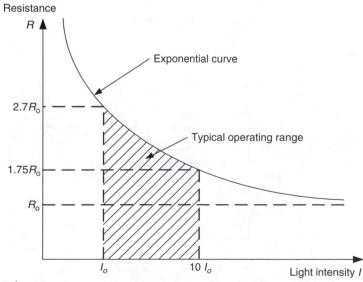

FIGURE 3.16
Characteristic curve of the sensor.

Note: For larger values of I the absolute error in R_o would be smaller. For example, for
$$I = 10 \, I_o \text{ we have,}$$
$$e_R = 0.01 + \frac{1}{4}\left(\frac{1}{10}\right)^{1/4}[0.01 + 0.01] \simeq 0.013.$$

d. *Advantages*

- Noncontacting
- Small moving mass (low inertial loading)
- All advantages of a potentiometer (see Chapter 4).

e. *Disadvantages*

- Nonlinear and exponential variation of R
- Effect of ambient lighting
- Possible nonlinear behavior of the device (nonlinear input–output relation)
- Effect of variations in the supply to the light source
- Effect of aging of the light source.

Example 3.14

a. You are required to select a sensor for a position control application. List several important considerations that you have to take into account in this selection. Briefly indicate why each of them is important.

b. A schematic diagram of a chip refiner that is used in the pulp and paper industry is shown in Figure 3.17. This machine is used for mechanical pulping of wood chips. The refiner has one fixed disk and one rotating disk (typical diameter = 2 m). The plate is rotated by an ac induction motor. The plate

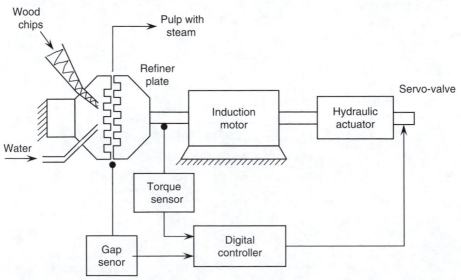

FIGURE 3.17
A single-disk chip refiner.

separation (typical gap $= 0.5$ mm) is controlled using a hydraulic actuator (piston-cylinder unit with servovalve; see Chapter 7). Wood chips are supplied to the eye of the refiner by a screw conveyor and are diluted with water. As the refiner plate rotates, the chips are ground into pulp within the internal grooves of the plates. This is accompanied by the generation of steam due to energy dissipation. The pulp is drawn and further processed for making paper.

An empirical formula relating the plate gap (h) and the motor torque (T) is given by

$$T = \frac{ah}{(1 + bh^2)}$$

with the model parameters a and b known to be positive.

i. Sketch the curve T vs. h. Express the maximum torque $T_{\max}$ and the plate gap (h_0) at this torque in terms of a and b only.

ii. Suppose that the motor torque is measured and the plate gap is adjusted by the hydraulic actuator according to the formula given previously. Show that the fractional error in h may be expressed as

$$e_h = \left[e_T + e_a + \frac{bh^2}{(1 + bh^2)} e_b \right] \frac{(1 + bh^2)}{|1 - bh^2|}$$

where e_T, e_a, and e_b are the fractional errors in T, a, and b, respectively, the latter two representing model error.

iii. The normal operating region of the refiner corresponds to $h > h_0$. The interval $0 < h < h_0$ is known as the "pad collapse region" and should be avoided. If the operating value of the plate gap is $h = 2/\sqrt{b}$ and if the error values are given as $e_T = \pm 0.05$, $e_a = \pm 0.02$, and $e_b = \pm 0.025$, compute the corresponding error in the plate gap estimate.

iv. Discuss why operation at $h = 1/\sqrt{b}$ is not desirable.

Solution

a. *Bandwidth*: Determines the useful (or flat) frequency range of operation, and also the speed of response (i.e., how quickly the plant reacts to a control action).

Accuracy: Low random error or high precision, low systematic error or reduced need for recalibration.

Dynamic Range: Range of operating amplitude.

Resolution: For a sensor, this determines the smallest signal change that could be correctly measured. In general, this is the smallest signal increment that is meaningful and representable.

Input Impedance: The impedance experienced by the input signal (the measurand, in the case of a sensor). High value means low loading error or distortion of the measurement.

Output Impedance: The impedance as present at the instrument output and experienced by the output signal (i.e., measurement, in the case of a sensor). Any device connected to the ouput of the instrument (e.g., signal conditioner) will experience this impedance at its input. Low value means high output level and low distortion due to subsequent signal-conditioning.

Linearity: Proportionality of the output with respect to the input. This is a measure of the ease of recalibration of the instrument (sensor).

Sensitivity: High value means high output for a given input.

Drift: Specifies output stability of the instrument under steady operating conditions.

Size: Smaller size typically means higher resolution and bandwidth, lower mechanical loading, and reduced space requirements.

Cost: The cheaper the better.

b. i. See the sketch in Figure 3.18.

$$T = \frac{ah}{1 + bh^2} \tag{i}$$

Differentiate with respect to h

$$\frac{\partial T}{\partial h} = \frac{(1 + bh^2)a - ah(2bh)}{(1 + bh^2)^2} = 0 \text{ at maximum } T.$$

Hence,

$$1 - bh^2 = 0$$

or,

$$h_o = 1/\sqrt{b}$$

Substitute in (i)

$$T_{max} = \frac{a}{2\sqrt{b}}$$

ii. The differential relation of Equation (i) is obtained by taking the differential of each term (i.e., slope times the increment). Thus,

$$\delta T = \frac{h}{(1 + bh^2)} \delta a + \frac{\partial T}{\partial h} \delta h - \frac{ah}{(1 + bh^2)^2} h^2 \delta b.$$

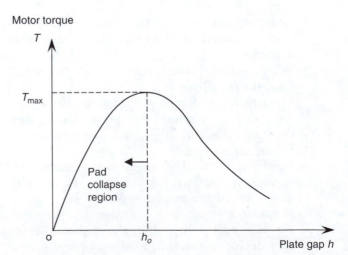

FIGURE 3.18
Characteristic curve of the chip refiner.

Substitute for $\partial T/\partial h$ from Part (i)

$$\delta T = \frac{h}{(1+bh^2)}\delta a + a\frac{(1-bh^2)}{(1+bh^2)^2}\delta h - \frac{ah^3}{(1+bh^2)^2}\delta b.$$

Divide throughout by Equation (i)

$$\frac{\delta T}{T} = \frac{\delta a}{a} + \left[\frac{1-bh^2}{1+bh^2}\right]\frac{\delta h}{h} - \frac{bh^2}{(1+bh^2)}\frac{\delta b}{b}$$

Or,

$$\frac{\delta h}{h} = \left[\frac{\delta T}{T} - \frac{\delta a}{a} + \frac{bh^2}{(1+bh^2)}\frac{\delta b}{b}\right]\left[\frac{1+bh^2}{1-bh^2}\right].$$

Now representing fractional errors by the fractional deviations (differentials), and using the absolute value method of error combination, we have

$$e_h = \left[e_T + e_a + \frac{bh^2}{(1+bh^2)}e_b\right]\frac{(1+bh^2)}{|1-bh^2|}. \tag{ii}$$

Note: The absolute values of the error terms are added. Hence, the minus sign in a term has become plus.

iii. With $h = 2/\sqrt{b}$ we have $bh^2 = 4$.

Substitute the given numerical values for fractional error in Equation (ii).

$$e_h = \left[0.05 + 0.02 + \frac{4}{5}\times 0.025\right]\frac{(1+4)}{|1-4|}$$
$$= \pm 0.15.$$

iv. When $h = 1/\sqrt{b}$ we see from Equation (ii) that $e_h \to \infty$. In addition, from the curve in Part (i) we see that at this point the motor torque is not sensitive to changes in plate gap. Hence, operation at this point is not appropriate, and should be avoided.

3.10 Statistical Process Control

In statistical process control (SPC), statistical analysis of process responses is used to generate control actions. This method of control is applicable in many situations of process control, including manufacturing quality control, control of chemical process plants, computerized office management systems, inventory control systems, and urban transit control systems. A major step in SPC is to compute control limits (or action lines) on the basis of measured data from the process.

3.10.1   Control Limits or Action Lines

Because a very high percentage of readings from an instrument should lie within $\pm 3\sigma$ about the mean value, according to the normal distribution (Gaussian distribution), these boundaries (-3σ and $+3\sigma$) drawn about the mean value may be considered *control limits* or *action lines* in SPC. If any measurements fall outside the action lines, corrective measures such as recalibration, controller adjustment, or redesign should be carried out.

3.10.2   Steps of SPC

The main steps of SPC are as follows:

1. Collect measurements of appropriate response variables of the process.
2. Compute the mean value of the data, the upper control limit, and the lower control limit.
3. Plot the measured data and draw the two control limits on a control chart.
4. During operation of the process, if the on-line measurements fall outside the control limits, take corrective action and repeat the control cycle.
5. When necessary (e.g., when the process performance under control is unsatisfactory) go to Step 1 and update the control limits.

If the measurements always fall within the control limits, the process is said to be in statistical control.

Example 3.15

Error in a satellite tracking system was monitored online for a period of 1 h to determine whether recalibration or gain adjustment of the tracking controller would be necessary. Four measurements of the tracking deviation were taken in a period of 5 min, and 12 such data groups were acquired during the 1 h period. Sample means and sample variance of the 12 groups of data were computed. The results are tabulated as follows:

Period i	1	2	3	4	5	6	7	8	9	10	11	12
Sample Mean $\overline{X}_i$	1.34	1.10	1.20	1.15	1.30	1.12	1.26	1.10	1.15	1.32	1.35	1.18
Sample Variance S_i^2	0.11	0.02	0.08	0.10	0.09	0.02	0.06	0.05	0.08	0.12	0.03	0.07

Draw a control chart for the error process, with control limits (action lines) at $\overline{X} \pm 3\sigma$. Establish whether the tracking controller is in statistical control or needs adjustment.

Solution

The overall mean tracking deviation,

$$\overline{X} = \frac{1}{12} \sum_{i=1}^{12} \overline{X}_i$$

is computed to be $\overline{X} = 1.214$. The average sample variance,

$$\overline{S}^2 = \frac{1}{12} \sum_{I=1}^{12} S_i^2$$

is computed to be $\overline{S}^2 = 0.069$. Since there are four readings within each period, the standard deviation σ of group mean $\overline{X}_i$ can be estimated as

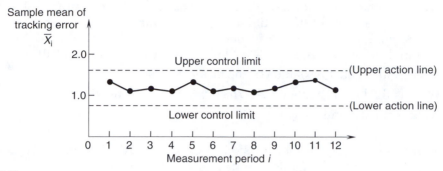

FIGURE 3.19
Control chart for the satellite tracking error example.

$$S = \frac{\overline{S}}{\sqrt{4}} = \frac{\sqrt{0.069}}{\sqrt{4}} = 0.131.$$

The upper control limit (action line) is at (approximately)

$$x = \overline{X} + 3S = 1.214 + 3 \times 0.131 = 1.607.$$

The lower control limit (action line) is at

$$x = \overline{X} - 3S = 0.821.$$

These two lines are shown on the control chart in Figure 3.19. Since the sample means lie within the two action lines, the process is considered to be in statistical control, and controller adjustments would not be necessary. Note that if better resolution is required in making this decision, individual readings, rather than group means should be plotted in Figure 3.19.

Problems

3.1 What do you consider a perfect measuring device? Suppose that you are asked to develop an analog device for measuring angular position in an application related to control of a kinematic linkage system (a robotic manipulator, for example). What instrument ratings (or specifications) would you consider crucial in this application? Discuss their significance.

3.2 List and explain some time–domain parameters and frequency–domain parameters that can be used to predominantly represent

a. Speed of response

b. Degree of stability of a control system. In addition, briefly discuss any conflicts that can arise in specifying these parameters.

3.3 A tactile (distributed touch) sensor (see Chapter 4) of the gripper of a robotic manipulator consists of a matrix of piezoelectric sensor elements placed 2 mm apart. Each element generates an electric charge when it is strained by an external load. Sensor elements are multiplexed at very high speed in order to avoid charge leakage and to read all data channels using a single high-performance charge amplifier. Load distribution on the surface of the tactile sensor is determined from the charge amplifier readings, since the multiplexing sequence is known. Each sensor element can read a maximum load of 50 N and can detect load changes in the order of 0.01 N.

a. What is the spatial resolution of the tactile sensor?

b. What is the load resolution (in N/m^2) of the tactile sensor?

c. What is the dynamic range?

3.4 A useful rating parameter for a robotic tool is *dexterity*. Though not complete, an appropriate analytical definition for dexterity of a device is

$$\text{dexterity} = \frac{\text{number of degrees of freedom}}{\text{motion resolution}},$$

where the number of degrees of freedom is equal to the number of independent variables that is required to completely define an arbitrary position increment of the tool (i.e., for an arbitrary change in its kinematic configuration).

a. Explain the physical significance of dexterity and give an example of a device for which the specification of dexterity would be very important.

b. The power rating of a tool may be defined as the product of maximum force that can be applied by it in a controlled manner and the corresponding maximum speed. Discuss why the power rating of a manipulating device is usually related to the dexterity of the device. Sketch a typical curve of power vs. dexterity.

3.5 Resolution of a feedback sensor (or resolution of a response measurement used in feedback) has a direct effect on the accuracy that is achievable in a control system. This is true because the controller cannot correct a deviation of the response from the desired value (set point) unless the response sensor can detect that change. It follows that the resolution of a feedback sensor will govern the minimum (best) possible deviation band (about the desired value) of the system response, under feedback control. An angular position servo uses a resolver as its feedback sensor. If peak-to-peak oscillations of the servo load (plant) under steady-state conditions have to be limited to no more than two degrees, what is the worst tolerable resolution of the resolver? Note that, in practice, the feedback sensor should have a resolution better (smaller) than this worst value.

3.6 Consider a simple mechanical dynamic device (single degree of freedom) that has low damping. An approximate design relationship between the two performance parameters T_r and f_b may be given as

$$T_r f_b = k,$$

where T_r is rise time in nanoseconds (ns) and f_b is the bandwidth in megahertz (MHz). Estimate a suitable value for k.

3.7 List several response characteristics of nonlinear control systems that are not exhibited by linear control systems in general. Additionally, determine the response y of the nonlinear system

$$\left[\frac{dy}{dt}\right]^{1/3} = u(t)$$

when excited by the input $u(t) = a_1 \sin \omega_1 t + a_2 \sin \omega_2 t$. What characteristic of a nonlinear system does this result illustrate?

3.8 Consider a mechanical component whose response x is governed by the relationship

$$f = f(x, \dot{x}),$$

where f denotes applied (input) force and $\dot{x}$ denotes velocity. Three special cases are

a. Linear spring:

$$f = kx$$

b. Linear spring with a viscous (linear) damper:

$$f = kx + b\dot{x}$$

c. Linear spring with coulomb friction:

$$f = kx + f_c \sin(\dot{x}).$$

Suppose that a harmonic excitation of the form $f = f_o \sin \omega t$ is applied in each case. Sketch the force-displacement curves for the three cases at steady state. Which of the three components exhibit hysteresis? Which components are nonlinear? Discuss your answers.

3.9 Discuss how the accuracy of a digital controller may be affected by

a. Stability and bandwidth of amplifier circuitry

b. Load impedance of the analog-to-digital conversion (ADC) circuitry. Moreover, what methods do you suggest to minimize problems associated with these parameters?

3.10 a. Sketch (not to scale) the magnitude vs. frequency curves of the following two transfer functions.

   i. $G_i(s) = \dfrac{1}{\tau_i s + 1}$

   ii. $G_d(s) = \dfrac{1}{1 + \dfrac{1}{\tau_d s}}$.

   Explain why these two transfer fractions may be used as an integrator, a low-pass filter, a differentiator, and a high-pass filter. In your magnitude vs. frequency curves, indicate in which frequency bands these four respective realizations are feasible. You may make appropriate assumptions for the time-constant parameters τ_i and τ_d.

b. Active vibration isolators, known as electronic mounts, have been considered for sophisticated automobile engines. The purpose is to actively filter out the cyclic excitation forces generated by the internal-combustion engines before they would adversely vibrate the components such as seats, floor, and steering column, which come into contact with the vehicle occupants. Consider a four-stroke, four-cylinder engine. It is known that the excitation frequency on the engine mounts is twice the crank-shaft speed, as a result of the firing cycles of the cylinders. A schematic representation of an active engine mount is shown in Figure P3.10a. The crank-shaft speed is measured and supplied to the controller of a valve actuator. The servovalve of a hydraulic cylinder is operated on the basis of this measurement. The hydraulic cylinder functions as an active

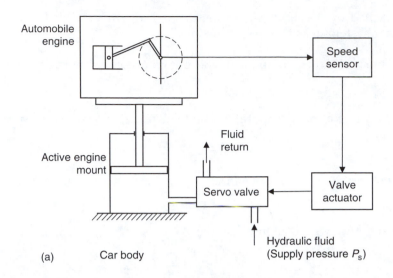

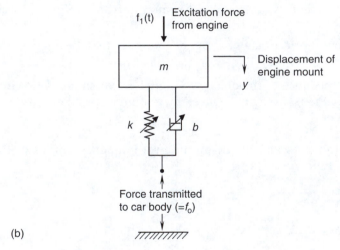

FIGURE P3.10
An active engine mount for an automobile. (a) Schematic diagram. (b) Approximate model.

suspension with a variable (active) spring and a damper. A simplified model of the mechanical interactions is shown in Figure P3.10b.

i. Neglecting gravity forces (which cancel out because of the static spring force) show that a linear model for system dynamics may be expressed as

$$m\ddot{y} + b\dot{y} + ky = f_i$$

$$b\dot{y} + ky - f_o = 0$$

where,

f_i = excitation force from the engine

f_o = force transmitted to the passenger compartment

y = displacement of the engine mount with respect to a frame fixed to the passenger compartment

m = mass of the engine unit

k = equivalent stiffness of the active mount

b = equivalent viscous damping constant of the active mount.

ii. Determine the transfer function (with the Laplace variable s) f_o/f_i for the system.

iii. Sketch the magnitude vs. frequency curve of the transfer function obtained in Part (ii) and show a suitable operating range for the active mount.

iv. For a damping ratio $\zeta = 0.2$, what is the magnitude of the transfer function when the excitation frequency ω is 5 times the natural frequency ω_n of the suspension (engine mount) system?

v. Suppose that the magnitude estimated in Part (iv) is satisfactory for the purpose of vibration isolation. If the engine speed varies from 600 to 1200 rpm, what is the range in which the spring stiffness k (N/m) should be varied by the control system in order to maintain this level of vibration isolation? Assume that the engine mass $m = 100$ kg and the damping ratio is approximately constant at $\zeta = 0.2$.

3.11 Consider the mechanical tachometer shown in Figure 3.8. Write expressions for sensitivity and bandwidth of the device. Using the example, show that the two performance ratings, sensitivity and bandwidth, generally conflict. Discuss ways to improve the sensitivity of this mechanical tachometer.

3.12 a. What is an antialiasing filter? In a particular application, the sensor signal is sampled at f_s Hz. Suggest a suitable cutoff frequency for an antialiasing filter to be used in this application.

3.13 a. Consider a multi-degree-of-freedom robotic arm with flexible joints and links. The purpose of the manipulator is to accurately place a payload. Suppose that the second natural frequency (i.e., the natural frequency of the second flexible mode) of bending of the robot, in the plane of its motion, is more than four times the first natural frequency. Discuss pertinent issues of sensing and control (e.g., types and locations of the sensors, types of control, operating bandwidth, control bandwidth, sampling rate of sensing information) if the primary frequency of the payload motion is:

   i. One-tenth of the first natural frequency of the robot

   ii. Very close to the first natural frequency of the robot

   iii. Twice the first natural frequency of the robot

   b. A single-link space robot is shown in Figure P3.13. The link is assumed to be uniform with length 10 m and mass 400 kg. The total mass of the end effector and the payload is also 400 kg. The robot link is assumed to be flexible, although the other components are rigid. The modulus of rigidity of bending deflection of the link in the plane of robot motion is known to be $EI = 8.25 \times 10^9$ N $\cdot$ m^2. The primary natural frequency of bending motion of a uniform cantilever beam with an end mass is given by

$$\omega_1 = \lambda_1^2 \sqrt{\frac{EI}{m}},$$

where m = mass per unit length, λ_1 = mode shape parameter for mode 1.

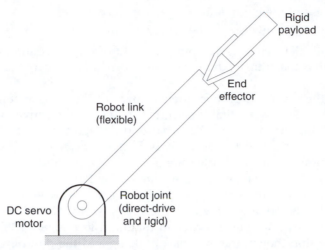

FIGURE P3.13
A single-link robotic manipulator.

For [beam mass/end mass] $= 1.0$, it is known that $\lambda_1 l = 1.875$, where $l =$ beam length. Give a suitable operating bandwidth for the robot manipulation. Estimate a suitable sampling rate for response measurements to be used in feedback control. What is the corresponding control bandwidth, assuming that the actuator and the signal-conditioning hardware can accommodate this bandwidth?

3.14   a. Define the following terms:

- Sensor
- Transducer
- Actuator
- Controller
- Control system
- Operating bandwidth of a control system
- Control bandwidth
- Nyquist frequency

b. Choose three practical dynamic systems each of which has at least one sensor, one actuator, and a feedback controller.

i. Briefly describe the purpose and operation of each dynamic system.

ii. For each system give a suitable value for the operating bandwidth, control bandwidth, operating frequency range of the sensor, and sampling rate for sensor signal for feedback control. Clearly justify the values that you have given.

3.15   Discuss and the contrast the following terms:

a. Measurement accuracy

b. Instrument accuracy

c. Measurement error

d. Precision

In addition, for an analog sensor-transducer unit of your choice, identify and discuss various sources of error and ways to minimize or account for their influence.

3.16 a. Explain why mechanical loading error due to tachometer inertia can be significantly higher when measuring transient speeds than when measuring constant speeds.

   b. A DC tachometer has an equivalent resistance $R_a = 20\ \Omega$ in its rotor windings. In a position plus velocity servo system, the tachometer signal is connected to a feedback control circuit with equivalent resistance 2 kΩ. Estimate the percentage error due to electrical loading of the tachometer at steady state.

   c. If the conditions were not steady, how would the electrical loading be affected in this application?

3.17 Briefly explain what is meant by the terms systematic error and random error of a measuring device. What statistical parameters may be used to quantify these two types of error? State, giving an example, how *precision* is related to error.

3.18 Four sets of measurements were taken on the same response variable of a process using four different sensors. The true value of the response was known to be constant. Suppose that the four sets of data are as shown in Figure P3.18a–d. Classify these data sets, and hence the corresponding sensors, with respect to precision and deterministic (repeatable) accuracy.

3.19 The damping constant b of the mounting structure of a machine is determined experimentally. First, the spring stiffness k is determined by applying a static load and measuring the resulting displacement. Further, mass m of the structure is directly measured. Finally, damping ratio ζ is determined using the logarithmic decrement method, by conducting an impact test and measuring the free response of the structure. A model for the structure is shown in Figure P3.19. Show that the damping constant is given by

$$b = 2\zeta\sqrt{km}.$$

If the allowable levels of error in the measurements of k, m, and ζ are $\pm 2\%$, $\pm 1\%$, and $\pm 6\%$, respectively, estimate a percentage absolute error limit for b.

3.20 Using the square-root-of-sum-of-squares (SRRSS) method for error combination determine the fractional error in each component x_i so that the contribution from each component to the overall error e_{SRSS} is the same.

3.21 A single-degree-of-freedom model of a robotic manipulator is shown in Figure P3.21a. The joint motor has rotor inertia J_m. It drives an inertial load that has moment of inertia J_l through a speed reducer of gear ratio 1:r (*Note: $r < 1$*). The control scheme used in this system is the so-called feedforward control (strictly, *computed-torque control*) method. Specifically, the motor torque T_m that is required to accelerate or decelerate the load is computed using a suitable dynamic model and a desired motion trajectory for the manipulator, and the motor windings are excited so as to generate that torque. A typical trajectory would consist of a constant angular acceleration segment followed by a constant angular velocity segment, and finally a constant deceleration segment, as shown in Figure P3.21b.

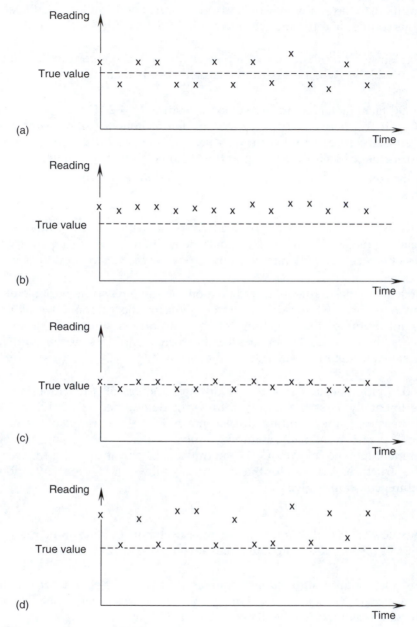

FIGURE P3.18
Four sets of measurements on the same response variable using different sensors.

a. Neglecting friction (particularly bearing friction) and inertia of the speed reducer, show that a dynamic model for torque computation during accelerating and decelerating segments of the motion trajectory would be

$$T_m = (J_m + r^2 J_l)\ddot{\theta}_l/r,$$

where $\ddot{\theta}_l$ is the angular acceleration of the load, hereafter denoted by α_l. Show that the overall system can be modeled as a single inertia rotating at the motor speed. Using this result, discuss the effect of gearing on a mechanical drive.

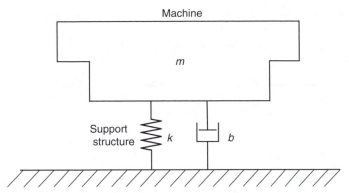

FIGURE P3.19
A model for the mounting structure of a machine.

b. Given that $r = 0.1$, $J_m = 0.1 \text{ kg m}^2$, $J_1 = 1.0 \text{ kg} \cdot \text{m}^2$, and $\alpha_1 = 5.0 \text{ rad}/\text{s}^2$, estimate the allowable error for these four quantities so that the combined error in the computed torque is limited to $\pm 4\%$ and that each of the four quantities contributes equally toward this error in the computed T_m. Use the absolute value method for error combination.

c. Arrange the four quantities r, J_m, J_1 and α_1 in the descending order of required accuracy for the numerical values given in the problem.

d. Suppose that $J_m = r^2 J_1$. Discuss the effect of error in r on the error in T_m.

3.22 An actuator (e.g., electric motor, hydraulic piston-cylinder) is used to drive a terminal device (e.g., gripper, hand, wrist with active remote center compliance)

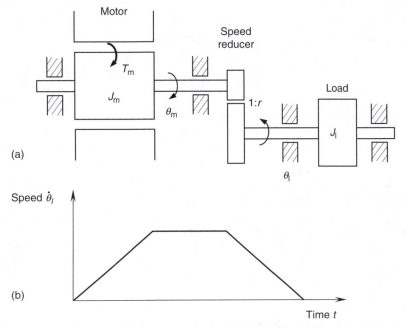

FIGURE P3.21
(a) A single-degree-of-freedom model of a robotic manipulator. (b) A typical reference (desired) speed trajectory for computed-torque control.

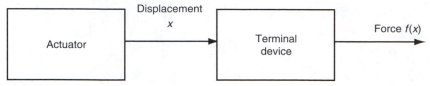

FIGURE P3.22
Block diagram of the terminal device of a robotic manipulator.

of a robotic manipulator. The terminal device functions as a force generator. A schematic diagram for the system is shown in Figure P3.22. Show that the displacement error e_x is related to the force error e_f through

$$e_f = \frac{x}{f} \frac{df}{dx} e_x.$$

The actuator is known to be 100% accurate for practical purposes, but there is an initial position error δx_o (at $x = x_o$). Obtain a suitable transfer relation $f(x)$ for the terminal device so that the force error e_f remains constant throughout the dynamic range of the device.

3.23 a. Clearly explain why the "square-root-of-sum-of-squares" (SRRSS) method of error combination is preferred to the "Absolute" method when the error parameters are assumed Gaussian and independent.

b. Hydraulic pulse generators (HPG) may be used in a variety of applications such as rock blasting, projectile driving, and seismic signal generation. In a typical HPG, water at very high pressure is supplied intermittently from an accumulator into the discharge gun, through a high-speed control valve. The pulsating water jet is discharged through a shock tube and may be used, for example, for blasting granite. A model for an HPG was found to be

$$E = aV\left(b + \frac{c}{V^{1/3}}\right),$$

where E is the hydraulic pulse energy (kJ), V is the volume of blast burden (m³), and, a, b, and c are model parameters whose values may be determined experimentally. Suppose that this model is used to estimate the blast volume of material (V) for a specific amount of pulse energy (E).

i. Assuming that the estimation error values in the model parameters a, b, and c are independent and may be represented by appropriate standard deviations, obtain an equation relating these fractional errors e_a, e_b, and e_c, to the fractional error e_v of the estimated blast volume.

ii. Assuming that $a = 2175.0$, $b = 0.3$, and $c = 0.07$ with consistent units, show that a pulse energy of $E = 219.0$ kJ can blast a material volume of approximately 0.6^3 m³. If $e_a = e_b = e_c = \pm 0.1$, estimate the fractional error e_v of this predicted volume.

3.24 The absolute method of error combination is suitable when the error contributions are additive (same sign). Under what circumstances would the square-root-of-sum-of-squares (SRRSS) method be more appropriate than the absolute method?

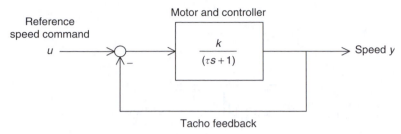

FIGURE P3.24
DC motor speed control system.

A simplified block diagram of a DC motor speed control system is shown in Figure P3.24. Show that in the Laplace domain, the fractional error e_y in the motor speed y is given by

$$e_y = -\frac{\tau s}{(\tau s + 1 + k)} e_\tau + \frac{(\tau s + 1)}{(\tau s + 1 + k)} e_k$$

where

e_τ = fractional error in the time constant τ

e_k = fractional error in the open-loop gain k.

The reference speed command u is assumed error free. Express the absolute error combination relation for this system in the frequency domain ($s = j\omega$). Using it show that

a. at low frequencies the contribution from the error in k will dominate and the error can be reduced by increasing the gain.

b. at high frequencies k and τ will make equal contributions toward the speed error and the error cannot be reduced by increasing the gain.

3.25 a. Compare and contrast the "Absolute Error" method with or against the "square-root-of-sum-of-squares" (SRRSS) method in analyzing error combination of multicomponent systems. Indicate situations where one method is preferred over the other.

  b. Figure P3.25 shows a schematic diagram of a machine that is used to produce steel billets. The molten steel in the vessel (called tundish) is poured into the copper mould having a rectangular cross section. The mould has a steel jacket with channels to carry cooling water upward around the copper mould. The mould, which is properly lubricated, is oscillated using a shaker (electro-mechanical or hydraulic) to facilitate stripping of the solidified steel inside it. A set of power-driven friction rollers is used to provide the withdrawal force for delivering the solidified steel strand to the cutting station. A billet cutter (torch or shear type) is used to cut the strand into billets of appropriate length.

The quality of the steel billets produced by this machine is determined on the basis of several factors, which include various types of cracks, deformation problems such as rhomboidity, and oscillation marks. It is known that the quality can be improved through proper control of the following variables: Q is the coolant (water) flow rate; v is the speed of the steel strand; s is the stroke of the mould oscillations; and f is the cyclic frequency of the mould oscillations. Specifically, these variables are measured and transmitted to the central controller of the billet casting machine, which in turn

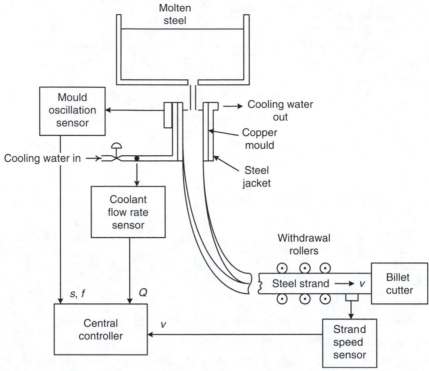

FIGURE P3.25
A steel-billet casting machine.

generates proper control commands for the coolant-valve controller, the drive controller of the withdrawal rollers, and the shaker controller.

A nondimensional quality index q has been expressed in terms of the measured variables, as

$$q = \left[1 + \frac{s}{s_o} \sin\frac{\pi}{2}\left(\frac{f}{f_o + f}\right)\right] \Big/ (1 + \beta v/Q)$$

where s_o, f_o, and β are operating parameters of the control system and are exactly known. Under normal operating conditions, the following conditions are (approximately) satisfied:

$$Q \approx \beta v$$

$$f \approx f_o$$

$$s \approx s_o.$$

Note that if the sensor readings are incorrect, the control system will not function properly, and the quality of the billets will deteriorate. It is proposed to use the "Absolute Error" method to determine the influence of the sensor errors on the billet quality.

  i. Obtain an expression for the quality deterioration δq in terms of the fractional errors $\delta v/v$, $\delta Q/Q$, $\delta s/s$, and $\delta f/f$ of the sensor readings.

 ii. If the sensor of the strand speed is known to have an error of $\pm1\%$, determine the allowable error percentages for the other three sensors so that there is equal contribution of error to the quality index from all four sensors, under normal operating conditions.

3.26 Consider the servo control system that is modeled as in Figure P3.24. Note that k is the equivalent gain and τ is the overall time constant of the motor and its controller.

a. Obtain an expression for the closed-loop transfer function y/u.

b. In the frequency domain, show that for equal contribution of parameter error toward the system response, we should have

$$\frac{e_k}{e_\tau} = \frac{\tau\omega}{\sqrt{\tau^2\omega^2 + 1}},$$

where fractional errors (or variations) are: for the gain,

$$e_k = \left|\frac{\delta k}{k}\right|;$$

and for the time constant,

$$e_\tau = \left|\frac{\delta\tau}{\tau}\right|.$$

Using this relation to explain why at low frequencies the control system has a larger tolerance to error in τ than to that in k. Show that, at very high frequencies the two error tolerance levels are almost equal.

3.27 Tension T at point P in a cable can be computed with the knowledge of the cable sag y, cable length s, cable weight w per unit length, and the minimum tension T_o at point O (see Figure P3.27). The applicable relationship is

$$1 + \frac{w}{T_o}y = \sqrt{1 + \frac{w^2}{T^2}s^2}.$$

For a particular arrangement, it is given that $T_o = 100$ lbf. The following parameter values were measured:

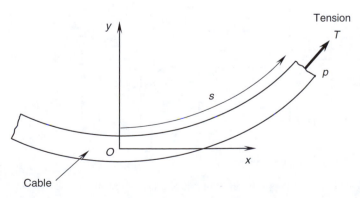

FIGURE P3.27
Cable tension example of error combination.

$$w = 11 \text{ b/ft}, \quad s = 10 \text{ ft}, \quad y = 0.412 \text{ ft}$$

Calculate the tension T.

In addition, if the measurements y and s each have 1% error and the measurement w has 2% error in this example, estimate the percentage error in T. Now suppose that equal contributions to error in T are made by y, s, and w. What are the corresponding percentage error values for y, s, and w so that the overall error in T is equal to the value computed in the previous part of the problem? Which of the three quantities y, s, and w should be measured most accurately, according to the equal contribution criterion?

3.28   In Problem 3.27, suppose that the percentage error values specified are in fact standard deviations in the measurements of y, s, and w. Estimate the standard deviation in the estimated value of tension T.

3.29   The quality control system in a steel rolling mill uses a proximity sensor to measure the thickness of rolled steel (steel gage) at every 1 m along the sheet, and the mill controller adjustments are made on the basis of the last 20 measurements. Specifically, the controller is adjusted unless the probability that the mean thickness lies within $\pm 1\%$ of the sample mean, exceeds 0.99. A typical set of 20 measurements in millimeters is as follows:

5.10,	5.05,	4.94,	4.98,	5.10,	5.12,	5.07,	4.96,	4.99,	4.95,
4.99,	4.97,	5.00	5.08,	5.10,	5.11,	4.99,	4.96,	4.90,	4.10,

Check whether adjustments would be made in the gage controller on the basis of these measurements.

3.30   Dynamics and control of inherently unstable systems, such as rockets, can be studied experimentally using simple scaled-down physical models of the prototype systems. One such study is the classic inverted pendulum problem. An experimental setup for the inverted pendulum is shown in Figure P3.30. The inverted pendulum is supported on a trolley that is driven on a tabletop along a straight line, using

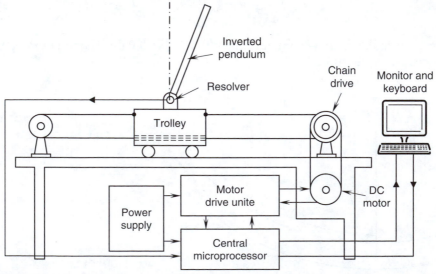

FIGURE P3.30
A microprocessor-controlled inverted pendulum—an application of statistical process control.

a chain-and-sprocket transmission operated by a dc motor. The motor is turned by commands from a microprocessor that is interfaced with the drive system of the motor. The angular position of the pendulum rod is measured using a resolver (see Chapter 4) and is transmitted (fed back) to the microprocessor. A strategy of SPC is used to balance the pendulum rod. Specifically, control limits are established from an initial set of measurement samples of the pendulum angle. Subsequently, if the angle exceeds one control limit, the trolley is accelerated in the opposite direction, using an automatic command to the motor. The control limits are also updated regularly. Suppose that the following 20 readings of the pendulum angle were measured (in degrees) after the system had operated for a few minutes:

0.5	−0.5	0.4	−0.3	0.3	0.1	−0.3	0.3	4.0	0.0
0.4	−0.4	0.5	−0.5	−5.0	0.4	− 0.4	0.3	−0.3	−0.1

Establish whether the system was in statistical control during the period in which the readings were taken. Comment on this method of control.

4

Analog Sensors and Transducers

Proper selection and integration of sensors and transducers are crucial in instrumenting a control system. Sensors may be used in a control system for a variety of purposes. In particular, output signals are measured for feedback control; input signals are measured for feedforward control; output signals are measured in system monitoring, tuning, and supervisory control; and input–output signal pairs are measured for experimental modeling and evaluation of a plant. Ideal characteristics of sensors and transducers are indicated in Chapter 3. Even though real sensors and transducers can behave quite differently in practice, when developing a control system we should use the ideal behavior as a reference for the design specifications. In this chapter, the significance of sensors and transducers in a control system is indicated; important criteria in selecting sensors and transducer for control applications are presented; and several representative sensors and transducers and their operating principles, characteristics, and applications are described.

4.1 Terminology

Potentiometers, differential transformers, resolvers, synchros, strain gages, tachometers, piezoelectric devices, bellows, diaphragms, flow meters, thermocouples, thermistors, and resistance temperature detectors (RTDs) are examples of sensors used in control systems. In a control system, sensors are used to measure the system response, and it enables the controller to take corrective actions if the system does not operate properly. A control system may have unknown excitations and disturbances, which can make the associated tasks (performance monitoring, experimental modeling, control, etc.), particularly difficult. Removing such excitations at the source level is desirable through proper design or system isolation. However, in the context of control, if these disturbances can be measured, or if some information about them is available, then they can be compensated for within the controller itself. This is in fact the approach of feedforward control. In summary, sensors may be used in a control system in several ways:

1. To measure the system outputs for feedback control.
2. To measure some types of system inputs (unknown inputs, disturbances, etc.) for feedforward control.
3. To measure output signals for system monitoring, diagnosis, evaluation, parameter adjustment, and supervisory control.
4. To measure input and output signal pairs for system testing and experimental modeling (i.e., for system identification).

The variable that is measured is termed the *measurand*. Examples are acceleration and velocity of a vehicle, torque into a robotic joint, temperature and pressure of a process plant, and current through an electric circuit. A measuring device passes through two stages while measuring a signal. First, the measurand is felt or *sensed* by the sensing element. Then, the sensed signal is *transduced* (or converted) into the form of the device output. In fact the sensor, which senses the response automatically, converts (i.e., transduces) this signal into the sensor output—the response of the sensor element. For example, a piezoelectric accelerometer senses acceleration and converts it into an electric charge; an electromagnetic tachometer senses velocity and converts it into a voltage; and a shaft encoder senses a rotation and converts it into a sequence of voltage pulses. Since sensing and transducing occur together, the terms sensor and transducer are used interchangeably to denote a sensor–transducer unit. Sensor and transducer stages are functional stages, and sometimes it is not easy or even feasible to draw a line to separate them or to separately identify physical elements associated with them. Furthermore, this separation is not very important in using existing devices. However, proper separation of sensor and transducer stages (physically as well as functionally) can be crucial, when designing new measuring devices.

Typically, the sensed signal is transduced (or converted) into a form that is particularly suitable for transmitting, recording, conditioning, processing, activating a controller, or driving an actuator. For this reason, the output of a transducer is often an electrical signal. The measurand is usually an analog signal because it represents the output of a dynamic system. For example, the charge signal generated in a piezoelectric accelerometer has to be converted into a voltage signal of appropriate level using a charge amplifier. To enable its use in a digital controller it has to be digitized using an analog-to-digital converter (ADC). In digital transducers the transducer output is discrete, and typically a sequence of pulses. Such discrete outputs can be counted and represented in a digital form. This facilitates the direct interface of a digital transducer with a digital processor.

A complex measuring device can have more than one sensing stage. Often, the measurand goes through several transducer stages before it is available for control and actuating purposes. Furthermore, filtering may be needed to remove measurement noise. Hence, signal conditioning is usually needed between the sensor and the controller as well as between the controller and the actuator. Charge amplifiers, lock-in amplifiers, power amplifiers, switching amplifiers, linear amplifiers, pulse-width modulation (PWM) amplifiers, tracking filters, low-pass filters, high-pass filters, bandpass filters, and notch filters are some of the signal-conditioning devices used in control systems. The subject of signal conditioning is discussed in Chapter 3, and typical signal condition devices are described there. In some literature, signal-conditioning devices such as electronic amplifiers are also classified as transducers. Since we are treating signal-conditioning and modification devices separately from measuring devices, this unified classification is avoided whenever possible, and the term transducer is used primarily in relation to measuring instruments. Note that it is somewhat redundant to consider electrical-to-electrical sensors–transducers as measuring devices because electrical signals need conditioning only before they are used to carry out a useful task. In this sense, electrical-to-electrical transduction should be considered a conditioning function rather than a measuring function. Additional components, such as power supplies, isolation devices, and surge-protection units are often needed in control systems, but they are only indirectly related to control functions. Relays and other switching devices and modulators and demodulators (see Chapter 3) may also be included.

Pure transducers depend on nondissipative coupling in the transduction stage. Passive transducers (sometimes called self-generating transducers) depend on their power transfer characteristics for operation and do not need an external power source. It follows that pure transducers are essentially passive devices. Some examples are electromagnetic, thermoelectric, radioactive, piezoelectric, and photovoltaic transducers. External power is required to operate active sensors/transducers, and they do not depend on their own power conversion characteristics for operation. A good example for an active device is a resistive transducer, such as a potentiometer, which depends on its power dissipation through a resistor to generate the output signal. Note that an active transducer requires a separate power source (power supply) for operation, whereas a passive transducer draws its power from a measured signal (measurand). Since passive transducers derive their energy almost entirely from the measurand, they generally tend to distort (or load) the measured signal to a greater extent than an active transducer would. Precautions can be taken to reduce such loading effects. On the other hand, passive transducers are generally simple in design, more reliable, and less costly. In the present classification of transducers, we are dealing with power in the immediate transducer stage associated with the measurand, and not the power used in subsequent signal conditioning. For example, a piezoelectric charge generation is a passive process. But, a charge amplifier, which uses an auxiliary power source, would be needed by a piezoelectric device in order to condition the generated charge.

Now, we analyze several analog sensor–transducer devices that are commonly used in control system instrumentation. The attempt here is not to present an exhaustive discussion of all types of sensors; rather, it is to consider a representative selection. Such an approach is reasonable because even though the scientific principles behind various sensors may differ, many other aspects (e.g., performance parameters and specification, signal conditioning, interfacing, and modeling procedures) can be common to a large extent.

4.1.1 Motion Transducers

By motion, we particularly mean one or more of the following four kinematic variables:

- Displacement (including position, distance, proximity, size or gage)
- Velocity (rate of change of displacement)
- Acceleration (rate of change of velocity)
- Jerk (rate of change of accleration)

Note that each variable is the time derivative of the preceding one. Motion measurements are extremely useful in controlling mechanical responses and interactions in control systems. Numerous examples can be cited: The rotating speed of a workpiece and the feed rate of a tool are measured in controlling machining operations. Displacements and speeds (both angular and translatory) at joints (revolute and prismatic) of robotic manipulators or kinematic linkages are used in controlling manipulator trajectory. In high-speed ground transit vehicles, acceleration and jerk measurements can be used for active suspension control to obtain improved ride quality. Angular speed is a crucial measurement that is used in the control of rotating machinery, such as turbines, pumps, compressors, motors, transmission units or gear boxes, and generators in power-generating plants. Proximity sensors (to measure displacement) and accelerometers (to measure acceleration) are the two most common types of measuring devices used in machine protection systems for condition monitoring, fault detection, diagnostic, and online (often real-time) control of large and complex machinery. The accelerometer is often the only measuring device used in controlling dynamic test rigs (e.g., in vibration testing). Displacement measurements are used for valve control in process applications.

Plate thickness (or gage) is continuously monitored by the automatic gage control (AGC) system in steel rolling mills.

A one-to-one relationship may not always exist between a measuring device and a measured variable. For example, although strain gages are devices that measure strains (and, hence, stresses and forces), they can be adapted to measure displacements by using a suitable front-end auxiliary sensor element, such as a cantilever (or spring). Furthermore, the same measuring device may be used to measure different variables through appropriate data interpretation techniques. For example, piezoelectric accelerometers with built-in microelectronic integrated circuitry (IC) are marketed as piezoelectric velocity transducers. Resolver signals, which provide angular displacements, are differentiated to obtain angular velocities. Pulse-generating (or digital) transducers, such as optical encoders and digital tachometers, can serve as both displacement transducers and velocity transducers, depending on whether the absolute number of pulses is counted or the pulse rate is measured. Note that pulse rate can be measured either by counting the number of pulses during a unit interval of time (i.e., pulse counting) or by gating a high-frequency clock signal through the pulse width (i.e., pulse timing). Furthermore, in principle, any force sensor can be used as an acceleration sensor, velocity sensor, or displacement sensor, depending on whether: (1) an inertia element (converting acceleration into force) (2) a damping element (converting velocity into force) or (3) a spring element (converting displacement into force) is used as the front-end auxiliary sensor.

We might question the need for separate transducers to measure the four kinematic variables—displacement, velocity, acceleration, and jerk—because any one variable is related to the other through simple integration or differentiation. It should be possible, in theory, to measure only one of these four variables and use either analog processing (through analog circuit hardware) or digital processing (through a dedicated processor) to obtain any one of the remaining motion variables. The feasibility of this approach is highly limited, however, and it depends crucially on several factors, including the following:

1. The nature of the measured signal (e.g., steady, highly transient, periodic, narrow-band, broadband)

2. The required frequency content of the processed signal (or the frequency range of interest)

3. The signal-to-noise ratio (SNR) of the measurement

4. Available processing capabilities (e.g., analog or digital processing, limitations of the digital processor and interface, such as the speed of processing, sampling rate, and buffer size)

5. Controller requirements and the nature of the plant (e.g., time constants, delays, complexity, hardware limitations)

6. Required accuracy as the end objective (on which processing requirements and hardware costs depend).

For instance, differentiation of a signal (in the time domain) is often unacceptable for noisy and high-frequency narrow-band signals. In any event, costly signal-conditioning hardware might be needed for preprocessing before differentiating a signal. As a rule of thumb, in low-frequency applications (in the order of 1 Hz), displacement measurements generally provide good accuracies. In intermediate-frequency applications (less than 1 kHz), velocity measurement is usually favored. In measuring high-frequency motions with high noise levels, acceleration measurement is preferred. Jerk is particularly useful in ground transit (ride quality), manufacturing (forging, rolling, cutting and similar impact-type operations), and shock isolation applications (for delicate and sensitive equipment).

4.2 Potentiometer

The potentiometer, or *pot*, is a displacement transducer. This active transducer consists of a uniform coil of wire or a film of high-resistance material–such as carbon, platinum, or conductive plastic, whose resistance is proportional to its length. A constant voltage v_{ref} is applied across the coil (or film) using an external dc voltage supply. The output signal v_o of the transducer is the dc voltage between the movable contact (wiper arm) sliding on the coil and the reference-voltage terminal of the coil, as shown schematically in Figure 4.1a. Slider displacement x is proportional to the output voltage:

$$v_o = kx. \tag{4.1}$$

This relationship assumes that the output terminals are in open-circuit; that is, a load of infinite impedance (or resistance in the present dc case) is present at the output terminals, so that the output current is zero. In actual practice, however, the load (the circuitry into which the pot signal is fed—e.g., conditioning, interfacing, processing, or control circuitry) has a finite impedance. Consequently, the output current (the current through the load) is nonzero, as shown in Figure 4.1b. The output voltage thus drops to $\tilde{v}_o$, even if the reference voltage v_{ref} is assumed to remain constant under load variations (i.e., even if the output impedance of the voltage source is zero); this consequence is known as the electrical *loading effect* of the transducer. Under these conditions, the linear relationship given by Equation 4.1 would no longer be valid, causing an error in the displacement reading. Loading can affect the transducer reading in two ways: by changing the reference voltage (i.e., loading the voltage source) and by loading the transducer. To reduce these effects, a voltage source that is not seriously affected by load variations (e.g., a regulated or stabilized power supply, which has a low output impedance), and data acquisition circuitry (including signal-conditioning circuitry) that has a high input impedance should be used.

The resistance of a potentiometer should be chosen with care. On the one hand, an element with high resistance is preferred because this results in reduced power dissipation for a given voltage, which has the added benefit of reduced thermal effects. On the

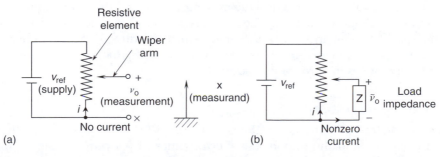

FIGURE 4.1
(a) Schematic diagram of a potentiometer. (b) Potentiometer loading.

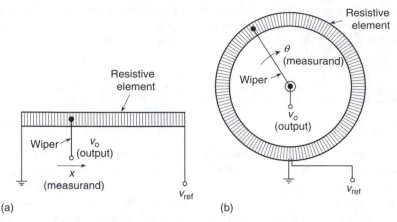

FIGURE 4.2
Practical potentiometer configurations for measuring. (a) Rectilinear motions. (b) Angular motions.

other hand, increased resistance increases the output impedance of the potentiometer and results in a corresponding increase in loading nonlinearity error unless the load resistance is also increased proportionately. Low-resistance pots have resistances less than 10 Ω. High-resistance pots can have resistances in the order of 100 kΩ. Conductive plastics can provide high resistances—typically about 100 Ω/mm—and are increasingly used in potentiometers. Reduced friction (low mechanical loading), reduced wear, reduced weight, and increased resolution are advantages of using conductive plastics in potentiometers.

4.2.1 Rotatory Potentiometers

Potentiometers that measure angular (rotatory) displacements are more common and convenient, because in conventional designs of rectilinear (translatory) potentiometers, the length of the resistive element has to be increased in proportion to the measurement range or stroke. Figure 4.2 presents schematic representations of translatory and rotatory potentiometers. Helix-type rotatory potentiometers are available for measuring absolute angles exceeding 360 degrees. The same function may be accomplished with a standard single-cycle rotatory pot simply by including a counter to record full 360 degree rotations.

Note that angular displacement transducers, such as rotatory potentiometers, can be used to measure large rectilinear displacements in the order of 3 m. A cable extension mechanism may be employed to accomplish this. A light cable wrapped around a spool, which moves with the rotary element of the transducer, is the cable extension mechanism. The free end of the cable is attached to the moving object, and the potentiometer housing is mounted on a stationary structure. The device is properly calibrated so that as the object moves, the rotation count and fractional rotation measure directly provide the rectilinear displacement. A spring-loaded recoil device, such as a spring motor, winds the cable back when the object moves toward the transducer.

4.2.1.1 Loading Nonlinearity

Consider the rotatory potentiometer shown in Figure 4.3. Let us now discuss the significance of the electrical loading nonlinearity error caused by a purely resistive load connected to the pot. For a general position θ of the pot slider arm, suppose that the resistance in the output (pick-off) segment of the coil is R_θ.

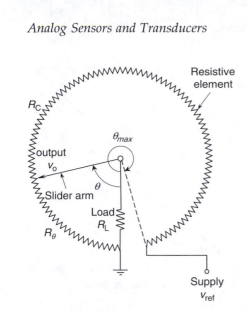

FIGURE 4.3
A rotatory potentiometer with a resistive load.

Assuming a uniform coil, one has

$$R_\theta = \frac{\theta}{\theta_{\max}} R_c, \tag{4.2}$$

where R_c is the total resistance of the potentiometer coil. The current balance at the sliding contact (node) point gives

$$\frac{v_{\text{ref}} - v_o}{R_c - R_\theta} = \frac{v_o}{R_\theta} + \frac{v_o}{R_L}, \tag{i}$$

where R_L is the load resistance. Multiply throughout Equation (i) by R_c and use Equation 4.2; thus,

$$\frac{v_{\text{ref}} - v_o}{1 - \theta/\theta_{\max}} = \frac{v_o}{\theta/\theta_{\max}} + \frac{v_o}{R_L/R_c}.$$

By using straightforward algebra, we have

$$\frac{v_o}{v_{\text{ref}}} = \left[\frac{(\theta/\theta_{\max})(R_L/R_C)}{\left(R_L/R_c + (\theta/\theta_{\max}) - (\theta/\theta_{\max})^2\right)} \right]. \tag{4.3}$$

Equation 4.3 is plotted in Figure 4.4. Loading error appears to be high for low values of the R_L/R_c ratio. Good accuracy is possible for $R_L/R_c > 10$, particularly for small values of $\theta/\theta_{\max}$.

It should be clear that the following actions can be taken to reduce loading error in pots:

1. Increase R_L/R_c (increase load impedance, reduce coil impedance).
2. Use pots to measure small values of $\theta/\theta_{\max}$ (or calibrate only a small segment of the resistance element, for linear reading).

The loading-nonlinearity error is defined by

$$e = \frac{(v_o/v_{\text{ref}} - \theta/\theta_{\max})}{\theta/\theta_{\max}} 100\%. \tag{4.4}$$

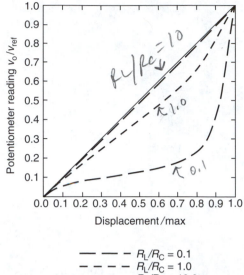

FIGURE 4.4
Electrical loading nonlinearity in a potentiometer.

$$
\begin{aligned}
&-\ -\ -\ -\quad R_L/R_C = 0.1\\
&-\ -\ -\ -\quad R_L/R_C = 1.0\\
&-\ -\ -\ -\quad R_L/R_C = 10.0
\end{aligned}
$$

The error at $\theta/\theta_{\max} = 0.5$ for three values of load resistance ratio is tabulated in Table 4.1. Note that this error is always negative. Using only a segment of the resistance element as the range of the potentiometer is similar to adding two end resistors to the elements. It is known that this tends to linearize the pot. If the load resistance is known to be small, a voltage follower may be used at the potentiometer output to virtually eliminate loading error, since this arrangement provides a high-load impedance to the pot and a low impedance at the output of the amplifier.

4.2.2 Performance Considerations

The potentiometer is a resistively coupled transducer. The force required to move the slider arm comes from the motion source, and the resulting energy is dissipated through friction. This energy conversion, unlike pure mechanical-to-electrical conversions, involves relatively high forces, and the energy is wasted rather than getting converted into the output signal of the transducer. Furthermore, the electrical energy from the reference source is also dissipated through the resistor element (coil or film), resulting in an undesirable temperature rise and coil degradation. These are two obvious disadvantages of a potentiometer. In coil-type pots there is another disadvantage, which is the finite resolution.

A coil, instead of a straight wire, is used as the resistance element of a pot to increase the resistance per unit travel of the slider arm. But the slider contact jumps from one turn

TABLE 4.1

Loading Nonlinearity Error in a Potentiometer

Load Resistance Ratio R_L/R_C	Loading Nonlinearity Error (e) at $\theta/\theta_{\max} = 0.5$
0.1	−71.4%
1.0	−20%
10.0	−2.4%

to the next in this case. Accordingly, the resolution of a coil-type potentiometer is determined by the number of turns in the coil. For a coil that has N turns, the resolution r, expressed as a percentage of the output range, is given by

$$r = \frac{100}{N}\%. \tag{4.5}$$

Resolutions better (smaller) than 0.1% (i.e., 1,000 turns) are available with coil potentiometers. Virtually infinitesimal (incorrectly termed infinite) resolutions are now possible with high-quality resistive film potentiometers, which use conductive plastics. In this case, the resolution is limited by other factors, such as mechanical limitations and signal-to-noise ratio (SNR). Nevertheless, resolutions in the order of 0.01 mm are possible with good rectilinear potentiometers.

Selection of a potentiometer involves many considerations. A primary factor is the required resolution for the specific application. Power consumption, loading, and size are also important factors. The following design example highlights some of these considerations.

Example 4.1

A high-precision mobile robot uses a potentiometer attached to the drive wheel to record its travel during autonomous navigation. The required resolution for robot motion is 1 mm, and the diameter of the drive wheel of the robot is 20 cm. Examine the design considerations for a standard (single-coil) rotary potentiometer to be used in this application.

Solution

Assume that the potentiometer is directly connected (without gears) to the drive wheel. The required resolution for the pot is

$$\frac{0.1}{\pi \times 20} \times 100\% = 0.16\%.$$

This resolution is feasible with a coil-type rotary pot. From Equation 4.5 the number of turns in the coil $= 100/0.16 = 625$ turns. Assuming an average pot diameter of 10 cm and denoting the wire diameter by d, since there are 625 turns covering the entire circumference of the pot we have:

$$\text{Potentiometer circumference} = \pi \times 10 = 625 \times d$$

or

$$d = 0.5 \text{ mm}.$$

Now, taking the resistance of the potentiometer to be 5 Ω and the resistivity of the wire to be 4 $\mu\Omega$ cm, the diameter D of the core of the coil is given by

$$\frac{4 \times 10^{-6} \times \pi D \times 625}{\pi(0.05/2)^2} = 5 \ \Omega.$$

Note: Resistivity $=$ (resistance) $\times$ (cross-section area)/(length).
Hence,

$$D = 1.25 \text{ cm}$$

The *sensitivity* of a potentiometer represents the change (Δv_o) in the output signal associated with a given small change ($\Delta\theta$) in the measurand (the displacement). The sensitivity is usually nondimensionalized, using the actual value of the output signal (v_o) and the actual value of the displacement (θ). For a rotatory potentiometer in particular, the sensitivity S is given by

$$S = \frac{\Delta v_o}{\Delta\theta} \qquad\qquad (4.6)$$

or, in the limit

$$S = \frac{\partial v_o}{\partial\theta}. \qquad\qquad (4.7)$$

These relations may be nondimensionalized by multiplying by θ/v_o. An expression for S may be obtained by simply substituting Equation 4.3 into Equation 4.7.

Some limitations and disadvantages of potentiometers as displacement measuring devices are given below:

1. The force needed to move the slider (against friction and arm inertia) is provided by the displacement source. This mechanical loading distorts the measured signal itself.

2. High-frequency (or highly transient) measurements are not feasible because of such factors as slider bounce, friction, and inertia resistance, and induced voltages in the wiper arm and primary coil.

3. Variations in the supply voltage cause error.

4. Electrical loading error can be significant when the load resistance is low.

5. Resolution is limited by the number of turns in the coil and by the coil uniformity. This limits small-displacement measurements.

6. Wear out and heating up (with associated oxidation) in the coil or film, and slider contact cause accelerated degradation.

There are several advantages associated with potentiometer devices, however, including the following:

1. They are relatively inexpensive.

2. Potentiometers provide high-voltage (low-impedance) output signals, requiring no amplification in most applications. Transducer impedance can be varied simply by changing the coil resistance and supply voltage.

Example 4.2

A rectilinear potentiometer was tested with its slider arm moving horizontally. It was found that at a speed of 1 cm/s, a driving force of 7×10^{-4} N was necessary to maintain the speed. At 10 cm/s, a force of 3×10^{-3} N was necessary. The slider weighs 5 gm, and the potentiometer stroke is ± 8 cm. If this potentiometer is used to measure the damped natural frequency of a simple mechanical oscillator of mass 10 kg, stiffness 10 N/m, and damping constant 2 N/m/s, estimate the percentage error due to mechanical loading. Justify this procedure for the estimation of damping.

Solution

Suppose that the mass, stiffness, and damping constant of the simple oscillator are denoted by M, K, and B, respectively. The equation of free motion of the simple oscillator is given by

$$M\ddot{y} + B\dot{y} + Ky = 0, \tag{i}$$

where y denotes the displacement of the mass from the static equilibrium position. This equation is of the form

$$\ddot{y} + 2\zeta\omega_n\dot{y} + \omega_n^2 y = 0, \tag{ii}$$

where ω_n is the undamped natural frequency of the oscillator and ζ is the damping ratio. By direct comparison of (i) and (ii), it is seen that

$$\omega_n = \sqrt{\frac{K}{M}} \quad \text{and} \quad \zeta = \frac{B}{2\sqrt{MK}}. \tag{iii}$$

The damped natural frequency is

$$\omega_d = \sqrt{1 - \zeta^2}\,\omega_n \quad \text{for } 0 < \zeta < 1. \tag{iv}$$

Hence,

$$\omega_d = \sqrt{\left(1 - \frac{B^2}{4MK}\right)\frac{K}{M}}. \tag{v}$$

Now, if the wiper arm mass and the damping constant of the potentiometer are denoted by m and b, respectively, the measured damped natural frequency (using the potentiometer) is given by

$$\tilde{\omega}_d = \sqrt{\left[1 - \frac{(B+b)^2}{4(M+m)K}\right]\frac{K}{(M+m)}}. \tag{vi}$$

Assuming linear viscous friction, which is not quite realistic, the equivalent damping constant b of the potentiometer may be estimated as

$$b = \text{damping force/steady state velocity of the wiper.}$$

For the present example,

$$b_1 = 7 \times 10^{-4}/1 \times 10^{-2} \text{ N/m/s} = 7 \times 10^{-2} \text{ N/m/s at 1 cm/s,}$$
$$b_2 = 3 \times 10^{-3}/10 \times 10^{-2} \text{ N/m/s} = 3 \times 10^{-2} \text{ N/m/s at 10 cm/s.}$$

We should use some form of interpolation to estimate b for the actual measuring conditions. Let us now estimate the average velocity of the wiper. The natural frequency of the oscillator is

$$\omega_n = \sqrt{\frac{10}{10}} = 1 \text{ rad/s} = \frac{1}{2\pi} \text{Hz.}$$

Since one cycle of oscillation corresponds to a motion of 4 strokes, the wiper travels a maximum distance of 4×8 cm $= 32$ cm in one cycle. Hence, the average operating speed of the wiper may be estimated as $32/(2\pi)$ cm/s, which is approximately equal to 5 cm/s. Therefore, the operating damping constant may be estimated as the average of b_1 (at 1 cm/s) and b_2 (at 10 cm/s):

$$b = 5 \times 10^{-2} \text{ N/m/s}.$$

With the foregoing numerical values

$$\omega_d = \sqrt{\left(1 - \frac{2^2}{4 \times 10 \times 10}\right) \frac{10}{10}} = 0.99499 \text{ rad/s},$$

$$\tilde{\omega}_d = \sqrt{\left(1 - \frac{2.05^2}{4 \times 10.005 \times 10}\right) \frac{10}{10.005}} = 0.99449 \text{ rad/s}. \cdot$$

$$\text{Percentage error} = \left[\frac{\tilde{\omega}_d - \omega_d}{\omega_d}\right] \times 100\% = 0.05\%$$

∎

Although pots are primarily used as displacement transducers, they can be adapted to measure other types of signals, such as pressure and force, using appropriate auxiliary sensor (front-end) elements. For instance, a bourdon tube or bellows may be used to convert pressure into displacement, and a cantilever element may be used to convert force or moment into displacement.

4.2.3   Optical Potentiometer

The optical potentiometer, shown schematically in Figure 4.5a, is a displacement sensor. A layer of photoresistive material is sandwiched between a layer of ordinary

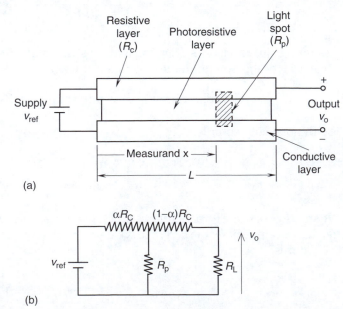

FIGURE 4.5
(a) An optical potentiometer.
(b) Equivalent circuit ($\alpha = x/L$).

resistive material and a layer of conductive material. The layer of resistive material has a total resistance of R_c, and it is uniform (i.e., it has a constant resistance per unit length). This corresponds to the coil resistance of a conventional potentiometer. The photoresistive layer is practically an electrical insulator when no light is projected on it. The moving object, whose displacement is measured, causes a moving light beam to be projected on a rectangular area of the photoresistive layer. This light-activated area attains a resistance of R_p, which links the resistive layer that is above the photoresistive layer and the conductive layer that is below the photoresistive layer. The supply voltage to the potentiometer is v_{ref}, and the length of the resistive layer is L. The light spot is projected at a distance x from the reference end of the resistive element, as shown in the figure.

An equivalent circuit for the optical potentiometer is shown in Figure 4.5b. Here it is assumed that a load of resistance R_L is present at the output of the potentiometer, with v_o as voltage across. Current through the load is v_o/R_L. Hence, the voltage drop across $(1 - \alpha) R_c + R_L$, which is also the voltage across R_p, is given by $[(1 - \alpha) R_c + R_L] v_o/R_L$. Note that $\alpha = x/L$, is the fractional position of the light spot. The current balance at the junction of the three resistors in Figure 4.5b is

$$\frac{v_{ref} - [(1 - \alpha)R_c + R_L]v_o/R_L}{\alpha R_c} = \frac{v_o}{R_L} + \frac{[(1 - \alpha)R_c + R_L]v_o/R_L}{R_p},$$

which can be written as

$$\frac{v_o}{v_{ref}} \left\{ \frac{R_c}{R_L} + 1 + \frac{x}{L}\frac{R_c}{R_p}\left[\left(1 - \frac{x}{L}\right)\frac{R_c}{R_L} + 1\right] \right\} = 1. \tag{4.8}$$

When the load resistance R_L is quite large in comparison with the element resistance R_c we have $R_c/R_L \simeq 0$. Hence, Equation 4.8 becomes

$$\frac{v_o}{v_{ref}} = \frac{1}{\left[\dfrac{x}{L}\dfrac{R_c}{R_p} + 1\right]}. \tag{4.9}$$

This relationship is still nonlinear in v_o/v_{ref} vs. x/L. The nonlinearity decreases, however, with decreasing R_c/R_p. This is also seen from Figure 4.6 where Equation 4.9 is plotted for several values of R_c/R_p. Then, for the case of $R_c/R_p = 0.1$, the original Equation 4.8 is

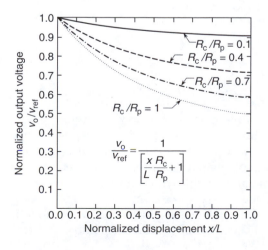

FIGURE 4.6

Behavior of the optical potentiometer at high load resistance.

$$\frac{v_o}{v_{ref}}\left\{\frac{R_c}{R_L} + 1 + 0.1\frac{x}{L}\left[\left(1-\frac{x}{L}\right)\frac{R_c}{R_L} + 1\right]\right\} = 1$$

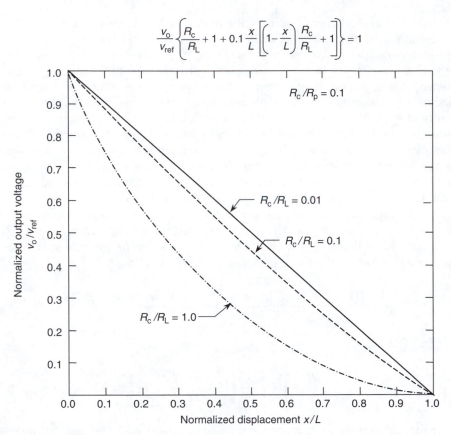

FIGURE 4.7
Behavior of the optical potentiometer for $R_c/R_p = 0.1$.

plotted in Figure 4.7, for several values of load resistance ratio. As expected, the behavior of the optical potentiometer becomes more linear for higher values of load resistance.

The potentiometer has disadvantages such as loading problems (both mechanical and electrical), limited speed of operation, considerable time constants, wear, noise, and thermal effects. Many of these problems arise from the fact that it is a contact device where its slider has to be in intimate contact with the resistance element of the pot, and also has to be an integral part of the moving object whose displacements need to be measured. Next we consider several noncontact motion sensors.

4.3   Variable-Inductance Transducers

Motion transducers that employ the principle of electromagnetic induction are termed variable-inductance transducers. When the flux linkage (defined as magnetic flux density times the number of turns in the conductor) through an electrical conductor changes, a voltage in proportion to the rate of change of flux is induced in the conductor. This voltage in turn, generates a magnetic field, which opposes the original (primary) field. Hence, a mechanical force is necessary to sustain the change of flux linkage. If the change in flux linkage is brought about by a relative motion, the associated mechanical energy is directly converted (induced) into electrical energy. This is the basis of electromagnetic induction, and it is the principle of operation of electrical generators and variable-inductance transducers. Note that in these devices, the change of flux linkage is caused by a mechanical motion, and mechanical-to-electrical energy transfer takes place under near-ideal conditions. The induced voltage or change in inductance may be used as a measure of the motion. Variable-inductance transducers are generally electromechanical devices coupled by a magnetic field.

There are many different types of *variable-inductance transducers*. Three primary types can be identified:

1. Mutual-induction transducers
2. Self-induction transducers
3. Permanent-magnet transducers.

Those variable-inductance transducers that use a nonmagnetized ferromagnetic medium to alter the reluctance (magnetic resistance) of the magnetic flux path are known as variable-reluctance transducers. Some of the mutual-induction transducers and most of the self-induction transducers are of this type. Permanent-magnet transducers are not considered variable-reluctance transducers.

4.3.1 Mutual-Induction Transducers

The basic arrangement of a mutual-induction transducer constitutes two coils, the primary winding and the secondary winding. One of the coils (primary winding) carries an alternating-current (ac) excitation, which induces a steady ac voltage in the other coil (secondary winding). The level (amplitude, rms value, etc.) of the induced voltage depends on the flux linkage between the coils. None of these transducers employ contact sliders or slip-rings and brushes as do resistively coupled transducers (potentiometer). Consequently, they have an increased design life and low mechanical loading. In mutual-induction transducers, a change in the flux linkage is effected by one of two common techniques. One technique is to move an object made of ferromagnetic material within the flux path between the primary coil and the secondary coil. This changes the reluctance of the flux path, with an associated change of the flux linkage in the secondary coil. This is, for example, the operating principle of the linear-variable differential transformer/ transducer (LVDT), the rotatory-variable differential transformer/transducer (RVDT), and the mutual-induction proximity probe. All of these are, in fact, variable-reluctance transducers as well. The other common way to change the flux linkage is to move one coil with respect to the other. This is the operating principle of the resolver, the synchro-transformer, and some types of ac tachometer. These are not variable-reluctance transducers, however, because a moving ferromagnetic element is not involved.

Motion can be measured by using the secondary signal (i.e., induced voltage in the secondary coil) in several ways. For example, the ac signal in the secondary coil may be demodulated by rejecting the carrier signal (i.e., the signal component at the excitation frequency) and the resulting signal, which represents the motion, is directly measured. This method is particularly suitable for measuring transient motions. Alternatively, the amplitude or the rms (root-mean-square) value of the secondary (induced) voltage may be measured. Another method is to measure the change of inductance (or reactance, which is equal to $Lj\omega$, since voltage $v = L\,di/dt$) where L is the inductance and i is the current in the secondary circuit directly, by using a device such as an inductance bridge circuits. See Chapter 3).

4.3.2   Linear-Variable Differential Transformer/Transducer

Differential transformer is a noncontact displacement sensor, which does not possess many of the shortcomings of the potentiometer. It falls into the general category of a variable-inductance transducer, and is also a variable-reluctance transducer and a mutual-induction transducer. Furthermore, unlike the potentiometer, the differential transformer is a passive device. Now we discuss the linear-variable differential transformer or transducer (LVDT), which is used for measuring rectilinear (or translatory) displacements. Next we describe the rotatory-variable differential transformer or transducer (RVDT), which is used for measuring angular (or rotatory) displacements.

The LVDT is considered a passive transducer because the measured displacement provides energy for changing the induced voltage in the secondary coil, even though an external power supply is used to energize the primary coil, which in turn induces a steady voltage at the carrier frequency in the secondary coil. In its simplest form (sse, Figure 4.8), the LVDT consists of an insulating, nonmagnetic form (a cylindrical structure on which a coil is wound and is integral with the housing), which has a primary coil in the midsegment and a secondary coil symmetrically wound in the two end segments, as depicted schematically in Figure 4.8b. The housing is made of magnetized stainless steel to shield the sensor from outside fields. The primary coil is energized by an ac supply of voltage v_{ref}. This generates, by mutual induction, an ac of the same frequency in the secondary coil. A core made of ferromagnetic material is inserted coaxially through the cylindrical form without actually touching it, as shown. As the core moves, the reluctance of the flux path between the primary and the secondary coils changes. The degree of flux linkage depends on the axial position of the core. Since the two secondary coils are connected in series opposition (as shown in Figure 4.9), so that the potentials induced in the two secondary coil segments oppose each other, it is seen that the net induced voltage is zero when the core is centered between the two secondary winding segments. This is known as the *null position*. When the core is displaced from this position, a nonzero induced voltage is generated. At steady state, the amplitude v_o of this induced voltage is proportional to the core displacement x in the linear (operating) region (see Figure 4.8c). Consequently, v_o may be used as a measure of the displacement. Note that because of opposed secondary windings, the LVDT provides the direction as well as the magnitude of displacement. If the output signal is not demodulated, the direction is determined by the phase angle between the primary (reference) voltage and the secondary (output) voltage, which includes the carrier signal.

For an LVDT to measure transient motions accurately, the frequency of the reference voltage (the carrier frequency) has to be at least 10 times larger than the largest significant (useful) frequency component in the measured motion, and typically can be as high as 20 kHz. For quasi-dynamic displacements and slow transients of the order of a few hertz, a standard ac supply (at 60 Hz line frequency) is adequate. The performance (particularly sensitivity and accuracy) is known to improve with the excitation frequency, however. Since the amplitude of the output signal is proportional to the amplitude of the primary signal, the reference voltage should be regulated to get accurate results. In particular, the power source should have a low output impedance.

4.3.2.1   *Phase Shift and Null Voltage*

An error known as *null voltage* is present in some differential transformers. This manifests itself as a nonzero reading at the null position (i.e., at zero displacement). This is usually 90° out of phase from the main output signal and, hence, is known as *quadrature error*. Nonuniformities in the windings (unequal impedances in the two segments of the

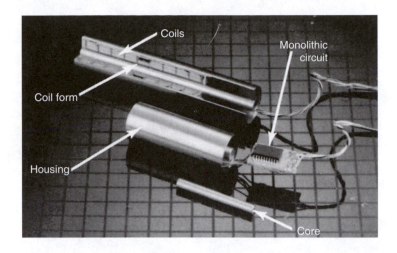

(a)

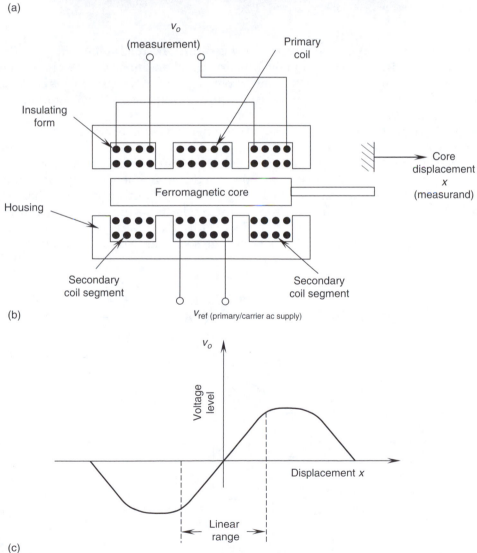

(b)

(c)

FIGURE 4.8
LVDT. (a) A commercial unit (Scheavitz Sensors, Measurement Specialties, Inc. With permission). (b) Schematic diagram. (c) A typical operating curve.

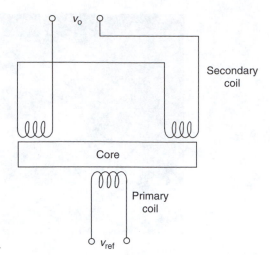

FIGURE 4.9
Series opposition connection of secondary windings.

secondary winding) are a major reason for this error. The null voltage may also result from harmonic noise components in the primary signal and nonlinearities in the device. Null voltage is usually negligible (typically about 0.1% of the full scale). This error can be eliminated from the measurements by employing appropriate signal-conditioning and calibration practices.

The output signal from a differential transformer is normally not in phase with the reference voltage. Inductance in the primary coil and the leakage inductance in the secondary coil are mainly responsible for this phase shift. Since demodulation involves extraction of the modulating signal by rejecting the carrier frequency component from the secondary signal, it is important to understand the size of this phase shift. This topic is addressed now. An equivalent circuit for a differential transformer is shown in Figure 4.10. The resistance in the primary coil is denoted by R_p and the corresponding inductance is denoted by L_p. The total resistance of the secondary coil is R_s. The net leakage

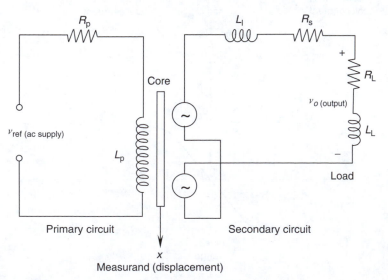

FIGURE 4.10
Equivalent circuit for a differential transformer/transducer.

inductance, due to magnetic flux leakage, in the two segments is denoted by L_1. The load resistance is R_L and the load inductance is L_L. First, let us derive an expression for the phase shift in the output signal.

The magnetizing voltage in the primary coil is given by

$$v_p = v_{ref} \left[\frac{j\omega L_p}{R_p + j\omega L_p} \right] \tag{4.10}$$

in the frequency domain. Now suppose that the core, length L, is moved through a distance x from the null position. The induced voltage in one segment (a) of the secondary coil would be

$$v_a = v_p k_a (L/2 + x) \tag{4.11}$$

and the induced voltage in the other segment (b) would be

$$v_b = v_p k_b (L/2 - x). \tag{4.12}$$

Here, k_a and k_b are nonlinear functions of the position of the core, and are also complex functions of the frequency variable ω. Furthermore, each function depends on the mutual-induction properties between the primary coil and the corresponding secondary-coil segment, through the core element. Due to series opposition connection of the two secondary segments, the net secondary voltage induced would be

$$v_s = v_a - v_b = v_p [k_a (L/2 + x) - k_b (L/2 - x)]. \tag{4.13}$$

In the ideal case, the two functions $k_a(.)$ and $k_b(.)$ would be identical. Then, at $x = 0$ we have $v_s = 0$. Hence, the null voltage would be zero in the ideal case. Suppose that, at $x = 0$, the magnitudes of $k_a(.)$ and $k_b(.)$ are equal, but there is a slight phase difference. Then the "difference vector", $k_a(L/2) - k_b(L/2)$ will have a small magnitude value, but its phase will be almost $90°$ with respect to both k_a and k_b. This is the *quadrature error*.

For small x, the Taylor series expansion of Equation 4.13 gives

$$v_s = v_p \left[k_a (L/2) + \frac{\partial k_a}{\partial x} (L/2) x - k_b (L/2) + \frac{\partial k_b}{\partial x} (L/2) x \right].$$

Then, assuming that $k_a(.) = k_b(.)$ is denoted by $k_o(.)$ we have

$$v_s = 2 v_p \frac{\partial k_o}{\partial x} (L/2) x;$$

or

$$v_s = v_p k x, \tag{4.14}$$

where,

$$k = 2 \frac{\partial k_o}{\partial x} (L/2). \tag{4.15}$$

In this case, the net induced voltage is proportional to x and is given by

$$v_s = v_{ref} \left[\frac{j\omega L_p}{R_p + j\omega L_p} \right] k x. \tag{4.16}$$

It follows that the output voltage v_o at the load is given by

$$v_o = \left[\frac{j\omega L_p}{R_p + j\omega L_p}\right]\left[\frac{R_L + j\omega L_L}{(R_L + R_s) + j\omega(L_L + L_l)}\right]kx. \tag{4.17}$$

Hence, for small displacements, the amplitude of the net output voltage of the LVDT is proportional to the displacement x. The phase lead at the output is, given by

$$\phi = 90° - \tan^{-1}\frac{\omega L_p}{R_p} + \tan^{-1}\frac{\omega L_L}{R_L} - \tan^{-1}\frac{\omega(L_L + L_l)}{R_L + R_s}. \tag{4.18}$$

Note that the level of dependence of the phase shift on the load (including the secondary circuit) can be reduced by increasing the load impedance.

4.3.2.2   Signal Conditioning

Signal conditioning associated with differential transformers includes filtering and amplification. Filtering is needed to improve the signal-to-noise ratio (SNR) of the output signal. Amplification is necessary to increase the signal strength for data acquisition, transmission, and processing. Since the reference frequency (carrier frequency) is induced into (and embedded in) the output signal, it is also necessary to interpret the output signal properly, particularly for transient motions.

The secondary (output) signal of an LVDT is an amplitude-modulated signal, where the signal component at the carrier frequency is modulated by the lower-frequency transient signal produced as a result of the core motion (x). Two methods are commonly used to interpret the crude output signal from a differential transformer: rectification and demodulation. Block diagram representations of these two procedures are given in Figure 4.11. In the first method (rectification) the ac output from the differential transformer is rectified to obtain a dc signal. This signal is amplified and then low-pass filtered to eliminate any high-frequency noise components. The amplitude of the resulting signal provides the transducer reading. In this method, phase shift in the LVDT output has to be checked separately to determine the direction of motion. In the second method (demodulation) the carrier frequency component is rejected from the output signal by comparing it with a phase-shifted and amplitude-adjusted version of the primary (reference) signal. Note that phase shifting is necessary because, as discussed earlier, the output signal is not in phase with the reference signal. The result is the modulating signal (proportional to x), which is subsequently amplified and filtered.

As a result of advances in miniature integrated circuit technology, differential transformers with built-in microelectronics for signal conditioning are commonly available today. A dc differential transformer uses a dc power supply (typically, ± 15 V) to activate it. A built-in oscillator circuit generates the carrier signal. The rest of the device is identical to an ac differential transformer. The amplified full-scale output voltage can be as high as ± 10 V. Let us now illustrate the demodulation approach of signal conditioning for an LVDT, using an example.

Example 4.3

Figure 4.12 shows a schematic diagram of a simplified signal-conditioning system for an LVDT. The system variables and parameters are as indicated in the figure.

In particular,

$x(t)$ = displacement of the LVDT core (measurand, to be measured)
ω_c = frequency of the carrier voltage
v_o = output signal of the system (measurement).

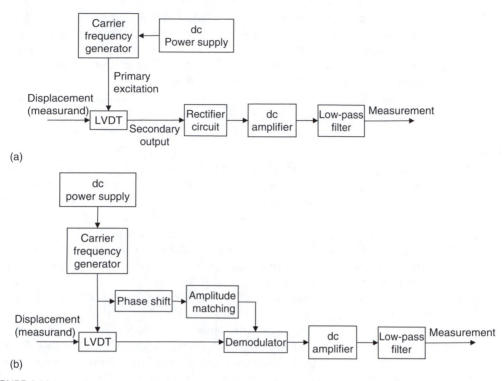

FIGURE 4.11
Signal-conditioning methods for a differential transformer. (a) Rectification. (b) Demodulation.

The resistances R_1, R_2, and R, and the capacitance C are as marked. In addition, we may introduce a transformer parameter r for the LVDT, as required.

  i. Explain the functions of the various components of the system shown in Figure 4.12.

  ii. Write equations for the amplifier and filter circuits and, using them, give expressions for the voltage signals v_1, v_2, v_3, and v_o, which are marked in Figure 4.12. Note that the excitation in the primary coil is $v_p \sin \omega_c t$.

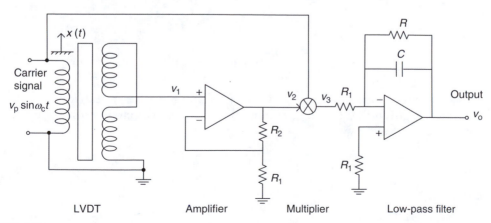

FIGURE 4.12
Signal-conditioning system for an LVDT.

iii. Suppose that the carrier frequency is $\boldsymbol{\omega}_c = 500$ rad/s and the filter resistance $R = 100$ kΩ. If no more than 5% of the carrier component should pass through the filter, estimate the required value of the filter capacitance C. Also, what is the useful frequency range (measurement bandwidth) of the measuring device in radians per second, with these parameter values?

iv. If the displacement $x(t)$ is linearly increasing (i.e., speed is constant), sketch the signals $u(t)$, v_1, v_2, v_3, and v_o as functions of time.

Solution

i. The LVDT has a primary coil, which is excited by an ac voltage of $v_p \sin \boldsymbol{\omega}_c t$. The ferromagnetic core is attached to the moving object whose displacement $x(t)$ is to be measured. The two secondary coils are connected in series opposition so that the LVDT output is zero at the null position, and the direction of motion can be detected as well. The amplifier is a noninverting type. It amplifies the output of the LVDT, which is an ac (carrier) signal of frequency $\boldsymbol{\omega}_c$, that is modulated by the core displacement $x(t)$. The multiplier circuit generates the product of the primary (carrier) signal and the secondary (LVDT output) signal. This is an important step in demodulating the LVDT output (see Chapter 3 for amplitude modulation and demodulation).

The product signal from the multiplier has a high-frequency ($2\boldsymbol{\omega}_c$) carrier component, added to the modulating component ($x(t)$). The low-pass filter removes this unnecessary high-frequency component, to obtain the demodulated signal, which is proportional to the core displacement $x(t)$.

ii. **Noninverting Amplifier**

Potentials at the $+$ and $-$ terminals of the op-amp are nearly equal. Also, currents through these leads are nearly zero. (These are the two common assumptions used for an op-amp; see Chapter 3). Then, the current balance at node A gives,

$$\frac{v_2 - v_1}{R_2} = \frac{v_1}{R_1}.$$

Hence,

$$v_2 = kv_1 \tag{i}$$

with

$$k = \frac{R_1 + R_2}{R_1} = \text{amplifier gain.} \tag{ii}$$

Loss-pass filter: Since the $+$ lead of the op-amp has approximately zero potential (ground), the voltage at point B is also approximately zero. The current balance for node B gives

$$\frac{v_3}{R_1} + \frac{v_o}{R} + C\dot{v}_o = 0.$$

Hence,

$$\tau \frac{dv_o}{dt} + v_o = -\frac{R}{R_1} v_3, \tag{iii}$$

where

$$\tau = RC = \text{filter time constant} \qquad \text{(iv)}$$

The transfer function of the filter is

$$\frac{v_o}{v_3} = -\frac{k_o}{(1 + \tau s)} \qquad \text{(v)}$$

with the filter gain

$$k_o = R/R_1. \qquad \text{(vi)}$$

In the frequency domain,

$$\frac{v_o}{v_3} = -\frac{k_o}{(1 + \tau j \omega)}. \qquad \text{(vii)}$$

Finally, neglecting the phase shift in the LVDT, we have

$$v_1 = v_p r\, x(t) \sin \omega_c t,$$
$$v_2 = v_p rk\, x(t) \sin \omega_c t,$$
$$v_3 = v_p^2 rk\, x(t) \sin^2 \omega_c t$$

or

$$v_3 = \frac{v_p^2 rk}{2} x(t)[1 - \cos 2\,\omega_c t]. \qquad \text{(viii)}$$

The carrier signal will be filtered out by the low-pass filter with an appropriate cutoff frequency. Then,

$$v_o = \frac{v_p^2 rk_o}{2} x(t). \qquad \text{(ix)}$$

iii.   Filter magnitude $= \dfrac{k_o}{\sqrt{1 + \tau^2 \omega^2}}.$ (4.19)

For no more than 5% of the carrier ($2\omega_c$) component to pass through, we must have

$$\frac{k_o}{\sqrt{1 + \tau^2 (2\omega_c)^2}} \le \frac{5}{100} k_o \qquad \text{(4.20)}$$

or, $\tau\omega_c \ge 10$ (approximately).

Pick $\tau\omega_c = 10$.
With $R = 100\ \text{k}\Omega$, $\omega_c = 500\ \text{rad/s}$, we have: $C \times 100 \times 10^3 \times 500 = 10$.
Hence, $C = 0.2\ \mu\text{F}$.

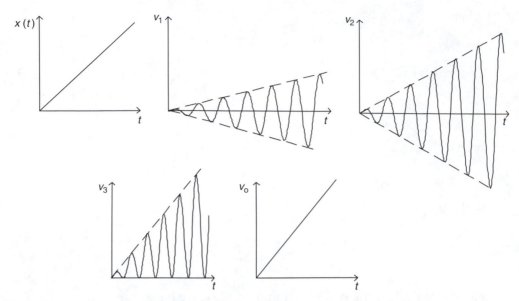

FIGURE 4.13
Nature of the signals at various locations in an LVDT measurement circuit.

According to the carrier frequency value (500 rad/s) we should be able to measure displacements $x(t)$ up to about 50 rad/s. But the flat region of the filter is about $\omega\tau = 0.1$, which, with the present value of $\tau = 0.02$ s, gives a bandwidth of only 5 rad/s for the overall measuring device (LVDT).

(iv) See Figure 4.13 for a sketch of various signals in the LVDT measurement system.

■

Advantages of the LVDT include the following:

1. It is essentially a noncontacting device with no frictional resistance. Near-ideal electromechanical energy conversion and lightweight core will result in very small resistive forces. Hysteresis (both magnetic hysteresis and mechanical backlash) is negligible.

2. It has low output impedance, typically in the order of 100 Ω. (Signal amplification is usually not needed beyond what is provided by the conditioning circuit.)

3. Directional measurements (positive/negative) are provided by it.

4. It is available in small sizes (e.g., 1 cm long with maximum travel or "stroke"of 2 mm).

5. It has a simple and robust construction (inexpensive and durable).

6. Fine resolutions are possible (theoretically, infinitesimal resolution; practically, much better than a coil potentiometer).

4.3.3   Rotatory-Variable Differential Transformer/Transducer

The rotatory-variable differential transformer or transducer (RVDT) operates using the same principle as the LVDT, except that in an RVDT, a rotating ferromagnetic core is used. The RVDT is used for measuring angular displacements. A schematic diagram of the device is shown in Figure 4.14a, and a typical operating curve is shown in Figure 4.14b.

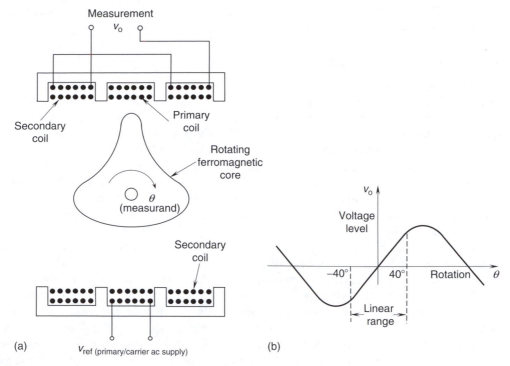

FIGURE 4.14

(a) Schematic diagram of an RVDT. (b) Operating curve.

The rotating core is shaped such that a reasonably wide linear operating region is obtained. Advantages of the RVDT are essentially the same as those cited for the LVDT. Since the RVDT measures angular motions directly, without requiring nonlinear transformations (which is the case in resolvers, as discussed subsequently), its use is convenient in angular position servos. The linear range is typically $\pm 40^\circ$ with a nonlinearity error less than $\pm 0.5\%$ of full scale.

As noted before, in variable-inductance devices, the induced voltage is generated through the rate of change of the magnetic flux linkage. Therefore, displacement readings are distorted by velocity of the moving member; similarly, velocity readings are affected by acceleration of the moving member; and so on. For the same displacement value, the transducer reading depends on the velocity of the measured object at that displacement (position). This error is known as the *rate error*, which increases with the ratio: (cyclic velocity of the core)/(carrier frequency), for an LVDT. Hence, the rate error can be reduced by increasing carrier frequency. The reason for this is discussed now.

At high carrier frequencies, the induced voltage due to the transformer effect (having frequency of the primary signal) is greater than the induced voltage due to the rate (velocity) effect of the moving member. Hence, the error is small. To estimate a lower limit for the carrier frequency in order to reduce rate effects to an acceptable level, we may proceed as follows:

1. For an LVDT: Let $\dfrac{\text{Maximum speed of operation}}{\text{Stroke of LVDT}} = \omega_o.$ (4.21)

The excitation frequency of the primary coil (i.e., carrier frequency) should be chosen as $5\omega_o$ or more.

2. For an RVDT: For the parameter ω_o in the above specification, use the maximum angular frequency of operation (of the rotor) of the RVDT.

4.3.4 Mutual-Induction Proximity Sensor

This displacement transducer also operates on the mutual-induction principle. A simplified schematic diagram of such a device is shown in Figure 4.15a . The insulating E core carries the primary winding in its middle limb. The two end limbs carry secondary windings, which are connected in series. Unlike the LVDT and the RVDT, the two voltages induced in the secondary winding segments are additive in this case. The region of the moving surface (target object) that faces the coils has to be made of ferromagnetic material so that as the object moves, the magnetic reluctance and the flux linkage between the primary and the secondary coils change. This, in turn, changes the induced voltage in the secondary coil, and this voltage change is a measure of the displacement.

Note that, unlike the LVDT, which has an axial displacement configuration, the proximity probe has a transverse (or, lateral) displacement configuration. Hence, it is particularly suitable for measuring transverse displacements or proximities of moving objects (e.g., transverse motion of a beam or whirling shaft). We can see from the operating curve shown in Figure 4.15b that the displacement–voltage relation of a proximity probe is nonlinear. Hence, these proximity sensors should be used only for measuring small displacements (e.g., in a typical linear range of 5.0 mm or 0.2 in.), unless accurate non-linear typical calibration curves are available. Since the proximity sensor is a noncontacting device, mechanical loading is small and the product life is high. Because a ferromagnetic object is used to alter the reluctance of the flux path, the mutual-induction proximity sensor is a variable-reluctance device as well. The operating frequency limit is about 1/10th the excitation frequency of the primary coil (carrier frequency). As for an LVDT, demodulation of the induced voltage (secondary voltage) is required to obtain direct (dc) output readings.

Proximity sensors are used in a wide variety of applications pertaining to noncontacting displacement sensing and dimensional gaging. Some typical applications are

1. Measurement and control of the gap between a robotic welding torch head and the work surface

2. Gaging the thickness of metal plates in manufacturing operations (e.g., rolling and forming)

3. Detecting surface irregularities in machined parts

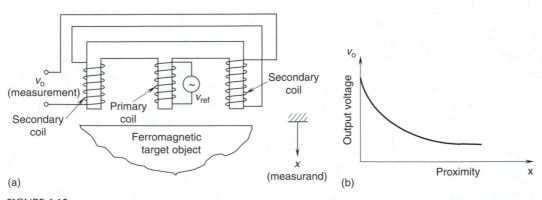

(a)

(b)

FIGURE 4.15

(a) Schematic diagram of a mutual-induction proximity sensor. (b) Operating curve.

4. Angular speed measurement at steady state, by counting the number of rotations per unit time

5. Measurement of vibration in rotating machinery and structures

6. Level detection (e.g., in the filling, bottling, and chemical process industries)

7. Monitoring of bearing assembly processes

Some mutual-induction displacement transducers use the relative motion between the primary coil and the secondary coil to produce a change in flux linkage. Two such devices are the resolver and the synchro-transformer, which are described next. These are not variable-reluctance transducers because they do not employ a ferromagnetic moving element.

4.3.5 Resolver

This mutual-induction transducer is widely used for measuring angular displacements. A simplified schematic diagram of the resolver is shown in Figure 4.16. The *rotor* contains the primary coil. It consists of a single two-pole winding element energized by an ac supply voltage v_{ref}. The rotor is directly attached to the object whose rotation is measured. The *stator* consists of two sets of windings placed 90° apart. If the angular position of the rotor with respect to one pair of stator windings is denoted by θ, the induced voltage in this pair of windings is given by

$$v_{o1} = a v_{ref} \cos \theta. \tag{4.22}$$

The induced voltage in the other pair of windings is given by

$$v_{o2} = a v_{ref} \sin \theta. \tag{4.23}$$

Note that these are amplitude-modulated signals—the carrier signal v_{ref}, which is a sinusoidal function of time, is modulated by the motion θ. The constant parameter a depends primarily on geometric and material characteristics of the device, for example, the ratio of the number of turns in the rotor and stator windings.

Either of the two output signals v_{o1} and v_{o2} may be used to determine the angular position in the first quadrant (i.e., $0 \leq \theta \leq 90°$). Both signals are needed, however, to

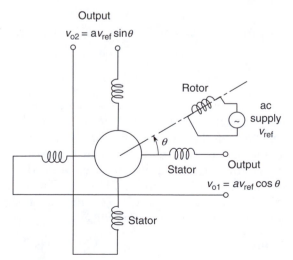

FIGURE 4.16
Schematic diagram of a resolver.

determine the displacement (direction as well as magnitude) in all four quadrants (i.e., in $0 \leq \theta \leq 360°$) without causing any ambiguity. For instance, the same sine value is obtained for both $90° + \theta$ and $90° - \theta$ (i.e., a positive rotation and a negative rotation from the $90°$ position), but the corresponding cosine values have opposite signs, thus providing the proper direction.

4.3.5.1 Demodulation

As for differential transformers (i.e., LVDT and RVDT) transient displacement signals of a resolver can be extracted by demodulating its (modulated) outputs. As usual, this is accomplished by filtering out the carrier signal, thereby extracting the modulating signal. The two output signals v_{o1} and v_{o2} of a resolver are termed quadrature signals. Suppose that the carrier (primary) signal is

$$v_{ref} = v_a \sin \omega t. \tag{4.24}$$

The induced quadrate signals are:

$$v_{o1} = a v_a \cos \theta \sin \omega t, \tag{4.25}$$

$$v_{o2} = a v_a \sin \theta \sin \omega t. \tag{4.26}$$

Multiply each quadrature signal by v_{ref} to get

$$v_{m1} = v_{o1} v_{ref} = a v_a^2 \cos \theta \sin^2 \omega t = \frac{1}{2} a v_a^2 \cos \theta [1 - \cos 2\omega t], \tag{4.27}$$

$$v_{m2} = v_{o2} v_{ref} = a v_a^2 \sin \theta \sin^2 \omega t = \frac{1}{2} a v_a^2 \sin \theta [1 - \cos 2\omega t]. \tag{4.28}$$

Since the carrier frequency ω should be about 10 times the maximum frequency content of interest in the angular displacement θ, one can use a low-pass filter with a cutoff set at $\omega/10$ to remove the carrier components in v_{m1} and v_{m2}. This gives the demodulated outputs

$$v_{f1} = \frac{1}{2} a v_a^2 \cos \theta, \tag{4.29}$$

$$v_{f2} = \frac{1}{2} a v_a^2 \sin \theta. \tag{4.30}$$

Note that Equation 4.29 and Equation 4.30 provide both $\cos \theta$ and $\sin \theta$, and hence magnitude and sign of θ.

4.3.5.2 Resolver with Rotor Output

An alternative form of resolver uses two ac voltages $90°$ out of phase, generated from a digital signal-generator board, to power the two coils of the stator. The rotor has the secondary winding in this case. The phase shift of the induced voltage determines the angular position of the rotor. An advantage of this arrangement is that it does not require slip-rings and brushes to energize the windings (which are stationary), as needed in the previous arrangement where the rotor has the primary winding. However, it will need some mechanism to pick-off the output signal from the rotor. To illustrate this alternative design, suppose that the excitation signals in the two stator coils are

$$v_1 = v_a \sin \omega t \tag{4.31}$$

$$v_2 = v_a \cos \omega t. \tag{4.32}$$

When the rotor coil is oriented at angular position θ with respect to the stator-coil 2, it will be at an angular position $\pi/2 - \theta$ from the stator-coil 1 (assuming that the rotor coil is in the first quadrant: $0 \leq \theta \leq \pi/2$). Hence, the voltage induced by stator coil 1 in the rotor coil would be $v_a \sin \omega t \sin \theta$, and the voltage induced by the stator coil 2 in the rotor coil would be $v_a \cos \omega t \cos \theta$. It follows that the total induced voltage in the rotor coil is given by

$$v_r = v_a \sin \omega t \sin \theta + v_a \cos \omega t \cos \theta$$

or

$$v_r = v_a \cos (\omega t - \theta). \tag{4.33}$$

It is seen that the phase angle of the rotor output signal with respect to the stator excitation signals v_1 and v_2 provides both magnitude and sign of the rotor position θ.

The output signals of a resolver are nonlinear (trigonometric) functions of the angle of rotation. (Historically, resolvers were used to compute trigonometric functions or to resolve a vector into orthogonal components.) In robot control applications, this is some-times viewed as a blessing. For computed torque control of robotic manipulators, for example, trigonometric functions of the joint angles are needed in order to compute the required input signals (reference joint torque values). Consequently, when resolvers are used to measure joint angles in manipulators, there is an associated reduction in processing time of the control input signals because the trigonometric functions them-selves are available as direct measurements.

The primary advantages of the resolver include

1. Fine resolution and high accuracy
2. Low output impedance (high signal levels)
3. Small size (e.g., 10 mm diameter)
4. Direct availability of the sine and cos functions of the measured angles

Its main limitations are

1. Nonlinear output signals (an advantage in some applications where trigono-metric functions of the rotations are needed);
2. Bandwidth limited by supply frequency;
3. Slip-rings and brushes would be needed if complete and multiple rotations have to be measured (which adds mechanical loading and also creates component wear, oxidation, and thermal and noise problems).

4.3.6 Synchro Transformer

The "synchro" is somewhat similar in operation to the resolver. The main differences are that the synchro employs two identical rotor–stator pairs, and each stator has three sets of windings, which are placed 120° apart around the rotor shaft. A schematic diagram for this arrangement is shown in Figure 4.17. Both rotors have single-phase windings. One of the rotors is energized with an ac supply voltage v_{ref}. This induces voltages in the three winding segments of the corresponding stator. These voltages have different amplitudes,

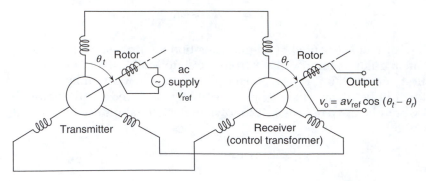

FIGURE 4.17
Schematic diagram of a synchro-transformer.

which depend on the angular position of the rotor. (Note: The resultant magnetic field
from the induced currents in these three stator winding sets must be in the same direction
as the rotor magnetic field.) This drive rotor–stator pair is known as the *transmitter*. The
other rotor–stator pair is known as the *receiver* or the *control transformer*. Windings of the
transmitter stator are connected correspondingly to the windings of the receiving stator, as
shown in Figure 4.17. Accordingly, the resultant magnetic field of the receiver stator must
be in the same direction as the resultant magnetic field of the transmitter stator (and of
course the transmitter rotor). This resultant magnetic field in the receiver stator induces a
voltage v_o in the rotor of the receiver. Suppose that the angle between the transmitter rotor
and one set of windings in its stator (the same reference winding set as what is used to
measure the angle of the transmitter rotor) is denoted by θ_t. If the receiver rotor is aligned
with this direction (i.e., $\theta_r = \theta_t$), then the induced voltage v_o becomes maximum. If the
receiver rotor is placed at $90°$ to this resultant magnetic field, then $v_o = 0$. Therefore, an
appropriate expression for the synchro output is

$$v_o = av_{ref} \cos(\theta_t - \theta_r). \tag{4.34}$$

Synchros are operated near $\theta_r = \theta_t + 90°$, where the output voltage is zero. Hence, we
define a new angle θ such that

$$\theta_r = \theta_t + 90° - \theta. \tag{4.35}$$

As a result, Equation (4.34) becomes

$$v_o = av_{ref} \sin\theta. \tag{4.36}$$

Synchro-transformers can be used to measure relative displacements between two rotating objects. When measuring absolute displacements, one of the rotors is attached to the
rotating member (e.g., the shaft), while the other rotor is fixed to a stationary member
(e.g., the bearing). As is clear from the previous discussion, a zero reading corresponds to
the case where the two rotors are $90°$ apart.

Synchros have been used extensively in position servos, particularly for the position
control of rotating objects. Typically, the input command is applied to the transmitter
rotor. The receiver rotor is attached to the object that is controlled. The initial physical
orientations of the two rotors should ensure that for a given command, the desired
position of the object corresponds to zero output voltage v_o; that is, when the two rotors
are $90°$ apart. In this manner, v_o can be used as the position error signal, which is fed
into the control circuitry that generates a drive signal so as to compensate for the error

(e.g., using proportional plus derivative control). For small angles θ, the output voltage may be assumed proportional to the angle. For large angles, inverse sine should be taken. Note that ambiguities arise when the angle θ exceeds 90°. Hence, synchro readings should be limited to $\pm 90°$. In this range, the synchro provides directional measurements. As for a resolver or LVDT, demodulation is required to extract transient measurements from the output signal. This is accomplished, as usual, by suppressing the carrier from the modulated signal, as demonstrated for the resolver and the LVDT.

The advantages and disadvantages of the synchro are essentially the same as those of the resolver. In particular, quadrature error (at null voltage) may be present because of impedance nonuniformities in the winding segments. Furthermore, velocity error or rate error (i.e., velocity-dependent displacement readings) is also a possibility. This may be reduced by increasing the carrier frequency, as in the case of a differential transformer and a resolver.

4.3.7 Self-Induction Transducers

These transducers are based on the principle of self-induction. Unlike mutual-induction transducers, only a single coil is employed. This coil is activated by an ac supply voltage v_{ref} of sufficiently high frequency. The current produces a magnetic flux, which is linked back with the coil. The level of flux linkage (or self-inductance) can be varied by moving a ferromagnetic object within the magnetic field. This movement changes the reluctance of the flux linkage path and also the inductance in the coil. The change in self-inductance, which can be measured using an inductance-measuring circuit (e.g., an inductance bridge; see Chapter 3), represents the measurand (displacement of the object). Note that self-induction transducers are usually variable-reluctance devices as well.

A typical self-induction transducer is a self-induction proximity sensor. A schematic diagram of this device is shown in Figure 4.18. This device can be used as a displacement sensor for transverse displacements. For instance, the distance between the sensor tip and ferromagnetic surface of a moving object, such as a beam or shaft, can be measured. Other applications include those mentioned for mutual-induction proximity sensors. High-speed displacement measurements can result in velocity error (rate error) when variable-inductance displacement sensors (including self-induction transducers) are used. This effect may be reduced, as in other ac-activated variable-inductance sensors, by increasing the carrier frequency.

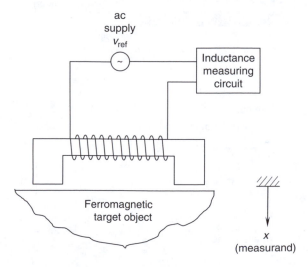

FIGURE 4.18
Schematic diagram of a self-induction proximity sensor.

4.4 Permanent-Magnet Transducers

In discussing this third category of variable-inductance transducer, we present several types of velocity transducers termed tachometers. A distinctive feature of permanent-magnet transducers is that they have a permanent magnet to generate a uniform and steady magnetic field. A relative motion between the magnetic field and an electrical conductor induces a voltage, which is proportional to the speed at which the conductor crosses the magnetic field (i.e., the rate of change of flux linkage). In some designs, a unidirectional magnetic field generated by a dc supply (i.e., an electromagnet) is used in place of a permanent magnet. Nevertheless, they are generally termed permanent-magnet transducers. Permanent-magnet transducers are not variable-reluctance devices in general.

4.4.1 DC Tachometer

This is a permanent-magnet dc velocity sensor in which the principle of electromagnetic induction in a conducting coil due to variations in the magnetic field of a permanent magnet is used. Depending on the configuration, either rectilinear speeds or angular speeds can be measured. Schematic diagrams of the two configurations are shown in Figure 4.19. These are passive transducers, because the energy for the output signal v_o is derived from the motion (i.e., measured signal) itself. The entire device is usually enclosed in a steel casing to shield (isolate) it from ambient magnetic fields.

In the rectilinear velocity transducer (Figure 4.19a), the conductor coil is wound on a core and placed centrally between two magnetic poles, which produce a cross-magnetic field. The core is attached to the moving object whose velocity v must be measured. This

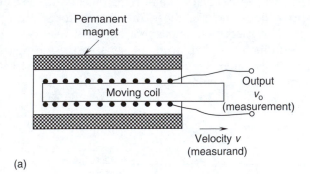

(a)

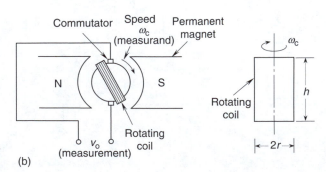

FIGURE 4.19
Permanent-magnet dc transducers. (a) Rectilinear velocity transducer. (b) DC tachometer.   (b)

velocity is proportional to the induced voltage v_o. Alternatively, a moving magnet and a fixed coil may be used as a velocity transducer (rectilinear or rotatory). This arrangement is perhaps more desirable since it eliminates the need for any sliding contacts (slip-rings and brushes) for the output leads, thereby reducing mechanical loading error, wear, and related problems.

The dc tachometer (or tachogenerator) is a common transducer for measuring angular velocities. Its principle of operation is the same as that for a dc generator (or back-driving of a dc motor). This principle of operation is illustrated in Figure 4.19b. The rotor is directly connected to the rotating object. The output signal that is induced in the rotating coil is picked up as dc voltage v_o using a suitable commutator device—typically consisting of a pair of low-resistance carbon brushes—that is stationary but makes contact with the rotating coil through split slip-rings so as to maintain the direction of the induced voltage the same throughout each revolution (see commutation in dc motors—Chapter 7). According to Faraday's law, the induced voltage is proportional to the rate of change of magnetic flux linkage. For a coil of height h and width $2r$ that has n turns, moving at an angular speed ω_c in a uniform magnetic field of flux density β, this is given by

$$v_o = (2nhr\ \beta)\omega_c = k\omega_c \tag{4.37}$$

This proportionality between v_o and ω_c is used to measure the angular speed ω_c. The proportionality constant k is known as the *back-emf constant* or the *voltage constant*.

4.4.1.1 Electronic Commutation

Slip-rings and brushes and associated drawbacks can be eliminated in a dc tachometer by using electronic commutation. In this case, a permanent-magnet rotor together with a set of stator windings are used. The output of the tachometer is drawn from the stationary (stator) coil. It has to be converted to a dc signal using an electronic switching mechanism, which has to be synchronized with the rotation of the tachometer (see Chapter 7, under brushless dc motors). Because of switching and associated changes in the magnetic field of the output signal, induced voltages known as *switching transients* will result. This is a drawback in electronic commutation.

4.4.1.2 Modeling and Design Example

A dc tachometer is shown schematically in Figure 4.20a. The field windings are powered by dc voltage v_f. The across variable at the input port is the measured angular speed ω_i.

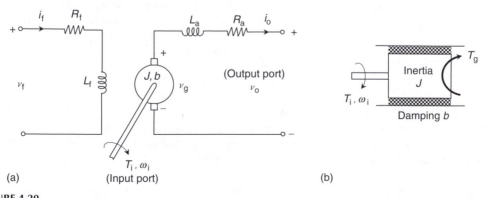

FIGURE 4.20
A dc tachometer example. (a) Equivalent circuit with an impedance load; (b) Armature free-body diagram.

The corresponding torque T_i is the through variable at the input port. The output voltage v_o of the armature circuit is the across variable at the output port. The corresponding current i_o is the through variable at the output port. Obtain a transfer-function model for this device. Discuss the assumptions needed to decouple this result into a practical input–output model for a tachometer. What are the corresponding design implications? In particular, discuss the significance of the mechanical time constant and the electrical time constant of the tachometer.

Solution

The generated voltage v_g at the armature (rotor) is proportional to the magnetic field strength of field windings, (which, in turn, is proportional to the filed current i_f) and the speed of the armature ω_i. Hence, $v_g = K'i_f\,\omega_i$. Now assuming constant field current, we have

$$v_g = K\omega_i \tag{i}$$

The rotor magnetic torque T_g, which resists the applied torque T_i, is proportional to the magnetic field strengths of the field windings and armature windings. Consequently, $T_g = K'i_f i_o$. Since i_f is assumed constant, we get

$$T_g = Ki_o. \tag{ii}$$

Note that the same constant K is used both in Equation (i) and Equation (ii). This is valid when the same units are used to measure mechanical power (N.m/s) and electrical power (W) and when the internal energy-dissipation mechanisms are not significant in the associated internal coupling (i.e., ideal energy conversion is assumed). The equation for the armature circuit is

$$v_o = v_g - R_a i_o - L_a \frac{di_o}{dt}, \tag{iii}$$

where R_a is the armature resistance and L_a is the leakage inductance in the armature circuit. With reference to Figure 4.20b, Newton's second law for a tachometer armature having inertia J and damping constant b is expressed as

$$J\frac{d\omega_i}{dt} = T_i - T_g - b\omega_i. \tag{iv}$$

Now Equation (i) is substituted into Equation (iii) in order to eliminate v_g. Similarly, Equation (ii) is substituted into Equation (iv) in order to eliminate T_g. Next, the time derivatives are replaced by the Laplace variable s. This results in the two algebraic relations:

$$v_o = K\omega_i - (R_a + sL_a)i_o, \tag{v}$$

$$(b + sJ)\omega_i = T_i - Ki_o. \tag{vi}$$

Note that the variables v_i, i_o, ω_i, and T_i in Equation (v) and Equation (vi) are actually Laplace transforms (functions of s), not functions of t, as in Equation (i) through Equation (iv). Finally, i_o in Equation (v) is eliminated using Equation (vi). This gives the matrix transfer function relation

$$\begin{bmatrix} v_o \\ i_o \end{bmatrix} = \begin{bmatrix} K + (R_a + sL_a)(b + sJ)/K & -(R_a + sL_a)/K \\ -(b + sJ)/K & 1/K \end{bmatrix} \begin{pmatrix} \omega_i \\ T_i \end{pmatrix}. \tag{vii}$$

The corresponding frequency domain relations are obtained by replacing s with $j\omega$, where ω represents the angular frequency (radians per second) in the frequency spectrum of an input signal.

Even though transducers are more accurately modeled as two-port elements, which have two variables associated with each port (see Figure 4.20), it is useful and often essential, for practical reasons, to relate just one input variable (measurand) and one output variable (measurement) so that only one (scalar) transfer function relating these two variables need be specified. This is particularly true for a transducer, as in the present example, and assumes some form of decoupling in the true model. If this assumption does not hold in the range of operation of the transducer, a measurement error would result. In particular, in the present tachometer example, we like to express the output voltage v_o in terms of the measured speed ω_i. In this case, the off-diagonal term—$(R_a + sL_a)/K$—in Equation (vii) has to be neglected. This is valid when the tachometer gain parameter K is large and the armature resistance R_a is negligible, since the leakage inductance L_a is negligible in any case for most practical purposes. Note from Equation (i) and Equation (ii) that the tachometer gain K can be increased by increasing the field current i_f. This will not be feasible if the field windings are already saturated, however. Furthermore, K (or K') depends on parameters such as number of turns and dimensions of the stator windings and magnetic properties of the stator core. Since there is a limitation on the physical size of the tachometer and the types of materials used in the construction, it is clear that K cannot be increased arbitrarily. The instrument designer should take such factors into consideration in developing a design that is optimal in many respects. In practical transducers, the operating range is specified in order to minimize the effect of coupling terms (and nonlinearities, frequency dependence, etc.), and the residual errors are accounted for by using correction curves. This approach is more convenient than using the coupled model Equation (vii), which introduces three more (scalar) transfer functions (in general) into the model.

Another desirable feature for practical transducers is to have a static (algebraic, non-dynamic) input–output relationship so that the output instantly reaches the input value (or the measured variable), and the frequency dependence of the transducer characteristic is eliminated. Then the transducer transfer function becomes a pure gain (i.e., independent of frequency). This happens when the transducer time constants are small (i.e., the transducer bandwidth is high). Returning to the tachometer example, it is clear from Equation (vii) that the transfer-function relations become static (frequency-independent) when both electrical time constant

$$\tau_e = \frac{L_a}{R_a} \tag{4.38}$$

and mechanical time constant

$$\tau_m = \frac{J}{b} \tag{4.39}$$

are negligibly small. The electrical time constant of a motor/generator is usually an order of magnitude smaller than the mechanical time constant (see Chapter 7). Hence, one must first concentrate on the mechanical time constant. Note from Equation (4.39) that τ_m can be reduced by decreasing rotor inertia and increasing rotor damping. Unfortunately, rotor inertia depends on rotor dimensions, and this determines the gain parameter K, as we saw earlier. Hence, we face some constraint in reducing K. Furthermore, when the rotor size is reduced (in order to reduce J), the number of turns in the windings should be reduced as well. Then, the air gap between the rotor and the stator becomes less uniform, which creates a voltage ripple in the induced voltage (tachometer output). The resulting measurement error can be significant. Next turning to damping, it is intuitively clear that if we increase b, a larger torque T_i will be required to drive the tachometer. This will load the measured object, which generates the measurand ω_i, possibly affecting the measurand itself. Hence, increasing b also has to be done cautiously. Now, going back to Equation (vii), we note that the dynamic terms in the transfer function between ω_i and v_o decrease as K is increased. So we note that increasing K has two benefits: reduction of coupling and reduction of dynamic effects (i.e., reduction of the frequency dependence of the system, thereby increasing the useful frequency range and bandwidth or speed of response).

4.4.1.3 Loading Considerations

The torque required to drive a tachometer is proportional to the current generated (in the dc output). The associated proportionality constant is the *torque constant*. With consistent units, in the case of ideal energy conversion, this constant is equal to the voltage constant. Since the tachometer torque acts on the moving object whose speed is measured, high torque corresponds to high mechanical loading, which is not desirable. Hence, it is desirable to reduce the tachometer current as much as possible. This can be realized by making the input impedance of the signal-acquisition device (i.e., voltage reading and interface hardware) for the tachometer as large as possible. Furthermore, distortion in the tachometer output signal (voltage) can result because of the reactive (inductive and capacitive) loading of the tachometer. When dc tachometers are used to measure transient velocities, some error results from the rate (acceleration) effect. This error generally increases with the maximum significant frequency that must be retained in the transient velocity signal, which in turn depends on the maximum speed that has to be measured. All these types of error can be reduced by increasing the load impedance.

For illustration, consider the equivalent circuit of a tachometer with an impedance load connected to the output port of the armature circuit shown in Figure 4.20. The induced voltage $k\omega_c$ is represented by a voltage source. The constant k depends on the coil geometry, the number of turns, and the magnetic flux density (see Equation 4.37). Coil resistance is denoted by R, and leakage inductance is denoted by L_1. The load impedance is Z_L. From straightforward circuit analysis in the frequency domain, the output voltage at the load is given by

$$v_o = \left[\frac{Z_L}{R + j\omega L_1 + Z_L}\right] k\omega_c. \tag{4.40}$$

It can be seen that because of the leakage inductance, the output signal attenuates more at higher frequencies ω of the velocity transient. In addition, a loading error is present. If Z_L is much larger than the coil impedance, however, the ideal proportionality, as given by $v_o = k\omega_c$ is achieved.

A *digital tachometer* is a velocity transducer, which is governed by somewhat different principles. It generates voltage pulses at a frequency proportional to the angular speed. Hence, it is considered a digital transducer, as discussed in Chapter 5.

4.4.2 Permanent-Magnet AC Tachometer

This device has a permanent magnet rotor and two separate sets of stator windings as schematically shown in Figure 4.21a. One set of windings is energized using an ac reference (carrier) voltage. Induced voltage in the other set of windings is the tachometer output. When the rotor is stationary or moving in a quasi-static manner, the output voltage is a constant-amplitude signal much like the reference voltage, as in an electrical transformer. As the rotor moves at a finite speed, an additional induced voltage, which is proportional to the rotor speed, is generated in the secondary coil. This is due to the rate of change of flux linkage into the secondary coil as a result of the rotating magnet. The overall output from the secondary coil is an amplitude-modulated signal whose amplitude is proportional to the rotor speed. For transient velocities, it becomes necessary to demodulate this signal in order to extract the transient velocity signal (i.e., the modulating signal) from the overall (modulated) output. The direction of velocity is determined from the phase angle of the modulated signal with respect to the carrier signal. Note that in an LVDT, the amplitude of the ac magnetic flux (linkage) is altered by the position of the ferromagnetic core. But in an ac permanent-magnet tachometer, a dc magnetic flux is generated by the magnetic rotor, and when the rotor is stationary it does not induce a voltage in the coils. The flux linked with the stator windings changes because of the rotation of the rotor, and the rate of change of linked flux is proportional to the speed of the rotor.

For low-frequency applications (5 Hz or less), a standard ac supply at line frequency (60 Hz) may be adequate to power an ac tachometer. For moderate-frequency applications, a 400 Hz supply may be used. For high-frequency (high-bandwidth) applications a high-frequency signal generator (oscillator) may be used as the primary signal. In high-bandwidth applications, carrier frequencies as high as 1.5 kHz are commonly used. Typical sensitivity of an ac permanent-magnet tachometer is of the order of 50–100 mV/rad/s.

4.4.3 AC Induction Tachometer

This tachometer is similar in construction to a two-phase induction motor (see Chapter 7). The stator arrangement is identical to that of the ac permanent-magnet tachometer, as presented before. The rotor has windings, which are shorted and not energized by an external source, as shown in Figure 4.21b. The primary stator coil is powered by an ac supply. This induces a voltage in the rotor coil and it is a modulated signal. The high-frequency (carrier) component of this induced signal is due to the direct transformer

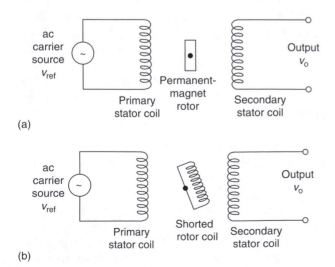

FIGURE 4.21
(a) An ac permanent-magnet tachometer.
(b) An ac induction tachometer.

action of the primary ac. The other (modulating) component is induced by the speed of rotation of the rotor, and its magnitude is proportional to the speed of rotation. The nonenergized stator (secondary) coil provides the output of the tachometer. This voltage output is a result of both the stator (primary) field and the speed of rotor coil. As a result, the tachometer output has a carrier ac component whose frequency is the same as the primary signal frequency, and a modulating component, which is proportional to the speed of rotation. Demodulation would be needed to extract the component that is proportional to the angular speed of the rotor.

The main advantage of ac tachometers over their conventional dc counterparts is the absence of slip-ring and brush devices, since the output is obtained from the stator. In particular, the signal from a dc tachometer usually has a voltage ripple, known as the *commutator ripple* or *brush noise*, which is generated as the split ends of the slip-ring pass over the brushes, and as a result of contact bounce, and so forth. The frequency of the commutator ripple is proportional to the speed of operation; consequently, filtering it out using a notch filter is difficult (because a speed-tracking notch filter would be needed). Also, there are problems with frictional loading and contact bounce in dc tachometers, and these problems are absent in ac tachometers. Note, however, that a dc tachometer with electronic commutation does not use slip-rings and brushes. But they produce switching transients, which are also undesirable.

As for any sensor, the noise components dominate at low levels of output signal. In particular, since the output of a tachometer is proportional to the measured speed, at low speeds, the level of noise, as a fraction of the output signal, can be large. Hence, removal of noise takes an increased importance at low speeds.

It is known that at high speeds the output from an ac tachometer is somewhat non-linear (primarily because of the saturation effect). Furthermore, signal demodulation is necessary, particularly for measuring transient speeds. Another disadvantage of ac tachometer is that the output signal level depends on the supply voltage; hence, a stabilized voltage source, which has a very small output-impedance, is necessary for accurate measurements.

4.4.4 Eddy Current Transducers

If a conducting (i.e., low-resistivity) medium is subjected to a fluctuating magnetic field, eddy currents are generated in the medium. The strength of eddy currents increases with the strength of the magnetic field and the frequency of the magnetic flux. This principle is used in eddy current proximity sensors. Eddy current sensors may be used as either dimensional gaging devices or displacement sensors.

A schematic diagram of an eddy current proximity sensor is shown in Figure 4.22a. Unlike variable-inductance proximity sensors, the target object of the eddy current sensor does not have to be made of a ferromagnetic material. A conducting target object is needed, but a thin film of conducting material, such as household aluminum foil glued to a nonconducting target object, would be adequate. The probe head has two identical coils, which form two arms of an impedance bridge (see Chapter 3). The coil closer to the probe face is the *active coil*. The other coil is the *compensating coil*. It compensates for ambient changes, particularly thermal effects. The remaining two arms of the bridge consist of purely resistive elements (see Figure 4.22b). The bridge is excited by a radio-frequency voltage supply. The frequency may range from 1 to 100 MHz. This signal is generated from a radio-frequency converter (an oscillator) that is typically powered by a 20 V dc supply. When the target (sensed) object is absent, the output of the impedance bridge is zero, which corresponds to the balanced condition. When the target object is moved close to the sensor, eddy currents are generated in the conducting medium

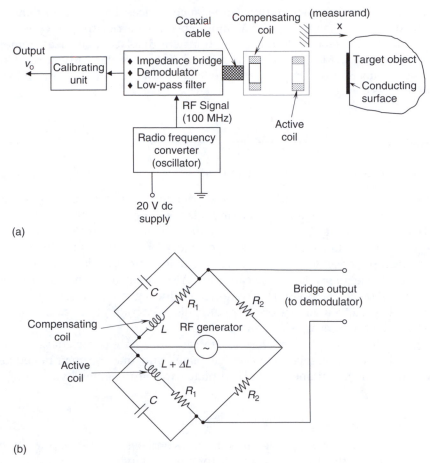

FIGURE 4.22
Eddy current proximity sensor. (a) Schematic diagram. (b) Impedance bridge.

because of the radio-frequency magnetic flux from the active coil. The magnetic field of the eddy currents opposes the primary field, which generates these currents. Hence, the inductance of the active coil increases, creating an imbalance in the bridge. The resulting output from the bridge is an amplitude-modulated signal containing the radio-frequency carrier. This signal can be demodulated by removing the carrier. The resulting signal (modulating signal) measures transient displacement of the target object. Low-pass filtering is used to remove high-frequency leftover noise in the output signal once the carrier is removed. For large displacements, the output is not linearly related to the displacement. Furthermore, the sensitivity of an eddy current probe depends nonlinearly on the nature of the conducting medium, particularly the resistivity. For example, for low resistivities, sensitivity increases with resistivity; for high resistivities, it decreases. A calibrating unit is usually available with commercial eddy current sensors to accommodate various target objects and nonlinearities. The gage factor is usually expressed in volts per millimeter. Note that eddy current probes can also be used to measure resistivity and surface hardness, which affects resistivity, in metals.

The facial area of the conducting medium on the target object has to be slightly larger than the frontal area of the eddy current probe head. If the target object has a curved surface, its radius of curvature has to be at least four times the diameter of the probe. These are not serious restrictions, because the typical diameter of a probe head is about

2 mm. Eddy current sensors are medium-impedance devices; 1000 Ω output impedance is typical. Sensitivity is in the order of 5 V/mm. Since the carrier frequency is very high, eddy current devices are suitable for highly transient displacement measurements—for example, bandwidths up to 100 kHz. Another advantage of the eddy current sensor is that it is a noncontacting device; hence, there it does not present mechanical loading on the moving (target) object.

4.5 Variable-Capacitance Transducers

Variable-inductance devices and variable-capacitance devices are *variable-reactance* devices. [Note that the reactance of an inductance L is given by $j\omega L$ and that of a capacitance C is given by $1/(j\omega C)$, since $v = L\,(di/dt)$ and $i = C\,(dv/dt)$]. For this reason, capacitive transducers fall into the general category of reactive transducers. They are typically high-impedance sensors, particularly at low frequencies, as clear from the impedance (reactance) expression for a capacitor. Also, capacitive sensors are noncontacting devices in the common usage. They require specific signal-conditioning hardware. In addition to analog capacitive sensors, digital (pulse-generating) capacitive transducers such as digital tachometers are also available.

A capacitor is formed by two plates, which can store an electric charge. The charge generates a potential difference between the plates and may be maintained using an external voltage. The capacitance C of a two-plate capacitor is given by

$$C = \frac{kA}{x}, \tag{4.41}$$

where A is the common (overlapping) area of the two plates, x is the gap width between the two plates, and k is the dielectric constant (or permittivity $k = \varepsilon = \varepsilon_r\,\varepsilon_o$; $\varepsilon_r =$ relative permittivity, $\varepsilon_o =$ permittivity in vacuum), which depends on dielectric properties of the medium between the two plates. A change in any one of the three parameters in Equation 4.41 may be used in the sensing process; for example, to measure small transverse displacements, large rotations, and fluid levels. Schematic diagrams for measuring devices that use this feature are shown in Figure 4.23. In Figure 4.23a, angular displacement of one of the plates causes a change in A. In Figure 4.23b, a transverse displacement of one of the plates results in a change in x. Finally, in Figure 4.23c, a change in k is produced as the fluid level between the capacitor plates changes. In all three cases, the associated change in capacitance is measured directly or indirectly and is used to estimate the measurand. A popular method is to use a capacitance bridge circuit to measure the change in capacitance, in a manner similar to how an inductance bridge (see Chapter 3) is used to measure changes in inductance. Other methods include measuring a change in such quantities as charge (using a charge amplifier), voltage (using a high input-impedance device in parallel), and current (using a very low-impedance device in series) that result from the change in capacitance in a suitable circuit. An alternative method is to make the capacitor a part of an inductance–capacitance (L–C) oscillator circuit—the natural frequency of the oscillator ($1/\sqrt{LC}$) measures the capacitance. (Incidentally, this method may be used to measure inductance as well.)

4.5.1 Capacitive Rotation Sensor

In the arrangement shown in Figure 4.23a, one plate of the capacitor rotates with (i.e., is attached to) a rotating object (shaft) and the other plate is kept stationary. Since the common area A is proportional to the angle of rotation θ, Equation 4.41 may be written as

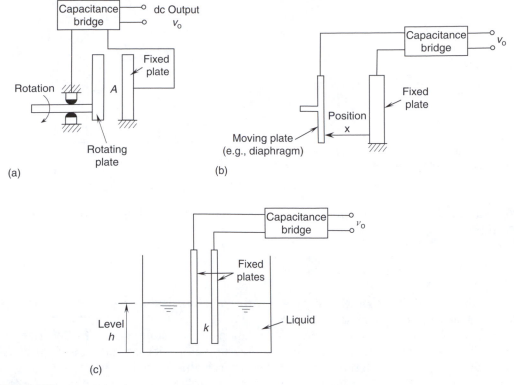

FIGURE 4.23
Schematic diagrams of capacitive sensors. (a) Capacitive rotation sensor. (b) Capacitive displacement sensor. (c) Capacitive liquid level sensor.

$$C = K\theta, \tag{4.42}$$

where K is a sensor constant. This is a linear relationship between C and θ. The capacitance may be measured by any convenient method. The sensor is linearly calibrated to give the angle of rotation.

The sensitivity of this angular displacement sensor is

$$S = \frac{\partial C}{\partial \theta} = K, \tag{4.43}$$

which is constant throughout the measurement. This is expected because the sensor relationship is linear. Note that in the nondimensional form, the sensitivity of the sensor is unity, implying direct sensitivity.

4.5.2 Capacitive Displacement Sensor

The arrangement shown in Figure 4.23b provides a sensor for measuring transverse displacements and proximities. One of the capacitor plates is attached to the moving object and the other is kept stationary. The sensor relationship is

$$C = \frac{K}{x}. \tag{4.44}$$

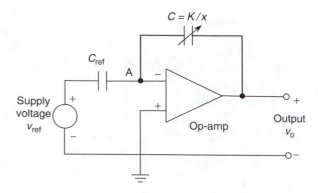

FIGURE 4.24
Linearizing amplifier circuit for a capacitive transverse displacement sensor.

The constant K has a different meaning here. The corresponding sensor sensitivity is given by

$$S = \frac{\partial C}{\partial x} = -\frac{K}{x^2}. \tag{4.45}$$

Again, the sensitivity is unity (negative) in the nondimensional form, which indicates direct sensitivity of the sensor.

Note that Equation 4.44 is a nonlinear relationship. A simple way to linearize this transverse displacement sensor is to use an inverting amplifier, as shown in Figure 4.24. Note that C_{ref} is a fixed reference capacitance, whose value is accurately known. Since the gain of the operational amplifier is very high, the voltage at the negative lead (node point A) is zero for most practical purposes (because the positive lead is grounded). Furthermore, since the input impedance of the op-amp is also very high, the current through the input leads is negligible. These are the two common assumptions used in op-amp analysis (See Chapter 3). Accordingly, the charge balance equation for node point A is: $v_{ref} C_{ref} + v_o C = 0$. Now, in view of Equation 4.44, we get the following linear relationship for the output voltage v_o in terms of the displacement x:

$$v_o = -\frac{v_{ref} C_{ref}}{K} x. \tag{4.46}$$

Hence, measurement of v_o gives the displacement through a linear relationship. The sensitivity of the device can be increased by increasing v_{ref} and C_{ref}. The reference voltage may be either dc or ac with frequency as high as 25 kHz (for high-bandwidth measurements). With an ac reference voltage, the output voltage is a modulated signal, which has to be demodulated to measure transient displacements, as discussed before in the context of variable-inductance sensors.

Example 4.4
Consider the circuit shown in Figure 4.25. Examine how this arrangement could be used to measure displacements.

Solution
Assuming that a very high-impedance device is used to measure the output voltage v_o, the current through the capacitor is the same as that through the resistor. Thus,

$$i = \frac{d}{dt}(Cv_o) = \frac{v_{ref} - v_o}{R}. \tag{4.47}$$

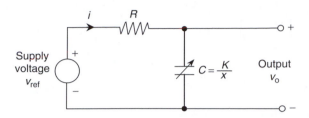

FIGURE 4.25

Capacitive displacement sensor example.

If a transverse displacement capacitor is considered, for example, from Equation 4.44 we have

$$x = RKv_o \Big/ \int_0^t (v_{ref} - v_o)dt. \tag{4.48}$$

This is a nonlinear differential relationship. To measure x, we need to measure the output voltage and perform an integration by either analog or digital means. That introduces a delay and hence reduces the operating speed (frequency range, bandwidth). Furthermore, since $v_o = v_{ref}$ and $v_o = 0$ at steady state, it follows that this approach cannot be used to make steady-state (or quasi-static) measurements. This situation can be corrected by using an ac source as the supply. If the supply frequency is ω, the frequency-domain transfer function between the supply and the output is given by

$$\frac{v_o}{v_{ref}} = \frac{1}{[1 + RKj\omega/x]}. \tag{4.49}$$

Now the displacement x may be determined by measuring either the signal amplification (i.e., amplitude ratio or magnitude) M at the output or the phase lag ϕ of the output signal. The corresponding relations are

$$x = \frac{RK\omega}{\sqrt{1/M^2 - 1}} \tag{4.50}$$

and

$$x = RK\omega/\tan\phi. \tag{4.51}$$

Note that the differential equation of the circuit is not linear unless x is constant. The foregoing transfer-function relations do not strictly hold if the displacement is transient. Nevertheless, reasonably accurate results are obtained when the measured displacement is varying slowly.

∎

The arrangement shown in Figure 4.23c can be used as well for displacement sensing. In this case a solid dielectric element, which is free to move in the longitudinal direction of the capacitor plates, is attached to the moving object whose displacement is to be measured. The dielectric constant of the capacitor changes as the common area between the dielectric element and the capacitor plates varies due to the motion. The same arrangement may be used as a liquid level sensor, in which case the dielectric medium is the measured liquid, as shown in Figure 4.23c.

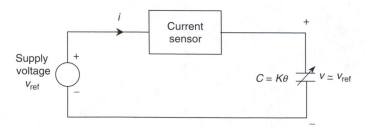

FIGURE 4.26
Rotating-plate capacitive angular velocity sensor.

4.5.3 Capacitive Angular Velocity Sensor

The schematic diagram for an angular velocity sensor that uses a rotating-plate capacitor is shown in Figure 4.26. Since the current sensor must have a negligible resistance, the voltage across the capacitor is almost equal to the supply voltage v_{ref}, which is kept constant. It follows that the current in the circuit is given by

$$i = \frac{\mathrm{d}}{\mathrm{d}t}(Cv_{\text{ref}}) = v_{\text{ref}}\frac{\mathrm{d}C}{\mathrm{d}t},$$

which in view of Equation 4.42, may be expressed as

$$\frac{\mathrm{d}\theta}{\mathrm{d}t} = \frac{i}{Kv_{\text{ref}}}. \tag{4.52}$$

This is a linear relationship for angular velocity in terms of the measured current i. Care must be exercised to ensure that the current-measuring device does not interfere with (e.g., does not load) the basic circuit.

An advantage of capacitance transducers is that because they are noncontacting devices, mechanical loading effects are negligible. There is some loading as a result of inertial forces of the moving plate and frictional resistance in associated sliding mechanisms, bearings, and so on. Such influences can be eliminated by using the moving object itself as the moving plate. Variations in the dielectric properties because of humidity, temperature, pressure, and impurities introduce errors. A capacitance bridge circuit can compensate for these effects. Extraneous capacitances, such as cable capacitance, can produce erroneous readings in capacitive sensors. This problem can be overcome by using proper conditioning circuitry such as a charge amplifier for the sensor signal. Another drawback of capacitance displacement sensors is low sensitivity. For a transverse displacement transducer, the sensitivity is typically less than one picofarad (pF) per millimeter (1 pF = 10^{-12} F). This problem is not serious because high supply voltages and amplifier circuitry can be used to increase the sensor sensitivity.

4.5.4 Capacitance Bridge Circuit

Sensors that are based on the change in capacitance (reactance) require some means of measuring that change. Furthermore, a change in capacitance, which is not caused by a change in measurand (for example, due to change in humidity, temperature, and so on), causes errors and should be compensated for. Both these goals are accomplished using a capacitance bridge circuit. An example is shown in Figure 4.27.

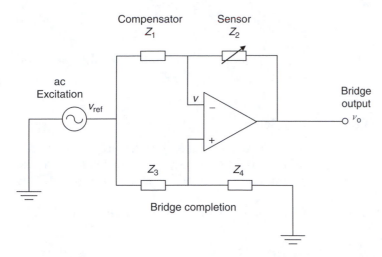

Compensator Z_1 Sensor Z_2

ac Excitation

v_{ref}

Bridge output

v_{o}

v

Z_3 Z_4

Bridge completion

FIGURE 4.27
A bridge circuit for capacitive sensors.

In this circuit,

$$Z_2 = \frac{1}{j\omega C_2} = \text{reactance (i.e., capacitive impedance) of the capacitive sensor (of capaci-}$$

tance C_2)

$$Z_1 = \frac{1}{j\omega C_1} = \text{reactance of the compensating capacitor } C_1$$

$Z_4, Z_3 = $ bridge completing impedances (typically, reactances)
$v_{\text{ref}} = v_a \sin \omega t = $ excitation ac voltage
$v_{\text{o}} = v_b \sin (\omega t - \phi) = $ bridge output
$\phi = $ phase lag of the output with respect to the excitation.

Using the two assumptions for an op-amp (potentials at the negative and positive leads are equal and the current through these leads is zero; see Chapter 3) we can write the current balance equations:

$$\frac{v_{\text{ref}} - v}{Z_1} + \frac{v_{\text{o}} - v}{Z_2} = 0, \tag{i}$$

$$\frac{v_{\text{ref}} - v}{Z_3} + \frac{0 - v}{Z_4} = 0, \tag{ii}$$

where v is the common voltage at the opamp leads. Next, eliminate v in Equation (i) and Equation (ii) to obtain

$$v_{\text{o}} = \frac{(Z_4/Z_3 - Z_2/Z_1)}{1 + Z_4/Z_3} v_{\text{ref}}. \tag{4.53}$$

It is noted that when

$$\frac{Z_2}{Z_1} = \frac{Z_4}{Z_3}, \tag{4.54}$$

the bridge output $v_{\text{o}} = 0$, and the bridge is said to be balanced. Since all capacitors in the bridge are similarly affected by ambient changes, a balanced bridge will maintain that condition even under ambient changes, unless the sensor reactance Z_2 is changed because of the measurand itself. It follows that the ambient effects are compensated (at least up to the first order) by the bridge circuit. From Equation 4.53 it is clear that the bridge output due to a sensor change of δZ, starting from a balanced state, is given by

$$\delta v_{\mathrm{o}} = -\frac{v_{\mathrm{ref}}}{Z_1(1 + Z_4/Z_3)}\delta Z. \tag{4.55}$$

The amplitude and phase angle of δv_{o} with respect to v_{ref} will determine δZ, assuming that Z_1 and Z_4/Z_3 are known.

4.5.5 Differential (Push–Pull) Displacement Sensor

Consider the capacitor shown in Figure 4.28 where the two end plates are fixed and the middle plate is attached to a moving object whose displacement (δx) needs to be measured. Suppose that the capacitor plates are connected to the bridge circuit of Figure 4.27 as shown, forming the reactances Z_3 and Z_4.

If initially the middle plate is placed at an equal separation of x from either end plate, and if it is moved by δx, we have

$$Z_3 = \frac{1}{j\omega C_3} = \frac{x - \delta x}{j\omega K}, \tag{i}$$

$$Z_4 = \frac{1}{j\omega C_4} = \frac{x + \delta x}{j\omega K}, \tag{ii}$$

where K is a capacitor constant as in Equation 4.44. Also assume that Z_1 and Z_2 are bridge-completion impedances and that they are equal. Then Equation 4.53 becomes

$$v_{\mathrm{o}} = \left[\frac{Z_4 - Z_3}{Z_4 + Z_3}\right]v_{\mathrm{ref}}. \tag{4.56}$$

This, in view of the results (i) and (ii) above, becomes:

$$v_{\mathrm{o}} = \frac{\delta x}{x}v_{\mathrm{ref}}. \tag{4.57}$$

This is a convenient linear relation for measuring the incremental displacement.

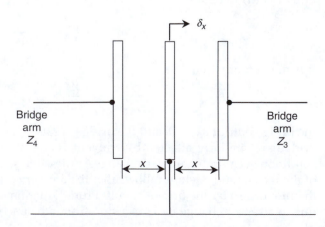

FIGURE 4.28
A linear push–pull displacement sensor.

4.6   Piezoelectric Sensors

Some substances, such as barium titanate, single-crystal quartz, and lead zirconate-titanate (PZT) can generate an electrical charge and an associated potential difference when they are subjected to mechanical stress or strain. This piezoelectric effect is used in piezoelectric transducers. Direct application of the piezoelectric effect is found in pressure and strain-measuring devices, touch screens of computer monitors, and a variety of microsensors. Many indirect applications also exist. They include piezoelectric accelerometers and velocity sensors, and piezoelectric torque sensors and force sensors. It is also interesting to note that piezoelectric materials deform when subjected to a potential difference (or charge or electric field), and can serve as actuators. Some delicate test equipment (e.g., in vibration testing) use such piezoelectric actuating elements (which undergo reverse piezoelectric action) to create fine motions. Also, piezoelectric valves (e.g., flapper valves), with direct actuation using voltage signals, are used in pneumatic and hydraulic control applications and in ink-jet printers. Miniature stepper motors based on the reverse piezoelectric action are available as well. Microactuators based on the piezoelectric effect are found in a number of applications including hard-disk drives (HDD). Modern piezoelectric materials include lanthanum modified PZT (or, PLZT) and piezoelectric polymeric polyvinylidene fluoride (PVDF).

The piezoelectric effect is caused by charge polarization in an anisotropic material (having nonsymmetric molecular structure), as a result of an applied strain. This is a reversible effect. In particular, when an electric field is applied to the material to change the ionic polarization, the material will shed the strain and regain its original shape. Natural piezoelectric materials are by and large crystalline, whereas synthetic piezoelectric materials tend to be ceramics. When the direction of the electric field and the direction of strain (or stress) are the same, we have direct sensitivity. Also crosssensitivities can be defined in a 6×6 matrix with reference to three orthogonal direct axes and three rotations about these axes.

Consider a piezoelectric crystal in the form of a disk with two electrodes plated on the two opposite faces. Since the crystal is a dielectric medium, this device is essentially a capacitor, which may be modeled by a capacitance C, as in Equation 4.41. Accordingly, a piezoelectric sensor may be represented as a charge source with a capacitive impedance in parallel (Figure 4.29). An equivalent circuit (Thevenin equivalent representation) can be given as well, where the capacitor is in series with an equivalent voltage source. The impedance from the capacitor is given by

$$Z = \frac{1}{j\omega C}. \tag{4.58}$$

As is clear from Equation 4.58, the output impedance of piezoelectric sensors is very high, particularly at low frequencies. For example, a quartz crystal may present an impedance

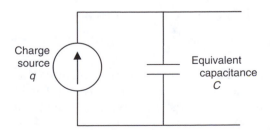

FIGURE 4.29
Equivalent circuit representation of a piezoelectric sensor.

of several megaohms at 100 Hz, increasing hyperbolically with decreasing frequencies. This is one reason why piezoelectric sensors have a limitation on the useful lower frequency. The other reason is the charge leakage.

4.6.1 Sensitivity

The sensitivity of a piezoelectric crystal may be represented either by its charge sensitivity or by its voltage sensitivity. Charge sensitivity is defined as

$$S_q = \frac{\partial q}{\partial F}, \tag{4.59}$$

where q denotes the generated charge and F denotes the applied force. For a crystal with surface area A, Equation 4.59 may be expressed as

$$S_q = \frac{1}{A} \frac{\partial q}{\partial p}, \tag{4.60}$$

where p is the stress (normal or shear) or pressure applied to the crystal surface. Voltage sensitivity S_v is given by the change in voltage due to a unit increment in pressure (or stress) per unit thickness of the crystal. Thus, in the limit, we have

$$S_v = \frac{1}{d} \frac{\partial v}{\partial p}, \tag{4.61}$$

where d denotes the crystal thickness. Now, since

$$\delta q = C \delta v \tag{4.62}$$

by using Equation 4.41 for a capacitor element, the following relationship between charge sensitivity and voltage sensitivity is obtained:

$$S_q = k S_v. \tag{4.63}$$

Note that k is the dielectric constant (permittivity) of the crystal capacitor, as defined in Equation 4.41. The overall sensitivity of a piezoelectric device can be increased through the use of properly designed multielement structures (i.e., bimorphs).

Example 4.5

A barium titanate crystal has a charge sensitivity of 150.0 picocoulombs per newton (pC/N). (Note: 1 pC $= 1 \times 10^{-12}$ coulombs; coulombs $=$ farads $\times$ volts). The dielectric constant for the crystal is 1.25×10^{-8} farads per meter (F/m). From Equation 4.63, the voltage sensitivity of the crystal is computed as

$$S_v = \frac{150.0 \text{ pC/N}}{1.25 \times 10^{-8} \text{ F/m}} = \frac{150.0 \times 10^{-12} \text{ C/N}}{1.25 \times 10^{-8} \text{ F/m}} = 12.0 \times 10^{-3} \text{ V.m/N} = 12.0 \text{ mV.m/N}.$$

■

The sensitivity of a piezoelectric element is dependent on the direction of loading. This is because the sensitivity depends on the molecular structure (e.g., crystal axis). Direct sensitivities of several piezoelectric materials along their most sensitive crystal axis are listed in Table 4.2.

TABLE 4.2

Sensitivities of Several Piezoelectric Materials

Material	Charge Sensitivity S_q (pC/N)	Voltage Sensitivity S_v (mV.m/N)
Lead Zirconate Titanate (PZT)	110	10
Barium Titanate	140	6
Quartz	2.5	50
Rochelle Salt	275	90

4.6.2 Accelerometers

It is known from Newton's second law that a force (f) is necessary to accelerate a mass (or inertia element), and its magnitude is given by the product of mass (M) and acceleration (a). This product (Ma) is commonly termed *inertia force*. The rationale for this terminology is that if a force of magnitude Ma were applied to the accelerating mass in the direction opposing the acceleration, then the system could be analyzed using static equilibrium considerations. This is known as *d'Alembert's principle* (Figure 4.30). The force that causes acceleration is itself a measure of the acceleration (mass is kept constant). Accordingly, mass can serve as a front-end element to convert acceleration into force. This is the principle of operation of common accelerometers. There are many different types of accelerometers, ranging from strain-gage devices to those that use electromagnetic induction. For example, the force which causes acceleration may be converted into a proportional displacement using a spring element, and this displacement may be measured using a convenient displacement sensor. Examples of this type are differential-transformer accelerometers, potentiometer accelerometers, and variable-capacitance accelerometers. Alternatively, the strain at a suitable location of a member that was deflected due to inertia force may be determined using a strain gage. This method is used in strain-gage accelerometers. Vibrating-wire accelerometers use the accelerating force to tension a wire. The force is measured by detecting the natural frequency of vibration of the wire, which is proportional to the square root of tension. In servo force-balance (or null-balance) accelerometers, the inertia element is restrained from accelerating by detecting its motion and feeding back a force (or torque) to exactly cancel out the accelerating force (torque). This feedback force is determined, for instance, by knowing the motor current, and it is a measure of the acceleration.

4.6.3 Piezoelectric Accelerometer

The piezoelectric accelerometer (or *crystal accelerometer*) is an acceleration sensor, which uses a piezoelectric element to measure the inertia force caused by acceleration.

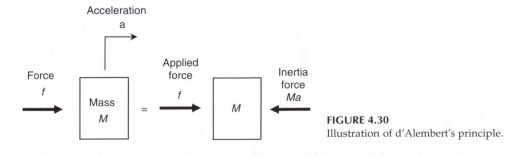

FIGURE 4.30
Illustration of d'Alembert's principle.

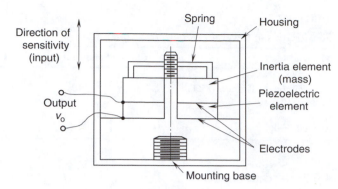

FIGURE 4.31
A compression-type piezoelectric accelerometer.

A piezoelectric velocity transducer is simply a piezoelectric accelerometer with a built-in integrating amplifier in the form of a miniature integrated circuit.

The advantages of piezoelectric accelerometers over other types of accelerometers are their light weight and high-frequency response (up to about 1 MHz). However, piezoelectric transducers are inherently high output-impedance devices, which generate small voltages (in the order of 1 mV). For this reason, special impedance-transforming amplifiers (e.g., charge amplifiers) have to be employed to condition the output signal and to reduce loading error.

A schematic diagram for a compression-type piezoelectric accelerometer is shown in Figure 4.31. The crystal and the inertia element (mass) are restrained by a spring of very high stiffness. Consequently, the fundamental natural frequency or resonant frequency of the device becomes high (typically 20 kHz). This gives a rather wide useful frequency range or operating range (typically up to 5 kHz). The lower limit of the useful frequency range (typically 1 Hz) is set by factors such as the limitations of the signal-conditioning system, the mounting methods, the charge leakage in the piezoelectric element, the time constant of the charge-generating dynamics, and the SNR. A typical frequency response curve of a piezoelectric accelerometer is shown in Figure 4.32.

In a compression-type crystal accelerometer, the inertia force is sensed as a compressive normal stress in the piezoelectric element. There are also piezoelectric accelerometers where the inertia force is applied to the piezoelectric element as a shear strain or as a tensile strain.

For an accelerometer, acceleration is the signal that is measured (the measurand). Hence, accelerometer sensitivity is commonly expressed in terms of electrical charge

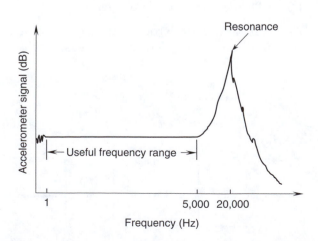

FIGURE 4.32
A typical frequency response curve for a piezoelectric accelerometer.

per unit acceleration or voltage per unit acceleration (compare this with Equation 4.60 and Equation 4.61). Acceleration is measured in units of acceleration due to gravity (g), and charge is measured in picocoulombs (pC), which are units of 10^{-12} coulombs (C). Typical accelerometer sensitivities are 10 pC/g and 5 mV/g. Sensitivity depends on the piezoelectric properties, the way in which the inertia force is applied to the piezoelectric element (e.g., compressive, tensile, shear), and the mass of the inertia element. If a large mass is used, the reaction inertia force on the crystal becomes large for a given acceleration, thus generating a relatively large output signal. Large accelerometer mass results in several disadvantages, however. In particular:

1. The accelerometer mass distorts the measured motion variable (mechanical loading effect).
2. A heavy accelerometer has a lower resonant frequency and hence a lower useful frequency range (Figure 4.32).

For a given accelerometer size, improved sensitivity can be obtained by using the shear-strain configuration. In this configuration, several shear layers can be used (e.g., in a delta arrangement) within the accelerometer housing, thereby increasing the effective shear area and hence the sensitivity in proportion to the shear area. Another factor that should be considered in selecting an accelerometer is its cross-sensitivity or transverse sensitivity. Cross-sensitivity is present because a piezoelectric element can generate a charge in response to forces and moments (or, torques) in orthogonal directions as well. The problem can be aggravated due to manufacturing irregularities of the piezoelectric element, including material unevenness and incorrect orientation of the sensing element, and due to poor design. Cross-sensitivity should be less than the maximum error (percentage) that is allowed for the device (typically 1%).

The technique employed to mount the accelerometer on an object can significantly affect the useful frequency range of the accelerometer. Some common mounting techniques are

1. Screw-in base
2. Glue, cement, or wax
3. Magnetic base
4. Spring-base mount
5. Hand-held probe

Drilling holes in the object can be avoided by using the second through fifth methods, but the useful frequency range can decrease significantly when spring-base mounts or hand-held probes are used (typical upper limit of 500 Hz). The first two methods usually maintain the full useful range (e.g., 5 kHz), whereas the magnetic attachment method reduces the upper frequency limit to some extent (typically 3 kHz).

4.6.4 Charge Amplifier

Piezoelectric signals cannot be read using low-impedance devices. The two primary reasons for this are:

1. High output impedance in the sensor results in small output signal levels and large loading errors.
2. The charge can quickly leak out through the load.

A charge amplifier is commonly used as the signal-conditioning device for piezoelectric sensors, in order to overcome these problems to a great extent. Because of impedance transformation, the impedance at the output of the charge amplifier becomes much smaller than the output impedance of the piezoelectric sensor. This virtually eliminates loading error and provides a low-impedance output for purposes such as signal communication, acquisition, recording, processing, and control. Also, by using a charge amplifier circuit with a relatively large time constant, the speed of charge leakage can be decreased. For example, consider a piezoelectric sensor and charge amplifier combination, as represented by the circuit in Figure 4.33. Let us examine how the rate of charge leakage is reduced by using this arrangement. Sensor capacitance, feedback capacitance of the charge amplifier, and feedback resistance of the charge amplifier are denoted by C, C_f, and R_f, respectively. The capacitance of the cable, which connects the sensor to the charge amplifier, is denoted by C_c.

For an opamp of gain K, the voltage at its inverting (negative) input is $-v_o/K$, where v_o is the voltage at the amplifier output. The noninverting (positive) input of the opamp is grounded (i.e., maintained at zero potential). Owing to very high input impedance of the opamp, the currents through its input leads are negligible (see Chapter 3). Current balance at point A gives:

$$\dot{q} + C\frac{\dot{v}_o}{K} + C_c\frac{\dot{v}_o}{K} + C_f\left(\dot{v}_o + \frac{\dot{v}_o}{K}\right) + \frac{v_o + v_o/K}{R_f} = 0. \tag{4.64}$$

Since gain K is very large (typically 10^5 to 10^9) compared to unity, this differential equation may be approximated as

$$R_f C_f \frac{dv_o}{dt} + v_o = -R_f \frac{dq}{dt} \tag{4.65}$$

Alternatively, instead of using Equation 4.64, it is possible to directly obtain Equation 4.65 from the two common assumptions (equal inverting and noninverting lead potentials and zero lead currents; see Chapter 3) for an opamp. Then, the potential at the negative (inverting) lead would be zero, as the positive lead is grounded. Also, as a result, the voltage across C_c would be zero. Hence, the current balance at point A gives:

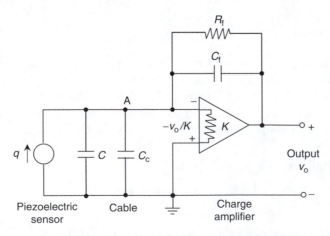

FIGURE 4.33
A piezoelectric sensor and charge amplifier combination.

$$\dot{q} + \frac{v_o}{R_f} + C_E \dot{v}_o = 0,$$

which is identical to Equation 4.65. The corresponding transfer function is

$$\frac{v_o(s)}{q(s)} = -\frac{R_f s}{[R_f C_f s + 1]}, \tag{4.66)*}$$

where s is the Laplace variable. Now, in the frequency domain ($s = j\omega$), we have

$$\frac{v_o(j\omega)}{q(j\omega)} = -\frac{R_f j\omega}{[R_f C_f j\omega + 1]}. \tag{4.66)**}$$

Note that the output is zero at zero frequency ($\omega = 0$). Hence, a piezoelectric sensor cannot be used for measuring constant (dc) signals. At very high frequencies, on the other hand, the transfer function approaches the constant value $-1/C_f$, which is the calibration constant for the device.

From Equation 4.65 or Equation 4.66* which represent a first-order system, it is clear that the time constant τ_c of the sensor-amplifier unit is

$$\tau_c = R_f C_f. \tag{4.66}$$

Suppose that the charge amplifier is properly calibrated (by the factor $-1/C_f$) so that the frequency transfer function (Equation 4.66**) can be written as

$$G(j\omega) = \frac{j\tau_c\omega}{[j\tau_c\omega + 1]}. \tag{4.66)***}$$

Magnitude M of this transfer function is given by

$$M = \frac{\tau_c\omega}{\sqrt{\tau_c^2\omega^2 + 1}}. \tag{4.68}$$

As $\omega \to \infty$, note that $M \to 1$. Hence, at infinite frequency there is no sensor error. Measurement accuracy depends on the closeness of M to 1. Suppose that we want the accuracy to be better than a specified value M_o. Accordingly, we must have

$$\frac{\tau_c\omega}{\sqrt{\tau_c^2\omega^2 + 1}} > M_o \tag{4.69)*}$$

or

$$\tau_c\omega > \frac{M_o}{\sqrt{1 - M_o^2}}. \tag{4.69}$$

If the required lower frequency limit is ω_{min}, the time constant requirement is

$$\tau_c > \frac{M_o}{\omega_{min}\sqrt{1 - M_o^2}} \tag{4.70)*}$$

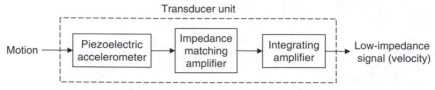

FIGURE 4.34
Schematic diagram of a piezoelectric velocity transducer.

or

$$R_f C_f > \frac{M_o}{\omega_{min} \sqrt{1 - M_o^2}}. \tag{4.70}$$

It follows that, for a specified level of accuracy, a specified lower limit on frequency of operation may be achieved by increasing the time constant (i.e., by increasing R_f, C_f, or both). The feasible lower limit on the frequency of operation (ω_{min}) can be set by adjusting the time constant.

Example 4.6
For a piezoelectric accelerometer with a charge amplifier, an accuracy level better than 99% is obtained if

$$\frac{\tau_c \omega}{\sqrt{\tau_c^2 \omega^2 + 1}} > 0.99$$

or: $\tau_c \, \omega > 7.0$. The minimum frequency of a transient signal, which can tolerate this level of accuracy, is

$$\omega_{min} = \frac{7.0}{\tau_c}.$$

■

In theory, it is possible to measure velocity by first converting velocity into a force using a viscous damping element and measuring the resulting force using a piezoelectric sensor. This principle may be used to develop a piezoelectric velocity transducer. However the practical implementation of an ideal velocity-force transducer is quite difficult, primarily due to nonlinearities in damping elements. Hence, commercial piezoelectric velocity transducers use a piezoelectric accelerometer and a built-in (miniature) integrating amplifier. A schematic diagram of the arrangement of such a piezoelectric velocity transducer is shown in Figure 4.34. The overall size of the unit can be as small as 1 cm. With double integration hardware, a piezoelectric displacement transducer is obtained using the same principle. Alternatively, an ideal spring element (or, cantilever), which converts displacement into a force (or bending moment or strain), may be employed to stress the piezoelectric element, resulting in a displacement transducer. Such devices are usually not practical for low-frequency (few hertz) applications because of the poor low-frequency characteristics of piezoelectric elements.

4.7 Effort Sensors

The response of a mechanical system depends on effort excitations (i.e., forces and torques) applied to the system. Many applications exist in which process performance is specified in terms of forces and torques. Examples include machine-tool operations, such as grinding, cutting, forging, extrusion, and rolling; manipulator tasks, such as parts handling, assembly, engraving, and robotic fine manipulation; and actuation tasks, such as locomotion. The forces and torques present in a dynamic system are generally functions of time. Performance monitoring and evaluation, failure detection and diagnosis, testing, and control of dynamic systems can depend heavily on accurate measurement of associated forces and torques. One example in which force (and torque) sensing can be very useful is a drilling robot. The drill bit is held at the end effector by the gripper of the robot, and the workpiece is rigidly fixed to a support structure by clamps. Although a displacement sensor such as a potentiometer or a differential transformer can be used to measure drill motion in the axial direction, this alone does not determine the drill performance. Depending on the material properties of the workpiece (e.g., hardness) and the nature of the drill bit (e.g., degree of wear), a small misalignment or slight deviation in feed (axial movement) or speed (rotational speed of the drill) can create large normal (axial) and lateral forces and resistance torques. This can create problems such as excessive vibrations and chattering uneven drilling, excessive tool wear, and poor product quality and may eventually lead to a major mechanical failure. Sensing the axial force or motor torque, for example, and using the information to adjust process variables (speed, feed rate, etc.), or even to provide warning signals and eventually stop the process, can significantly improve the system performance. Another example in which force sensing is useful is in nonlinear feedback control (or feedback linearization technique or FLT) of mechanical systems such as robotic manipulators.

Since both force and torque are effort variables, the term force may be used to represent both these variables. This generalization is adopted here except when discrimination might be necessary—for example, when discussing specific applications.

4.7.1 Force Causality Issues

One important application of force (and torque) sensing is in the area of control. Since forces are variables in a mechanical system, their measurement can lead to effective control. There are applications in which force control is invaluable. This is particularly evident in situations where a small error in motion can lead to the generation of large forces, which is the case, for example, in parts assembly operations. In assembly, a slight misalignment (or position error) can cause jamming and generation of damaging forces. As another example, consider precision machining of a hard workpiece. A slight error in motion could generate large cutting forces, which might lead to unacceptable product quality or even to rapid degradation of the machine tool. In such situations, measurement and control of forces seem to be an effective way to improve the system performance. First, we shall address the force control problem from a generalized and unified point of view. The concepts introduced here are illustrated further by examples.

4.7.1.1 Force–Motion Causality

The response of a mechanical control system does not necessarily have to be a motion variable. When the objective of a control system is to produce a desired motion, the response variables (outputs) are the associated motion variables. A good example is the response of a spray-painting robot whose end effector is expected to follow a specific trajectory of motion. On the other hand, when the objective is to exert a desired set of forces (or torques)—which is the case in some tasks of machining, forging, gripping, engraving, and assembly—the outputs are the associated force variables. The choice of inputs and outputs cannot be done arbitrarily, however, for a physical system. The conditions of physical realizability have to be satisfied, as we illustrate in the examples to follow.

A lumped-parameter mechanical system can be treated as a set of basic mechanical elements (springs, inertia elements, dampers, levers, gyros, force sources, velocity sources, and so on) that are interconnected through ports (or bonds) through which power (or energy) flows. Each port actually consists of two terminals. There is an effort variable (force) f and a flow variable (motion, such as velocity) v associated with each port. In particular, consider a port that connects two subsystems A and B, as shown in Figure 4.35a. The two subsystems interact, and power flows through the connecting port. Hence, for example, if we assume that A pushes B with a force f, then f is considered input to B. Now, B responds with motion v—the output of B—which, in turn causes A to respond with force f. Consequently, the input to A is v and the output of A is f. This cause-effect relationship (causality) is clearly shown by the block diagram representation in Figure 4.35b. A similar argument would lead to the opposite causality if we had started with the assumption that B is pushing A with force f. It follows that one has to make some causal decisions for a dynamic system, based on system requirements and physical realizability, and the rest of the causalities are decided automatically. It can be shown that conflicts in causality indicate that the energy storage elements (e.g., springs, masses) in the particular system model are not independent. Note that causality is not governed by the positive direction assigned to the variables f and v. This is clear from the free-body diagram representation in Figure 4.35c. In particular, if the force f on A (the action) is taken as positive in one direction then according to Newton's third law, the force f on B (the reaction) is positive in the opposite direction. Similarly, if the velocity v of B relative to A is positive in one direction, then the velocity v of A relative to B is positive in the opposite direction. These directions are unrelated to which variable is the input and which variable is the output (causality). The foregoing discussion shows that forces can be considered inputs (excitations) as well as outputs (responses), depending on the functional requirements of a particular system. In particular, when force is considered as an output, we should imagine that there is an interacting subsystem, which receives this force as an input. In these considerations, however, the requirements of physical realizability have to be satisfied.

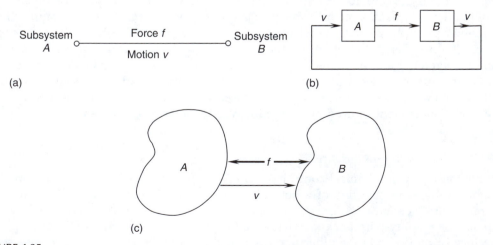

FIGURE 4.35
Force–motion causality. (a) Bond or port representation. (b) Block diagram representation. (c) Free-body diagram representation.

4.7.1.2 Physical Realizability

A primary requirement for physical realizability of a linear system is that its transfer function

$$G(s) = \frac{N(s)}{\Delta(s)} \qquad (4.71)$$

must satisfy

$$O(N) \leq O(\Delta), \qquad (4.72)$$

where N and Δ are the numerator polynomial and the denominator (characteristic) polynomial, respectively, of the transfer function, and $O(N)$ and $O(\Delta)$ denote their orders (Note: polynomial order = highest power among the terms). It should be intuitively clear why this condition must be satisfied. In particular, if the condition in Equation 4.72 is violated, then we would encounter the following unrealistic characteristics in the system, particularly at very high frequencies:

1. A finite input would produce an infinite output.
2. Infinite levels of power would be needed to drive the system.
3. Saturation would not be possible.
4. Future inputs would affect the present outputs.

Example 4.7

In the linear system given by

$$\frac{dy}{dt} + a_0 y = b_0 u + b_1 \frac{du}{dt} + b_2 \frac{d^2 u}{dt^2},$$

suppose that u denotes the input and y denotes the output. Show that the physical realizability is violated by this system. Using the same reasoning show that if u and y were switched the physical realizability would be satisfied.

Solution

Integrate the system equation. We get

$$y = -a_0 \int y dt + b_0 \int u dt + b_1 u + b_2 \frac{du}{dt}.$$

If u is a step input, the derivative term on the RHS becomes infinite instantaneously, at $t = 0$. That means (to satisfy the system equation) the output also becomes infinite, which is not feasible. Hence, we conclude that the system is not physically realizable. Also note that the system transfer function is

$$G(s) = \frac{b_2 s^2 + b_1 s + b_0}{s + a_0},$$

where $O(N) = 2$ and $O(\Delta) = 1$. Hence the condition (Equation 4.72) is not satisfied.

Now if we switch u and y in the system equation and integrate twice to express the output, there will not be any derivatives of the input term. Specifically:

$$b_2 y = -b_1 \int y dt - b_0 \int \int y dt + a_1 \int u dt + a_0 \int \int u dt.$$

Hence, in this case, a finite input will not produce an infinite output instantaneously. Furthermore, the system transfer function becomes

$$G(s) = \frac{s + a_0}{b_2 s^2 + b_1 s + b_0},$$

which satisfies Equation 4.72.

Example 4.8

A mechanical system is modeled as in Figure 4.36. The system parameters are the masses m_1 and m_2, damping constants b_1 and b_2, and stiffness terms k_1 and k_2. Forces f_1 and f_2 exist at masses m_1 and m_2, respectively. The displacements of the corresponding masses are y_1 and y_2. In general, a displacement y_i may be considered either as an input or as an output. Similarly, a force f_i may be considered either as an input or as an output. In particular, when a displacement is treated as an input at a mass, we may assume it to be generated by an internal actuator of adequate capability, and also may produce an external force which acts through the environment. When a force is treated as an output, we may consider it to be applied on a system that is external to the original system (i.e., on the environment of the original system).

It is easy to verify, by applying Newton's second law to each mass that the equations of motion of the given system are:

$$m_1 \ddot{y}_1 = -b_1 \dot{y}_1 - k_1 y_1 + b_2(\dot{y}_2 - \dot{y}_1) + k_2(y_2 - y_1) + f_1$$
$$m_2 \ddot{y}_2 = -b_2(\dot{y}_2 - \dot{y}_1) - k_2(y_2 - y_1) + f_2.$$

Using the Laplace variable s, these equations may be expressed as the transfer-function relations:

$$[m_1 s^2 + (b_1 + b_2)s + (k_1 + k_2)]y_1 - [b_2 s + k_2]y_2 = f_1$$
$$-[b_2 s + k_2]y_1 + [m_2 s^2 + b_2 s + k_2]y_2 = f_2.$$

(i) Express in the Laplace domain, y_1 in terms of f_1 and f_2, and also y_2 in terms of f_1 and f_2, together with the system parameters.

(ii) Indicate, giving reasons, whether f_1 and f_2 can be treated as inputs and y_1 and y_2 as outputs of the system.

(iii) Also indicate, giving reasons, whether y_1 and y_2 can be treated as inputs and f_1 and f_2 as outputs of the system.

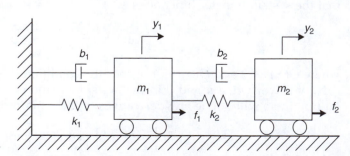

FIGURE 4.36
A model of a mechanical system.

Solution

(i)   The system transfer equations may be expressed in the vector-matrix form as:

$$\begin{bmatrix} a_{11} & a_{12} \\ a_{21} & a_{22} \end{bmatrix} \begin{bmatrix} y_1 \\ y_2 \end{bmatrix} = \begin{bmatrix} f_1 \\ f_2 \end{bmatrix}, \tag{i}$$

where

$$\begin{aligned} a_{11} &= m_1 s^2 + (b_1 + b_2)s + k_1 + k_2 \\ a_{22} &= m_2 s^2 + b_2 s + k_2 \\ a_{12} &= a_{21} = -(b_2 s + k_2). \end{aligned} \tag{ii}$$

Then, by matrix inversion, we get

$$\begin{bmatrix} y_1 \\ y_2 \end{bmatrix} = \frac{1}{(a_{11}a_{22} - a_{12}a_{21})} \begin{bmatrix} a_{22} & -a_{12} \\ -a_{21} & a_{11} \end{bmatrix} \begin{bmatrix} f_1 \\ f_2 \end{bmatrix}$$

or

$$\begin{aligned} y_1 &= \frac{a_{22}}{\Delta} f_1 - \frac{a_{12}}{\Delta} f_2 \\ y_2 &= -\frac{a_{21}}{\Delta} f_1 + \frac{a_{11}}{\Delta} f_2, \end{aligned} \tag{iii}$$

where

$$\Delta = (a_{11}a_{22} - a_{12}a_{21}) = \text{characteristic polynomial.}$$

(ii)   From Equation (iii) note that it is feasible to treat f_1 and f_2 as inputs and y_1 and y_2 as the resulting outputs. The reason is, the associated transfer functions

$$\frac{a_{22}}{\Delta}, -\frac{a_{12}}{\Delta}, -\frac{a_{21}}{\Delta}, \text{ and } \frac{a_{11}}{\Delta}$$

are all physically realizable. This is true because the order of the characteristic polynomial Δ (i.e., the denominator of the transfer functions) is 4, whereas the orders of the numerator polynomials are: for a_{22} it is 2; for a_{12} it is 1, for a_{21} it is 1, and for a_{11} it is 2. Accordingly, for each transfer function we have,

Numerator Order < Denominator Order, which satisfies the condition of physical realizability (4.72).

(iii)   From Equation (i)

$$\begin{aligned} f_1 &= a_{11}y_1 + a_{12}y_2 \\ f_2 &= a_{21}y_1 + a_{22}y_2. \end{aligned}$$

Then, for y_1 and y_2 to be inputs and f_1 and f_2 to be the resulting outputs, the associated transfer functions $a_{11}, a_{12}, a_{21},$ and a_{22} all must be physically realizable. However, this is not the case because for these transfer functions, the denominator order is equal to 0 and the numerator order is ≥ 1.

4.7.2 Force Control Problems

Some forces in a control system are actuating or excitation or input forces, and some others are response or output forces. For example, torques (or forces) driving the joints of a robotic manipulator are considered inputs to the robot. From the control point of view, however, joint input is the voltage applied to the motor drive amplifier, which produces a field current that generates the motor (magnetic) torque (see Chapter 7), which in turn is resisted by the load torque (transmitted to the next link of the robot) at the joint. Gripping or tactile (touch) forces at the end effector of a manipulator, tool tip forces in a milling machine, and forces at the die in a forging machine can be considered as output forces, which are exerted on objects external to the system, while the corresponding motions are externally "constrained" by the environment (i.e., these motions are inputs to the system). However, output forces should be completely determined by the inputs (motions as well as forces) to the system. Unknown input forces and output forces can be measured using appropriate force sensors, and force control (feedforward and feedback) may be implemented using these measurements.

4.7.2.1 Force Feedback Control

Consider the system shown in Figure 4.37, which is connected to its environment through two ports. Force variable f_A and motion variable v_A (both in the same direction) are associated with port A, and force variable f_B and motion variable v_B (in the same direction) are associated with port B. Suppose that f_A is an input to the system and f_B is an output. It follows that v_B is an input to the system and v_A is an output of the system. Specifically, the motion at B is constrained. For example, the port might be completely restrained in the direction of f_B, resulting in the constraint $v_B = 0$. Ideally, if we knew the dynamic behavior of the system—assuming that there are no extraneous inputs (disturbances and noise)—we would be able to analytically determine the input f_A that generates a desired outputs f_B and v_A for a specified v_B. Then, we could control the output force f_B and the output motion v_A simply by supplying the predetermined f_A and by subjecting B to the specified motion v_B. The corresponding open-loop configuration is shown in Figure 4.38a. Since it is practically impossible to achieve accurate system performance with this open-loop control arrangement (inappropriately known as feedforward control), except in a few simple situations, the feedback control loop shown in Figure 4.38b has to be added. In this case, the response force f_B is measured and fed back into the controller, which will modify the control input signals according to a suitable control law, to correct any deviations in f_B from the desired value.

4.7.2.2 Feedforward Force Control

Once again, consider the system shown in Figure 4.37, which has two ports, A and B. Here again, suppose that f_A is an actuating (input) force that can be generated according to a given specification (i.e., a "known" input). But suppose that f_B is an unknown input force.

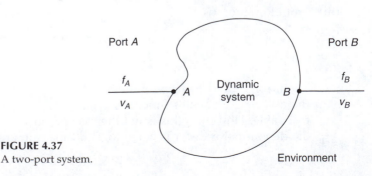

FIGURE 4.37
A two-port system.

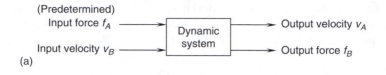

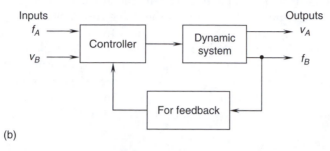

(b)

FIGURE 4.38
(a) An open-loop system. (b) A force feedback control system.

It could be a disturbance force, such as that resulting from a collision, or a useful force, such as a suspension force whose value is not known. This configuration is shown in Figure 4.39a. Since f_B is unknown, it might not be possible to accurately control the system response (v_A and v_B). One solution is to measure the unknown force f_B, using a suitable force sensor, and feed it forward into the controller (Figure 4.39b). The controller can use this additional information to compensate for the influence of f_B on the system and produce the desired response. This is an example of feedforward control. Feedforward control of this type "anticipates" adverse effects of an unknown input and takes corrective actions before the system response deteriorates. Consequently, feedforward provides a relatively fast control system. Sometimes, if an input force to a system (e.g., a joint force or

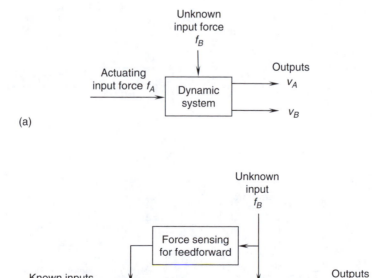

FIGURE 4.39
(a) A system with an unknown input force. (b) Feedforward force control.

torque of a robotic manipulator) is computed using an analytical model and is supplied to the actuator, the associated control is inappropriately termed feedforward control. It should be termed *computed force/torque control* or *computed input control*, to be exact.

Example 4.9

A schematic representation of a single joint of a direct-drive robotic manipulator is shown in Figure 4.40a. Direct-drive joints have no speed transmission devices such as gears or harmonic drives (see Chapter 8); the stator of the drive motor is rigidly attached to one link, and the rotor is rigidly attached to the next link. Motor (magnetic) torque is T_m, and the joint torque, which is transmitted to the driven link (link 2) is T_J. Draw a block diagram for the joint and show that T_J is represented as an input to the joint control system. If T_J is measured directly, using a semiconductor (SC) strain-gage torque sensor, what type of control would you recommend for improving the manipulator performance? Extend the discussion to the case in which the joint actuator is a hydraulic piston-cylinder mechanism.

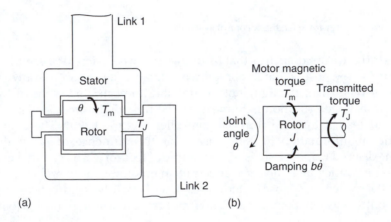

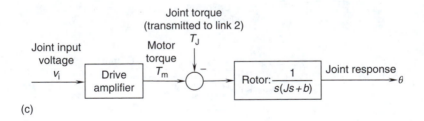

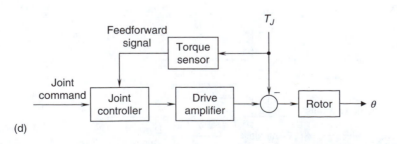

FIGURE 4.40
(a) A single joint of a direct-drive arm. (b) Free-body diagram of the motor rotor. (c) Block diagram of the joint. (d) Feedforward control using the joint torque.

Solution

Let us assume linear viscous friction at the joint, with damping constant b. If θ denotes the relative rotation of Link 2 with respect to Link 1, the equation of motion for the motor rotor (with inertia J) may be written using the free-body diagram shown in Figure 4.40b; thus,

$$J\ddot{\theta} = T_m - b\dot{\theta} - T_J.$$

Here, for simplicity we have assumed that Link 1 is at rest (or moving with constant velocity).

The motor magnetic torque T_m is generated from the field current provided by the drive amplifier, in response to a command voltage v_i to the joint (see Chapter 7). Figure 4.40c shows a block diagram for the joint. Note that T_J enters as an input. Hence, an appropriate control method for the robot joint would be to measure the joint torque T_J directly and feed it forward so as to correct any deviations in the joint response θ. This feedforward control structure is shown by the block diagram in Figure 4.40d.

The same concepts are true in the case of a hydraulic actuator. Note that the control input is the voltage signal to the hydraulic valve actuator. The pressure in the hydraulic fluid is analogous to the motor torque. Joint torque/force is provided by the force exerted on the driven link by the piston. This force should be measured for feedforward control.

4.7.3 Impedance Control

Consider a mechanical operation where we push against a spring that has constant stiffness. Here, the value of the force completely determines the displacement; similarly the value of the displacement completely determines the force. It follows that, in this example, we are unable to control force and displacement independently at the same time. Also, it is not possible, in this example, to apply a command force that has an arbitrarily specified relationship with displacement. In other words, stiffness control is not possible. Now suppose that we push against a complex dynamic system, not a simple spring element. In this case, we should be able to command a pushing force in response to the displacement of the dynamic system so that the ratio of force to displacement varies in a specified manner. This is a stiffness control (or compliance control) action. Dynamic stiffness is defined as the ratio: (output force)/(input displacement), expressed in the frequency domain. Dynamic flexibility or compliance or receptance is the inverse of dynamic stiffness. Mechanical impedance is defined as the ratio: (output force)/(input velocity), in the frequency domain. Mobility is the inverse of mechanical impedance. Note that stiffness and impedance both relate force and motion variables in a mechanical system. The objective of impedance control is to make the impedance function equal to some specified function (without separately controlling or independently constraining the associated force variable and velocity variable). Force control and motion control can be considered extreme cases of impedance control (and stiffness control). Since the objective of force control is to keep the force variable from deviating from a desired level, in the presence of independent variations of the associated motion variable (an input), force is the output variable, whose deviation (increment) from the desired value must be made zero, under control. Hence, force control can be considered zero-impedance control, when velocity is chosen as the motion variable (or zero-stiffness control, when displacement is chosen as the motion variable). Conversely, displacement control can be considered infinite-stiffness control and velocity control can be considered infinite-impedance control.

Impedance control has to be accomplished through active means, generally by generating forces as specified functions of associated displacements. Impedance control is particularly useful in mechanical manipulation against physical constraints that are not "hard", which is the case in compliant assembly and machining tasks. In particular, very high impedance is naturally present in the direction of a motion constraint and very low impedance in the direction of a free motion. Problems that arise using motion control in applications where small motion errors would create large forces, can be avoided to some extent if stiffness control or impedance control is used. Furthermore, the stability of the overall system can be guaranteed and the robustness of the system improved by properly bounding the values of impedance parameters.

Impedance control can be particularly useful in tasks of fine and flexible manipulation; for example, in the processing of flexible and inhomogeneous natural material such as meat. In this case, the mechanical impedance of the task interface (i.e., in the region where the mechanical processor or cutting tool interacts with the processed object) provides valuable characteristics of the process, which can be used in fine control of the processing task. Since impedance relates the input velocity to the output force, it is a transfer function. The concepts of impedance control can be applied to situations where the input is not a velocity and the output is not a force. Still, the term impedance control is used in literature, even though the corresponding transfer function is, strictly speaking, not an impedance.

Example 4.10
The control of processes such as machine tools and robotic manipulators may be addressed from the point of view of impedance control. For example, consider a milling machine that performs a straight cut on a workpiece, as shown in Figure 4.41a. The tool position is stationary, and the machine table, which holds the workpiece, moves along a horizontal axis at speed v—the feed rate. The cutting force in the direction of feed is f. Suppose that the machine table is driven using the speed error. Then according to the law,

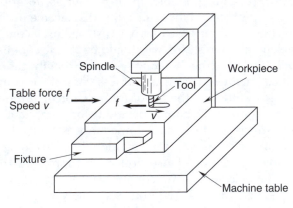

(a)

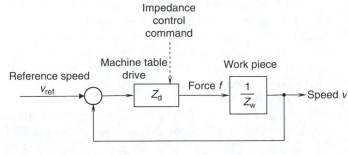

FIGURE 4.41
(a) A straight-cut milling operation.
(b) Impedance block diagram representation.

(b)

$$F = Z_d(V_{ref} - V), \tag{i}$$

where Z_d denotes the drive impedance of the table and V_{ref} is the reference (command) feed rate. (The uppercase letters are used to represent frequency domain variables of the system). Cutting impedance Z_w, of the workpiece satisfies the relation

$$F = Z_w V. \tag{ii}$$

Note that Z_w depends on system properties, and we usually do not have direct control over it. The overall system is represented by the block diagram in Figure 4.41b. An impedance control problem would be to adjust (or adapt) the drive impedance Z_d so as to maintain the feed rate near V_{ref} and the cutting force near F_{ref}. We wish to determine an adaptive control law for Z_d.

Solution
The control objective is satisfied by minimizing the objective function

$$J = \frac{1}{2}\left[\frac{F - F_{ref}}{f_o}\right]^2 + \frac{1}{2}\left[\frac{V - V_{ref}}{v_o}\right]^2, \tag{iii}$$

where f_o denotes the force tolerance and v_o denotes the speed tolerance. For example, if we desire stringent control of the feed rate, we need to choose a small value for v_o, which corresponds to a heavy weighting on the feed rate term in J. Hence, these two tolerance parameters are weighting parameters as well, in the cost function.

The optimal solution is given by

$$\frac{\partial J}{\partial Z_d} = 0 = \frac{(F - F_{ref})}{f_0^2}\frac{\partial F}{\partial Z_d} + \frac{(V - V_{ref})}{v_0^2}\frac{\partial V}{\partial Z_d}. \tag{iv}$$

Now, from Equation (i) and Equation (ii), we obtain

$$V = \left[\frac{Z_d}{Z_d + Z_w}\right]V_{ref}, \tag{v}$$

$$F = \left[\frac{Z_d Z_w}{Z_d + Z_w}\right]V_{ref}. \tag{vi}$$

On differentiating Equation (v) and Equation (vi), we get

$$\frac{\partial V}{\partial Z_d} = \frac{Z_w}{(Z_d + Z_w)^2}V_{ref}. \tag{vii}$$

and

$$\frac{\partial F}{\partial Z_d} = \frac{Z_w^2}{(Z_d + Z_w)^2}V_{ref}. \tag{viii}$$

Next, we substitute Equation (vii) and Equation (viii) in (iv) and divide by the common term; thus,

$$\frac{(F - F_{ref})}{f_0^2}Z_w + \frac{(V - V_{ref})}{v_0^2} = 0. \tag{ix}$$

Equation (ix) is expanded after substituting Equation (v) and Equation (vi) in order to get the required expression for Z_d:

$$Z_d = \left[\frac{Z_0^2 + Z_w Z_{ref}}{Z_w - Z_{ref}}\right], \tag{x}$$

where

$$Z_o = \frac{f_o}{v_o} \quad \text{and} \quad Z_{ref} = \frac{F_{ref}}{V_{ref}}.$$

Equation (x) is the impedance control law for the table drive. Specifically, since Z_w—which depends on workpiece characteristics, tool bit characteristics, and the rotating speed of the tool bit—is known through a suitable model or might be experimentally determined (identified) by monitoring v and f, and since Z_d and Z_{ref} are specified, we are able to determine the necessary drive impedance Z_d using Equation (x). Parameters of the table drive controller—particularly gain—can be adjusted to match this optimal impedance. Unfortunately, exact matching is virtually impossible, because Z_d is generally a function of frequency. If the component bandwidths are high, we may assume that the impedance functions are independent of frequency, and this somewhat simplifies the impedance control task.

Note from Equation (ii) that for the ideal case of $V = V_{ref}$ and $F = F_{ref}$, we have $Z_w = Z_{ref}$. Then, from Equation (x), it follows that a drive impedance of infinite magnitude is needed for exact control. This is impossible to achieve in practice, however. Of course, an upper limit for the drive impedance should be set in any practical scheme of impedance control.

4.7.4 Force Sensor Location

In force feedback control, the location of the force sensor with respect to the location of actuation can have a crucial effect on system performance, stability in particular. For example, in robotic manipulator applications, it has been experienced that with some locations and configurations of a force-sensing wrist at the robot end effector, dynamic instabilities were present in the manipulator response for some (large) values of control gains in the force feedback loop. These instabilities were found to be limit-cycle-type motions in most cases. Generally, it is known that when the force sensors are more remotely located with respect to the drive actuators of a mechanical system, the system is more likely to exhibit instabilities under force feedback control. Hence, it is desirable to make force measurements very close to the actuator locations when force feedback is used.

Consider a mechanical processing task. The tool actuator generates the processing force, which is applied to the workpiece. The force transmitted to the workpiece by the tool is measured by a force sensor and is used by a feedback controller to generate the correct actuator force. The machine tool is a dynamic system, which consists of a tool subsystem (dynamic) and a tool actuator (dynamic). The workpiece is also a dynamic system.

Relative location of the tool actuator with respect to the force sensor (at the tool–workpiece interface) can affect the stability of the feedback control system. In general, the closer the actuator to the sensor the more stable the feedback control system. Two scenarios are shown in Figure 4.42, which can be used to study the stability of the overall control system. In both cases, the processing force at the interface between the tool and the workpiece is measured using a force sensor, and is used by the feedback controller to generate the actuator drive signal. In Figure 4.42a the tool actuator, which generates the drive signal of the actuator, is located next to the force sensor. In Figure 4.42b the tool

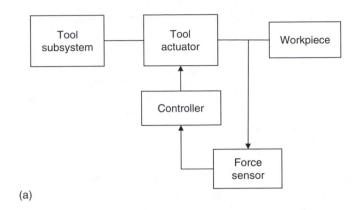

(a)

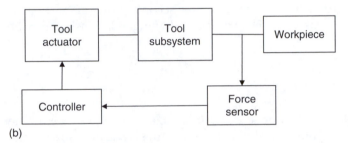

(b)

FIGURE 4.42
(a) Force sensor located next to the actuator. (b) Force sensor separated from the actuator by a dynamic subsystem.

actuator is separated from the force sensor by a dynamic system of the processing machine. It is known that the arrangement (b) is less stable than the arrangement (a). The reason is simple. Arrangement (b) introduces more dynamic delay into the feedback control loop. It is well known that time delay has a destabilizing effect on feedback control systems, particularly at high control gains.

4.8 Strain Gages

Many types of force and torque sensors are based on strain-gage measurements. Although strain gages measure strain, the measurements can be directly related to stress and force. Hence, it is appropriate to discuss strain gages under force and torque sensors. Note, however, that strain gages may be used in a somewhat indirect manner (using auxiliary front-end elements) to measure other types of variables, including displacement, acceleration, pressure, and temperature. Two common types of resistance strain gages are discussed next. Specific types of force and torque sensors are dealt in the subsequent sections.

4.8.1 Equations for Strain-Gage Measurements

The change of electrical resistance in material when mechanically deformed is the property used in resistance-type strain gages. The resistance R of a conductor that has length ℓ and area of cross-section A is given by

$$R = \rho \frac{\ell}{A},$$

(4.73)

where ρ denotes the resistivity of the material. Taking the logarithm of Equation 4.73, we have

$$\log R = \log \rho + \log (\ell/A).$$

Now, taking the differential of each term, we obtain

$$\frac{\mathrm{d}R}{R} = \frac{\mathrm{d}\rho}{\rho} + \frac{\mathrm{d}(\ell/A)}{\ell/A}. \tag{4.74}$$

The first term on the RHS of Equation 4.74 is the fractional change in resistivity, and the fractional second term represents deformation. It follows that the change in resistance in the material comes from the change in shape as well as from the change in resistivity of the material. For linear deformations, the two terms on the RHS of Equation 4.74 are linear functions of strain ε; the proportionality constant of the second term, in particular, depends on Poisson's ratio of the material. Hence, the following relationship can be written for a strain-gage element:

$$\frac{\delta R}{R} = S_\mathrm{s}\varepsilon. \tag{4.75}$$

The constant S_s is known as the gage factor or sensitivity of the strain-gage element. The numerical value of this parameter ranges from 2 to 6 for most metallic strain-gage elements and from 40 to 200 for semiconductor strain gages. These two types of strain gages are discussed later.

The change in resistance of a strain-gage element, which determines the associated strain (Equation 4.75), is measured using a suitable electrical circuit. Many variables—including displacement, acceleration, pressure, temperature, liquid level, stress, force, and torque—can be determined using strain measurements. Some variables (e.g., stress, force, and torque) can be determined by measuring the strain of the dynamic object itself at suitable locations. In other situations, an auxiliary front-end device may be required to convert the measurand into a proportional strain. For instance, pressure or displacement may be measured by converting them to a measurable strain using a diaphragm, bellows, or bending element. Acceleration may be measured by first converting it into an inertia force of a suitable mass (seismic mass) element, then subjecting a cantilever (strain member) to that inertia force and, finally, measuring the strain at a high-sensitivity location of the cantilever element (see Figure 4.43). Temperature may be measured by

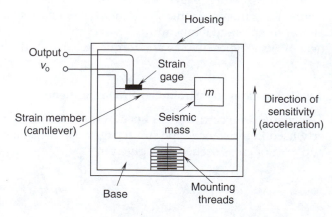

FIGURE 4.43
A strain-gage accelerometer.

measuring the thermal expansion or deformation in a bimetallic element. *Thermistors* are temperature sensors made of semiconductor material whose resistance changes with temperature. Resistance temperature detectors (RTDs) operate by the same principle, except that they are made of metals, not of semiconductor material. These temperature sensors, and the piezoelectric sensors discussed previously, should not be confused with strain gages. Resistance strain gages are based on resistance change as a result of strain, or the piezoresistive property of materials.

Early strain gages were fine metal filaments. Modern strain gages are manufactured primarily as metallic foil (e.g., using the copper–nickel alloy known as constantan) or semiconductor elements (e.g., silicon with trace impurity boron). They are manufactured by first forming a thin film (foil) of metal or a single crystal of semiconductor material and then cutting it into a suitable grid pattern, either mechanically or by using photo etching (opto-chemical) techniques. This process is much more economical and is more precise than making strain gages with metal filaments. The strain-gage element is formed on a backing film of electrically insulated material (e.g., polymide plastic). This element is cemented or bonded using epoxy, onto the member whose strain is to be measured. Alternatively, a thin film of insulating ceramic substrate is melted onto the measurement surface, on which the strain gage is mounted directly. The direction of sensitivity is the major direction of elongation of the strain-gage element (Figure 4.44a). To measure strains in more than one direction, multiple strain gages (e.g., various rosette configurations) are available as single units. These units have more than one direction of sensitivity. Principal strains in a given plane (the surface of the object on which the strain gage is mounted) can be determined by using these multiple strain-gage units. Typical foil-type gages are shown in Figure 4.44b, and a semiconductor strain gage is shown in Figure 4.44c.

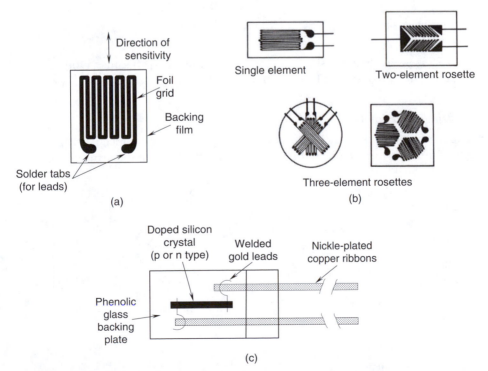

FIGURE 4.44
(a) Strain-gage nomenclature. (b) Typical foil-type strain gages. (c) A semiconductor strain-gage.

A direct way to obtain strain-gage measurement is to apply a constant dc voltage across a series-connected pair of strain-gage element (of resistance R) and a suitable (complementary) resistor R_c, and to measure the output voltage v_o across the strain gage under open-circuit conditions (using a voltmeter with high input impedance). It is known as a *potentiometer circuit* or *ballast circuit*. This arrangement has several weaknesses. Any ambient temperature variation directly introduces some error because of associated change in the strain-gage resistance and the resistance of the connecting circuitry. Also, measurement accuracy will be affected by possible variations in the supply voltage v_{ref}. Furthermore, the electrical loading error will be significant unless the load impedance is very high. Perhaps the most serious disadvantage of this circuit is that the change in signal due to strain is usually a small fraction of the total signal level in the circuit output. This problem can be reduced to some extent by decreasing v_o, which may be accomplished by increasing the resistance R_c. This, however, reduces the sensitivity of the circuit. Any changes in the strain-gage resistance due to ambient changes will directly affect enter the strain-gage reading unless R and R_c have identical coefficients with respect to ambient changes.

A more favorable circuit for use in strain-gage measurements is the Wheatstone bridge, as discussed in Chapter 3. One or more of the four resistors R_1, R_2, R_3, and R_4 in the bridge (Figure 4.45) may represent strain gages. The output relationship for the Wheatstone bridge circuit is given by (see Chapter 3)

$$v_o = \frac{R_1 v_{ref}}{(R_1 + R_2)} - \frac{R_3 v_{ref}}{(R_3 + R_4)} = \frac{(R_1 R_4 - R_2 R_3)}{(R_1 + R_2)(R_3 + R_4)} v_{ref}. \tag{4.76}$$

When this output voltage is zero, the bridge is balanced. It follows from Equation 4.76 that for a balanced bridge,

$$\frac{R_1}{R_2} = \frac{R_3}{R_4}. \tag{4.77}$$

Equation 4.77 is valid for any value of R_L, not just for large R_L, because when the bridge is balanced, current i through the load becomes zero, even for small R_L.

4.8.1.1 Bridge Sensitivity

Strain-gage measurements are calibrated with respect to a balanced bridge. When the strain gages in the bridge deform, the balance is upset. If one of the arms of the bridge has

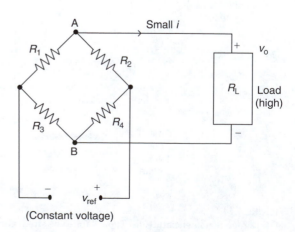

FIGURE 4.45
Wheatstone bridge circuit.

a variable resistor, it can be changed to restore balance. The amount of this change measures the amount by which the resistance of the strain gages changed, thereby measuring the applied strain. This is known as the *null-balance method* of strain measurement. This method is inherently slow because of the time required to balance the bridge each time a reading is taken. A more common method, which is particularly suitable for making dynamic readings from a strain-gage bridge, is to measure the output voltage resulting from the imbalance caused by the deformation of active strain gages in the bridge. To determine the calibration constant of a strain-gage bridge, the sensitivity of the bridge output to changes in the four resistors in the bridge should be known. For small changes in resistance, using straightforward calculus, this may be determined as

$$\frac{\delta v_0}{v_{ref}} = \frac{(R_2 \delta R_1 - R_1 \delta R_2)}{(R_1 + R_2)^2} - \frac{(R_4 \delta R_3 - R_3 \delta R_4)}{(R_3 + R_4)^2}. \tag{4.78}$$

This result is subject to Equation 4.77, because changes are measured from the balanced condition. Note from Equation 4.78 that if all four resistors are identical (in value and material), the changes in resistance due to ambient effects cancel out among the first-order terms (δR_1, δR_2, δR_3, δR_4), producing no net effect on the output voltage from the bridge. Closer examination of Equation 4.78 reveals that only the adjacent pairs of resistors (e.g., R_1 with R_2 and R_3 with R_4) have to be identical in order to achieve this environmental compensation. Even this requirement can be relaxed. In fact, compensation is achieved if R_1 and R_2 have the same temperature coefficient and if R_3 and R_4 have the same temperature coefficient.

Example 4.11

Suppose that R_1 represents the only active strain-gage and R_2 represents an identical dummy gage in Figure 4.45. The other two elements of the bridge are bridge-completion resistors, which do not have to be identical to the strain gages. For a balanced bridge, we must have $R_3 = R_4$, but not necessarily equal to the resistance of the strain-gage. Let us determine the output of the bridge.

In this example, only R_1 changes. Hence, from Equation 4.78, we have

$$\frac{\delta v_0}{v_{ref}} = \frac{\delta R}{4R}, \tag{4.79}*$$

where R denotes the strain-gage resistance.

4.8.1.2 The Bridge Constant

Equation (4.79)* assumes that only one resistance (strain-gage) in the Wheatstone bridge (Figure 4.45) is active. Numerous other activating combinations are possible, however; for example, tension in R_1 and compression in R_2, as in the case of two strain gages mounted symmetrically at 45° about the axis of a shaft in torsion. In this manner, the overall sensitivity of a strain-gage bridge can be increased. It is clear from Equation 4.78 that if all four resistors in the bridge are active, the best sensitivity is obtained if, for example, R_1 and R_4 are in tension and R_2 and R_3 are in compression, so that all four differential terms have the same sign. If more than one strain-gage is active, the bridge output may be expressed as

$$\frac{\delta v_0}{v_{ref}} = k \frac{\delta R}{4R}, \tag{4.79}$$

where

$$k = \frac{\text{bridge output in the general case}}{\text{bridge output if only one strain gage is active}}.$$

This constant is known as the *bridge constant*. The larger the bridge constant, the better the sensitivity of the bridge.

Example 4.12

A strain-gage load cell (force sensor) consists of four identical strain gages, forming a Wheatstone bridge, and are mounted on a rod that has a square cross-section. One opposite pair of strain gages is mounted axially and the other pair is mounted in the transverse direction, as shown in Figure 4.46a. To maximize the bridge sensitivity, the strain gages are connected to the bridge as shown in Figure 4.46b. Determine the bridge constant k in terms of Poisson's ratio ν of the rod material.

Solution

Suppose that $\delta R_1 = \delta R$. Then, for the given configuration, we have

$$\delta R_2 = -\nu \delta R$$
$$\delta R_3 = -\nu \delta R.$$
$$\delta R_4 = \delta R$$

Note that from the definition of Poisson's ratio:

$$\text{Transverse strain} = (-\nu) \times \text{longitudinal strain}.$$

Now, it follows from Equation 4.78 that

$$\frac{\delta v_o}{v_{\text{ref}}} = 2(1 + \nu)\frac{\delta R}{4R}$$

according to which the bridge constant is given by

$$k = 2(1 + \nu).$$

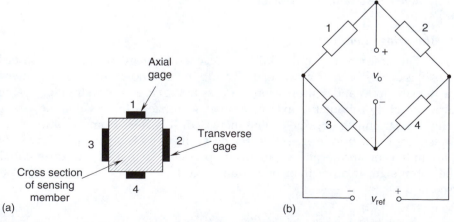

FIGURE 4.46
An example of four active strain gages. (a) Mounting configuration on the load cell. (b) Bridge circuit.

4.8.1.3   The Calibration Constant

The calibration constant C of a strain-gage bridge relates the strain that is measured to the output of the bridge. Specifically,

$$\frac{\delta v_o}{v_{\text{ref}}} = C\varepsilon. \tag{4.80}$$

Now, in view of Equation 4.75 and Equation 4.79, the calibration constant may be expressed as

$$C = \frac{k}{4}S_s, \tag{4.81}$$

where k is the bridge constant and S_s is the sensitivity (or *gage factor*) of the strain gage. Ideally, the calibration constant should remain constant over the measurement range of the bridge (i.e., should be independent of strain ε and time t) and should be stable with respect to ambient conditions. In particular, there should not be any creep, nonlinearities such as hysteresis or thermal effects.

Example 4.13

A schematic diagram of a strain-gage accelerometer is shown in Figure 4.47a. A point mass of weight W is used as the acceleration sensing element, and a light cantilever with rectangular cross-section, mounted inside the accelerometer casing, converts the inertia force of the mass into a strain. The maximum bending strain at the root of the cantilever is measured using four identical active semiconductor strain gages. Two of the strain gages (A and B) are mounted axially on the top surface of the cantilever, and the remaining two (C and D) are mounted on the bottom surface, as shown in Figure 4.47b. In order to maximize the sensitivity of the accelerometer, indicate the manner in which the four strain gages—A, B, C, and D—should be connected to a Wheatstone bridge circuit. What is the bridge constant of the resulting circuit?

Obtain an expression relating applied acceleration a (in units of g, which denotes acceleration due to gravity) to bridge output δv_o (measured using a bridge balanced at zero acceleration) in terms of the following parameters:

$W = Mg =$ weight of the seismic mass at the free end of the cantilever element
$E \;=$ Young's modulus of the cantilever
$\ell \;=$ length of the cantilever
$b \;=$ cross-section width of the cantilever
$h \;=$ cross-section height of the cantilever
$S_s =$ gage factor (sensitivity) of each strain-gage
$v_{\text{ref}} =$ supply voltage to the bridge.

If $M = 5$ gm, $E = 5 \times 10^{10}$ N/m^2, $\ell = 1$ cm, $b = 1$ mm, $h = 0.5$ mm, $S_s = 200$, and $v_{\text{ref}} = 20$ V, determine the sensitivity of the accelerometer in microvolts per gram.

If the yield strength of the cantilever element is 5×10^7 N/m^2, what is the maximum acceleration that could be measured using the accelerometer? If the analog-to-digital converter (ADC) which reads the strain signal into a process computer has the range 0 to 10 V, how much amplification (bridge amplifier gain) would be needed at the bridge output so that this maximum acceleration corresponds to the upper limit of the ADC (10 V)?

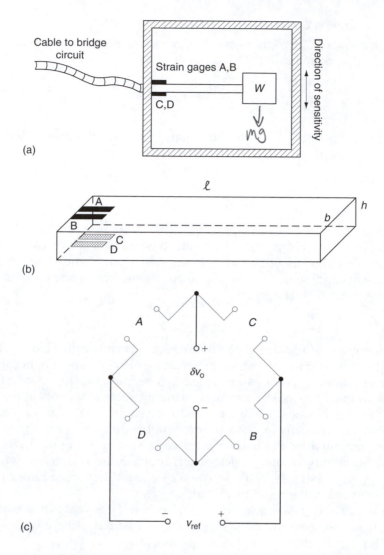

FIGURE 4.47
A miniature accelerometer using strain gages: (a) Schematic diagram; (b) Mounting configuration of the strain gages; (c) Bridge connection.

Is the cross-sensitivity (i.e., the sensitivity in the two directions orthogonal to the direction of sensitivity shown in Figure 4.47a) small with your arrangement of the strain-gage bridge? Explain.

Hint: For a cantilever subjected to force F at the free end, the maximum stress at the root is given by

$$\sigma = \frac{6F\ell}{bh^2}$$

with the present notation.

Note: Microelectromechanical systems (MEMS) accelerometers where the cantilever member, inertia element, and the strain gage are all integrated into a single semiconductor (silicon) unit are available in commercial applications such as air bag activation sensors for automobiles.

Solution

Clearly, the bridge sensitivity is maximized by connecting the strain gages *A, B, C,* and *D* to the bridge as shown in Figure 4.47c. This follows from Equation 4.78, noting that the contributions from the four strain gages are positive when δR_1 and δR_4 are positive, and δR_2 and δR_3 are negative. The bridge constant for the resulting arrangement is $k = 4$. Hence, from Equation 4.79,

$$\frac{\delta v_o}{v_{\text{ref}}} = \frac{\delta R}{R}$$

or from Equation 4.80 and Equation 4.81,

$$\frac{\delta v_o}{v_{\text{ref}}} = S_s \varepsilon.$$

Also,

$$\varepsilon = \frac{\sigma}{E} = \frac{6F\ell}{Ebh^2}$$

where *F* denotes the inertia force;

$$F = \frac{W}{g}\ddot{x} = Wa.$$

Note that $\ddot{x}$ is the acceleration in the direction of sensitivity and $\ddot{x}/g = a$ is the acceleration in units of *g*.

Thus,

$$\varepsilon = \frac{6W\ell}{Ebh^2}a$$

or

$$\delta v_o = \frac{6W\ell}{Ebh^2}S_s v_{\text{ref}}a.$$

Now, with the given values,

$$\frac{\delta v_o}{a} = \frac{6 \times 5 \times 10^{-3} \times 9.81 \times 1 \times 10^{-2} \times 200 \times 20}{5 \times 10^{10} \times 1 \times 10^{-3} \times (0.5 \times 10^{-3})^2}\text{V/g}$$

$$= 0.94 \text{ V/g}$$

$$\frac{\varepsilon}{a} = \frac{1}{S_s v_{\text{ref}}}\frac{\delta v_o}{a} = \frac{0.94}{200 \times 20}\text{strain/g}$$

$$= 2.35 \times 10^{-4}\ \varepsilon/g = 235.0\ \mu\varepsilon/g$$

$$\text{Yield strain} = \frac{\text{Yield strength}}{E} = \frac{5 \times 10^7}{5 \times 10^{10}} = 1 \times 10^{-3}\ \text{strain.}$$

Hence,

$$\text{Number of } g\text{'s to yield point} = \frac{1 \times 10^{-3}}{2.35 \times 10^{-4}}g = 4.26 \text{ g}$$

$$\text{Corresponding voltage} = 0.94 \times 4.26 \text{ V} = 4.0 \text{ V}$$

$$\text{Hence, the amplifier gain} = 10.0/4.0 = 2.25.$$

Cross-sensitivity comes from accelerations in the two directions y and z, which are orthogonal to the direction of sensitivity (x). In the lateral (y) direction, the inertia force causes lateral bending. This produces equal tensile (or compressive) strains in B and D and equal compressive (or tensile) strains in A and C. According to the bridge circuit, we see that these contributions cancel each other. In the axial (z) direction, the inertia force causes equal tensile (or compressive) stresses in all four strain gages. These also cancel out, as is clear from the relationship in Equation 4.78 for the bridge, which gives

$$\frac{\delta v_o}{v_{\text{ref}}} = \frac{(\delta R_A - \delta R_C - \delta R_D + \delta R_B)}{4R}.$$

It follows that this arrangement compensates for cross-sensitivity problems.

4.8.1.4 Data Acquisition

For measuring dynamic strains, either the servo null-balance method or the imbalance output method should be employed (see Chapter 3). A schematic diagram for the imbalance output method is shown in Figure 4.48. In this method, the output from the active bridge is directly measured as a voltage signal and calibrated to provide the measured strain. Figure 4.48 shows the use of an ac bridge. In this case, the bridge is powered by an ac voltage. The supply frequency should be about 10 times the maximum frequency of interest in the dynamic strain signal (bandwidth). A supply frequency in the order of 1 kHz is typical. This signal is generated by an oscillator and is fed into the bridge. The transient component of the output from the bridge is very small (typically less than 1 mV and possibly a few microvolts). This signal has to be amplified, demodulated (especially if the signals are transient), and filtered to provide the strain reading. The calibration constant of the bridge should be known in order to convert the output voltage to strain.

Strain-gage bridges powered by dc voltages are common. They have the advantages of simplicity with regard to necessary circuitry and portability. The advantages of ac bridges include improved stability (reduced drift) and accuracy, and reduced power consumption.

4.8.1.5 Accuracy Considerations

Foil gages are available with resistances as low as 50 Ω and as high as several kilohms. The power consumption of a bridge circuit decreases with increased resistance. This has the added advantage of decreased heat generation. Bridges with a high range of

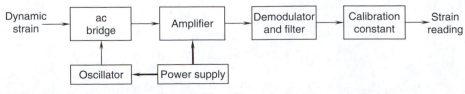

FIGURE 4.48
Measurement of dynamic strains using an ac bridge.

measurement (e.g., a maximum strain of 0.04 m/m) are available. The accuracy depends on the linearity of the bridge, environmental effects (particularly temperature), and mounting techniques. For example, zero shifts, due to strains produced when the cement or epoxy that is used to mount the strain-gage dries, results in calibration error. Creep introduces errors during static and low-frequency measurements. Flexibility and hysteresis of the bonding cement (or epoxy) bring about errors during high-frequency strain measurements. Resolutions in the order of 1 μm/m (i.e., one *microstrain*) are common.

As noted earlier, the cross-sensitivity of a strain gage is the sensitivity to strains that are orthogonal to the measured strain. This cross-sensitivity should be small (say, less than 1% of the direct sensitivity). Manufacturers usually provide cross-sensitivity factors for their strain gages. This factor, when multiplied by the cross strain present in a given application, gives the error in the strain reading due to cross-sensitivity.

Often, strains in moving members are sensed for control purposes. Examples include real-time monitoring and failure detection in machine tools, measurement of power, measurement of force and torque for feedforward and feedback control in dynamic systems, biomechanical devices, and tactile sensing using instrumented hands in industrial robots. If the motion is small or the device has a limited stroke, strain gages mounted on the moving member can be connected to the signal-conditioning circuitry and power source, using coiled flexible cables. For large motions, particularly in rotating shafts, some form of commutating arrangement has to be used. Slip-rings and brushes are commonly used for this purpose. When ac bridges are used, a mutual-induction device (rotary transformer) may be used, with one coil located on the moving member and the other coil stationary. To accommodate and compensate for errors (e.g., losses and glitches in the output signal) caused by commutation, it is desirable to place all four arms of the bridge, rather than just the active arms, on the moving member.

4.8.2 Semiconductor Strain Gages

In some low-strain applications (e.g., dynamic torque measurement), the sensitivity of foil gages is not adequate to produce an acceptable strain-gage signal. Semiconductor (SC) strain gages are particularly useful in such situations. The strain element of an SC strain-gage is made of a single crystal of piezoresistive material such as silicon, doped with a trace impurity such as boron. A typical construction is shown in Figure 4.49. The gage

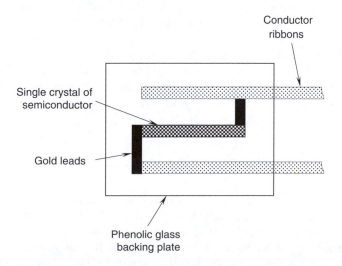

Conductor ribbons

Single crystal of semiconductor

Gold leads

Phenolic glass backing plate

FIGURE 4.49
Component details of a semiconductor strain gage.

TABLE 4.3

Properties of Common Strain-Gage Material

Material	Composition	Gage Factor (Sensitivity)	Temperature Coefficient of Resistance ($10^{-6}/°C$)
Constantan	45% Ni, 55% Cu	2.0	15
Isoelastic	36% Ni, 52% Fe, 8% Cr, 4% (Mn, Si, Mo)	3.5	200
Karma	74% Ni, 20% Cr, 3% Fe, 3% Al	2.3	20
Monel	67% Ni, 33% Cu	1.9	2000
Silicon	p-type	100 to 170	70 to 700
Silicon	n-type	−140 to −100	70 to 700

factor (sensitivity) of an SC strain gage is about two orders of magnitude higher than that of a metallic foil gage (typically, 40–200), as seen for Silicon, from the data given in Table 4.3. The resistivity is also higher, providing reduced power consumption and lower heat generation. Another advantage of SC strain gages is that they deform elastically to fracture. In particular, mechanical hysteresis is negligible. Furthermore, they are smaller and lighter, providing less cross-sensitivity, reduced distribution error (i.e., improved spatial resolution), and negligible error from mechanical loading. The maximum strain that is measurable using a semiconductor strain-gage is typically 0.003 m/m (i.e., 3000 $\mu\varepsilon$). Strain-gage resistance can be an order of magnitude greater for an SC strain gage; for example, several hundred ohms for a metal foil strain gage (typically, 120 Ω or 350 Ω), while several thousand ohms (5000 Ω) for an SC strain gage. There are several disadvantages associated with semiconductor strain gages, however, which can be interpreted as advantages of foil gages. Undesirable characteristics of SC gages include the following:

1. The strain–resistance relationship is more nonlinear.

2. They are brittle and difficult to mount on curved surfaces.

3. The maximum strain that can be measured is one to two orders of magnitude smaller (typically, less than 0.001 m/m).

4. They are more costly.

5. They have much larger temperature sensitivity.

The first disadvantage is illustrated in Figure 4.50. There are two types of semiconductor strain gages: the p-type, which are made of a semiconductor (e.g., silicon) doped with an acceptor impurity (e.g., boron), and the n-type, which are made of a semiconductor doped with a donor impurity (e.g., arsenic). In p-type strain gages, the direction of sensitivity is along the (1, 1, 1) crystal axis, and the element produces a positive (p) change in resistance in response to a positive strain. In n-type strain gages, the direction of sensitivity is along the (1, 0, 0) crystal axis, and the element responds with a negative (n) change in resistance to a positive strain. In both types, the response is nonlinear and can be approximated by the quadratic relationship

$$\frac{\delta R}{R} = S_1\varepsilon + S_2\varepsilon^2. \tag{4.82}$$

The parameter S_1 represents the linear gage factor (linear sensitivity), which is positive for p-type gages and negative for n-type gages. Its magnitude is usually somewhat larger for p-type gages, corresponding to better sensitivity. The parameter S_2 represents the degree of nonlinearity, which is usually positive for both types of gages. Its magnitude, however, is typically somewhat smaller for p-type gages. It follows that p-type gages are

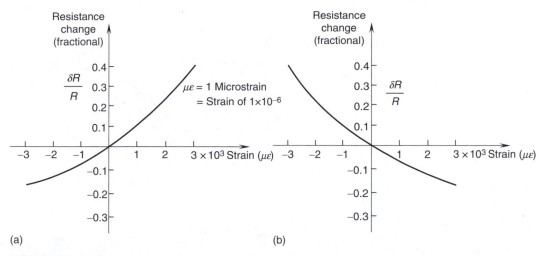

FIGURE 4.50
Nonlinear behavior of a semiconductor (silicon/boron) strain gage. (a) A p-type gage. (b) An n-type gage.

less nonlinear and have higher strain sensitivities. The nonlinear relationship given by Equation 4.82 or the nonlinear characteristic curve (Figure 4.50) should be used when measuring moderate to large strains with semiconductor strain gages. Otherwise, the nonlinearity error would be excessive.

Example 4.14
For a semiconductor strain gage characterized by the quadratic strain–resistance relationship, Equation (4.82), obtain an expression for the equivalent gage factor (sensitivity) S_s, using least squares error linear approximation and assuming that strains in the range $\pm \varepsilon_{max}$, have to be measured. Derive an expression for the percentage nonlinearity.

Taking $S_1 = 117$, $S_2 = 3600$, and $\varepsilon_{max} = 1 \times 10^{-2}$, calculate S_s and the percentage nonlinearity.

Solution
The linear approximation of Equation 4.82 may be expressed as

$$\left[\frac{\delta R}{R}\right]_L = S_s \varepsilon.$$

The error is given by

$$e = \frac{\delta R}{R} - \left[\frac{\delta R}{R}\right]_L = S_1 \varepsilon + S_2 \varepsilon^2 - S_s \varepsilon = (S_1 - S_s)\varepsilon + S_2 \varepsilon^2. \tag{i}$$

The quadratic integral error is

$$J = \int_{-\varepsilon_{max}}^{\varepsilon_{max}} e^2 d\varepsilon = \int_{-\varepsilon_{max}}^{\varepsilon_{max}} \left[(S_1 - S_s)\varepsilon + S_2 \varepsilon^2\right]^2 d\varepsilon. \tag{ii}$$

We have to determine S_s that results in minimum J. Hence, we use $\partial J / \partial S_s = 0$. Thus, from Equation (ii):

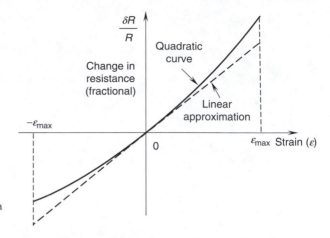

FIGURE 4.51
Least squares linear approximation for a semiconductor strain gage.

$$\int_{-\varepsilon_{max}}^{\varepsilon_{max}} (-2\varepsilon)\left[(S_1 - S_s)\varepsilon + S_2\varepsilon^2\right]^2 d\varepsilon = 0$$

On performing the integration and solving the equation, we get

$$S_s = S_1. \tag{4.83}$$

The quadratic curve and the linear approximation are shown in Figure 4.51. Note that the maximum error is at $\varepsilon = \pm\,\varepsilon_{max}$. The maximum error value is obtained from Equation (i), with $S_s = S_1$ and $\varepsilon = \pm\,\varepsilon_{max}$, as

$$e_{max} = S_2\varepsilon_{max}^2.$$

The true change in resistance (nondimensional) from $-\varepsilon_{max}$ to $+\varepsilon_{max}$ is obtained using Equation 4.82; thus,

$$\frac{\Delta R}{R} = (S_1\varepsilon_{max} + S_2\varepsilon_{max}^2) - (-S_1\varepsilon_{max} + S_2\varepsilon_{max}^2)$$

$$= 2S_1\varepsilon_{max}.$$

Hence, the percentage nonlinearity is given by

$$N_p = \frac{\text{max error}}{\text{range}} \times 100\% = \frac{S_2\varepsilon_{max}^2}{2S_1\varepsilon_{max}} \times 100\%$$

or

$$N_p = 50S_2\varepsilon_{max}/S_1\%. \tag{4.84}$$

Now, with the given numerical values, we have

$$S_s = 117$$

and

$$N_p = 50 \times 3600 \times 1 \times 10^{-2}/117\% = 15.4\%.$$

We obtained this high value for nonlinearity because the given strain limits were high. Usually, the linear approximation is adequate for strains up to $\pm 1 \times 10^{-3}$.

∎

The higher temperature sensitivity, which is listed as a disadvantage of semiconductor strain gages may be considered an advantage in some situations. For instance, it is this property of high temperature sensitivity that is used in piezoresistive temperature sensors. Furthermore, using the fact that the temperature sensitivity of a semiconductor strain-gage can be determined very accurately, precise methods can be employed for temperature compensation in strain-gage circuitry, and temperature calibration can also be done accurately. In particular, a passive SC strain gage may be used as an accurate temperature sensor for compensation purposes.

4.8.3 Automatic (Self) Compensation for Temperature

In foil gages the change in resistance due to temperature variations is typically small. Then the linear (first-order) approximation for the contribution from each arm of the bridge to the output signal, as given by Equation 4.78, would be adequate. These contributions cancel out if we pick strain-gage elements and bridge completion resistors properly—for example, R_1 identical to R_2 and R_3 identical R_4. If this is the case, the only remaining effect of temperature change on the bridge output signal is because of changes in the parameter values k and S_s (see Equation 4.80 and Equation 4.81). For foil gages, such changes are also typically negligible. Hence, for small to moderate temperature changes, additional compensation is not required when foil gage bridge circuits are employed.

In semiconductor gages, the change in resistance with temperature (and with strain) is not only larger, but the change in S_s with temperature is also large compared with the corresponding values for foil gages. Hence, the linear approximation given by Equation 4.78 might not be accurate for SC gages under variable temperature conditions; furthermore, the bridge sensitivity could change significantly with temperature. Under such conditions, temperature compensation becomes necessary.

A straightforward way to account for temperature changes is by directly measuring temperature and correcting strain-gage readings, using calibration data. Another method of temperature compensation is described here. This method assumes that the linear approximation given by Equation 4.78 is valid; hence, Equation 4.80 is applicable.

The resistance R and strain sensitivity (or gage factor) S_s of a semiconductor strain-gage are highly dependent on the concentration of the trace impurity, in a nonlinear manner. The typical behavior of the temperature coefficients of these two parameters for a p-type semiconductor strain gage is shown in Figure 4.52. The *temperature coefficient of resistance* α and the *temperature coefficient of sensitivity* β are defined by

$$R = R_o(1 + \alpha.\Delta T), \tag{4.85}$$

$$S_s = S_{so}(1 + \beta.\Delta T), \tag{4.86}$$

where ΔT denotes the temperature increase. Note from Figure 4.52 that β is a negative quantity and that for some dope concentrations, its magnitude is less than the value of the temperature coefficient of resistance (α). This property can be used in self-compensation with regard to temperature for a p-type semiconductor (silicon) strain gage.

Consider a constant-voltage bridge circuit with a compensating resistor R_c connected to the supply lead, as shown in Figure 4.53a. It can be shown that self-compensation can result if R_c is set to a value predetermined on the basis of the temperature coefficients of

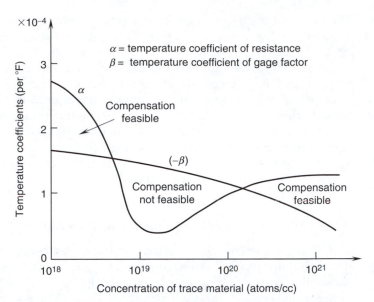

FIGURE 4.52
Temperature coefficients of
resistance and gage factor.

the strain gages. Consider the case where load impedance is very high and the bridge has
four identical SC strain gages, which have resistance R. In this case, the bridge can be
represented by the circuit shown in Figure 4.53b.

Since series impedances and parallel admittances (inverse of impedance) are additive,
the equivalent resistance of the bridge is R. Hence, the voltage supplied to the bridge,
allowing for the voltage drop across R_c, is not v_{ref} but v_i, as given by

$$v_i = \frac{R}{(R + R_c)} v_{ref}. \tag{4.87}$$

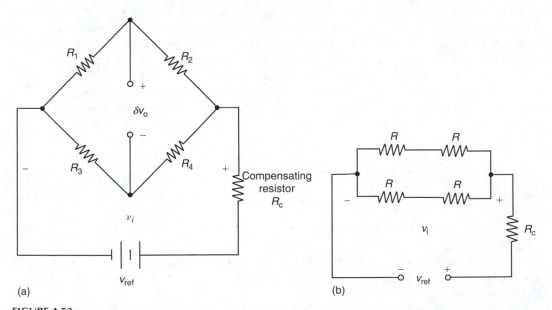

FIGURE 4.53
A strain-gage bridge with a compensating resistor. (a) Constant voltage dc bridge. (b) Equivalent circuit with
high load impedance.

Now, from Equation 4.80, we have

$$\frac{\delta v_o}{v_{ref}} = \frac{R}{(R + R_c)} \frac{kS_s}{4} \varepsilon. \tag{4.88}$$

We assume that the bridge constant k does not change with temperature. Otherwise, the following procedure still holds, provided that the calibration constant C is used in place of the gage factor S_s (see Equation 4.81). For self-compensation, we must have the same output after the temperature has changed through ΔT. Hence, from Equation 4.88, we have

$$\frac{R_o}{(R_o + R_c)} S_s o = \frac{R_o(1 + \alpha.\Delta T)}{[R_o(1 + \alpha.\Delta T) + R_c]} S_{so}(1 + \beta.\Delta T),$$

where the subscript o denotes values before the temperature change. Cancellation of the common terms and cross-multiplication gives

$$R_o\beta + R_c(\alpha + \beta) = (R_o + R_c)\alpha\beta\Delta T.$$

Now, since both $\alpha.\Delta T$ and $\beta.\Delta T$ are usually much smaller than unity, we may neglect the RHS (second-order) term in the preceding equation. This gives the following expression for the compensating resistance:

$$R_c = -\left[\frac{\beta}{\alpha + \beta}\right] R_o. \tag{4.89}$$

Note that compensation is possible because the temperature coefficient of the strain-gage sensitivity (β) is negative. The feasible ranges of operation, which correspond to positive R_c are indicated in Figure 4.52. This method requires that R_c be maintained constant at the chosen value under changing temperature conditions. One way to accomplish this is by selecting a material with negligible temperature coefficient of resistance for R_c. Another way is to locate R_c in a separate, temperature-controlled environment (e.g., ice bath).

4.9 Torque Sensors

Sensing of torque and force is useful in many applications, including the following:

1. In robotic tactile (distributed touch) and manufacturing applications such as gripping, surface gaging, and material forming, where exerting an adequate load on an object is a primary purpose of the task.

2. In the control of fine motions (e.g., fine manipulation and micromanipulation) and in assembly tasks, where a small motion error can cause large damaging forces or performance degradation.

3. In control systems that are not fast enough when motion feedback alone is employed, where force feedback and feedforward force control can be used to improve accuracy and bandwidth.

4. In process testing, monitoring, and diagnostic applications, where torque sensing can detect, predict, and identify abnormal operation, malfunction,

component failure, or excessive wear (e.g., in monitoring of machine tools such as milling machines and drills)

5. In the measurement of power transmitted through a rotating device, where power is given by the product of torque and angular velocity in the same direction.

6. In controlling complex nonlinear mechanical systems, where measurement of force and acceleration can be used to estimate unknown nonlinear terms. Non-linear feedback of the estimated terms will linearize or simplify the system (nonlinear feedback control or linearizing feedback technique or LFT).

In most applications, sensing is done by detecting an effect of torque or the cause of torque. As well, there are methods for measuring torque directly. Common methods of torque sensing include the following:

1. Measuring strain in a sensing member between the drive element and the driven load, using a strain-gage bridge

2. Measuring displacement in a sensing member (as in the first method)—either directly, using a displacement sensor, or indirectly, by measuring a variable such as magnetic inductance or capacitance that varies with displacement

3. Measuring reaction in support structure or housing (by measuring the force and the associated lever arm length that is required to hold it down)

4. In electric motors, measuring the field current or armature current, which produces motor torque; in hydraulic or pneumatic actuators, measuring the actuator pressure

5. Measuring torque directly, using piezoelectric sensors, for example

6. Employing a servo method—balancing the unknown torque with a feedback torque generated by an active device (say, a servomotor) whose torque characteristics are precisely known

7. Measuring the angular acceleration caused by the unknown torque in a known inertia element.

The remainder of this section is devoted to a discussion of torque measurement using some of these methods. Force sensing may be accomplished by essentially the same techniques. For the sake of brevity, however, we limit our treatment primarily to torque sensing, which may be interpreted as sensing of a "generalized force." The extension of torque-sensing techniques to force sensing is rather challenging.

4.9.1 Strain-Gage Torque Sensors

The most straightforward method of torque sensing is to connect a torsion member between the drive unit and the (driven) load in series, as shown in Figure 4.54, and to measure the torque in the torsion member.

If a circular shaft (solid or hollow) is used as the torsion member, the torque–strain relationship becomes relatively simple, and is given by:

$$\varepsilon = \frac{r}{2GJ}T, \tag{4.90}$$

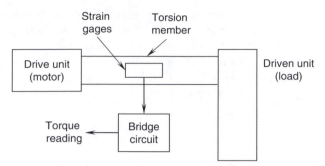

FIGURE 4.54
Torque sensing using a torsion member.

where T is the torque transmitted through the member, ε is the principal strain (which is at 45° to shaft axis) at radius r within the member, J is the polar moment of area of cross-section of the member, and G is the shear modulus of the material.

Moreover, the shear stress τ at a radius r of the shaft is given by

$$\tau = \frac{Tr}{J}. \tag{4.91}$$

It follows from Equation 4.90 that torque T can determined by measuring direct strain ε on the shaft surface along a principal stress direction (i.e., at 45° to the shaft axis). This is the basis of torque sensing using strain measurements. Using the general bridge Equation 4.80 along with Equation 4.81 in Equation 4.90, we can obtain torque T from bridge output δv_o:

$$T = \frac{8GJ}{kS_s r} \frac{\delta v_o}{v_{ref}}, \tag{4.92}$$

where S_s is the gage factor (or sensitivity) of the strain gages. The bridge constant k depends on the number of active strain-gages used. Strain gages are assumed to be mounted along a principal direction. Three possible configurations are shown in Figure 4.55. In configurations (a) and (b) only two strain gages are used, and the bridge constant $k = 2$. Note

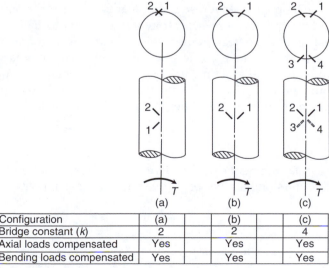

Strain-gage bridge

Configuration	(a)	(b)	(c)
Bridge constant (k)	2	2	4
Axial loads compensated	Yes	Yes	Yes
Bending loads compensated	Yes	Yes	Yes

FIGURE 4.55
Strain-gage configurations for a circular shaft torque sensor.

that both axial and bending loads are compensated with the given configurations because resistance in both gages are changed by the same amount (same sign and same magnitude), which cancels out up to first order, for the bridge circuit connection shown in Figure 4.55. Configuration (c) has two pairs of gages mounted on the two opposite surfaces of the shaft. The bridge constant is doubled in this configuration, and here again, the sensor self-compensates for axial and bending loads up to first order [$O(\delta R)$].

4.9.2 Design Considerations

Two conflicting requirements in the design of a torsion element for torque sensing are sensitivity and bandwidth. The element has to be sufficiently flexible in order to get an acceptable level of sensor sensitivity (i.e., a sufficiently large output signal). According to Equation 4.90, this requires a small torsional rigidity GJ, to produce a large strain for a given torque. Unfortunately, since the torsion-sensing element is connected in series between a drive element and a driven element, an increase in flexibility of the torsion element results in reduction of the overall stiffness of the system. Specifically, with reference to Figure 4.56, the overall stiffness K_{old} before connecting the torsion element is given by

$$\frac{1}{K_{old}} = \frac{1}{K_m} + \frac{1}{K_L}, \tag{4.93}$$

and the stiffness K_{new} after connecting the torsion member is given by

$$\frac{1}{K_{new}} = \frac{1}{K_m} + \frac{1}{K_L} + \frac{1}{K_s} \tag{4.94}$$

where K_m is the equivalent stiffness of the drive unit (motor), K_L is the equivalent stiffness of the load, and K_s is the stiffness of the torque-sensing element. It is clear from Equation 4.93 and Equation 4.94 that $1/K_{new} > 1/K_{old}$. Hence, $K_{new} < K_{old}$. This reduction in stiffness is associated with a reduction in natural frequency and bandwidth, resulting in slower response to control commands in the overall system. Furthermore, a reduction in stiffness causes a reduction in the loop gain. As a result, the steady-state error in some motion variables can increase, which demands more effort from the controller to achieve a required level of accuracy. One aspect in the design of the torsion element is to guarantee that the element stiffness is small enough to provide adequate sensitivity but large enough to maintain adequate bandwidth and system gain. In situations where K_s cannot be increased adequately without seriously jeopardizing the sensor sensitivity, the system bandwidth can be improved by decreasing either the load inertia or the drive unit (motor) inertia.

Example 4.15

Consider a rigid load, which has a polar moment of inertia J_L, and driven by a motor with a rigid rotor, which has inertia J_m. A torsion member of stiffness K_s is connected between

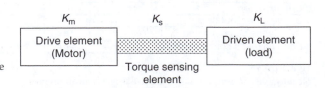

FIGURE 4.56
Stiffness degradation due to flexibility of the torsion-sensing element.

the rotor and the load, as shown in Figure 4.57a , to measure the torque transmitted to the load. Determine the transfer function between the motor torque T_m and the twist angle θ of the torsion member. What is the torsional natural frequency ω_n of the system? Discuss why the system bandwidth depends on ω_n. Show that the bandwidth can be improved by increasing K_s, by decreasing J_m, or by decreasing J_L. Mention some advantages and disadvantages of introducing a gearbox at the motor output.

Solution
From the free-body diagram shown in Figure 4.57b, the equations motion can be written:

$$\text{For motor: } J_m \ddot{\theta}_m = T_m - K_s(\theta_m - \theta_L) \tag{i}$$

$$\text{For load: } J_L \ddot{\theta}_L = K_s(\theta_m - \theta_L). \tag{ii}$$

Note that θ_m is the motor rotation and θ_L is the load rotation. Divide Equation (i) by J_m, divide Equation (ii) by J_L, and subtract the second equation from the first; thus,

$$\ddot{\theta}_m - \ddot{\theta}_L = \frac{T_m}{J_m} - \frac{K_s}{J_m}(\theta_m - \theta_L) - \frac{K_s}{J_L}(\theta_m - \theta_L).$$

This equation can be expressed in terms of the twist angle:

$$\theta = \theta_m - \theta_L \tag{iii}$$

$$\ddot{\theta} + K_s\left(\frac{1}{J_m} + \frac{1}{J_L}\right)\theta = \frac{T_m}{J_m}. \tag{iv}$$

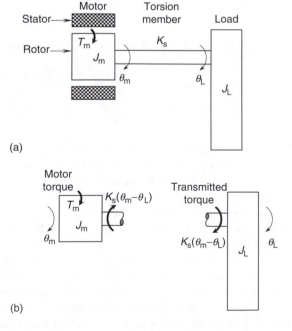

(a)

(b)

FIGURE 4.57
An example of bandwidth analysis of a system with a torque sensor. (a) System model. (b) Free-body diagram.

Hence, the transfer function $G(s)$ between input T_m and output θ is obtained by introducing the Laplace variable s in place of the time derivative d/dt. Specifically, we have:

$$G(s) = \frac{1/J_m}{s^2 + K_s(1/J_m + 1/J_L)}. \tag{v}$$

The characteristic equation of the twisting system is

$$s^2 + K_s\left(\frac{1}{J_m} + \frac{1}{J_L}\right) = 0. \tag{vi}$$

It follows that the torsional (twisting) natural frequency ω_n is given by

$$\omega_n = \sqrt{K_s\left(\frac{1}{J_m} + \frac{1}{J_L}\right)}. \tag{vii}$$

In addition to this natural frequency, there is a zero natural frequency in the overall system, which corresponds to rotation of the entire system as a rigid body without any twisting in the torsion member (i.e., the rigid-body mode). Both natural frequencies are obtained if the output is taken as either θ_m or θ_L, not the twist angle θ. When the output is taken as the twist angle θ, the response is measured relative to the rigid-body mode; hence, the zero-frequency term disappears from the characteristic equation.

The transfer function given by the Equation (v) may be written as

$$G(s) = \frac{1/J_m}{s^2 + \omega_n^2}. \tag{viii}$$

In the frequency domain $s = j\omega$, and the resulting frequency transfer function is

$$G(j\omega) = \frac{1/J_m}{\omega_n^2 - \omega^2}. \tag{ix}$$

It follows that if ω is small in comparison to ω_n, the transfer function can be approximated by

$$G(j\omega) = \frac{1/J_m}{\omega_n^2}, \tag{x}$$

which is a static relationship, implying an instantaneous response without any dynamic delay. Since, system bandwidth represents the excitation frequency range ω within which the system responds sufficiently fast (which corresponds to the sufficiently flat region of the transfer function magnitude), it follows that system bandwidth improves when ω_n is increased. Hence, ω_n is a measure of system bandwidth.

Now, observe from Equation (vii) that ω_n (and the system bandwidth) increases when K_s is increased, when J_m is decreased, or when J_L is decreased. If a gearbox is added to the system (see Chapter 8), the equivalent inertia increases and the equivalent stiffness decreases. This reduces the system bandwidth, resulting in a slower response. Another disadvantage of a gearbox is the backlash and friction, which are nonlinearities, that enter the system. The main advantage, however, is that torque transmitted to the load is amplified through speed reduction between motor and load. However, high torques and low speeds can be achieved by using torque motors without employing any speed reducers or by using backlash-free transmissions such as harmonic drives and traction (friction) drives (see Chapter 8).

■

The design of a torsion element for torque sensing can be viewed as the selection of the polar moment of area J of the element to meet the following four requirements:

1. The strain capacity limit specified by the strain-gage manufacturer is not exceeded.
2. A specified upper limit on nonlinearity for the strain-gage is not exceeded, for linear operation.
3. Sensor sensitivity is acceptable in terms of the output signal level of the differential amplifier (see Chapter 3) in the bridge circuit.
4. The overall stiffness (bandwidth, steady-state error, and so on) of the system is acceptable.

Now we develop design criteria for each of these requirements.

4.9.2.1 Strain Capacity of the Gage

The maximum strain handled by a strain-gage element is limited by factors such as strength, creep problems associated with the bonding material (epoxy), and hysteresis. This limit ε_{max} is specified by the strain-gage manufacturer. For a typical semiconductor gage, the maximum strain limit is in the order of 3000 $\mu\varepsilon$. If the maximum torque that needs to be handled by the sensor is T_{max}, we have, from Equation (4.90):

$$\frac{r}{2GJ}T_{max} < \varepsilon_{max},$$

which gives

$$J > \frac{r}{2G}\frac{T_{max}}{\varepsilon_{max}}, \tag{4.95}$$

where ε_{max} and T_{max} are specified.

4.9.2.2 Strain-Gage Nonlinearity Limit

For large strains, the characteristic equation of a strain-gage becomes increasingly nonlinear. This is particularly true for semiconductor gages. If we assume the quadratic Equation 4.82, the percentage nonlinearity N_p is given by Equation 4.84. For a specified nonlinearity, an upper limit for strain can be determined using this result; thus,

$$\frac{r}{2GJ}T_{max} = \varepsilon_{max} \leq \frac{N_p S_1}{50 S_2}. \tag{4.96}$$

The corresponding J is given by,

$$J \geq \frac{25 S_2}{G S_1}\frac{T_{max}}{N_p}, \tag{4.97}$$

where N_p and T_{max} are specified.

4.9.2.3 Sensitivity Requirement

The output signal from the strain-gage bridge is provided by a differential amplifier (see Chapter 3), which detects the voltages at the two output nodes of the bridge (A and B in Figure 4.45), takes the difference, and amplifies it by a gain K_a. This output signal is supplied to an analog-to-digital converter (ADC), which provides a digital signal to the computer for performing further processing and control. The signal level of the amplifier output has to be sufficiently high so that the SNR is adequate. Otherwise, serious noise problems can result. Typically, a maximum voltage in the order of ± 10 V is desired.

Amplifier output v is given by

$$v = K_a \delta v_o, \tag{4.98}$$

where δv_o is the bridge output before amplification. It follows that the desired signal level can be obtained by simply increasing the amplifier gain. There are limits to this approach, however. In particular, a large gain increases the susceptibility of the amplifier to saturation and instability problems such as drift, and errors as a result of parameter changes. Hence, sensitivity has to be improved as much as possible through mechanical considerations.

By substituting Equation 4.92 into 4.98, we get the signal level requirement as:

$$v_o \leq \frac{K_a k S_s r v_{\text{ref}}}{8GJ} T_{\text{max}},$$

where v_o is the specified lower limit on the output signal from the bridge amplifier, and T_{max} is also specified. Then the limiting design value for J is given by:

$$J \leq \frac{K_a k S_s r v_{\text{ref}}}{8G} \frac{T_{\text{max}}}{v_o}, \tag{4.99}$$

where v_o and T_{max} are specified.

4.9.2.4 Stiffness Requirement

The lower limit of the overall stiffness of the system is constrained by factors such as speed of response (represented by system bandwidth) and steady-state error (represented by system gain). The polar moment of area J should be chosen such that the stiffness of the torsion element does not fall below a specified limit K. First, we have to obtain an expression for the torsional stiffness of a circular shaft. For a shaft of length L and radius r, a twist angle of θ corresponds to a shear strain of

$$\gamma = \frac{r\theta}{L} \tag{4.100}$$

on the outer surface. Accordingly, shear stress is given by

$$\tau = \frac{Gr\theta}{L}. \tag{4.101}$$

Now in view of Equation 4.91, the torsional stiffness of the shaft is given by

$$K_s = \frac{T}{\theta} = \frac{GJ}{L}. \tag{4.102}$$

Note that the stiffness can be increased by increasing GJ. However, this decreases the sensor sensitivity because, in view of Equation 4.90, measured direct strain ε decreases for a given torque when GJ is increased. There are two other parameters—outer radius r and length L of the torsion element—which we can manipulate. Although for a solid shaft J increases (to the fourth power) with r, for hollow shafts it is possible to manipulate J and r independently, with practical limitations. For this reason, hollow members are commonly used as torque-sensing elements. With these design freedoms, for a given value of GJ, we can increase r to increase the sensitivity of the strain-gage bridge without changing the system stiffness, and we can decrease L to increase the system stiffness without affecting the bridge sensitivity.

Assuming that the shortest possible length L is used in the sensor, for a specified stiffness limit K we should have $GJ/L \geq K$. Then, the limiting design value for J is given by

$$J \geq \frac{L}{G}K, \tag{4.103}$$

where K is specified.

The governing formulas for the polar moment of area J of a torque sensor, based on the four criteria discussed earlier, are summarized in Table 4.4.

Example 4.16

A joint of a direct-drive robotic arm is sketched in Figure 4.58. The rotor of the drive motor is an integral part of the driven link, and there are no gears or any other speed reducers. Also, the motor stator is an integral part of the drive link. A tachometer measures the joint speed (relative), and a resolver measures the joint rotation (relative). Gearing is used to improve the performance of the resolver, and it does not affect the load transfer characteristics of the joint. Neglecting mechanical loading from sensors and gearing, but including bearing friction, sketch the torque distribution along the joint axis. Suggest a location (or locations) for measuring the net torque transmitted to the driven link using a strain-gage torque sensor.

Solution

For simplicity, assume point torques. Denoting the motor (magnetic) torque by T_m; the total rotor inertia torque and frictional torque in the motor by T_1; and the frictional torques at the two bearings by T_{f1}, and T_{f2}; the torque distribution can be sketched as shown in Figure 4.59. The net torque transmitted to the driven link is T_L. Locations available to install strain gages include, A, B, C, and D; note that T_L is given by the

TABLE 4.4

Design Criteria for a Strain-Gage Torque-Sensing Element

Criterion	Specification	Governing Formula for Polar Moment of Area (J)
Strain capacity of strain-gage element	ε_{max} and T_{max}	$> \dfrac{r}{2G} \cdot \dfrac{T_{max}}{\varepsilon_{max}}$
Strain-gage nonlinearity	N_p and T_{max}	$> \dfrac{25rS_2}{GS_1} \cdot \dfrac{T_{max}}{N_p}$
Sensor sensitivity	v_o and T_{max}	$\leq \dfrac{K_a k S_s r v_{ref}}{8G} \cdot \dfrac{T_{max}}{v_o}$
Sensor stiffness (system bandwidth and gain)	K	$\geq \dfrac{L}{G} \cdot K$

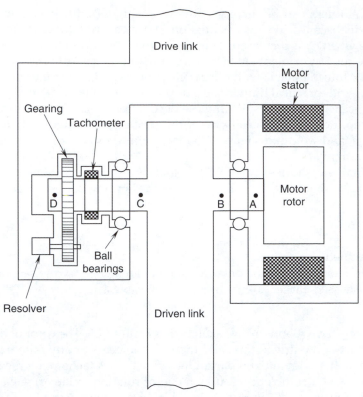

FIGURE 4.58
A joint of a direct-drive robotic arm.

difference between the torques at B and C. Hence, strain-gage torque sensors should be mounted at B and C and the difference of the readings should be taken for accurate measurement of T_L. Since bearing friction is small for most practical purposes, a single torque sensor located at B provides reasonably accurate results. The motor torque T_m is also approximately equal to the transmitted torque when the effects of bearing friction and motor loading (inertia and friction) are negligible. This is the reason behind using motor current (field or armature) to measure joint torque in some robotic applications.

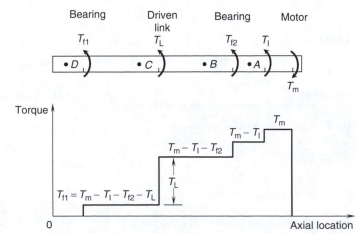

FIGURE 4.59
Torque distribution along the axis of a direct-drive manipulator joint.

Example 4.17

Consider the design of a tubular torsion element. Using the notation of Table 4.4, the following design specifications are given: $\varepsilon_{max} = 3000\ \mu\varepsilon$; $N_p = 5\%$; $v_o = 10$ V; and for a system bandwidth of 50 Hz, $K = 2.5 \times 10^3$ N·m/rad. A bridge with four active strain gages is used to measure torque in the torsion element. The following parameter values are provided:

1. For strain gages:
   $S_s = S_1 = 115,\ S_2 = 3500$.
2. For the torsion element:
   Outer radius $r = 2$ cm
   Shear modulus $G = 3 \times 10^{10}$ N/m^2
   Length $L = 2$ cm.
3. For the bridge circuitry:
   $v_{ref} = 20$ V and $K_a = 100$.

The maximum torque that is expected is $T_{max} = 10$ N m.
Using these values, design a torsion element for the sensor. Compute the operating parameter limits for the designed sensor.

Solution

Let us assume a safety factor of 1 (i.e., use the limiting values of the design formulas). We can compute the polar moment of area J using each of the four criteria given in Table 4.4:

1. For $\varepsilon_{max} = 3000\ \mu\varepsilon$:

$$J = \frac{0.02 \times 10}{2 \times 3 \times 10^{10} \times 3 \times 10^{-3}}\ m^4 = 1.11 \times 10^{-9}\ m^4.$$

2. For $N_p = 5$:

$$J = \frac{25 \times 0.02 \times 3500 \times 10}{3 \times 10^{10} \times 115 \times 5}\ m^4 = 1.01 \times 10^{-9}\ m^4.$$

3. For $v_o = 10$ V:

$$J = \frac{100 \times 4 \times 115 \times 0.02 \times 20 \times 10}{8 \times 3 \times 10^{10} \times 10}\ m^4 = 7.67 \times 10^{-8}\ m^4.$$

4. For $K = 2.5 \times 10^3$ N m/rad:

$$J = \frac{0.02 \times 2.5 \times 10^3}{3 \times 10^{10}}\ m^4 = 1.67 \times 10^{-9}\ m^4.$$

It follows that for an acceptable sensor, we should satisfy

$$J \geq (1.11 \times 10^{-9}) \quad \text{and} \quad (1.01 \times 10^{-9}) \quad \text{and} \quad (1.67 \times 10^{-9}) \quad \text{and} \quad J \leq 7.67 \times 10^{-8}\ m^4.$$

We pick $J = 7.67 \times 10^{-8}\ m^4$ so that the tube thickness is sufficiently large to transmit load without buckling or yielding. Since, for a tubular shaft,

$$J = \frac{\pi}{2}(r_0^4 - r_i^4),$$

where r_o is the outer radius and r_i is the inner radius, we have

$$7.67 \times 10^{-8} = \frac{\pi}{2}(0.02^4 - r_i^4)$$

or

$$r_i = 1.8 \text{ cm}.$$

Now, with the chosen value for J:

$$\varepsilon_{max} = \frac{7.67 \times 10^{-8}}{1.11 \times 10^{-9}} \times 3000 \ \mu\varepsilon = 2.07 \times 10^5 \ \mu\varepsilon$$

$$N_P = \frac{1.01 \times 10^{-9}}{7.67 \times 10^{-8}} \times 5\% = 0.07\%$$

$$v_o = 10 \text{ V}$$

$$K = \frac{7.67 \times 10^{-8}}{1.67 \times 10^{-9}} \times 2.5 \times 10^3 = 1.15 \times 10^5 \text{ N m/rad}$$

Since natural frequency is proportional to the square root of stiffness, for a given inertia, we note that a bandwidth of

$$50\sqrt{\frac{1.15 \times 10^5}{2.5 \times 10^3}} = 339 \text{ Hz}$$

is possible with this design.

∎

Although the manner in which strain gages are configured on a torque sensor can be exploited to compensate for cross-sensitivity effects arising from factors such as tensile and bending loads, it is advisable to use a torque-sensing element that inherently possesses low sensitivity to these factors, which cause error in a torque measurement. The tubular torsion element discussed in this section is convenient for analytical purposes because of the simplicity of the associated expressions for design parameters. Its mechanical design and integration into a practical system are convenient as well. Unfortunately, this member is not optimal with respect to rigidity (stiffness) for the transmission of both bending and tensile loads. Alternative shapes and structural arrangements have to be considered when inherent rigidity (insensitivity) to cross-loads is needed. Furthermore, a tubular element has the same principal strain at all locations on the element surface. This does not give us a choice with respect to mounting locations of strain gages in order to maximize the torque sensor sensitivity. Another disadvantage of the basic tubular torsion member is that, due to curved surface, much care is needed in mounting fragile semi-conductor gages, which could be easily damaged even with slight bending. Hence, a sensor element that has flat surfaces to mount the strain gages would be desirable.

A torque-sensing element that has the foregoing desirable characteristics (i.e., inherent insensitivity to cross-loading, nonuniform strain distribution on the surface, and availability of flat surfaces to mount strain gages) is shown in Figure 4.60. Note that two sensing elements are connected radially between the drive unit and the driven member. In this design, the sensing elements undergo bending to transmit a torque between the driver and the driven member. Bending strains are measured at locations of high sensitivity and are taken to be proportional to the transmitted torque. Analytical determination

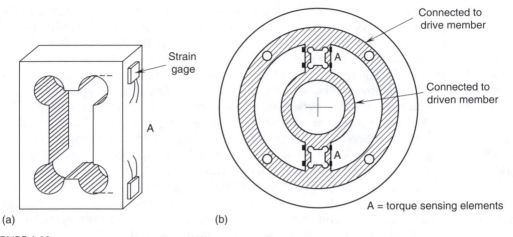

FIGURE 4.60
A bending element for torque sensing. (a) Shape of the sensing element. (b) Element locations (two radially placed elements).

of the calibration constant is not easy for such complex sensing elements, but experimental determination is straightforward. Finite element analysis may be used as well for this purpose. Note that the strain-gage torque sensors measure the direction as well as the magnitude of the torque transmitted through it.

4.9.3 Deflection Torque Sensors

Instead of measuring strain in the sensor element, the actual deflection (twisting or bending) can be measured and used to determine torque, through a suitable calibration constant. For a circular shaft (solid or hollow) torsion element, the governing relationship is given by Equation 4.102, which can be written in the form

$$T = \frac{GJ}{L}\theta. \tag{4.104}$$

The calibration constant GJ/L has to be small in order to achieve high sensitivity. This means that the element stiffness should be low. This limits the bandwidth, which measures the speed of response; and the gain, which determines the steady-state error, of the overall system. The twist angle θ is very small (e.g., a fraction of a degree) in systems with high bandwidth. This requires very accurate measurement of θ, for accurate determination of torque T. Two types of displacement torque sensors are described next. One sensor directly measures the angle of twist, and the other sensor uses the change in magnetic induction associated with sensor deformation.

4.9.3.1 Direct-Deflection Torque Sensor

Direct measurement of the twist angle between two axial locations in a torsion member, using an angular displacement sensor, can be used to determine torque. The difficulty in this case is that under dynamic conditions, relative deflection has to be measured while the torsion element is rotating. One type of displacement sensor that could be used here is a synchro transformer. Suppose that the two rotors of the synchro are mounted at the two ends of the torsion member. The synchro output gives the relative angle of rotation of the two rotors. Another type of displacement sensor that could be used for the same objective is shown in Figure 4.61a. Two ferromagnetic gear wheels are splined at two axial locations of the torsion element. Two stationary proximity probes of the magnetic induction type (self-induction or mutual induction) are placed radially, facing the gear teeth, at the two locations. As the shaft rotates, the gear teeth cause a change in flux linkage with the proximity sensor coils. The resulting output signals of the two probes are pulse sequences, shaped somewhat like sine waves. The phase shift of one signal with respect to the other determines the relative angular deflection of one gear wheel with respect to the other, assuming that the two probes are synchronized under no-torque conditions. Both the magnitude and the direction of the transmitted torque are determined using this method. A $360°$ phase shift corresponds to a relative deflection by an integer multiple of the gear pitch. It follows that deflections less than half the gear-tooth pitch can be measured without ambiguity. Assuming that the output signals of the two probes are sine waves (narrow-band filtering can be used to achieve this), the phase shift ϕ is proportional to the angular twist θ. If the gear wheel has n teeth, a primary phase shift of 2π corresponds to a twist angle of $2\pi/n$ radians. Hence, $\theta = \phi/n$ and from Equation 4.104, we get

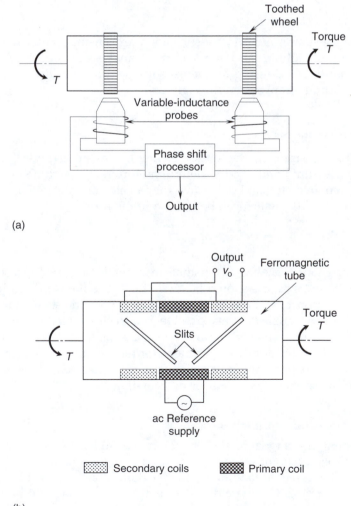

(a)

(b)

FIGURE 4.61
Deflection torque sensors. (a) A direct-deflection torque sensor. (b) A variable-reluctance torque sensor.

$$T = \frac{GJ\phi}{Ln},$$ (4.105)

where G is the shear modulus of the torsion element, J the polar moment of area of the torsion element, ϕ the phase shift between the two proximity probe signals, L the axial separation of the proximity probes, and n is the number of teeth in each gear wheel.

Note that the proximity probes are noncontact devices, unlike dc strain-gage sensors. In addition, note that eddy current proximity probes and Hall effect proximity probes could be used instead of magnetic induction probes in this method of torque sensing.

4.9.3.2 Variable-Reluctance Torque Sensor

A torque sensor that is based on sensor element deformation and that does not require a contacting commutator is shown in Figure 4.61b. This is a variable-reluctance device, which operates like a differential transformer (RVDT or LVDT). The torque-sensing element is a ferromagnetic tube, which has two sets of slits, typically oriented along the two principal stress directions of the tube (i.e., at $45°$ to the axial direction) under torsion. When a torque is applied to the torsion member, one set of gaps closes and the other set opens as a result of the principal stresses normal to the slit axes. Primary and secondary coils are placed around the slit tube, and they remain stationary. One segment of the secondary coil is placed around one set of slits, and the second segment is placed around the other (perpendicular) set. The primary coil is excited by an ac supply, and the induced voltage v_o in the secondary coil is measured. As the tube deforms, it changes the magnetic reluctance in the flux linkage path, thus changing the induced voltage. To obtain the best sensitivity, the two segments of the secondary coil, as shown in Figure 4.61b, should be connected so that the induced voltages are absolutely additive (algebraically subtractive), because one voltage increases and the other decreases. The output signal should be demodulated (by removing the carrier frequency component) to effectively measure transient torques. Note that the direction of torque is given by the sign of the demodulated signal.

4.9.4 Reaction Torque Sensors

The foregoing methods of torque sensing use a sensing element that is connected between the drive member and the driven member. A major drawback of such an arrangement is that the sensing element modifies the original system in an undesirable manner, particularly by decreasing the system stiffness and adding inertia. Not only does the overall bandwidth of the system decrease, but the original torque is also changed (mechanical loading) because of the inclusion of an auxiliary sensing element. Furthermore, under dynamic conditions, the sensing element is in motion, thereby making torque measurement more difficult. The reaction method of torque sensing eliminates these problems to a large degree. This method can be used to measure torque in a rotating machine. The supporting structure (or housing) of the rotating machine (e.g., motor, pump, compressor, turbine, generator) is cradled by releasing the fixtures, and the effort necessary to keep the structure from moving (i.e., to hold down) is measured. A schematic representation of the method is shown in Figure 4.62a. Ideally, a lever arm is mounted on the cradled housing, and the force required to maintain the housing stationary is measured using a force sensor (load cell). The reaction torque on the housing is given by

$$T_R = F_R \cdot L,$$ (4.106)

where F_R is the reaction force measured using load cell and L is the lever arm length.

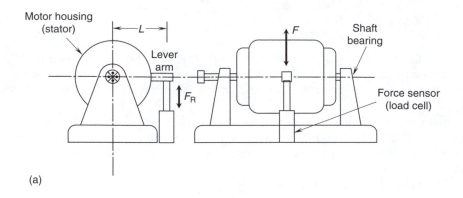

(a)

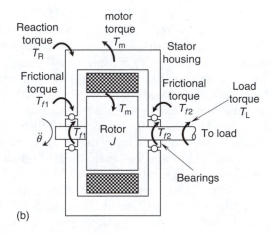

(b)

FIGURE 4.62
(a) Schematic representation of a reaction torque sensor setup (reaction dynamometer). (b) The relationship between reaction torque and load torque.

Alternatively, strain gages or other types of force sensors may be mounted directly at the fixture locations (e.g., on the mounting bolts) of the housing, to measure the reaction forces without actually having to cradle the housing. Then the reaction torque is determined with knowledge of the distance of the fixture locations from the shaft axis.

The reaction-torque method of torque sensing is widely used in dynamometers (reaction dynamometers), which determine the transmitted power in rotating machinery through the measurement of torque and shaft speed. A drawback of reaction-type torque sensors can be explained using Figure 4.62b. A motor of rotor inertia J, which rotates at angular acceleration $\ddot{\theta}$, is shown. By Newton's third law (action = reaction), the electromagnetic torque generated at the rotor of the motor T_m is reacted back onto the stator and housing. In the figure, T_{f1} and T_{f2} denote the frictional torques at the two bearings and T_L is the torque transmitted to the driven load.

When applying Newton's second law to the entire system, note that the friction torques and the motor (magnetic) torque all cancel out, giving $J\ddot{\theta} = T_R - T_L$, or

$$T_L = T_R - J\ddot{\theta}. \qquad (4.107)$$

Note that T_L is what must be measured. Under accelerating or decelerating conditions, the reaction torque T_R, which is measured, is not equal to the actual torque T_L that is transmitted. One method of compensating for this error is to measure shaft acceleration, compute the inertia torque, and adjust the measured reaction torque using this inertia torque. Note that the frictional torque in the bearings does not enter the final equation, which is an other advantage of this method.

4.9.5 Motor Current Torque Sensors

Torque in an electric motor is generated as a result of the electromagnetic interaction between the armature windings and the field windings of the motor (see Chapter 7 for details). For a dc motor, the armature windings are located on the rotor and the field windings are on the stator. The (magnetic) torque T_m in a dc motor is given by

$$T_m = k i_f i_a, \tag{4.108}$$

where i_f is the field current, i_a the armature current, and k is the torque constant.

It is seen from Equation 4.108 that the motor torque can be determined by measuring either i_a or i_f while the other is kept constant at a known value. In particular, note that i_f is assumed constant in armature control and i_a is assumed constant in field control.

As noted before (e.g., see Figure 4.62b), the magnetic torque of a motor is not quite equal to the transmitted torque, which is what needs to be sensed in most applications. It follows that the motor current provides only an approximation for the needed torque. The actual torque that is transmitted through the motor shaft (the load torque) is different from the motor torque generated at the stator–rotor interface of the motor. This difference is necessary for overcoming the inertia torque of the moving parts of the motor unit (particularly rotor inertia) and frictional torque (particularly bearing friction). Methods are available to adjust (compensate for) the readings of magnetic toque, so as to estimate the transmitted torque at reasonable accuracy. One approach is to incorporate a suitable dynamic model for the electromechanical system of the motor and the load, into a Kalman filter whose input is the measured current and the estimated output is the transmitted load. A detailed presentation of this approach is beyond the present scope. The current can be measured by sensing the voltage across a known resistor (of low resistance) placed in series with the current circuit.

In the past, dc motors were predominantly used in complex control applications. Although ac synchronous motors were limited mainly to constant-speed applications in the past, they are finding numerous uses in variable-speed applications (e.g., robotic manipulators) and servo systems, because of rapid advances in solid-state drives. Today, ac motor drive systems incorporate both frequency control and voltage control using advanced semiconductor technologies (see Chapter 7). Torque in an ac motor may also be determined by sensing the motor current. For example, consider the three-phase synchronous motor shown schematically in Figure 4.63.

The armature windings of a conventional synchronous motor are carried by the stator (in contrast to the case of a dc motor). Suppose that the currents in the three phases (armature currents) are denoted by i_1, i_2, and i_3. The dc field current in the rotor windings is denoted by i_f. Then, the motor torque T_m can be expressed as

$$T_m = k i_f \left[i_1 \sin\theta + i_2 \sin\left(\theta - \frac{2\pi}{3}\right) + i_3 \sin\left(\theta - \frac{4\pi}{3}\right) \right], \tag{4.109}$$

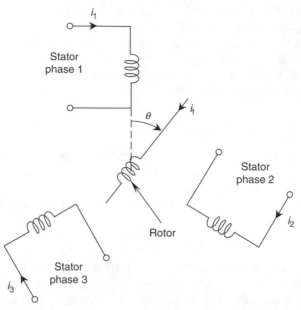

FIGURE 4.63
Schematic representation of a three-phase synchronous motor.

where θ denotes the angular rotation of the rotor and k is the torque constant of the synchronous motor. Since i_f is assumed fixed, the motor torque can be determined by measuring the phase currents. For the special case of a balanced three-phase supply, we have

$$i_1 = i_a \sin \omega t, \tag{4.110}$$

$$i_2 = i_a \sin \left(\omega t - \frac{2\pi}{3} \right), \tag{4.111}$$

$$i_3 = i_a \sin \left(\omega t - \frac{4\pi}{3} \right), \tag{4.112}$$

where ω denotes the line frequency (frequency of the current in each supply phase) and i_a is the amplitude of the phase current. Substituting Equation 4.110 through Equation 4.112 into Equation 4.109 and simplifying using well-known trigonometric identities, we get

$$T_m = 1.5 k i_f i_a \cos (\theta - \omega t). \tag{4.113}$$

We know that the angular speed of a three-phase synchronous motor with one pole pair per phase is equal to the line frequency ω (see Chapter 7). Accordingly, we have

$$\theta = \theta_0 + \omega t, \tag{4.114}$$

where θ_0 denotes the angular position of the rotor at $t = 0$. It follows that with a balanced three-phase supply, the torque of a synchronous motor is given by

$$T_m = 1.5 k i_f i_a \cos \theta_0. \tag{4.115}$$

This expression is quite similar to the one for a dc motor, as given by Equation 4.108.

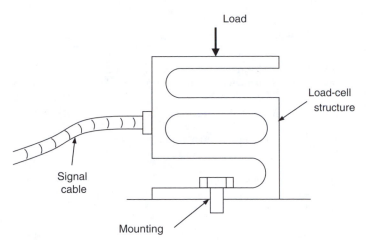

FIGURE 4.64
An industrial force sensor (load cell).

4.9.6 Force Sensors

Force sensors are useful in numerous applications. For example, cutting forces generated by a machine tool may be monitored to detect tool wear and an impending failure and to diagnose the causes of it; to control the machine tool, with feedback; and to evaluate the product quality. In vehicle testing, force sensors are used to monitor impact forces on the vehicles and crash-test dummies. Robotic handling and assembly tasks are controlled by measuring the forces generated at the end effector. Measurement of excitation forces and corresponding responses is employed in experimental modeling (model identification) of mechanical systems. Direct measurement of forces is useful in nonlinear feedback control of mechanical systems.

Force sensors that employ strain-gage elements or piezoelectric (quartz) crystals with built-in microelectronics are common. For example, thin-film and foil sensors that employ the strain-gage principle for measuring forces and pressures are commercially available. A sketch of an industrial load cell, which uses strain-gage method, is shown in Figure 4.64. Both impulsive forces and slowly varying forces can be monitored using this sensor. Some types of force sensors are based on measuring a deflection caused by the force. Relatively high deflections (fraction of a mm) would be necessary for this technique to be feasible. Commercially available sensors range from sensitive devices, which can detect forces in the order of thousandth of a newton to heavy-duty load cells, which can handle very large forces (e.g., 10,000 N). Since the techniques of torque sensing can be extended in a straightforward manner to force sensing, further discussion of the topic is not undertaken here. Typical rating parameters for several types of sensors are given in Table 4.5.

4.10 Tactile Sensing

Tactile sensing is usually interpreted as touch sensing, but tactile sensing is different from a simple clamping where very few discrete force measurements are made. In tactile sensing, a force distribution is measured, using a closely spaced array of force sensors and usually exploiting the skin-like properties of the sensor array.

TABLE 4.5

Rating Parameters of Several Sensors and Transducers

Transducer	Measurand	Measurand Frequency Max/Min	Output Impedance	Typical Resolution	Accuracy	Sensitivity
Potentiometer	Displacement	10 Hz/DC	Low	0.1 mm or less	0.1%	200 mV/ mm
LVDT	Displacement	2500 Hz/DC (Max, limited by carrier frequency)	Moderate	0.001 mm or less	0.1%	50 mV/ mm
Resolver	Angular displacement	500 Hz/DC (Max, limited by carrier frequency)	Low	2 min.	0.2%	10 mV/ deg
DC Tachometer	Velocity	700 Hz/DC	Moderate (50 Ω)	0.2 mm/s	0.5%	5 mV/ mm/s 75 mV/ rad/s
Eddy current proximity sensor	Displacement	100 kHz/DC	Moderate	0.001 mm 0.05% full scale	0.5%	5 V/mm
Piezoelectric accelerometer	Acceleration (and velocity, etc.)	25 kHz/1Hz	High	1 mm/s^2	0.1%	0.5 mV/ m/s^2
Semiconductor strain-gage	Strain (displacement, acceleration, etc.)	1 kHz/DC (limited by fatigue)	200 Ω	1–10 $\mu\varepsilon$ (1 $\mu\varepsilon = 10^{-6}$ unity strain)	0.1%	1 V/ε, 2000 $\mu\varepsilon$ max
Loadcell	Force (1–1000 *N*)	500 Hz/DC	Moderate	0.01 N	0.05%	1 mV/N
Laser	Displacement/ Shape	1 kHz/DC	100 Ω	1.0 μm	0.5%	1 V/mm
Optical encoder	Motion	100 kHz/DC	500 Ω	10 bit	$\pm\frac{1}{2}$ bit	10^4 pulses/ rev.

Tactile sensing is particularly important in two types of operations: (1) grasping and (2) object identification. In grasping, the object has to be held in a stable manner without being allowed to slip and without being damaged. Object identification includes recognizing or determining the shape, location, and orientation of an object as well as detecting or identifying surface properties (e.g., density, hardness, texture, flexibility), and defects. Ideally, these tasks would require two types of sensing:

1. Continuous spatial sensing of time-varying contact forces
2. Sensing of surface deformation profiles (time-varying).

These two types of measured data are generally related through the constitutive relations (e.g., stress–strain relations) of the touch surface of the tactile sensor or of the object that is grasped. As a result, either the almost-continuous spatial sensing of tactile forces or the sensing of a tactile deflection profile, separately, is often termed tactile sensing. Note that the learning experience is also an important part of tactile sensing. For example, picking up a fragile object such as an egg or a light bulb and picking up an object that has the same shape but is made of a flexible material (e.g., tennis ball), are not identical processes; they require some learning through touch, particularly when vision capability is not available.

4.10.1 Tactile Sensor Requirements

Significant advances in tactile sensing have taken place in the robotics area. Applications, which are very general and numerous, include: automated inspection of surface profiles and joints (e.g., welded or glued parts) for defects; material handling or parts transfer (e.g., pick and place); parts assembly (e.g., parts mating); parts identification and gaging in manufacturing applications (e.g., determining the size and shape of a turbine blade picked from a bin); and fine-manipulation tasks (e.g., production of arts and craft, robotic engraving, and robotic microsurgery). Some of these applications might need only simple touch (force or torque) sensing if the grasped parts are properly oriented and if adequate information about the process and the objects is already available.

Naturally, the frequently expressed design objective for tactile sensing devices has been to mimic the capabilities of human fingers. Specifically, the tactile sensor should have a compliant covering with skin-like properties, along with enough degrees of freedom for flexibility and dexterity, adequate sensitivity and resolution for information acquisition, adequate robustness and stability to accomplish various tasks, and some local intelligence for identification and learning purposes. Although the spatial resolution of a human fingertip is about 2 mm, still finer spatial resolutions (e.g., less than 1 mm) can be realized if information through other senses (e.g., vision), prior experience, and intelligence are used simultaneously during the touch. The force resolution (or sensitivity) of a human fingertip is in the order of 1 gm. Moreover, human fingers can predict impending slip during grasping, so that corrective actions can be taken before the object actually slips. At an elementary level, this requires the knowledge of shear stress distribution and friction properties at the common surface between the object and the hand (see Chapter 8). Additional information and an intelligent processing capability are also needed to accurately predict slip and to take corrective actions that would prevent slipping. These are, of course, ideal goals for a tactile sensor, but they are not unrealistic in the long run. Typical specifications for an industrial tactile sensor are as follows:

1. Spatial resolution of about 2 mm
2. Force resolution (sensitivity) of about 2 gm
3. Force capacity (maximum touch force) of about 1 kg
4. Response time of 5 ms or less
5. Low hysteresis (low energy dissipation)
6. Durability under harsh working conditions
7. Robustness and insensitivity to change in environmental conditions (temperature, dust, humidity, vibration, and so on)
8. Capability to detect and even predict slip.

Although the technology of tactile sensing has not peaked yet, and the widespread use of tactile sensors in industrial applications is still to come, several types of tactile sensors that meet and even exceed the foregoing specifications are commercially available. In future developments of these sensors, two separate groups of issues need to be addressed:

1. Ways to improve the mechanical characteristics and design of a tactile sensor so that accurate data with high resolution can be acquired quickly using the sensor

2. Ways to improve signal analysis and processing capabilities so that useful information can be extracted accurately and quickly from the data acquired through tactile sensing.

Under the second category, we also have to consider techniques for using tactile information in the feedback control of dynamic processes. In this context, the development of control algorithms, rules, and inference techniques for intelligent controllers that use tactile information, has to be addressed. Advances in micro-electromechanical systems (MEMS) and network communication technologies have contributed to the development of advanced sensor networks, distributed sensor arrays, and tactile sensors.

4.10.2 Construction and Operation of Tactile Sensors

The touch surface of a tactile sensor is usually made of an elastomeric pad or flexible membrane. While starting from this common basis, the principle of operation of a tactile sensor differs primarily depending on whether the distributed force is sensed or the deflection of the tactile surface is measured. The common methods of tactile sensing include the following:

1. Use a closely spaced set of strain gages or other types of force sensors to sense the distributed force.

2. Use a conductive elastomer as the tactile surface. The change in its resistance as it deforms determines the distributed force.

3. Use a closely spaced array of deflection sensors or proximity sensors (e.g., optical sensors) to determine the deflection profile of the tactile surface.

Since force and deflection are related through a constitutive law for the tactile sensor (touch pad), only one type of measurement, not both—force and deflection—is needed in tactile sensing. A force distribution profile or a deflection profile obtained in this manner may be treated as a 2-D array or an image and may be processed (filtered, function-fitted, etc.) and displayed as a tactile image, or used in applications (object identification, manipulation control, etc.).

The contact force distribution in a tactile sensor is commonly measured using an array of force sensors located under the flexible membrane. Arrays of piezoelectric sensors and metallic or semiconductor strain gages (piezoresistive sensors) in sufficient density (number of elements per unit area) may be used for the measurement of the tactile force distribution. In particular, semiconductor elements are poor in mechanical strength but have good sensitivity. Alternatively, the skin-like membrane itself can be made from a conductive elastomer (e.g., graphite-leaded neoprene rubber) whose changes in resistance can be sensed and used in determining the force and deflection distribution. In particular, as the tactile pressure increases, the resistance of the particular elastomer segment decreases and the current conducted through it (because of an applied constant voltage) increases. Conductors can be etched underneath the elastomeric pad to detect the current distribution in the pad, through proper signal acquisition circuitry. Common problems with conductive elastomers are electrical noise, nonlinearity, hysteresis, low sensitivity, drift, low bandwidth, and poor material strength.

On the other hand the deflection profile of a tactile surface may be determined using a matrix of proximity sensors or deflection sensors. Electromagnetic and capacitive sensors may be used in obtaining this information. The principles of operation of these types of sensors have been discussed previously, in this chapter. Optical tactile sensors use light-sensitive elements (photosensors) to sense the intensity of light (or laser beams) reflected from the tactile surface. In one approach (extrinsic), the proximity of a light-reflecting surface, which is attached to the back of a transparent tactile pad, is measured. Since the light intensity depends on the distance from the light-reflecting surface to the photosensor, the deflection profile can be determined. In another approach (intrinsic), the deformation of the tactile pad alters the light transmission characteristics of the pad. As a result, the intensity distribution of the transmitted light, as detected by an array of photosensors, determines the deflection profile. Optical methods have the advantages of freedom from electromagnetic noise and safety in explosive environments, but they can have errors as a result of stray light reaching the sensor, variation in intensity of the light source, and changes in environmental conditions (e.g., dirt, humidity, and smoke).

Example 4.18

A tactile sensor pad consists of a matrix of conductive elastomer elements. The resistance R_t in each tactile element is given by

$$R_t = \frac{a}{F_t},$$

where F_t is the tactile force applied to the element and a is a constant. The circuit shown in Figure 4.65 is used to acquire the tactile sensor signal v_o, which measures the local tactile force F_t. The entire matrix of tactile elements may be scanned by addressing the corresponding elements through appropriate switching electronics.

For the signal acquisition circuit shown in Figure 4.65 obtain a relationship for the output voltage v_o in terms of the parameters a, R_o, and others if necessary, and the force variable $F_{t(\text{the measurand})}$.

Show that $v_o = 0$ when the tactile element is not addressed (i.e., when the circuit is switched to the reference voltage 2.5 V).

Solution

Define,

v_i = input to the circuit (2.5 V or 0.0 V)

v_{o1} = output of the first opamp.

According to the properties of an opamp (see Chapter 3):

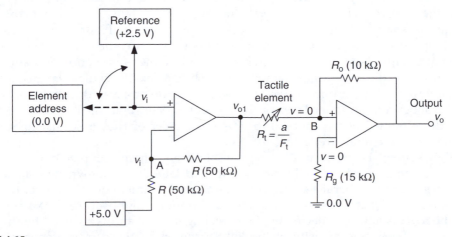

FIGURE 4.65
A signal acquisition circuit for a conductive-elastomer tactile sensor.

1. Voltages at the two input leads are equal
2. Currents through the two input leads are zero.

Hence, note the same v_i at both input leads of the first op-amp (and at node A); and the same zero voltage at both input leads of the second op-amp (and at node B), because one of the leads is grounded.

$$\text{Current balance at A: } \frac{5.0 - v_i}{R} = \frac{v_i - v_{o1}}{R} \Rightarrow v_{o1} = 2v_i - 5.0. \tag{i}$$

$$\text{Current balance at B: } \frac{v_{o1} - 0}{R_t} = \frac{0 - v_o}{R_o} \Rightarrow v_o = -v_{o1} \frac{R_o}{R_t}. \tag{ii}$$

Substitute Equation (i) into Equation (ii) and also substitute the given expression for R_t. We get,

$$v_o = \frac{R_o}{a} F_t (5.0 - 2v_i). \tag{4.116}$$

Substitute the two switching values for v_i. We have,

$$v_o = \frac{5R_o}{a} F_t \text{ when addressed.}$$
$$= 0 \qquad \text{when at reference. This is the required result.} \tag{4.116*}$$

4.10.3 Optical Tactile Sensors

A schematic representation of an optical tactile sensor (built at the Man-Machine Systems Laboratory of Massachusetts Institute of Technology—MIT) is shown in Figure 4.66. If a beam of light (or laser) is projected onto a reflecting surface, the intensity of light reflected back and received by a light receiver depends on the distance (proximity) of the reflecting surface. For example, in Figure 4.66a, more light is received by the light receiver when the reflecting surface is at Position 2 than when it is at Position 1. But if the reflecting surface actually touches the light source, light becomes completely blocked off, and no light reaches the receiver. Hence, in general, the proximity–intensity curve for an optical proximity sensor is nonlinear and has the shape shown in Figure 4.66a. Using this (calibration) curve, we can determine the position (x) once the intensity of the light received at the photosensor is known. This is the principle of operation of many optical tactile sensors. In the system shown in Figure 4.66b, the flexible tactile element consists of a thin, light-reflecting surface embedded within an outer layer (touch pad) of high-strength rubber and an inner layer of transparent rubber. Optical fibers are uniformly and rigidly mounted across (normal to) this inner layer of rubber so that light can be projected directly onto the reflecting surface.

The light source, the beam splitter, and the solid-state (charge-coupled device, or CCD) camera form an integral unit, which can be moved laterally in known steps to scan the entire array of optical fiber if a single image frame of the camera does not cover the entire array. The splitter plate reflects part of the light from the light source onto a bundle of optical fiber. This light is reflected by the reflecting surface and is received by the solid-state camera. Since the intensity of the light received by the camera depends on the proximity of the reflecting surface, the gray-scale intensity image detected by the camera determines the deflection profile of the tactile surface. Using appropriate constitutive relations for the tactile sensor pad, the tactile force distribution can be determined from the deflection profile. The image processor carries out the standard operations (filtering, segmenting, etc.) on the successive image frames captured by the frame grabber and computes the deflection profile and the associated tactile force distribution in this manner. The image resolution depends on the pixel (picture-element) size of each image frame (e.g., 512×512 pixels, 1024×1024 pixels, etc.) as well as the spacing of the fiber-optic matrix. The force resolution or sensitivity of the tactile sensor can be improved at the expense of the thickness of the elastomeric layer, which determines the robustness of the sensor.

In the described fiber-optic tactile sensor (Figure 4.66), the optical fibers serve as the medium through which light or laser rays are transmitted to the tactile site. This is an extrinsic use of fiber-optics for sensing. Alternatively, an intrinsic application can be developed where an optical fiber serves as the sensing element itself. Specifically, the tactile pressure is directly applied to a mesh of optical fibers. Since the amount of light transmitted through a fiber decreases due to deformation caused by the tactile pressure, the light intensity that reaches the receiver can be used to determine the tactile pressure distribution.

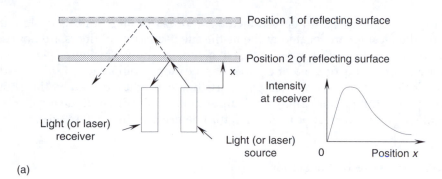

(a)

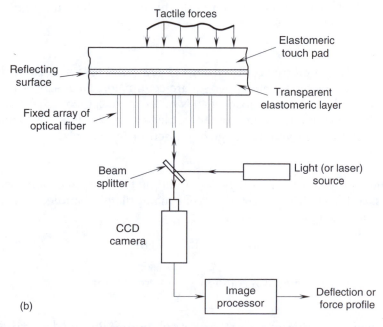

(b)

FIGURE 4.66

(a) The principle of an optical proximity sensor. (b) Schematic representation of a fiber-optic tactile sensor.

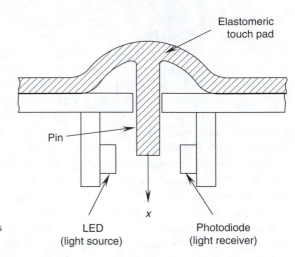

FIGURE 4.67
An optical tactile sensor with localized light sources
and photosensors.

Yet another alternative of an optical tactile sensor is available. In this design, the light source and the receiver are located at the tactile site itself; optical fibers are not used. The principle of operation of this type of tactile sensor is shown in Figure 4.67. When the elastomeric touch pad is pressed at a particular location, a pin attached to the pad at that point moves (in the x direction), thereby obstructing the light received by the photodiode from the light-emitting diode (LED). The output signal of the photodiode measures the pin movement (i.e., tactile deflection), which in turn determines the tactile force.

4.10.4 Piezoresistive Tactile Sensors

One type of piezoresistive tactile sensor uses an array of semiconductor strain gages mounted under the touch pad on a rigid base. In this manner, the force distribution on the touch pad is measured directly.

Example 4.19
When is tactile sensing preferred over sensing of a few point forces? A piezoelectric tactile sensor has 25 force-sensing elements per square centimeter. Each sensor element in the sensor can withstand a maximum load of 40 N and can detect load changes in the order of 0.01 N. What is the force resolution of the tactile sensor? What is the spatial resolution of the sensor? What is the dynamic range of the sensor in decibels?

Solution
Tactile sensing is preferred when it is not a simple-touch application. Shape, surface characteristics, and flexibility characteristics of a manipulated (handled or grasped) object can be determined using tactile sensing.

$$\text{Force resolution} = 0.01 \text{ N}$$

$$\text{Spatial resolution} = \frac{\sqrt{1}}{\sqrt{25}} \text{ cm} = 2 \text{ mm}$$

$$\text{Dynamic range} = (\text{maximum force})/(\text{force resolution}) = 20 \log_{10}\left(\frac{40}{0.01}\right) \text{dB} = 72 \text{ dB}.$$

4.10.5 Dexterity

Dexterity is an important consideration in sophisticated manipulators and robotic hands, which employ tactile sensing. The dexterity of a device is conventionally defined as the ratio: (number of degrees of freedom in the device)/(motion resolution of the device). In fact we call this *motion dexterity*.

We can define another type of dexterity called *force dexterity*, as follows:

$$\text{Force dexterity} = \frac{\text{number of degrees of freedom}}{\text{force resolution}}.$$

Both types of dexterity are useful in mechanical manipulation where tactile sensing is used.

4.10.6 A Strain-Gage Tactile Sensor

A strain-gage tactile sensor has been developed by the Eaton Corporation in Troy, Michigan. The concept behind it can be employed to determine the size and location of a point-contact force, which is useful, for example, in parts-mating applications. A square plate of length a is simply supported by frictionless hinges at its four corners on strain-gage load cells, as shown in Figure 4.68a. The magnitude, direction, and location of a point force P applied normally to the plate can be determined using the readings of the four (strain-gage) load cells.

To illustrate this principle, consider the free-body diagram shown in Figure 4.68b. The location of force P is given by the coordinates (x, y) in the Cartesian coordinate system (x, y, z), with the origin located at 1, as shown. The load cell reading at location i is denoted by R_i. Equilibrium in the z direction gives the force balance

$$P = R_1 + R_2 + R_3 + R_4. \tag{4.117}$$

Equilibrium about the y-axis gives the moment balance

$$Px = R_2a + R_3a \tag{i}$$

or

$$x = \frac{a}{P}(R_2 + R_3). \tag{4.118}$$

Similarly, equilibrium about the x-axis gives

$$y = \frac{a}{P}(R_3 + R_4). \tag{4.119}$$

It follows from Equation 4.117 through Equation 4.119 that the force P (direction as well as magnitude) and its location (x, y) are completely determined by the load cell readings. Typical values for the plate length a, and the maximum force P are 5 cm and 10 kg, respectively.

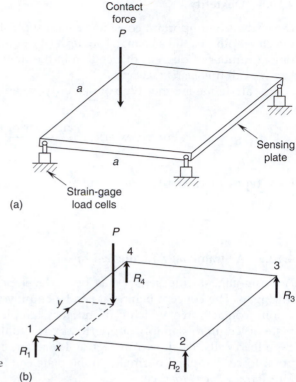

FIGURE 4.68
(a) Schematic representation of a strain-gage point-contact sensor. (b) Free-body diagram.

Example 4.20

In a particular parts-mating process using the principle of strain-gage tactile sensor described earlier, suppose that the tolerance on the measurement error of the force location is limited to δr. Determine the tolerance δF on the load-cell error.

Solution

Take the differentials of Equation 4.117 and Equation ii:

$$\delta P = \delta R_1 + \delta R_2 + \delta R_3 + \delta R_4$$
$$P\delta x + x\delta P = a\delta R_2 + a\delta R_3.$$

Direct substitution gives

$$\delta x = \frac{a}{P}(\delta R_2 + \delta R_3) - \frac{x}{P}(\delta R_1 + \delta R_2 + \delta R_3 + \delta R_4).$$

Note that x lies between 0 and a, and each δR_i can vary up to $\pm \delta F$. Hence, the largest error in x is given by $(2a/P)\,\delta F$. This is limited to δr. Hence, we have

$$\delta r = \frac{2a}{P}\delta F$$

or

$$\delta F = \frac{P}{2a} \delta r,$$

which gives the tolerance on the force error. The same result is obtained by considering y instead of x.

4.10.7 Other Types of Tactile Sensors

Ultrasonic tactile sensors are based, for example, on pulse-echo ranging. In this method, the tactile surface consists of two membranes separated by an air gap. The time taken for an ultrasonic pulse to travel through the gap and be reflected back onto a receiver depends, in particular on the thickness of the air gap. Since this time interval changes with deformation of the tactile surface, it can serve as a measure of the deformation of the tactile surface at a given location. Other possibilities for tactile sensors include the use of chemical effects that might be present when an object is touched and the influence of grasping on the natural frequencies of an array of sensing elements.

Sensor density or resolution, dynamic range, response time or bandwidth, strength and physical robustness, size, stability (dynamic robustness), linearity, flexibility, and localized intelligence (including data processing, learning and reorganization) are important factors, which require consideration in the analysis, design, or selection of a tactile sensor. The chosen specifications depend on the particular application. Typical values are 100 sensor elements spaced at 1 mm, a dynamic range of 60 dB, and a bandwidth of over 100 Hz (i.e., a response time of 10 ms or less). Because of the large number of sensor elements, signal conditioning and processing for tactile sensors present enormous difficulties. For instance, in piezoelectric tactile sensors, it is usually impractical to use a separate charge amplifier (or a voltage amplifier) for each piezoelectric element, even when built-in microelectronic amplifiers are available. Instead, signal multiplexing could be employed, along with a few high-performance signal amplifiers. The sensor signals could then be serially transmitted for conditioning and digital processing. The obvious disadvantages here are the increase in data acquisition time and the resulting reduction in sensor bandwidth.

4.10.8 Passive Compliance

Tactile sensing may be used for active control of processes. This is particularly useful in robotic applications that call for fine manipulation (e.g., microsurgery, assembly of delicate instruments, and robotic artistry). Heavy-duty industrial manipulators are often not suitable for fine manipulation because of errors that arise from such factors as backlash, friction, drift, errors in control hardware and algorithms, and generally poor dexterity. This situation can be improved to a great extent by using a heavy-duty robot for gross manipulations and using a miniature robot (or hand or gripper or finger manipulator) to serve as an end effector for fine-manipulation purposes. The end effector uses tactile sensing and localized control to realize required levels of accuracy in fine manipulation. This approach to accurate manipulation is expensive, primarily because of the sophisticated instrumentation and local processing needed at the end effector. A more cost-effective approach is to use passive remote-center compliance (RCC) at the end effector. With this method, passive devices (particularly, linear or nonlinear springs) are incorporated in the end-effector design (typically, at the wrist) so that some compliance (flexibility) is present. Errors in manipulation (e.g., jamming during parts mating) generate forces and moments, which deflect the end effector so as to self-correct the situation. Active compliance, which employs local sensors to adaptively change the end-effector compliance, is used in operations where passive compliance alone is not adequate. Impedance control is useful here.

4.11   Gyroscopic Sensors

Gyroscopic sensors are used for measuring angular orientations and angular speeds of aircraft, ships, vehicles, and various mechanical devices. These sensors are commonly used in control systems for stabilizing various vehicle systems. Since a spinning body (a gyroscope) requires an external torque to turn (precess) its axis of spin, it is clear that if this gyro is mounted (in a frictionless manner) on a rigid vehicle so that there are a sufficient number of degrees of freedom (at most three) between the gyro and the vehicle, the spin axis remains unchanged in space, regardless of the motion of the vehicle. Hence, the axis of spin of the gyro provides a reference with respect to which the vehicle orientation (e.g., azimuth or yaw, pitch, and roll angles) and angular speed can be measured. The orientation can be measured by using angular sensors at the pivots of the structure, which mounts the gyro on the vehicle. The angular speed about an orthogonal axis can be determined; for example, by measuring the precession torque (which is proportional to the angular speed) using a strain-gage sensor; or by measuring using a resolver, the deflection of a torsional spring that restrains the precession. The angular deflection in the latter case is proportional to the precession torque and hence the angular speed.

Consider a rigid disk spinning about an axis at angular speed ω. If the moment of inertia of the disk about that axis (polar moment of inertia) is J, the angular momentum H about the same axis is given by

$$H = J\omega. \tag{4.120}$$

Newton's second law (torque = rate of change of angular momentum) tells us that to rotate (precess) the spinning axis slightly, a torque has to be applied, because precession causes a change in the spinning angular momentum vector (the magnitude remains constant but the direction changes), as shown in Figure 4.69a . This explains the principle of operation of a gyroscope.

In the gyroscope shown in Figure 4.69b, the disk is spun about frictionless bearings using a constant-speed motor. Since the gimbal (the framework on which the disc is supported) is free to turn about frictionless bearings in the vertical axis, it remains fixed with respect to an inertial frame, even if the bearing housing (the main structure in which the gyroscope is located) rotates about the same vertical axis. Hence, the relative angle between the gimbal and the bearing housing (angle θ in the figure) can be measured (using a resolver, RVDT, encoder, etc.) and this gives the angle of rotation of the main structure. Bearing friction introduces an error into this measurement, which has to be compensated for, perhaps by recalibration before a reading is taken or by active feedback using a motor (torquer).

Figure 4.69b illustrates the case of a single-axis gyro sensor. The idea can be extended to the three-axis (3 d.o.f.) case, by providing two further frames, which are mounted on

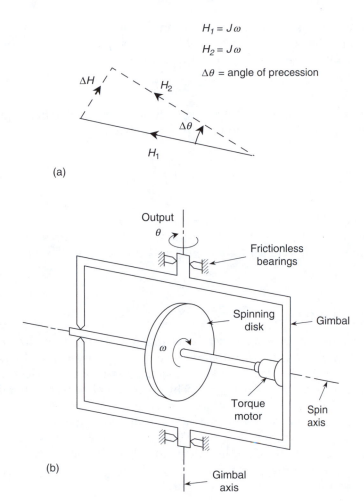

FIGURE 4.69
(a) Illustration of the gyroscopic torque needed to change the direction of an angular momentum vector. (b) A simple single-axis gyroscope for sensing angular displacement.

gimbals with their axes orthogonal to each other. The angular displacement sensors are mounted at all three gimbal bearings. For small rotations, these three angles can be considered uncoupled and will provide the orientation of the body on which the gyro unit is mounted. For large rotations, proper coordinate transformation has to be applied in converting the sensor readings to the orientation of the moving body (vehicle).

4.11.1 Rate Gyro

A rate gyro is used to measure angular speeds. The same arrangement shown in Figure 4.69b with a slight modification can be used. In this case, the gimbal is not free; and may be restrained by a torsional spring. A viscous damper is provided to suppress any oscillations. By analyzing this gyro, we note that the relative angle of rotation θ gives the angular speed Ω of the structure (vehicle) about the axis that is orthogonal to both gimbal axis and spin axis.

For a simplified analysis, assume that the angles of rotation are small and that the moment of inertia of the gimbal frame is negligible (compared with J). Newton's second law of motion for the unit about the gimbal axis gives:

$$J\omega\Omega = K\theta + B\dot{\theta},$$

where K is the torsional stiffness of the spring restraint at the gimbal bearings, and B is the damping constant of rotational motion about the gimbal axis.
We get

$$\Omega = \frac{K\theta + B\dot{\theta}}{J\omega}. \tag{4.121}$$

In particular, when B is very small, angular rotation at the gimbal bearings (e.g., as measured by a resolver) become proportional to the angular speed of the system including the main body (vehicle) about the axis orthogonal to both gimbal axis and spin axis.

4.11.2 Coriolis Force Devices

Consider a mass m moving at velocity v relative to a rigid frame. If the frame itself rotates at an angular velocity ω, it is known that the acceleration of m is given by $2\omega \times v$. This is known as the *Coriolis acceleration* (Note: the vector-cross-product is denoted by x). The associated force $2m\omega \times v$ is the *Coriolis force*. This force can be sensed either directly using a force sensor or by measuring a resulting deflection in a flexible element, and may be used to determine the variables (ω or v) in the Coriolis force. Note that Coriolis force is somewhat similar to gyroscopic force even though the concepts are different. For this reason, devices based on the Coriolis effect are also commonly termed gyroscopes. Coriolis concepts are gaining popularity in MEMS-based sensors, which use MEMS (micro-electromechanical systems) technologies.

4.12   Optical Sensors and Lasers

The laser (*l*ight *a*mplification by *s*timulated *e*mission of *r*adiation) produces electromagnetic radiation in the ultraviolet, visible, or infrared bands of the spectrum. A laser can provide a single-frequency (i.e., monochromatic) light source. Furthermore, the electromagnetic radiation in a laser is *coherent* in the sense that all the generated waves have constant phase angles. The laser uses oscillations of atoms or molecules of various elements. It is useful in fiber optics, and it can also be used directly in sensing and gaging applications. The helium–neon (He–Ne) laser and the semiconductor laser are commonly used in optical sensor applications.

The characteristic component in a fiber-optic sensor is a bundle of glass fibers (typically a few hundred) that can carry light. Each optical fiber may have a diameter in the order of a few micrometers to about 0.01 mm. There are two basic types of fiber-optic sensors. In one type—the indirect or the extrinsic type—the optical fiber acts only as the medium in which the sensor light is transmitted. In this type, the sensing element itself does not consist of optical fibers. In the second type—the direct or the intrinsic type—the optical fiber itself acts as the sensing element. When the conditions of the sensed medium change,

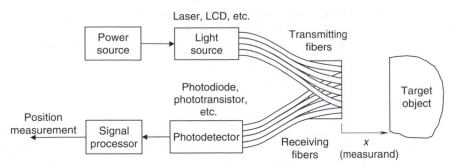

FIGURE 4.70
A fiber-optic position sensor.

the light-propagation properties of the optical fibers change (e.g., due to microbending of a straight fiber as a result of an applied force), providing a measurement of the change in conditions. Examples of the first (extrinsic) type of sensor include fiber-optic position sensors, proximity sensors, and tactile sensors. The second (intrinsic) type of sensor is found, for example, in fiber-optic gyroscopes, fiber-optic hydrophones, and some types of microdisplacement or force sensors MEMS devices).

4.12.1 Fiber-Optic Position Sensor

A schematic representation of a fiber-optic position sensor (or proximity sensor or displacement sensor) is shown in Figure 4.70.

The optical fiber bundle is divided into two groups: transmitting fibers and receiving fibers. Light from the light source is transmitted along the first bundle of fibers to the target object whose position is measured. Light reflected (or diffused) onto the receiving fibers by the surface of the target object is carried to a photodetector. The intensity of the light received by the photodetector depends on position x of the target object. In particular, if $x = 0$, the transmitting bundle is completely blocked off and the light intensity at the receiver becomes zero. As x is increased, the intensity of the received light increases, because more and more light is reflected on the tip of the receiving bundle. This reaches a peak at some value of x. When x is increased beyond that value, more and more light is reflected outside the receiving bundle; hence, the intensity of the received light drops. In general then, the proximity–intensity curve for an optical proximity sensor will be non-linear and will have the shape shown in Figure 4.71. Using this (calibration) curve, we can determine the position (x) once the intensity of the light received at the photosensor is known. The light source could be a laser (structured light), infrared light-source, or some

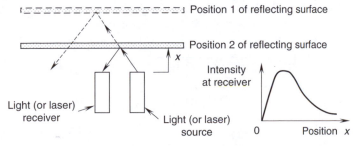

FIGURE 4.71
The principle of a fiber-optic proximity sensor.

other type, such as a light-emitting diode (LED). The light sensor (photodetector) could be some device such as a photodiode or a photo field effect transistor (photo FET). This type of fiber-optic sensors can be used, with a suitable front-end device (such as bellows, springs, etc.) to measure pressure, and force as well.

4.12.2   Laser Interferometer

This sensor is useful in the accurate measurement of small displacements. In this fiber-optic position sensor, the same bundle of fibers is used for sending and receiving a monochromatic beam of light (typically, laser). Alternatively, monomode fibers, which transmit only monochromatic light (of a specific wavelength), may be used for this purpose. In either case, as shown in Figure 4.72, a beam splitter (Λ) is used so that part of the light is directly reflected back to the bundle tip and the other part reaches the target object (as in Figure 4.70) and reflected back from it (using a reflector mounted on the object) on to the bundle tip. In this manner, part of the light returning through the bundle had not traveled beyond the beam splitter, whereas the other part had traveled between the beam splitter (A) and the object (through an extra distance equal to twice the separation between the beam splitter and the object). As a result, the two components of light will have a phase difference ϕ, which is given by

$$\phi = \frac{2x}{\lambda} \times 2\pi, \tag{4.122}$$

where x is the distance of the target object from the beam splitter and λ is the wavelength of monochromatic light.

The returning light is directed to a light sensor using a beam splitter (B). The sensed signal is processed using principles of interferometry to determine ϕ, and from Equation 4.122, the distance x. Very fine resolutions better than a fraction of a micrometer (μm) can be obtained using this type of fiber-optic position sensors.

The advantages of fiber optics include insensitivity to electrical and magnetic noise (due to optical coupling), safe operation in explosive, high-temperature, corrosive, and hazardous environments, and high sensitivity. Furthermore, mechanical loading and wear problems do not exist because fiber-optic position sensors are noncontact devices

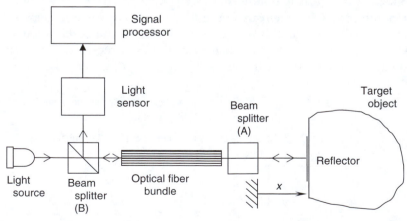

FIGURE 4.72
Laser interferometer position sensor.

with no moving parts. The disadvantages include direct sensitivity to variations in the intensity of the light source and dependence on ambient conditions (temperature, dirt, moisture, smoke, etc.). Compensation can be made, however, with respect to temperature. An *optical encoder* is a digital (or pulse-generating) motion transducer (see Chapter 5). Here, a light beam is intercepted by a moving disk that has a pattern of transparent windows. The light that passes through, as detected by a photosensor, provides the transducer output. These sensors may also be considered in the extrinsic category.

As an intrinsic application of fiber optics in sensing, consider a straight optical fiber element that is supported at its two ends. In this configuration almost 100% of the light at the source end transmits through the optical fiber and reaches the detector (receiver) end. Now, suppose that a slight load is applied to the optical fiber segment at its mid span. It deflects slightly due to the load, and as a result the amount of light received at the detector can drop significantly. For example, a microdeflection of just 50 μm can result in a drop in intensity at the detector by a factor of 25. Such an arrangement may be used in deflection, force, and tactile sensing. Another intrinsic application is the fiber-optic gyroscope, as described next.

4.12.3 Fiber-Optic Gyroscope

This is an angular speed sensor that uses fiber optics. Contrary to the implication of its name, however, it is not a gyroscope in the conventional sense. Two loops of optical fibers wrapped around a cylinder are used in this sensor, and they rotate with the cylinder, at the same angular speed, which is the sensed quantity (measurand). One loop carries a monochromatic light (or laser) beam in the clockwise direction; the other loop carries a beam from the same light (laser) source in the counterclockwise direction (see Figure 4.73). Since the laser beam traveling in the direction of rotation of the cylinder attains a higher frequency than that of the other beam, the difference in frequencies (known as the Sagnac effect) of the two laser beams received at a common location, is used as a measure of the angular speed of the cylinder. This may be accomplished through interferometry. Because the combined signal is a sine beat, light and dark patterns (fringes) will be present in the detected light, and they will measure the frequency difference and hence the rotating speed of the optical fibers.

In a laser (ring) gyroscope, it is not necessary to have a circular path for the laser. Triangular and square paths are commonly used as well. In general the beat frequency $\Delta\omega$ of the combined light from the two laser beams traveling in opposite directions is given by

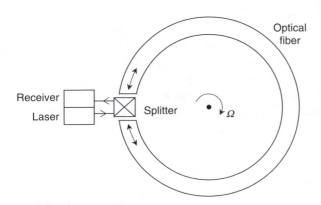

FIGURE 4.73
A fiber-optic laser gyroscope.

$$\Delta\omega = \frac{4A}{p\lambda}\Omega, \qquad (4.123)$$

where A is the area enclosed by the travel path (πr^2 for a cylinder of radius r), p the length (perimeter) of the traveled path ($2\pi r$ for a cylinder), λ the wavelength of the laser, and Ω is the angular speed of the object (or optical fiber).

The length of the optical fiber wound around the rotating object can exceed 100 m and can be about 1 km. Angular displacements can be measured with a laser gyro simply by counting the number of cycles and clocking fractions of cycles. Acceleration can be determined by digitally determining the rate of change of speed. In a laser gyro, there is an alternative to the use of two separate loops of optical fiber, wound in opposite direction. The same loop can be used to transmit light from the same laser from the opposite ends of the fiber. A beam splitter has to be used in this case, as shown in Figure 4.73.

4.12.4 Laser Doppler Interferometer

The laser Doppler interferometer is used for accurate measurement of speed. To understand the operation of this device, we should explain two phenomena: the Doppler effect and light wave interference. The latter phenomenon is used in the laser interferometer position sensor, which was discussed before. Consider a wave source (e.g., a light source or sound source) that is moving with respect to a receiver (observer). If the source moves toward the receiver, the frequency of the received wave appears to have increased; if the source moves away from the receiver, the frequency of the received wave appears to have decreased. The change in frequency is proportional to the velocity of the source relative to the receiver. This phenomenon is known as the *Doppler effect*. Now consider a monochromatic (single-frequency) light wave of frequency f (say, 5×10^{14} Hz) emitted by a laser source. If this ray is reflected by a target object and received by a light detector, the frequency of the received wave would be

$$f_2 = f + \Delta f. \qquad (4.124)$$

The frequency increase Δf is proportional to the velocity v of the target object, which is assumed positive when moving toward the light source. Specifically,

$$\Delta f = \frac{2f}{c}v = kv, \qquad (4.125)$$

where c is the speed of light in the particular medium (typically, air). Now by comparing the frequency f_2 of the reflected wave, with the frequency $f_1 = f$ of the original wave, we can determine Δf and, hence, the velocity v of the target object.

The change in frequency Δf due to the Doppler effect can be determined by observing the fringe pattern due to light wave interference. To understand this, consider the two waves

$$v_1 = a \sin 2\pi f_1 t \qquad (4.126)$$

and

$$v_2 = a \sin 2\pi f_2 t. \qquad (4.127)$$

If we add these two waves, the resulting wave would be

$$v = v_1 + v_2 = a(\sin 2\pi f_1 t + \sin 2\pi f_2 t), \qquad (4.128)$$

which can be expressed as

$$v = 2a \sin \pi(f_2 + f_1)t \cos \pi(f_2 - f_1)t. \qquad (4.129)$$

It follows that the combined signal will exhibit beats at the beat frequency $\Delta f/2$. Since f_2 is very close to f_1 (because Δf is small compared with f), the beats appear as dark and light lines (fringes) in the resulting light wave. This is known as *wave interference*. Note that Δf can be determined by two methods:

1. By measuring the spacing of the fringes
2. By counting the beats in a given time interval or by timing successive beats using a high-frequency clock signal.

The velocity of the target object is determined in this manner. Displacement can be obtained simply by digital integration (or by accumulating the count). A schematic diagram for the laser Doppler interferometer is shown in Figure 4.74. Industrial interferometers usually employ a He–Ne laser, which has waves of two frequencies close together. In that case, the arrangement shown in Figure 4.74 has to be modified to take into account the two frequency components.

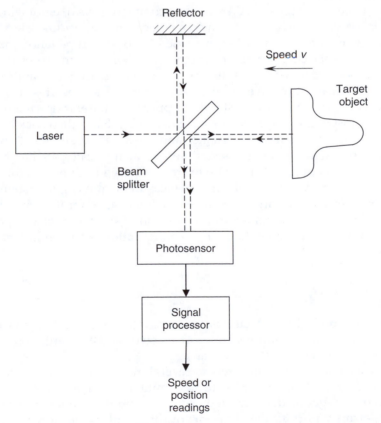

FIGURE 4.74
A laser-Doppler interferometer for measuring velocity and displacement.

Note that the laser interferometer discussed before (Figure 4.72) directly measures displacement rather than speed. It is based on measuring the phase difference between the direct and returning laser beams, not the Doppler effect (frequency difference).

4.13   Ultrasonic Sensors

Audible sound waves have frequencies in the range of 20 Hz to 20 kHz. Ultrasound waves are pressure waves, just like sound waves, but their frequencies are higher (ultra) than the audible frequencies. Ultrasonic sensors are used in many applications, including medical imaging, ranging systems for cameras with autofocusing capability, level sensing, and speed sensing. For example, in medical applications, ultrasound probes of frequencies 40 kHz, 75 kHz, 7.5 MHz and 10 MHz are commonly used. Ultrasound can be generated according to several principles. For example, high-frequency (gigahertz) oscillations in a piezoelectric crystal subjected to an electrical potential is used to generate very high-frequency ultrasound. Another method is to use the magnetostrictive property of ferromagnetic material. Ferromagnetic materials deform when subjected to magnetic fields. Respondent oscillations generated by this principle can produce ultrasonic waves. Another method of generating ultrasound is to apply a high-frequency voltage to a metal-film capacitor. A microphone can serve as an ultrasound detector (receiver).

Analogous to fiber-optic sensing, there are two common ways of employing ultrasound in a sensor. In one approach—the intrinsic method—the ultrasound signal undergoes changes as it passes through an object, because of acoustic impedance and absorption characteristics of the object. The resulting signal (image) may be interpreted to determine properties of the object, such as texture, firmness, and deformation. This approach has been used, for example, in an innovative firmness sensor for herring roe. It is also the principle used in medical ultrasonic imaging. In the other approach—the extrinsic method—the time of flight of an ultrasound burst from its source to an object and then back to a receiver is measured. This approach is used in distance and position measurement and in dimensional gaging. For example, an ultrasound sensor of this category has been used in thickness measurement of fish. This is also the method used in camera autofocusing.

In distance (range, proximity, displacement) measurement using ultrasound, a burst of ultrasound is projected at the target object, and the time taken for the echo to be received is clocked. A signal processor computes the position of the target object, possibly compensating for environmental conditions. This configuration is shown in Figure 4.75. The applicable relation is

$$x = \frac{ct}{2}, \qquad\qquad (4.130)$$

where t is the time of flight of the ultrasound pulse (from generator to receiver), x the distance between the ultrasound generator or receiver and the target object, and c the speed of sound in the medium (typically, air).

Distances as small as a few centimeters to several meters may be accurately measured by this approach, with fine resolution (e.g., a millimeter or less). Since the speed of ultrasonic wave propagation depends on the medium and the temperature of the medium (typically air), errors will enter the ultrasonic readings unless the sensor is compensated for the variations in the medium; particularly for temperature.

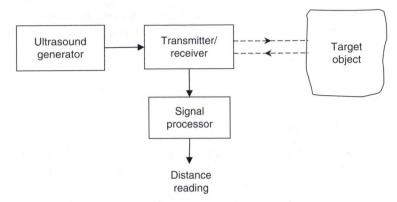

FIGURE 4.75
An ultrasonic position sensor.

Alternatively, the velocity of the target object can be measured, using the Doppler effect, by measuring (clocking) the change in frequency between the transmitted wave and the received wave. The beat phenomenon is employed here. The applicable relation is Equation 4.125, except, now f is the frequency of the ultrasound signal and c the speed of sound.

4.13.1 Magnetostrictive Displacement Sensors

The ultrasound-based time-of-flight method is used somewhat differently in a magnetostrictive displacement sensor (e.g., the sensor manufactured by Temposonics). The principle behind this method is illustrated in Figure 4.76. The sensor head generates an interrogation current pulse, which travels along the magnetostrictive wire. This pulse interacts with the magnetic field of the permanent magnet and generates an ultrasound pulse (by magnetostrictive action in the wire). This pulse is received (and timed) at the sensor head. The time of flight is proportional to the distance of the magnet from the sensor head. If the target object is attached to the magnet of the sensor, its position (x) can be determined using the time of flight as usual.

Strokes (maximum displacement) ranging from a few centimeters to 1 or 2 m, at resolutions better than 50 μm are possible with these sensors. With a 15 VDC power supply, the sensor can provide a dc output in the range ± 5 V. Since the sensor uses a

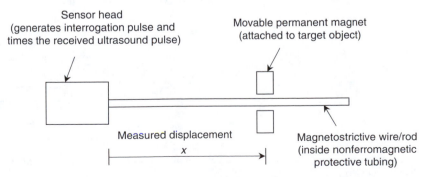

FIGURE 4.76
A magnetostrictive ultrasound displacement sensor.

magnetostrictive medium with a protective nonferromagnetic tubing, some of the common sources of error in ultrasonic sensors that use air as the medium of propagation can be avoided.

4.14 Thermofluid Sensors

Common thermofluid (mechanical engineering) sensors include those measuring pressure, fluid flow rate, temperature, and heat transfer rate. Such sensors are useful in mechatronic applications as well in view of the fact that the plant (i.e., the system to be controlled; e.g., automobile, machine tool, and aircraft) may involve these measurands. Several common types of sensors in this category are presented next.

4.14.1 Pressure Sensors

Common methods of pressure sensing are the following:

1. Balance the pressure with an opposing force (or head) and measure this force. Examples are liquid manometers and pistons.
2. Subject the pressure to a flexible front-end (auxiliary) member and measure the resulting deflection. Examples are Bourdon tube, bellows, and helical tube.
3. Subject the pressure to a front-end auxiliary member and measure the resulting strain (or stress). Examples are diaphragms and capsules.

Some of these devices are illustrated in Figure 4.77.

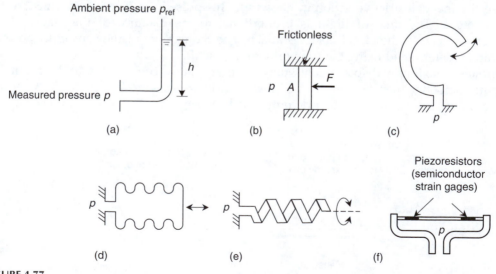

FIGURE 4.77
Typical pressure sensors. (a) Manometer. (b) Counterbalance piston. (c) Bourdon tube. (d) Bellows. (e) Helical tube; and (f) Diaphragm.

In the manometer shown in Figure 4.77a, the liquid column of height h and density ρ provides a counterbalancing pressure head to support the measured pressure p with respect to the reference (ambient) pressure p_{ref}. Accordingly, this device measures the gage pressure as given by

$$p - p_{ref} = \rho g h, \tag{4.131}$$

where g is the acceleration due to gravity. In the pressure sensor shown in Figure 4.77b, a frictionless piston of area A supports the pressure load by means of an external force F. The governing equation is

$$p = \frac{F}{A}. \tag{4.132}$$

The pressure is determined by measuring F using a force sensor. The Bourdon tube shown in Figure 4.77c deflects with a straightening motion as a result of internal pressure. This deflection can be measured using a displacement sensor (typically, a rotatory sensor) or indicated by a moving pointer. The bellows unit deflects as a result of the internal pressure, causing a linear motion, as shown in Figure 4.77d. The deflection can be measured using a sensor such as LVDT or a capacitive sensor, and can be calibrated to indicate pressure. The helical tube shown in Figure 4.77e undergoes a twisting (rotational) motion when deflected by internal pressure. This deflection can be measured by an angular displacement sensor (RVDT, resolver, potentiometer, etc.), to provide pressure reading through proper calibration. Figure 4.77f illustrates the use of a diaphragm to measure pressure. The membrane (typically metal) is strained due to pressure. The pressure can be measured by means of strain gages (piezoresistive sensors) mounted on the diaphragm. MEMS pressure sensors that use this principle are available. In one such device, the diaphragm has a silicon wafer substrate integral with it. Through proper doping (using boron, phosphorous, etc.) a microminiature semiconductor strain-gage can be formed. In fact, more than one piezoresistive sensor can be etched on the diaphragm, and used in a bridge circuit to provide the pressure reading, through proper calibration. The most sensitive locations for the piezoresistive sensors are closer to the edge of the diaphragm, where the strains reach the maximum.

4.14.2 Flow Sensors

The volume flow rate Q of a fluid is related to the mass flow rate Q_m through

$$Q_m = \rho Q, \tag{4.133}$$

where ρ is the mass density of the fluid. In addition, for a flow across an area A at average velocity v, we have

$$Q = Av. \tag{4.134}$$

When the flow is not uniform, a suitable correction factor has to be included in Equation 4.134, if a local velocity or the maximum velocity is used in Equation (4.134).

According to Bernoulli's equation for incompressible, ideal flow (no energy dissipation) we have

$$p + \frac{1}{2}\rho v^2 = \text{constant}. \tag{4.135}$$

This theorem may be interpreted as conservation of energy. Moreover, note that the pressure p due to a fluid head of height h is given by (gravitational potential energy) $\rho g h$. Using Equation 4.135 and allowing for dissipation (friction), the flow across a constriction (i.e., a fluid resistance element such as an orifice, nozzle, valve, and so on) of area A can be shown to obey the relation

$$Q = c_d A \sqrt{\frac{2\Delta p}{\rho}}, \tag{4.136}$$

where Δp is the pressure drop across the constriction and c_d is the discharge coefficient for the constriction.

Common methods of measuring fluid flow may be classified as follows:

1. Measure pressure across a known constriction or opening. Examples include nozzles, Venturi meters, and orifice plates.
2. Measure the pressure head, which brings the flow to static conditions. The examples include pitot tube, liquid level sensing using floats, and so on.
3. Measure the flow rate (volume or mass) directly. Turbine flow meter and angular-momentum flow meter are examples.
4. Measure the flow velocity. Coriolis meter, laser-Doppler velocimeter, and ultrasonic flow meter are examples.
5. Measure an effect of the flow and estimate the flow rate using that information. Hot-wire (or hot-film) anemometer and magnetic induction flow meter are examples.

Several examples of flow meters are shown in Figure 4.78.

For the orifice meter shown in Figure 4.78a, Equation 4.136 is applied to measure the volume flow rate. The pressure drop is measured using the techniques outlined earlier. For the pitot tube shown in Figure 4.78b, Bernoulli's Equation 4.135 is applicable, noting that the fluid velocity at the free surface of the tube is zero. This gives the flow velocity

$$v = \sqrt{2gh}. \tag{4.137}$$

Note that a correction factor is needed when determining the flow rate because the velocity is not uniform across the flow section. In the angular momentum method shown in Figure 4.78c, the tube bundle through which the fluid flows, is rotated by a motor. The motor torque τ and the angular speed ω are measured. As the fluid mass passes through the tube bundle, it imparts an angular momentum at a rate governed by the mass flow rate Q_m of the fluid. The motor torque provides the torque needed for this rate of change of angular momentum. Neglecting losses, the governing equation is

$$\tau = \omega r^2 Q_m, \tag{4.138}$$

where r is the radius of the centroid of the rotating fluid mass. In a turbine flow meter, the rotation of the turbine wheel located in the flow can be calibrated to directly give the flow rate. In the Coriolis method shown in Figure 4.78d, the fluid is made to flow through a "U" segment, which is hinged to oscillate out of plane (at angular velocity ω) and restrained by springs (with known stiffness) in the lateral direction. If the fluid velocity is v, the resulting Coriolis force (due to Coriolis acceleration $2\omega \times v$) is supported by the springs. The out-of-plane angular speed is measured by a motion sensor. In addition, the

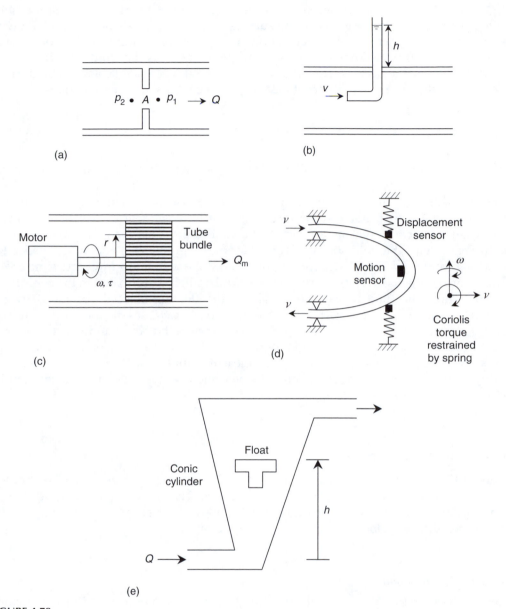

FIGURE 4.78
Several flow meters. (a) Orifice flow meter. (b) Pitot tube. (c) Angular-momentum flow meter. (d) Coriolis velocity meter. (e) Rotameter.

spring force is measured using a suitable sensor (e.g., displacement sensor). This information determines the Coriolis acceleration of the fluid particles and hence their velocity.

In the laser-Doppler velocimeter, a laser beam is projected on the fluid flow (through a window) and its frequency shift due to the Doppler effect is measured (see under optical sensors, as described before). This is a measure of the speed of the fluid particles. As another method of sensing velocity of a fluid, an ultrasonic burst is sent in the direction of flow and the time of flight is measured. Increase in the speed of propagation is due to the fluid velocity, and may be determined as usual (see under ultrasonic sensors, as outlined earlier).

In the hot-wire anemometer, a conductor carrying current (i) is placed in the fluid flow. The temperatures of the wire (T) and the surrounding fluid (T_f) are measured along with the current. The coefficient of heat transfer (forced convection) at the boundary of the wire and the moving fluid is known to vary with $\sqrt{v}$, where v is the fluid velocity. Under steady conditions, the heat loss from the wire into the fluid is exactly balanced by the heat generated by the wire due to its resistance (R). The heat balance equation gives

$$i^2 R = c(a + \sqrt{v})(T - T_f). \tag{4.139}$$

This relation can be used to determine v. Instead of a wire, a metal film (e.g., platinum plated glass tube) may be used.

There are other indirect methods of measuring fluid flow rate. In one method, the drag force on an object suspended in the flow using a cantilever arm is measured (using a strain-gage sensor at the clamped end of the cantilever). This force is known to vary quadratically with the fluid speed. A rotameter (see Figure 4.78e) is another device for measuring fluid flow. This device consists of a conic tube with uniformly increasing cross-sectional area, which is vertically oriented. A cylindrical object is floated in the conic tube, through which the fluid flows. The weight of the floating object is balanced by the pressure differential on the object. When the flow speed increases, the object rises within the conic tube, thereby allowing more clearance between the object and the tube for the fluid to pass. The pressure differential, however, still balances the weight of the object, and is constant. Equation 4.136 is used to measure fluid flow rate, since A increases quadratically with the height of the object. Consequently, the level of the object can be calibrated to give the flow rate.

4.14.3 Temperature Sensors

In most (if not all) temperature measuring devices, the temperature is sensed through heat transfer from the source to the measuring device. The physical (or chemical) change in the device caused by this heat transfer is the transducer stage. Several temperature sensors are outlined below.

4.14.3.1 *Thermocouple*

When the temperature changes at the junction formed by joining two unlike conductors, its electron configuration changes due to the resulting heat transfer. This electron reconfiguration produces a voltage (emf or electromotive force), and is known as the Seebeck effect. Two junctions (or more) of a thermocouple are made with two unlike conductors such as iron and constant an, copper and constantan, chrome and alumel, and so on. One junction is placed in a reference source (cold junction) and the other in the temperature source (hot junction), as shown in Figure 4.79. The voltage across the two junctions is measured to give the temperature of the hot junction with respect to the cold junction. The presence of any other junctions, such as the ones formed by the wiring to the voltage sensor, does not affect the reading as long as these leads are maintained at the same temperature. Very low temperatures (e.g., $-250°C$) as well as very hightemperatures (e.g., $3000°C$) can be measured using a thermocouple. Since the temperature–voltage relationship is nonlinear, correction has to be made when measuring changes in temperature; usually by using polynomial relations. Sensitivity is quite reasonable (e.g., $10\,mV/°C$), but

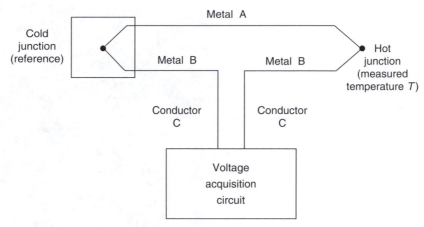

FIGURE 4.79
A thermocouple.

signal conditioning may be needed in some applications. Fast measurements are possible with small thermocouples having low time constants (e.g., 1 ms).

4.14.3.2 Resistance Temperature Detector

A resistance temperature detector (RTD) is simply a metal element (in a ceramic tube) whose resistance typically increases with temperature, according to a known function. A linear approximation, as given by Equation 4.140 is adequate when the temperature change is not too large. Temperature is measured using an RTD simply by measuring the change in resistance (say, using a bridge circuit; see Chapter 3). Metals used in RTDs include platinum, nickel, copper, and various alloys. The temperature coefficient of resistance (α) of several metals, which can be used in RTDs, is given in Table 4.6.

$$R = R_0(1 + \alpha T). \tag{4.140}$$

The useful temperature range of an RTD is about $-200°C$ to $+800°C$. At high temperatures, these devices may tend to be less accurate than thermocouples. The speed of response can be lower as well (e.g., fraction of a second). A commercial RTD unit is shown in Figure 4.80.

4.14.3.3 Thermistor

Unlike an RTD, a thermistor is made of a semiconductor material (e.g., metal oxides such as those of chromium, cobalt, copper, iron, manganese, and nickel), which usually has a

TABLE 4.6

Temperature Coefficients of Resistance of Some RTD Metals

Metal	Temperature Coefficient of Resistance α ($°K$)
Copper	0.0043
Nickel	0.0068
Platinum	0.0039

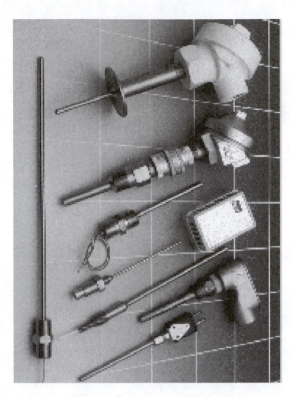

FIGURE 4.80
A commercial RTD unit (RdF
Corp. With permission).

negative change in resistance with temperature (i.e., negative α). The resistance change is detected through a bridge circuit or a voltage divider circuit. Even though the accuracy provided by a thermistor is usually better than that of an RTD, the temperature–resistance relation is far more nonlinear, as given by

$$R = R_0 \exp \beta \left(\frac{1}{T} - \frac{1}{T_0} \right). \tag{4.141}$$

The characteristic temperature β (about $4000°K$) itself is temperature dependent, thereby adding to the nonlinearity of the device. Hence, proper calibration is essential when operating in a wide temperature range (say, greater than $50°C$). Thermistors are quite robust and they provide a fast response and high sensitivity (compared with RTDs).

4.14.3.4 *Bi-Metal Strip Thermometer*

Unequal thermal expansion of different materials is used in this device. If strips of the two materials (typically metals) are firmly bonded, thermal expansion causes this element to bend toward the material with the lower expansion. This motion can be measured using a displacement sensor, or indicated using a needle and scale. Household thermostats commonly use this principle for temperature sensing and control (on–off).

4.15   Other Types of Sensors

There are many other types of sensors and transducers, which cannot be discussed here due to space limitation. But, the principles and techniques presented in this chapter may be extended to many of these devices. One area where a great variety of sensors are used is factory automation. Here, in applications of automated manufacturing and robotics it is important to use proper sensors for specific operations and needs. For example, mechanical and electronic switches (binary or two-state sensors), chemical sensors, camera-based vision systems, and ultrasonic motion detectors may be used for human safety requirements. Motion and force, power-line, debris, sound, vibration, temperature, pressure, flow and liquid-level sensing may be used in machine monitoring and diagnosis. Motion, force, torque, current, voltage, flow, and pressure sensing are important in machine control. Vision, motion, proximity, tactile, force, torque, and pressure sensing and dimensional gaging are useful in task monitoring and control.

Several areas can be identified where new developments and innovations are made in sensor technology:

1. Microminiature (MEMS and nano) Sensors: (integrated-circuit- or IC-based, with built-in signal processing)

2. Intelligent Sensors: (Built-in reasoning or information pre-processing to provide high-level knowledge and decision-making capability)

3. Embedded and Distributed Sensor Networks: (Sensors which are integral with the components; networks of sensors or agents, which communicate with each other through high-speed wireless communication networks in an overall multi-agent system. Networks of sensor arrays also have useful applications)

4. Hierarchical Sensory Architectures: (Low-level sensory information is pre-processed to match higher level requirements).

These four areas of activity are also representative of future trends in sensor technology development.

Problems

4.1 In each of the following examples, indicate at least one (unknown) input, which should be measured and used for feedforward control to improve the accuracy of the control system.

a. A servo system for positioning a mechanical load. The servomotor is a field-controlled dc motor, with position feedback using a potentiometer and velocity feedback using a tachometer.

b. An electric heating system for a pipeline carrying a liquid. The exit temperature of the liquid is measured using a thermocouple and is used to adjust the power of the heater.

c. A room heating system. Room temperature is measured and compared with the set point. If it is low, a valve of a steam radiator is opened; if it is high, the valve is shut.

d. An assembly robot, which grips a delicate part to pick it up without damaging the part.

e. A welding robot, which tracks the seam of a part to be welded.

4.2 A typical input variable is identified for each of the following examples of dynamic systems. Give at least one output variable for each system.

a. Human body: neuroelectric pulses

b. Company: information

    c. Power plant: fuel rate

    d. Automobile: steering wheel movement

    e. Robot: voltage to joint motor.

4.3 Measuring devices (sensors–transducers) are useful in measuring outputs of a process for feedback control.

    a. Give other situations in which signal measurement would be important.

    b. List at least five different sensors used in an automobile engine.

4.4 Give one situation where output measurement is needed and give another where input measurement is needed for proper control of the chosen system. In each case justify the need.

4.5 Giving examples, discuss situations in which measurement of more than one type of kinematic variables using the same measuring device is

    a. An advantage

    b. A disadvantage.

4.6 Giving examples for suitable auxiliary front-end elements, discuss the use of a force sensor to measure

    a. Displacement

    b. Velocity

    c. Acceleration.

4.7 Derive the expression for electrical-loading nonlinearity error (percentage) in a rotatory potentiometer in terms of the angular displacement, maximum displacement (stroke), potentiometer element resistance, and load resistance. Plot the percentage error as a function of the fractional displacement for the three cases $R_L/R_c = 0.1, 1.0,$ and 10.0.

4.8 Determine the angular displacement of a rotatory potentiometer at which the loading nonlinearity error is the largest.

4.9 A potentiometer circuit with element resistance R_c and equal end resistors R_e is shown in Figure P4.9. Derive the necessary input–output relations. Show that the end resistors can produce a linearizing effect in the potentiometer. At half the maximum reading of the potentiometer shown in Figure P4.9, calculate the percentage loading error for the three values of the resistance ratio: $R_c/R_e = 0.1, 1.0,$ and 10.0, assuming that the load resistance R_L is equal to the element resistance. Compare the results with the corresponding value for $R_e = 0$. Finally, choose a suitable value for R_c/R_e and plot the curve of percentage loading error versus fractional displacement x/x_{max}. From the graph, estimate the maximum loading error.

4.10 Derive an expression for the sensitivity (normalized) of a rotatory potentiometer as a function of displacement (normalized). Plot the corresponding curve in the non-dimensional form for the three load values given by $R_L/R_c = 0.1, 1.0,$ and 10.0. Where does the maximum sensitivity occur? Verify your observation using the analytical expression.

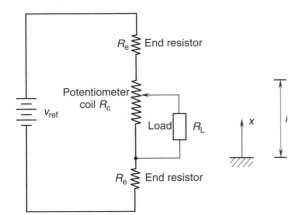

FIGURE P4.9
A potentiometer circuit with end resistors.

4.11 The range of a coil-type potentiometer is 10 cm. If the wire diameter is 0.1 mm, determine the resolution of the device.

4.12 The data acquisition system connected at the output of a differential transformer (say, an LVDT) has a very high resistive load. Obtain an expression for the phase lead of the output signal (at the load) of the differential transformer, with reference to the supply to the primary windings of the transformer, in terms of the impedance of the primary windings only.

4.13 At the null position, the impedances of the two secondary winding segments of an LVDT were found to be equal in magnitude but slightly unequal in phase. Show that the quadrature error (null voltage) is about 90° out of phase with reference to the predominant component of the output signal under open-circuit conditions.

Hint: This may be proved either analytically, or graphically by considering the difference between two rotating directed lines (phasors) that are separated by a very small angle.

4.14 A vibrating system has an effective mass M, an effective stiffness K, and an effective damping constant B in its primary mode of vibration at point A with respect to coordinate y. Write expressions for the undamped natural frequency, the damped natural frequency, and the damping ratio for this first mode of vibration of the system.
A displacement transducer is used to measure the fundamental undamped natural frequency and the damping ratio of the system by subjecting the system to an initial excitation and recording the displacement trace at a suitable location (point A along y in the Figure P4.14) in the system. This trace provides the period of damped oscillations and the logarithmic decrement of the exponential decay from which the required parameters can be computed using well-known relations. However, it was found that the mass m of the moving part of the displacement sensor and the associated equivalent viscous damping constant b are not negligible. Using the model shown in Figure P4.14, derive expressions for the measured undamped natural frequency and damping ratio. Suppose that $M = 10$ kg, $K = 10$ N/m, and $B = 2$ N/m/s. Consider an LVDT whose core weighs 5 gm and has negligible damping, and a potentiometer whose slider arm weighs 5 gm and has an equivalent viscous damping constant of 0.05 N/m/s. Estimate the percentage error

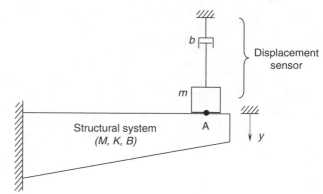

FIGURE P4.14

The use of a displacement sensor to measure the natural frequency and damping ratio of a structure.

of the results for the undamped natural frequency and damping ratio, as measured using each of these two displacement sensors.

4.15 Standard rectilinear displacement sensors such as the LVDT and the potentiometer are used to measure displacements up to 25 cm; within this limit, accuracies as high as $\pm 0.2\%$ can be obtained. For measuring large displacements in the order of 3 m, cable extension displacement sensors, which have an angular displacement sensor as the basic sensing unit, may be used. One type of rectilinear displacement sensor has a rotatory potentiometer and a light cable, which wraps around a spool that rotates with the wiper arm of the pot. In using this sensor, the free end of the cable is connected to the moving member whose displacement is to be measured. The sensor housing is mounted on a stationary platform, such as the support structure of the system that is monitored. A spring motor winds the cable back as the cable retracts. Using suitable sketches, describe the operation of this displacement sensor. Discuss the shortcomings of this device.

4.16 It is known that the factors that should be considered in selecting an LVDT for a particular application include: linearity, sensitivity, response time, size and weight of core, size of the housing, primary excitation frequency, output impedance, phase change between primary and secondary voltages, null voltage, stroke, and environmental effects (temperature compensation, magnetic shielding, etc.). Explain why and how each of these factors is an important consideration.

4.17 The signal-conditioning system for an LVDT has the following components: power supply, oscillator, synchronous demodulator, filter, and voltage amplifier. Using a schematic block diagram, show how these components are connected to the LVDT. Describe the purpose of each component. A high-performance LVDT has a linearity rating of 0.01% within its output range of 0.1–1.0 VAC. The response time of the LVDT is known to be 10 ms. What should be the frequency of the primary excitation?

4.18 List merits and shortcomings of a potentiometer (pot) as a displacement sensing device, in comparison with an LVDT. Give several ways to improve the measurement linearity of a potentiometer.

Suppose that a resistance R_l is added to the conventional potentiometer circuit as shown in Figure P4.18. With $R_l = R_L$ show that

$$\frac{v_o}{v_{ref}} = \frac{(R_L/R_c + 1 - x/x_{max})x/x_{max}}{[R_L/R_c + 2x/x_{max} - 2(x/x_{max})^2]},$$

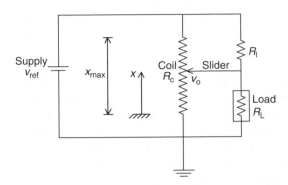

FIGURE P4.18
Potentiometer circuit with a linearizing resistor.

where
R_c = potentiometer coil resistance (total)
R_L = load resistance
v_{ref} = supply voltage to the coil
v_o = output voltage
x = slider displacement
x_{max} = slider stroke (maximum displacement).
Explain why R_l produces a linearizing effect.

4.19 Suppose that a sinusoidal carrier frequency is applied to the primary coil of an LVDT. Sketch the shape of the output voltage of the LVDT when the core is stationary at: (a) null position; (b) the left of null position; and (c) the right of null position.

4.20 For directional sensing using an LVDT it is necessary to determine the phase angle of the induced signal. In other words, phase-sensitive demodulation would be needed.

a. First consider a linear core displacement starting from a positive value, moving to zero, and then returning to the same position in an equal time period. Sketch the output of the LVDT for this triangular core displacement.

b. Next sketch the output if the core continued to move to the negative side at the same speed.

By comparing the two outputs show that phase-sensitive demodulation would be needed to distinguish between the two cases of displacement.

4.21 Joint angles and angular speeds are the two basic measurements used in the direct (low-level) control of robotic manipulators. One type of robot arm uses resolvers to measure angles and differentiates these signals (digitally) to obtain angular speeds. A gear system is used to step up the measurement (typical gear ratio, 1:8). Since the gear wheels are ferromagnetic, an alternative measuring device would be a self-induction or mutual-induction proximity sensor located at a gear wheel. This arrangement, known as a pulse tachometer, generates a pulse (or near-sine) signal, which can be used to determine both angular displacement and angular speed. Discuss the advantages and disadvantages of these two arrangements (resolver and pulse tachometer) in this particular application.

4.22 Why is motion sensing important in trajectory-following control of robotic manipulators? Identify five types of motion sensors that could be used in robotic manipulators.

4.23 Compare and contrast the principles of operation of DC tachometer and AC tachometer (both permanent-magnet and induction types). What are the advantages and disadvantages of these two types of tachometers?

4.24 Describe three different types of proximity sensors. In some applications, it may be required to sense only two-state values (e.g., presence or absence, go or no-go). Proximity sensors can be used in such applications, and in that context they are termed proximity switches (or limit switches). For example, consider a parts-handling application in automated manufacturing in which a robot end effector grips a part and picks it up to move it from a conveyor to a machine tool. We can identify four separate steps in the gripping process. Explain how proximity switches can be used for sensing in each of these four tasks:

a. Make sure that the part is at the expected location on the conveyor.

b. Make sure that the gripper is open.

c. Make sure that the end effector has moved to the correct location so that the part is in between the gripper fingers.

d. Make sure that the part did not slip when the gripper was closed.

Note: A similar use of limit switches is found in lumber mills, where tree logs are cut (bucked) into smaller logs; bark removed (de-barked); cut into a square or rectangular log using a chip-n-saw operation; and sawed into smaller dimensions (e.g., two by four cross-sections) for marketing.

4.25 Discuss the relationships among displacement sensing, distance sensing, position sensing, and proximity sensing. Explain why the following characteristics are important in using some types of motion sensors:

a. Material of the moving (or target) object

b. Shape of the moving object

c. Size (including mass) of the moving object

d. Distance (large or small) of the target object

e. Nature of motion (transient or not, what speed, etc.) of the moving object

f. Environmental conditions (humidity, temperature, magnetic fields, dirt, lighting conditions, shock, and vibration, etc.).

4.26 In some industrial processes, it is necessary to sense the condition of a system at one location and, depending on that condition, activate an operation at a location far from that location. For example, in a manufacturing environment, when the count of the finished parts exceeds some value, as sensed in the storage area, a milling machine could be shut down or started. A proximity switch could be used for sensing, and a networked (e.g., Ethernet-based) control system for process control. Since activation of the remote process usually requires a current that is larger than the rated load of a proximity switch, one would have to use a relay circuit, which is operated by the proximity switch. One such arrangement is shown in Figure P4.26. The relay circuit can be used to operate a device such as a valve, a motor, a pump, or a heavy-duty switch. Discuss an application of the arrangement shown in Figure P4.26 in the food-packaging industry. A mutual-induction proximity sensor with the following ratings is used in this application:

Sensor diameter = 1 cm

Sensing distance (proximity) = 1 mm

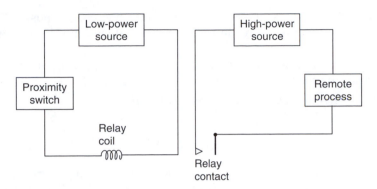

FIGURE P4.26
Proximity switch-operated relay circuit.

Supply to primary winding $= 110$ AC at 60 Hz
Load current rating (in secondary) $= 200$ mA.
Discuss the limitations of this proximity sensor.

4.27 Compression molding is used in making parts of complex shapes and varying sizes. Typically, the mold consists of two platens, the bottom platen fixed to the press table and the top platen operated by a hydraulic press. Metal or plastic sheets—for example, for the automotive industry—can be compression-molded in this manner. The main requirement in controlling the press is to position the top platen accurately with respect to the bottom platen (say, with a 0.001 in or 0.025 mm tolerance), and it has to be done quickly (say, in a few seconds). How many degrees of freedom have to be sensed (how many position sensors are needed) in controlling the mold? Suggest typical displacement measurements that would be made in this application and the types of sensors that could be employed. Indicate sources of error that cannot be perfectly compensated for in this application.

4.28 Seam tracking in robotic arc welding needs precise position control under dynamic conditions. The welding seam has to be accurately followed (tracked) by the welding torch. Typically, the position error should not exceed 0.2 mm. A proximity sensor could be used for sensing the gap between the welding torch and the welded part. The sensor has to be mounted on the robot end effector in such a way that it tracks the seam at some distance (typically 1 in or 2.5 cm) ahead of the welding torch. Explain why this is important. If the speed of welding is not constant and the distance between the torch and the proximity sensor is fixed, what kind of compensation would be necessary in controlling the end effector position? Sensor sensitivity of several volts per millimeter is required in this application of position control. What type of proximity sensor would you recommend?

4.29 An angular motion sensor, which operates somewhat like a conventional resolver, has been developed at Wright State University. The rotor of this resolver is a permanent magnet. A 2 cm diameter Alnico-2 disk magnet, diametrically magnetized as a two-pole rotor, has been used. Instead of the two sets of stationary windings placed at $90°$ in a conventional resolver, two Hall-effect sensors (see Chapter 5) placed at $90°$ around the permanent-magnet rotor are used for detecting quadrature signals. Note that Hall-effect sensors can detect moving magnetic sources. Describe the operation of this modified resolver and explain how this device could be used to measure angular motions continuously. Compare this device with a conventional resolver, giving advantages and disadvantages.

4.30 Discuss factors that limit the lower frequency and upper frequency limits of the output from the following sensors:

a. Potentiometer   d. Eddy current proximity sensor

b. LVDT   e. DC tachometer

c. Resolver   f. Piezoelectric transducer.

4.31 An active suspension system is proposed for a high-speed ground transit vehicle in order to achieve significant improvements in ride quality. The system senses jerk (rate of change of acceleration) as a result of road disturbances and adjusts system parameters accordingly.

a. Draw a suitable schematic diagram for the proposed control system and describe appropriate measuring devices.

b. Suggest a way to specify the desired ride quality for a given type of vehicle. (Would you specify one value of jerk, a jerk range, or a jerk curve with respect to time or frequency?)

c. Discuss the drawbacks and limitations of the proposed control system with respect to such factors as reliability, cost, feasibility, and accuracy.

4.32 A design objective in most control system applications is to achieve small time constants. An exception is the time constant requirements for a piezoelectric sensor. Explain why a large time constant, in the order of 1.0 s, is desirable for a piezoelectric sensor in combination with its signal-conditioning system.

An equivalent circuit for a piezoelectric accelerometer, which uses a quartz crystal as the sensing element, is shown in Figure P4.32. The generated charge is denoted by q, and the output voltage at the end of the accelerometer cable is v_o. The piezoelectric sensor capacitance is modeled by C_p, and the overall capacitance experienced at the sensor output, whose primary contribution is due to cable capacitance, is denoted by C_c. The resistance of the electric insulation in the accelerometer is denoted by R. Write a differential equation relating v_o to q. What is the corresponding transfer function? Using this result, show that the accuracy of accelerometer improves when the sensor time constant is large and when the frequency of the measured acceleration is high. For a quartz crystal sensor with $R = 1 \times 10^{11}$ Ω and $C_5 = 300$ pF, and a circuit with $C_c = 700$ pF compute the time constant.

4.33 Applications of accelerometers are found in the following areas:

a. Transit vehicles (automobiles—microsensors for airbag sensing in particular, aircraft, ships, etc.)

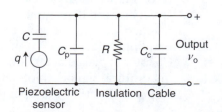

FIGURE P4.32
Equivalent circuit for a quartz crystal (piezoelectric) accelerometer.

  b. Power cable monitoring

  c. Robotic manipulator control

  d. Building structures

  e. Shock and vibration testing

  f. Position and velocity sensing

Describe one direct use of acceleration measurement in each application area.

4.34 a. A standard accelerometer that weighs 100 gm is mounted on a test object that has an equivalent mass of 3 kg. Estimate the accuracy in the first natural frequency of the object measured using this arrangement, considering mechanical loading due to accelerometer mass alone. If a miniature accelerometer that weighs 0.5 gm is used instead, what is the resulting accuracy?

  b. A strain-gage accelerometer uses a semiconductor strain gage mounted at the root of a cantilever element, with the seismic mass mounted at the free end of the cantilever. Suppose that the cantilever element has a square cross-section with dimensions 1.5×1.5 mm^2. The equivalent length of the cantilever element is 25 mm, and the equivalent seismic mass is 0.2 gm. If the cantilever is made of an aluminum alloy with Young's modulus $E = 69 \times 10^9$ N/m^2, estimate the useful frequency range of the accelerometer in Hertz.

  Hint: When force F is applied to the free end of a cantilever, the deflection y at that location may be approximated by the formula

$$y = \frac{Fl^3}{3EI}$$

  where l = cantilever length
    I = second moment area of the cantilever cross-section about the bending axis = $bh^3/12$
    b = cross-section width
    h = cross-section height.

4.35 Applications of piezoelectric sensors are numerous: push-button devices and switches, airbag MEMS sensors in vehicles, pressure and force sensing, robotic tactile sensing, accelerometers, glide testing of computer hard-disk-drive (HDD) heads, excitation sensing in dynamic testing, respiration sensing in medical diagnostics, and graphics input devices for computers. Discuss advantages and disadvantages of piezoelectric sensors.

  What is cross-sensitivity of a sensor? Indicate how the anisotropy of piezoelectric crystals (i.e., charge sensitivity quite large along one particular crystal axis) is useful in reducing cross-sensitivity problems in a piezoelectric sensor.

4.36 As a result of advances in microelectronics, piezoelectric sensors (such as accelerometers and impedance heads) are now available in miniature form with built-in charge amplifiers in a single integral package. When such units are employed, additional signal conditioning is usually not necessary. An external power supply unit is needed, however, to provide power for the amplifier circuitry. Discuss the advantages and disadvantages of a piezoelectric sensor with built-in microelectronics for signal conditioning.

A piezoelectric accelerometer is connected to a charge amplifier. An equivalent circuit for this arrangement is shown in Figure 4.33.

a. Obtain a differential equation for the output v_o of the charge amplifier, with acceleration a as the input, in terms of the following parameters: S_a = charge sensitivity of the accelerometer (charge/acceleration); R_f = feedback resistance of the charge amplifier; τ_c = time constant of the system (charge amplifier).

b. If an acceleration pulse of magnitude a_o and duration T is applied to the accelerometer, sketch the time response of the amplifier output v_o. Show how this response varies with τ_c. Using this result, show that the larger the τ_c the more accurate the measurement.

4.37 Give typical values for the output impedance and the time constant of the following measuring devices:

a. Potentiometer

b. Differential transformer

c. Resolver

d. Piezoelectric accelerometer

An RTD has an output impedance of 500 Ω. If the loading error has to be maintained near 5%, estimate a suitable value for the load impedance.

4.38 A signature verification pen has been developed by IBM Corporation. The purpose of the pen is to authenticate the signature, by detecting whether the user is trying to forge someone else's signature. The instrumented pen has analog sensors. Sensor signals are conditioned using microcircuitry built into the pen and sampled into a digital computer through a wireless communication link, at the rate of 80 samples/second. Typically, about 1000 data samples are collected per signature. Before the pen's use, authentic signatures are collected off-line and stored in a reference database. When a signature and the corresponding identification code are supplied to the computer for verification, a program in the processor retrieves the authentic signature from the database, by referring to the identification code, and then compares the two sets of data for authenticity. This process takes about 3 s. Discuss the types of sensors that could be used in the pen. Estimate the total time required for signal verification. What are the advantages and disadvantages of this method in comparison with the user punching in an identification code alone or providing the signature without the identification code?

4.39 Under what conditions can displacement control be treated as force control? Describe a situation in which this is not feasible.

4.40 Consider the joint of a robotic manipulator, shown schematically in Figure P4.40. Torque sensors are mounted at locations 1, 2, and 3. If the magnetic torque generated at the motor rotor is T_m write equations for the torque transmitted to link 2, the frictional torque at bearing A, the frictional torque at bearing B, and the reaction torque on link 1, in terms of the measured torques, the inertia torque of the rotor, and T_m.

4.41 A model for a machining operation is shown in Figure P4.41. The cutting force is denoted by f, and the cutting tool with its fixtures is modeled by a spring (stiffness k),

| = torque-sensing locations

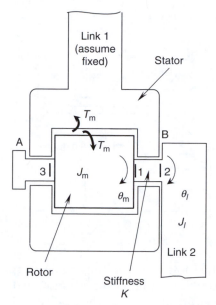

FIGURE P4.40
Torque sensing locations for a manipulator joint.

a viscous damper (damping constant b), and a mass m. The actuator (hydraulic) with its controller is represented by an active stiffness g. Assuming linear g, obtain a transfer relation between the actuator input u and the cutting force f. Now determine an approximate expression for the gradient $\partial g/\partial u$. Discuss a control strategy for counteracting effects from random variations in the cutting force. Note that this is important for controlling the product quality.

Hint: You may use a reference-adaptive feedforward control strategy where a reference g and u are the inputs to the machine tool. The reference g is adapted using the gradient $\partial g/\partial u$, as u changes by Δu.

4.42 A strain-gage sensor to measure the torque T_m generated by a motor is shown schematically in Figure P4.42. The motor is floated on frictionless bearings.

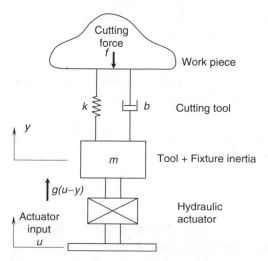

FIGURE P4.41
A model for a machining operation.

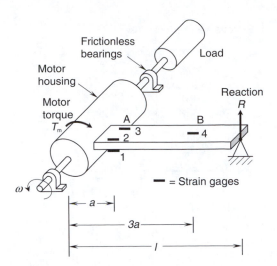

FIGURE P4.42
A strain-gage sensor for measuring motor torque.

A uniform rectangular lever arm is rigidly attached to the motor housing, and its projected end is restrained by a pin joint. Four identical strain gages are mounted on the lever arm, as shown. Three of the strain gages are at point A, which is located at a distance a from the motor shaft, and the fourth strain-gage is at point B, which is located at a distance $3a$ from the motor shaft. The pin joint is at a distance l from the motor shaft. Strain gages 2, 3, and 4 are on the top surface of the lever arm, and gage 1 is on the bottom surface. Obtain an expression for T_m in terms of the bridge output δv_o and the following additional parameters: S_s is the gage factor (strain-gage sensitivity), v_{ref} the supply voltage to the bridge, b the width of the lever arm cross-section, h the height of the lever arm cross-section, and E is the Young's modulus of the lever arm.

Verify that the bridge sensitivity does not depend on a and l. Describe means to improve the bridge sensitivity. Explain why the sensor reading is only an approximation to the torque transmitted to the load. Give a relation to determine the net normal reaction force at the bearings, using the bridge output.

4.43 The sensitivity S_s of a strain gage consists of two parts: the contribution from the change in resistivity of the material and the direct contribution due to the change in shape of the strain-gage when deformed. Show that the second part may be approximated by $(1 + 2v)$, where v denotes the Poisson's ratio of the strain-gage material.

4.44 Compare the potentiometer (ballast) circuit with the Wheatstone bridge circuit for strain-gage measurements, with respect to the following considerations:

a. Sensitivity to the measured strain

b. Error due to ambient effects (e.g., temperature changes)

c. Signal-to-noise ratio (SNR) of the output voltage

d. Circuit complexity and cost

e. Linearity

4.45 In the strain-gage bridge shown in Figure 4.45, suppose that the load current i is not negligible. Derive an expression for the output voltage v_o in terms of $R_1, R_2, R_3, R_4, R_L,$

and v_{ref}. Initially, the bridge was balanced, with equal resistances in the four arms. Then one of the resistances (say, R_1) was increased by 1%. Plot to scale the ratio (actual output from the bridge)/(output under open-circuit, or infinite-load-impedance, conditions) as a function of the nondimensionalized load resistance R_L/R in the range 0.1 to 10.0, where R denotes the initial resistance in each arm of the bridge.

4.46 What is meant by the term bridge sensitivity in strain-gage measurements? Describe methods of increasing bridge sensitivity. Assuming the load resistance to be very high in comparison with the arm resistances in the strain-gage bridge shown in Figure 4.45, obtain an expression for the power dissipation p in terms of the bridge resistances and the supply voltage. Discuss how the limitation on power dissipation can affect bridge sensitivity.

4.47 Consider the strain-gage bridge shown in Figure 4.45. Initially, the bridge is balanced, with $R_1 = R_2 = R$. (*Note*: R_3 may not be equal to R_1.) Then R_1 is changed by δR. Assuming the load current to be negligible, derive an expression for the percentage error as a result of neglecting the second-order and higher-order terms in δR. If $\delta R/R = 0.05$, estimate this nonlinearity error.

4.48 Discuss the advantages and disadvantages of the following techniques in the context of measuring transient signals.
  a. DC bridge circuits vs. ac bridge circuits
  b. Slip ring and brush commutators vs. ac transformer commutators
  c. Strain-gage torque sensors vs. variable-inductance torque sensors
  d. Piezoelectric accelerometers vs. strain-gage accelerometers
  e. Tachometer velocity transducers vs. piezoelectric velocity transducers

4.49 For a semiconductor strain gage characterized by the quadratic strain–resistance relationship

$$\frac{\delta R}{R} = S_1 \varepsilon + S_2 \varepsilon^2,$$

obtain an expression for the equivalent gage factor (sensitivity) S_s using the least squares error linear approximation. Assume that only positive strains up to ε_{max} are measured with the gage. Derive an expression for the percentage nonlinearity. Taking $S_1 = 117, S_2 = 3600$, and $\varepsilon_{max} = 0.01$ strain, compute S_s and the percentage nonlinearity.

4.50 Briefly describe how strain gages may be used to measure
  a. Force   d. Pressure
  b. Displacement   e. Temperature
  c. Acceleration

Show that if a compensating resistance R_c is connected in series with the supply voltage v_{ref} to a strain-gage bridge that has four identical members, each with resistance R, the output equation is given by

$$\frac{\delta v_o}{v_{ref}} = \frac{R}{(R + R_c)} \frac{kS_s}{4} \varepsilon$$

in the usual rotation.

A foil-gage load cell uses a simple (one-dimensional) tensile member to measure force. Suppose that k and S_s are insensitive to temperature change. If the temperature coefficient of R is α_1, that of the series compensating resistance R_c is α_2, and that of the Young's modulus of the tensile member is $(-\beta)$, determine an expression for R_c that would result in automatic (self) compensation for temperature effects. Under what conditions is this arrangement realizable?

4.51 Draw a block diagram for a single joint of a robot, identifying inputs and outputs. Using the diagram, explain the advantages of torque sensing in comparison to displacement and velocity sensing at the joint. What are the disadvantages of torque sensing?

4.52 Figure P4.52 shows a schematic diagram of a measuring device.

  a. Identify the various components in this device.

  b. Describe the operation of the device, explaining the function of each component and identifying the nature of the measurand and the output of the device.

  c. List the advantages and disadvantages of the device.

  d. Describe a possible application of this device.

4.53 Discuss the advantages and disadvantages of torque sensing by the motor current method. Show that for a synchronous motor with a balanced three-phase supply, the electromagnetic torque generated at the rotor–stator interface is given by

$$T_{\mathrm{m}} = k i_f i_a \cos(\theta - \omega t),$$

where i_f = dc current in the rotor (field) winding
  i_a = amplitude of the supply current to each phase in the stator (armature)
  θ = angle of rotation
  ω = frequency (angular) of the ac supply
  t = time
  k = motor torque constant.

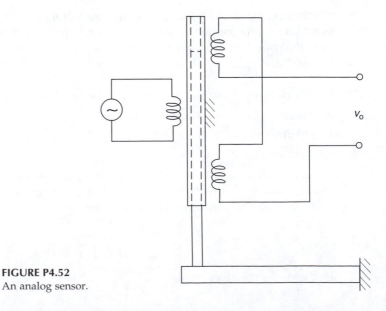

FIGURE P4.52
An analog sensor.

4.54 Discuss factors that limit the lower frequency and upper frequency limits of measurements obtained from the following devices:

    a. Strain gage

    b. Rotating shaft torque sensor

    c. Reaction torque sensor

4.55 Briefly describe a situation in which tension in a moving belt or cable has to be measured under transient conditions. What are some of the difficulties associated with measuring tension in a moving member? A strain-gage tension sensor for a belt-drive system is shown in Figure P4.55. Two identical active strain gages, G_1 and G_2, are mounted at the root of a cantilever element with rectangular cross-section, as shown. A light, frictionless pulley is mounted at the free end of the cantilever element. The belt makes a 90° turn when passing over this idler pulley.

    a. Using a circuit diagram, show the Wheatstone bridge connections necessary for the strain gages G_1 and G_2 so that strains as a result of axial forces in the cantilever member have no effect on the bridge output (i.e., effects of axial loads are compensated) and the sensitivity to bending loads is maximized.

    b. Obtain an equation relating the belt tension T and the bridge output δv_o in terms of the following additional parameters:

        S_s = gage factor (sensitivity) of each strain gage

        E = Young's modulus of the cantilever element

        L = length of the cantilever element

        b = width of the cantilever cross-section

        h = height of the cantilever cross-section.

    In particular, show that the radius of the pulley does not enter this equation. Give the main assumptions made in your derivation.

4.56 Consider a standard strain-gage bridge (Figure 4.45) with R_1 as the only active gage and $R_3 = R_4$. Obtain an expression for R_1 in terms of R_2, v_o and v_{ref}. Show that when $R_1 = R_2$, we get $v_o = 0$—a balanced bridge—as required. Note that the equation for R_1, assuming that v_o is measured using a high-impedance voltmeter, can be used to detect large resistance changes in R_1. Now suppose that the active gage R_1 is connected to the bridge using a long, twisted wire pair, with each wire having a resistance of R_c. The bridge circuit has to be modified as in Figure P4.56 in this case.

    Using the expression obtained earlier for R_1, show that the equation of the modified bridge is given by

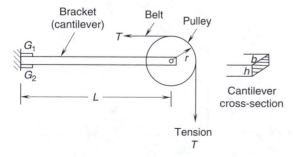

Bracket (cantilever)   Belt   Pulley

G_1

T

r

G_2

L

b

h

Cantilever cross-section

Tension T

FIGURE P4.55
A strain-gage tension sensor for a moving belt.

$$R_1 = R_2 \left[\frac{v_{\text{ref}} + 2v_0}{v_{\text{ref}} - 2v_0}\right] + 4R_c \frac{v_0}{[v_{\text{ref}} - 2v_0]}.$$

Obtain an expression for the fractional error in the R_1 measurement due to cable resistance R_c. Show that this error can be decreased by increasing R_2 and v_{ref}.

4.57 The read-write head in a hard-disk-drive (HDD) of a digital computer should float at a constant but small height (say, fraction of a μm) above the disk surface. Because of aerodynamics resulting from the surface roughness and the surface deformations of the disk, the head can be excited into vibrations that could cause head-disk contacts. These contacts, which are called head-disk interferences (HDIs), are clearly undesirable. They can occur at very high frequencies (say, 1 MHz). The purpose of a glide test is to detect HDIs and to determine the nature of these interferences. Glide testing can be used to determine the effect of parameters such as the flying height of the head and the speed of the disk, and to qualify (certify the quality of) disk drive units for specific types of operating conditions. Indicate the basic instrumentation needed in glide testing. In particular, suggest the types of sensors that could be used and their advantages and disadvantages.

4.58 What are the typical requirements for an industrial tactile sensor? Explain how a tactile sensor differs from a simple touch sensor. Define spatial resolution and force resolution (or sensitivity) of a tactile sensor.

The spatial resolution of your fingertip can be determined by a simple experiment using two pins and a helper. Close your eyes. Instruct the helper to apply one pin or both pins randomly to your fingertip so that you feel the pressure of the tip of the pins. You should respond by telling the helper whether you feel both pins or

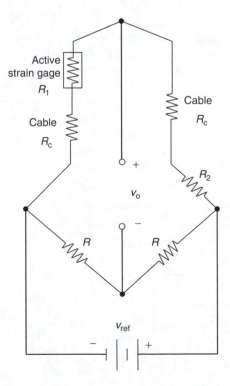

FIGURE P4.56
The influence of cable resistance on strain-gage
bridge measurements.

just one pin. If you feel both pins, the helper should decrease the spacing of the two pins in the next round of tests. The test should be repeated in this manner by successively decreasing the spacing between the pins until you feel only one pin when both pins are actually applied. Then measure the distance between the two pins in millimeters. The largest spacing between the two pins that will result in this incorrect sensation corresponds to the spatial resolution of your fingertip. Repeat this experiment for all your fingers, repeating the test several times on each finger. Compute the average and the standard deviation. Then perform the test on other subjects. Discuss your results. Do you notice large variations in the results?

4.59 Torque, force, and tactile sensing can be very useful in many applications, particularly in the manufacturing industry. For each of the following applications, indicate the types of sensors that would be useful for properly performing the task:

a. Controlling the operation of inserting printed circuit (PC) boards in card cages using a robotic end effector

b. Controlling a robotic end effector that screws a threaded part into a hole

c. Failure prediction and diagnosis of a drilling operation

d. Gripping a fragile, delicate, and somewhat flexible object by a robotic hand without damaging the object

e. Gripping a metal part using a two-fingered gripper

f. Quickly identifying and picking a complex part from a bin containing many different parts

4.60 The *motion dexterity* of a device is defined as the ratio: (Number of degrees of freedom in the device)/(Motion resolution of the device). The *force dexterity* may be defined as: (Number of degrees of freedom in the device)/(Force resolution of the device). Give a situation where both types of dexterity mean the same thing and a situation where the two terms mean different things. Outline how force dexterity of a device (say, an end effector) can be improved by using tactile sensors. Provide the dexterity requirements for the following tasks by indicating whether motion dexterity or force dexterity is preferred in each case:

a. Gripping a hammer and driving a nail with it

b. Threading a needle

c. Seam tracking of a complex part in robotic arc welding

d. Finishing the surface of a complex metal part using robotic grinding

4.61 Describe four advantages and four disadvantages of a semiconductor strain-gage weight sensor. A weight sensor is used in a robotic wrist. What would be the purpose of this sensor? How can the information obtained from the weight sensor be used in controlling the robotic manipulator?

4.62 Discuss whether there is any relationship between the dexterity and the stiffness of a manipulator hand. The stiffness of a robotic hand can be improved during gripping operations by temporarily decreasing the number of degrees of freedom of the hand using suitable fixtures. What purpose does this serve?

4.63 Using the usual equation for a dc strain-gage bridge (Figure 4.45) show that if the resistance elements R_1 and R_2 have the same temperature coefficient of resistance and if R_3 and R_4 have the same temperature coefficient of resistance, the temperature effects are compensated up to first order.

A microminiature (MEMS) strain-gage accelerometer uses two semiconductor strain-gages, one integral with the cantilever element near the fixed end (root) and the other mounted at an unstrained location of the accelerometer. The entire unit including the cantilever and the strain gages, has a silicon integrated-circuit (IC) construction, and measures smaller than 1 mm in size. Outline the operation of the accelerometer. What is the purpose of the second strain gage?

4.64 a. List three advantages and three disadvantages of a semiconductor strain-gage when compared with a foil strain-gage.

   b. A fly-wheel device is schematically shown in Figure P4.64. The wheel consists of four spokes which carry lumped masses at one end and are clamped to the rotating hub at the other end, as shown. Suppose that the inertia of the spokes can be neglected in comparison with that of the lumped masses.

Four active strain gages are used in a bridge circuit for measuring speed.

   (i) If the bridge can be calibrated to measure the tensile force F in each spoke, express the dynamic equation, which may be used to measure the rotating speed (ω). The following parameters may be used:

   m = mass of the lumped element at the end of a spoke

   r = radius of rotation of the center of mass of the lumped element.

   (ii) For good results with regard to high sensitivity of the bridge and also for compensation of secondary effects such as out-of-plane bending, indicate where the four strain gages (1, 2, 3, 4) should be located on the spokes and in what configuration they should be connected in a dc bridge.

   (iii) Compare this method of speed sensing to that using a tachometer and a potentiometer by giving three advantages and two disadvantages of the strain-gage method.

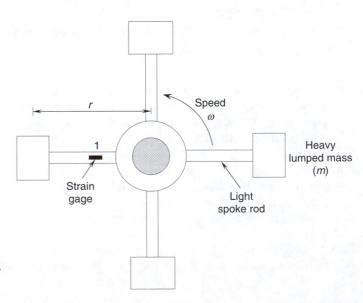

FIGURE P4.64
Strain-gage speed sensor for a fly wheel.

4.65 a. Consider a simple mechanical manipulator. Explain why in some types of manipulation tasks, motion sensing alone might not be adequate for accurate control, and torque or force sensing might be needed as well.

b. Discuss what factors should be considered when installing a torque sensor to measure the torque transmitted from an actuator to a rotating load.

c. A harmonic drive (see Chapter 8) consists of the following three main components:

1. Input shaft with the elliptical wave generator (cam)
2. Circular flexispline with external teeth
3. Rigid circular spline with internal teeth

Consider the free-body diagrams shown in Figure P4.65. The following variables are defined:

ω_i = speed of the input shaft (wave generator)

ω_o = speed of the output shaft (rigid spline)

T_o = torque transmitted to the driven load by the output shaft (rigid spline)

T_i = torque applied on the harmonic drive by the input shaft

T_f = torque transmitted by the flexispline to the rigid spline

T_r = reaction torque on the flexispline at the fixture

T_w = torque transmitted by the wave generator

If strain gages are to be used to measure the output torque T_o, suggest suitable locations for mounting them and discuss how the torque measurement can be obtained in this manner. Using a block diagram for the system, indicate whether you consider T_o to be an input to or an output of the harmonic drive. What are the implications of this consideration?

4.66 a. Describe three different principles of torque sensing. Discuss relative advantages and disadvantages of the three approaches.

b. A torque sensor is needed for measuring the drive torque that is transmitted to a link of a robot (i.e., joint torque). What characteristics and specifications of the

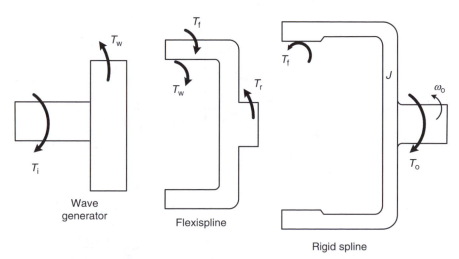

FIGURE P4.65
Free body diagram of a harmonic drive.

sensor and the requirement of the system should be considered in selecting a suitable torque sensor for this application?

4.67 A simple rate gyro, which may be used to measure angular speeds, is shown in Figure P4.67. The angular speed of spin is ω and is kept constant at a known value. The angle of rotation of the gyro about the gimbal axis (or the angle of twist of the torsional spring) is θ, and is measured using a displacement sensor. The angular speed of the gyro about the axis that is orthogonal to both gimbal axis and spin axis is Ω. This is the angular speed of the supporting structure (vehicle), which needs to be measured. Obtain a relationship between Ω and θ in terms of parameters such as the following: J, the moment of inertia of the spinning wheel; k, the torsional stiffness of the spring restraint at the gimbal bearings; and b, the damping constant of rotational motion about the gimbal axis; and the spinning speed. How would you improve the sensitivity of this device?

4.68 Level sensors are used in a wide variety of applications, including soft drink bottling, food packaging, monitoring of storage vessels, mixing tanks, and pipe-lines. Consider the following types of level sensors, and briefly explain the principle of operation of each type in level sensing. Additionally, what are the limitations of each type?

a. Capacitive sensors

b. Inductive sensors

c. Ultrasonic sensors

d. Vibration sensors

4.69 Consider the following types of position sensors: inductive, capacitive, eddy current, fiber optic, and ultrasonic. For the following conditions, indicate which of these types are not suitable and explain why:

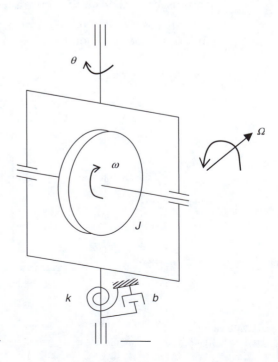

FIGURE P4.67
A rate gyro speed sensor.

    a. Environment with variable humidity

    b. Target object made of aluminum

    c. Target object made of steel

    d. Target object made of plastic

    e. Target object several feet away from the sensor location

    f. Environment with significant temperature fluctuations

    g. Smoke-filled environment

4.70 Discuss the advantages and disadvantages of fiber-optic sensors. Consider the fiber-optic position sensor. In the curve of intensity of received light versus x, in which region would you prefer to operate the sensor, and what are the corresponding limitations?

4.71 The manufacturer of an ultrasonic gage states that the device has applications in measuring cold roll steel thickness, determining parts positions in robotic assembly, lumber sorting, measurement of particle board and plywood thickness, ceramic tile dimensional inspection, sensing the fill level of food in a jar, pipe diameter gaging, rubber tire positioning during fabrication, gaging of fabricated automotive components, edge detection, location of flaws in products, and parts identification. Discuss whether the following types of sensors are also equally suitable for some or all of the foregoing applications. In each case where do a particular sensor is not suitable for a given application, give reasons to support that claim.

    a. Fiber-optic position sensors

    b. Self-induction proximity sensors

    c. Eddy current proximity sensors

    d. Capacitive gages

    e. Potentiometers

    f. Differential transformers

4.72 a. Consider the motion control system that is shown by the block diagram in Figure P4.72.

    (i) Giving examples of typical situations explain the meaning of the block represented as Load in this system.

    (ii) Indicate advantages and shortcomings of moving the motion sensors from the motor shaft to the load response point, as indicated by the broken lines in the figure.

    b. Indicate, giving reasons, what type of sensors will you recommend for the following applications:

    (i) In a soft drink bottling line, for on-line detection of improperly fitted metal caps on glass bottles.

    (ii) In a paper processing plant, to simultaneously measure both the diameter and eccentricity of rolls of newsprint.

    (iii) To measure the dynamic force transmitted from a robot to its support structure, during operation.

    (iv) In a plywood manufacturing machine, for on-line measurement of the thickness of plywood.

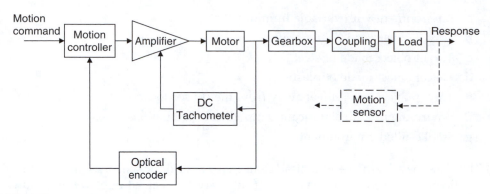

FIGURE P4.72
Block diagram of a motion control system.

    (v) In a food canning plant, to detect defective cans (with damage to flange and side seam, bulging of the lid, and so on,)

    (vi) To read codes on food packages

5

Digital Transducers

A digital transducer is a measuring device that produces a digital output. A transducer whose output is a pulse signal may be considered in this category since the pulses can be counted and presented in the digital form using a counter. Similarly, a transducer whose output is a frequency falls into the same category as it can use a frequency counter to generate a digital output.

Sensors and transducers are useful in many industrial applications within the general area of control engineering. Numerous examples are found in robotic manipulation, transit systems, digital computers and accessories, process monitoring and control, material processing, fabrication, finishing, handling, inspection, testing, grading, and packaging. In Chapter 4, we discussed analog sensors and transducers. In the present chapter, we study some useful types, concepts, operation, and use of digital transducers. Our discussion will be limited primarily to motion transducers. Note, however, that by using a suitable auxiliary front-end sensor, other measurands, such as force, torque, temperature, and pressure, may be converted into a motion and subsequently measured using a motion transducer. For example, altitude (or pressure) measurements in aircraft and aerospace applications are made using a pressure-sensing front-end, such as a bellows or diaphragm device, in conjunction with an optical encoder (which is a digital transducer) to measure the resulting displacement. Similarly, a bimetallic element may be used to convert temperature into a displacement, which may be measured using a displacement sensor.

As we have done, it is acceptable to call an analog sensor as an analog transducer, because both the sensor stage and the transducer stage are analog in this case. The sensor stage of a digital transducer is typically analog as well. For example, motion, as manifested in physical systems, is continuous in time. Therefore, we cannot generally speak of digital motion sensors. Actually, the transducer stage generates the discrete output signal in a digital motion-measuring device. Hence, we have termed the present category of devices as digital transducers rather than digital sensors. Commercially available digital transducers are not as numerous as analog sensors, but what is available has found extensive application.

5.1 Advantages of Digital Transducers

Any measuring device that presents information as discrete samples and does not introduce a quantization error when the reading is represented in the digital form may be classified as a digital transducer. According to this definition, for example, an analog sensor such as a thermocouple along with an analog-to-digital converter (ADC) is not a digital transducer. This is so because a quantization error is introduced by the ADC process (see Section 2.6). A digital processor plays the role of controller in a digital control system. This facilitates complex processing of measured signals and other known quantities, thereby generating control signals for the plant of the control system. If the measured signals are available in analog form, an ADC stage is necessary before processing within a digital controller.

Digital signals (or digital representation of information) have several advantages in comparison with analog signals. Notably

1. Digital signals are less susceptible to noise, disturbances, or parameter variation in instruments because data can be generated, represented, transmitted, and processed as binary words consisting of bits, which possess two identifiable states.

2. Complex signal processing with very high accuracy and speed is possible through digital means (hardware implementation is faster than software implementation).

3. High reliability in a system can be achieved by minimizing analog hardware components.

4. Large amounts of data can be stored using compact, high-density data storage methods.

5. Data can be stored or maintained for very long periods of time without any drift or disruption by adverse environmental conditions.

6. Fast data transmission is possible over long distances with no attenuation and with less dynamic delays, compared to analog signals.

7. Digital signals use low voltages (e.g., 0–12 V DC) and low power.

8. Digital devices typically have low overall cost.

These advantages help build a strong case in favor of digital measuring and signal transmission devices for control systems.

Digital measuring devices (or digital transducers, as they are commonly known) generate discrete output signals, such as pulse trains or encoded data, which can be directly read by a digital controller. Nevertheless, the sensor stage itself of a digital measuring device is usually quite similar to that of an analog counterpart. There are digital measuring devices that incorporate microprocessors to locally perform numerical manipulations and conditioning and provide output signals in either digital form or analog form. These measuring systems are particularly useful when the required variable is not directly measurable but could be computed using one or more measured outputs (e.g., power $=$ force $\times$ speed). Although a microprocessor is an integral part of the measuring device in this case, it performs a conditioning task rather than a measuring task. For our purposes, we consider the two tasks separately.

When the output of a digital transducer is a pulse signal, a common method of reading the signal is using a counter, either to count the pulses (for high-frequency pulses) or to count clock cycles over one pulse duration (for low-frequency pulses). The count is placed as a digital word in a buffer, which can be accessed by the host (control) computer, typically at a constant frequency (sampling rate). On the other hand, if the output of a digital transducer is automatically available in a coded form (e.g., natural binary code or gray code), it can be directly read by a computer. In the latter case, the coded signal is normally generated by a parallel set of pulse signals; each pulse transition generates one bit of the digital word, and the numerical value of the word is determined by the pattern of the generated pulses. This is the case, for example, with absolute encoders. Data acquisition from (i.e., computer interfacing) a digital transducer is commonly carried out using a general-purpose input/output (I/O) card; for example, a motion control (servo) card, which may be able to accommodate multiple transducers (e.g., 8 channels of encoder inputs with 24-bit counters) or by using a data acquisition card specific to the particular transducer.

5.2   Shaft Encoders

Any transducer that generates a coded (digital) reading of a measurement can be termed as an encoder. Shaft encoders are digital transducers that are used for measuring angular displacements and angular velocities. Applications of these devices include motion measurement in performance monitoring and control of robotic manipulators, machine tools, industrial processes (e.g., food processing and packaging, pulp and paper), digital data storage devices, positioning tables, satellite mirror positioning systems, vehicles, and rotating machinery such as motors, pumps, compressors, turbines, and generators. High resolution (which depends on the word size of the encoder output and the number of pulses generated per revolution of the encoder), high accuracy (particularly due to noise immunity and reliability of digital signals and superior construction), and relative ease of adoption in digital control systems (because transducer output can be read as a digital word), with associated reduction in system cost and improvement of system reliability, are some of the relative advantages of digital transducers in general and shaft encoders in particular, in comparison with their analog counterparts.

5.2.1   Encoder Types

Shaft encoders can be classified into two categories depending on the nature and the method of interpretation of the transducer output: (1) incremental encoders and (2) absolute encoders.

The output of an incremental encoder is a pulse signal, which is generated when the transducer disk rotates as a result of the motion that is measured. By counting the pulses or by timing the pulse width using a clock signal, both angular displacement and angular velocity can be determined. With an incremental encoder, displacement is obtained with respect to some reference point. The reference point can be the home position of the moving component (say, determined by a limit switch) or a reference point on the encoder disk, as indicated by a reference pulse (index pulse) generated at that location on the disk. Furthermore, the index pulse count determines the number of full revolutions.

An absolute encoder (or, whole-word encoder) has many pulse tracks on its transducer disk. When the disk of an absolute encoder rotates, several pulse trains—equal in number to the tracks on the disk—are generated simultaneously. At a given instant, the magnitude of each pulse signal will have one of two signal levels (i.e., a binary state), as determined by a level detector (or edge detector). This signal level corresponds to a binary digit (0 or 1). Hence, the set of pulse trains gives an encoded binary number at any instant. The pulse windows on the tracks can be organized into some pattern (code) so that the generated binary number at a particular instant corresponds to the specific angular position of the encoder disk at that time. The pulse voltage can be made compatible with some digital interface logic (e.g., transistor-to-transistor logic or TTL). Consequently, the direct digital readout of an angular position is possible with an absolute encoder, thereby expediting digital data acquisition and processing. Absolute encoders are commonly used to measure fractions of a revolution. However, complete revolutions can be measured using an additional track, which generates an index pulse, as in the case of an incremental encoder.

The same signal generation (and pick-off) mechanism may be used in both types (incremental and absolute) of transducers. Four techniques of transducer signal generation can be identified:

1. Optical (photosensor) method
2. Sliding contact (electrical conducting) method

3. Magnetic saturation (reluctance) method

4. Proximity sensor method

By far, the optical encoder is most popular and cost-effective. The other three approaches may be used in special circumstances, where the optical method may not be suitable (e.g., under extreme temperatures) or may be redundant (e.g., where a code disk such as a toothed wheel is already available as an integral part of the moving member). For a given type of encoder (incremental or absolute), the method of signal interpretation is identical for all four types of signal generation listed previously. We briefly describe the principle of signal generation for all four techniques and consider only the optical encoder in the context of signal interpretation and processing.

The optical encoder uses an opaque disk (code disk) that has one or more circular tracks, with some arrangement of identical transparent windows (slits) in each track. A parallel beam of light (e.g., from a set of light-emitting diodes or LEDs) is projected to all tracks from one side of the disk. The transmitted light is picked off using a bank of photosensors on the other side of the disk, which typically has one sensor for each track. This arrangement is shown in Figure 5.1a, which indicates just one track and one pick-off sensor. The light sensor could be a silicon photodiode or a phototransistor. Since the light from the source is interrupted by the opaque regions of the track, the output signal from the photosensor is a series of voltage pulses. This signal can be interpreted (e.g., through edge detection or level detection) to obtain the increments in the angular position and also angular velocity of the disk. Note that in standard terminology, the sensor element of such a measuring device is the encoder disk, which is coupled to the rotating object (directly or through a gear mechanism). The transducer stage is the conversion of disk motion (analog) into the pulse signals, which can be coded into a digital word. The opaque background of transparent windows (the window pattern) on an encoder disk may be produced by contact printing techniques. The precision of this production procedure is a major factor that determines the accuracy of optical encoders. Note that a transparent disk with a track of opaque spots will work equally well as the encoder disk of an optical encoder. In either form, the track has a 50% duty cycle (i.e., the length of the transparent region is equal to the length of the opaque region). Components of a commercially available optical encoder are shown in Figure 5.1b.

In a sliding contact encoder, the transducer disk is made of an electrically insulating material. Circular tracks on the disk are formed by implanting a pattern of conducting areas. These conducting regions correspond to the transparent windows on an optical encoder disk. All conducting areas are connected to a common slip ring on the encoder shaft. A constant voltage v_{ref} is applied to the slip ring using a brush mechanism. A sliding contact such as a brush touches each track, and as the disk rotates, a voltage pulse signal is picked off by it. The pulse pattern depends on the conducting–nonconducting pattern on each track, as well as the nature of rotation of the disk. The signal interpretation is done as it is for optical encoders. The advantages of sliding contact encoders include high sensitivity (depending on the supply voltage) and simplicity of construction (low cost). The disadvantages include the familiar drawbacks of contacting and commutating devices (e.g., friction, wear, brush bounce due to vibration, and signal glitches and metal oxidation due to electrical arcing). A transducer's accuracy is very much dependent on the precision of the conducting patterns of the encoder disk. One method of generating the conducting pattern on the disk is electroplating.

Magnetic encoders have high-strength magnetic regions imprinted on the encoder disk using techniques such as etching, stamping, or recording (similar to magnetic data

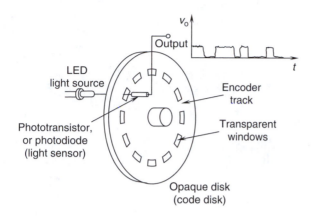

(a)

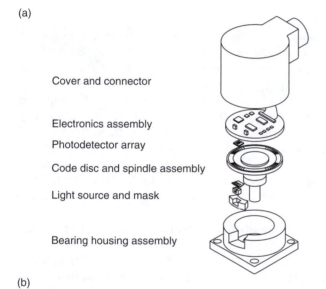

(b)

FIGURE 5.1

(a) Schematic representation of an (incremental) optical encoder; (b) components of a commercial incremental encoder. (BEI Electronics, Inc. With permission.)

recording). These magnetic regions correspond to the transparent windows on an optical encoder disk. The signal pick-off device is a microtransformer, which has primary and secondary windings on a circular ferromagnetic core. This pick-off sensor resembles a core storage element in historical mainframe computers. The encoder arrangement is illustrated schematically in Figure 5.2. A high-frequency (typically 100 kHz) primary voltage induces a voltage in the secondary winding of the sensing element at the same frequency, operating as a transformer. A magnetic field of sufficient strength can saturate the core, however, thereby significantly increasing the reluctance and dropping the induced voltage. By demodulating the induced voltage, a pulse signal is obtained. This signal can be interpreted in the usual manner. Note that a pulse peak corresponds to a nonmagnetic area and a pulse valley corresponds to a magnetic area on each track. Magnetic encoders have noncontacting pick-off sensors, which is an advantage. They are more costly than the contacting devices, however, primarily because of the cost of transformer elements and demodulating circuitry for generating the output signal.

Proximity sensor encoders use a proximity sensor as the signal pick-off element. Any type of proximity sensor may be used; for example, a magnetic induction probe

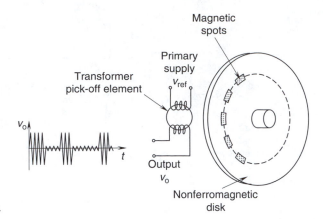

FIGURE 5.2
Schematic representation of a magnetic encoder.

or an eddy current probe, as discussed in Chapter 4. In the magnetic induction probe, for example, the disk is made of ferromagnetic material. The encoder tracks have raised spots of the same material, serving a purpose analogous to that of the windows on an optical encoder disk. As a raised spot approaches the probe, the flux linkage increases as a result of the associated decrease in reluctance, thereby raising the induced voltage level. The output voltage is a pulse-modulated signal at the frequency of the supply (primary) voltage of the proximity sensor. This is then demodulated, and the resulting pulse signal is interpreted. Instead of a disk with a track of raised regions, a ferromagnetic toothed wheel may be used along with a proximity sensor placed in a radial orientation. In principle, this device operates like a conventional digital tachometer. If an eddy current probe is used, pulse areas in the track have to be plated with a conducting material.

Note that an incremental encoder disk requires only one primary track that has equally spaced and identical window (pick-off) regions. The window area is equal to the area of the inter-window gap (i.e., 50% duty cycle). Usually, a reference track that has just one window is also present to generate a pulse (known as the index pulse) to initiate pulse counting for angular position measurement and to detect complete revolutions. In contrast, absolute encoder disks have several rows of tracks, equal in number to the bit size of the output data word. Furthermore, the windows in a track are not equally spaced but are arranged in a specific pattern to obtain a binary code (or a gray code) for the output data from the transducer. It follows that absolute encoders need at least as many signal pick-off sensors as there are tracks, whereas incremental encoders need just one pick-off sensor to detect the magnitude of rotation. As will be explained, they will also need a sensor at a quarter-pitch separation (pitch = center-to-center distance between adjacent windows) to generate a quadrature signal, which will identify the direction of rotation. Some designs of incremental encoders have two identical tracks, one at a quarter-pitch offset from the other, and the two pick-off sensors are placed radially without offset. The two (quadrature) signals obtained with this arrangement will be similar to those with the previous arrangement. A pick-off sensor for receiving a reference pulse is also used in some designs of incremental encoders (i.e., three-track incremental encoders).

In many control applications, encoders are built into the plant itself, rather than being externally fitted onto a rotating shaft. For instance, in a robot arm, the encoder might be an integral part of the joint motor and may be located within its housing. This reduces coupling errors (e.g., errors due to backlash, shaft flexibility, and resonances added by the transducer and fixtures), installation errors (e.g., misalignment and eccentricity), and overall cost. Encoders are available in sizes as small as 2 cm and as large as 15 cm in diameter.

Since the techniques of signal interpretation are quite similar for the various types of encoders with different principles of signal generation, we limit further discussion to optical encoders only. These encoders are in fact the most common types in practical applications. Signal interpretation differs depending on whether the particular optical encoder is an incremental device or an absolute device.

5.3 Incremental Optical Encoders

There are two possible configurations for an incremental encoder disk: (1) Offset sensor configuration and (2) Offset track configuration. The first configuration is schematically shown in Figure 5.3. The disk has a single circular track with identical and equally spaced transparent windows. The area of the opaque region between adjacent windows is equal to the window area. Two photodiode sensors (pick-offs 1 and 2 in Figure 5.3) are positioned facing the track at a quarter-pitch (half the window length) apart. The forms of their output signals (v_1 and v_2), after passing them through pulse-shaping circuitry (idealized), are shown in Figure 5.4a and Figure 5.4b for the two directions of rotation.

In the second configuration of incremental encoders, two identical tracks are used, one offset from the other by a quarter-pitch. Each track has its own pick-off sensor, oriented normally facing the track. The two pick-off sensors are positioned on a radial line facing the disk, without any circumferential offset unlike the previous configuration. The output signals from the two sensors are the same as before, however (Figure 5.4). Note that an output pulse signal is on for half the time and off for half the time, giving a 50% duty cycle.

In both configurations, an additional track with a lone window and associated sensor is also usually available. This track generates a reference pulse (index pulse) per revolution of the disk (see Figure 5.4c). This pulse is used to initiate the counting operation. Furthermore, the index pulse count gives the number of complete revolutions, which

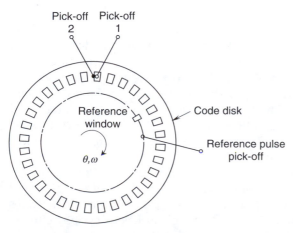

FIGURE 5.3
An incremental encoder disk (offset sensor configuration).

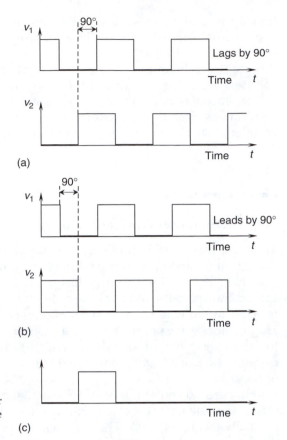

FIGURE 5.4
Shaped pulse signals from an incremental encoder (a) for clockwise rotation; (b) for counterclockwise rotation; (c) reference pulse signal.

is required in measuring absolute angular rotations. Note that when the disk rotates at constant angular speed, the pulse width and pulse-to-pulse period (encoder cycle) are constant (with respect to time) in each sensor output. When the disk accelerates, the pulse width decreases continuously; when the disk decelerates, the pulse width increases.

5.3.1 Direction of Rotation

The quarter-pitch offset in sensor location (or in track placement) is used to determine the direction of rotation of the disk. For example, Figure 5.4a shows the shaped (idealized) sensor outputs (v_1 and v_2) when the disk rotates in the clockwise (cw) direction; and Figure 5.4b shows the outputs when the disk rotates in the counterclockwise (ccw) direction. It is clear from these two figures that in cw rotation, v_1 lags v_2 by a quarter of a cycle (i.e., a phase lag of 90°) and in ccw rotation, v_1 leads v_2 by a quarter of a cycle. Hence, the direction of rotation is obtained by determining the phase difference of the two output signals, using phase-detecting circuitry.

One method for determining the phase difference is to time the pulses using a high-frequency clock signal. Suppose that the counting (timing) begins when the v_1 signal begins to rise (i.e., when a rising edge is detected). Let n_1 be the number of clock cycles (time) up to the time when v_2 begins to rise; and n_2 be the number of clock cycles up to the time when v_1 begins to rise again. Then, the following logic applies:

If $n_1 > n_2 - n_1 \Rightarrow$ cw rotation

If $n_1 < n_2 - n_1 \Rightarrow$ ccw rotation.

This logic for direction detection should be clear from Figure 5.4a and Figure 5.4b.

Another scheme can be given for direction detection. In this case, we first detect a high level (logic high or binary 1) in signal v_2 and then check whether the edge in signal v_1 rises or falls during this period. As shown in Figure 5.4a and Figure 5.4b, the following logic applies:

If rising edge in v_1 when v_2 is logic high $\Rightarrow$ cw rotation

If falling edge in v_1 when v_2 is logic high $\Rightarrow$ ccw rotation.

5.3.2 Hardware Features

The actual hardware of commercial encoders is not as simple as that suggested by Figure 5.3 (also see Figure 5.1b). A more detailed schematic diagram of the signal generation mechanism of an optical incremental encoder is shown in Figure 5.5a. The light generated by the LED is collimated (forming parallel rays) using a lens. This pencil of parallel light passes through a window of the rotating code disk. The masking (grating) disk is

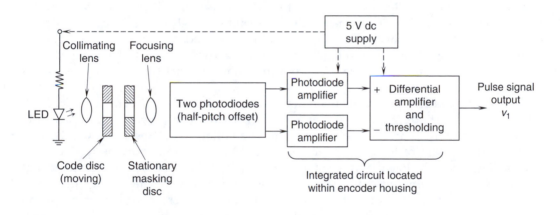

(a)

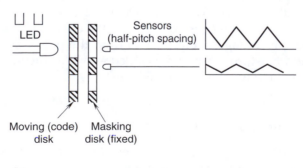

(b)

FIGURE 5.5
(a) Internal hardware of an optical incremental encoder; (b) use of two sensors at $180°$ spacing to generate a shaped pulse.

stationary and has a track of windows identical to that in the code disk. Because of the presence of the masking disk, light from the LED will pass through more than one window of the code disk, thereby improving the intensity of light received by the photosensor but not introducing any error due to the diameter of the pencil of light as it is larger than the window length. When the windows of the code disk face the opaque areas of the masking disk, virtually no light is received by the photosensor. When the windows of the code disk face the transparent areas of the masking disk, the maximum amount of light reaches the photosensor. Hence, as the code disk moves, a sequence of triangular (and positive) pulses of light is received by the photosensor. Pulse width in this case is a full cycle (i.e., it corresponds to the window pitch) and not a half cycle.

Fluctuation in the supply voltage to the encoder light source also directly influences the light level received by the photosensor. If the sensitivity of the photosensor is not high enough, a low light level might be interpreted as no light, which would result in measurement error. Such errors due to instabilities and changes in the supply voltage can be eliminated by using two photosensors, one placed half a pitch away from the other along the window track, as shown in Figure 5.5b. This arrangement is for contrast detection, and it should not be confused with the quarter-of-a-pitch offset arrangement that is required for direction detection. The sensor facing the opaque region of the masking disk will always read a low signal. The other sensor will read a triangular signal whose peak occurs when a moving window overlaps with a window of the masking disk and whose valley occurs when a moving window faces an opaque region of the masking disk. The two signals from these two sensors are amplified separately and fed into a differential amplifier (see Chapter 2). The result is a high-intensity triangular pulse signal. A shaped (or binary) pulse signal can be generated by subtracting a threshold value from this signal and identifying the resulting positive (or binary 1) and negative (or binary 0) regions. This procedure will produce a more distinct (or binary) pulse signal that is immune to noise.

Signal amplifiers are integrated circuit devices and are housed within the encoder itself. Additional pulse-shaping circuitry may also be present. The power supply has to be provided separately as an external component. The voltage level and pulse width of the output pulse signal are logic-compatible (e.g., TTL) so that they may be read directly using a digital board. Note that if the output level v_1 is positive high, we have a logic high (or binary 1) state. Otherwise, we have a logic low (or binary 0) state. In this manner, a stable and accurate digital output can be obtained even under unstable voltage supply conditions. The schematic diagram in Figure 5.5 shows the generation of only one (v_1) of the two quadrature pulse signals. The other pulse signal (v_2) is generated using identical hardware but at a quarter-of-a-pitch offset. The index pulse (reference pulse) signal is also generated in a similar manner. The cable of the encoder (usually a ribbon cable) has a multipin connector. Three of the pins provide the three output pulse signals. Another pin carries the DC supply voltage (typically 5 V) from the power supply into the encoder. Typically, a ground line (ground pin) is included as well. Note that the only moving part in the system shown in Figure 5.5 is the code disk.

5.3.3 Displacement Measurement

An incremental encoder measures displacement as a pulse count and it measures velocity as a pulse frequency. A digital processor is able to express these readings in engineering units (radians, degrees, rad/s, etc.) using pertinent parameter values of the physical system. Suppose that the maximum count possible is M pulses and the range of the encoder is $\pm\theta_{max}$.

The angular position θ corresponding to a count of n pulses is computed as

$$\theta = \frac{n}{M}\theta_{max}. \tag{5.1}$$

5.3.3.1 Digital Resolution

The resolution of an encoder represents the smallest change in measurement that can be measured realistically. Since an encoder can be used to measure both displacement and velocity, we can identify a resolution for each case. First, we consider displacement resolution, which is governed by the number of windows N in the code disk and the digital size (number of bits) of the buffer (counter output). In the following paragraphs, we discuss digital resolution.

Suppose that the encoder count is stored as digital data of r bits. Allowing for a sign bit, we have

$$M = 2^{r-1}. \tag{5.2}$$

The displacement resolution of an incremental encoder is given by the change in displacement corresponding to a unit change in the count (n). It follows from Equation 5.1 that the displacement resolution is given by

$$\Delta\theta = \frac{\theta_{max}}{M}. \tag{5.3}$$

In particular, the digital resolution corresponds to a unit change in the bit value. By substituting Equation 5.2 into Equation 5.3, we have the digital resolution

$$\Delta\theta_d = \frac{\theta_{max}}{2^{r-1}}. \tag{5.4}*$$

Typically, $\theta_{max} = \pm 180°$ or $360°$. Then,

$$\Delta\theta_d = \frac{180°}{2^{r-1}} = \frac{360°}{2^r}. \tag{5.4}$$

Note that the minimum count corresponds to the case where all the bits are zero and the maximum count corresponds to the case where all the bits are unity. Suppose that these two readings represent the angular displacements θ_{min} and θ_{max}. We have,

$$\theta_{max} = \theta_{min} + (M - 1)\Delta\theta \tag{5.5}$$

or substituting Equation 5.2,

$$\theta_{max} = \theta_{min} + (2^{r-1} - 1)\Delta\theta_d. \tag{5.6}*$$

Equation 5.5 leads to the conventional definition for digital resolution:

$$\Delta\theta_d = \frac{(\theta_{max} - \theta_{min})}{(2^{r-1} - 1)}. \tag{5.6}$$

This result is exactly the same as that given by Equation 5.4.

If θ_{max} is 2π and θ_{min} is 0, then θ_{max} and θ_{min} will correspond to the same position of the code disk. To avoid this ambiguity, we use

$$\theta_{min} = \frac{\theta_{max}}{2^{r-1}}. \tag{5.7}$$

Note that if we substitute Equation 5.7 into Equation 5.6 we get Equation 5.4* as required. Then, the digital resolution is given by

$$\frac{(360° - 360°/2^r)}{(2^r - 1)},$$

which is identical to Equation 5.4.

5.3.3.2 Physical Resolution

The physical resolution of an encoder is governed by the number of windows N in the code disk. If only one pulse signal is used (i.e., no direction sensing) and if only the rising edges of the pulses are detected (i.e., full cycles of the encoder signal are counted), the physical resolution is given by the pitch angle of the track (i.e., angular separation between adjacent windows), which is $(360/N)°$. However; quadrature signals (i.e., two pulse signals, one out of phase with the other by $90°$ or quarter-of-a-pitch angle) are available and the capability to detect both rising and falling edges of a pulse is also present, four counts can be made per encoder cycle, thereby improving the resolution by a factor of four. Under these conditions, the physical resolution of an encoder is given by

$$\Delta\theta_p = \frac{360°}{4N}. \tag{5.8}$$

To understand this, note in Figure 5.4a (or Figure 5.4b) that when the two signals v_1 and v_2 are added, the resulting signal has a transition at every quarter of the encoder cycle. This is illustrated in Figure 5.6. By detecting each transition (through edge detection or level detection), four pulses can be counted within every main cycle. It should be mentioned that each signal (v_1 or v_2) has a resolution of half a pitch separately, provided that all transitions (rising edges and falling edges) are detected and counted instead of counting pulses (or high signal levels). Accordingly, a disk with 10,000 windows has a resolution of $0.018°$ if only one pulse signal is used (and both transitions, rise and fall, are detected). When two signals (with a phase shift of a quarter of a cycle) are used, the resolution improves to $0.009°$. This resolution is achieved directly from the mechanics of the transducer; no interpolation is involved. It assumes, however, that the pulses are nearly ideal and, in particular, that the transitions are perfect. In practice, this cannot be realized if the pulse signals are noisy. Then, pulse shaping will be necessary as mentioned earlier.

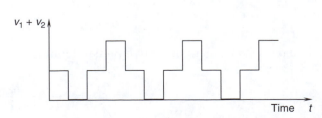

FIGURE 5.6
Quadrature signal addition to
improve physical resolution.

The larger value of the two resolutions given by Equation 5.4 and Equation 5.8 governs the displacement resolution of the encoder.

Example 5.1
For an ideal design of an incremental encoder, obtain an equation relating the parameters d, w, and r, where

$d =$ diameter of encoder disk
$w =$ number of windows per unit diameter of disk
$r =$ word size (bits) of the angle measurement.

Assume that quadrature signals are available. If $r = 12$ and $w = 500/\text{cm}$, determine a suitable disk diameter.

Solution
In this problem, we take the ideal design as the case where the physical resolution is equal to the digital resolution. The position resolution due to physical constraints (assuming that quadrature signals are available) is given by Equation 5.8. Hence,

$$\Delta\theta_{\text{P}} = \frac{1}{4}\left(\frac{360}{wd}\right)^{\circ}.$$

The resolution limited by the digital word size of the buffer is given by Equation 5.4:

$$\Delta\theta_{\text{d}} = \left(\frac{360}{2^r}\right)^{\circ}.$$

For an ideal design we need $\Delta\theta_{\text{p}} = \Delta\theta_{\text{d}}$, which gives

$$\frac{1}{4}\frac{360}{wd} = \frac{360}{2^r}.$$

Simplifying, we have

$$wd = 2^{r-2}. \tag{5.9}$$

Substitute $r = 12$ and $w = 500/\text{cm}$ to obtain

$$d = \left(\frac{2^{12-2}}{500}\right)\text{cm} = 2.05\text{ cm}.$$

5.3.3.3 Step-Up Gearing
The physical resolution of an encoder can be improved by using step-up gearing so that one rotation of the moving object that is monitored corresponds to several rotations of the code disk of the encoder. This improvement is directly proportional to the step-up gear ratio (p). Specifically, we have

$$\Delta\theta_{\text{p}} = \frac{360°}{4pN}. \tag{5.10}$$

Backlash in the gearing introduces a new error, however. For best results, this backlash error should be several times smaller than the resolution with no backlash.

The digital resolution will not improve by gearing if the maximum angle of rotation of the moving object (say, 360°) still corresponds to the buffer/register size. Then the change in the least significant bit (LSB) of the buffer corresponds to the same change in the angle of rotation of the moving object. In fact, the overall displacement resolution can be harmed in this case if excessive backlash is present. However, if the buffer or register size corresponds to a full rotation of the code disk (i.e., a rotation of $360°/p$ in the object) and if the output register (or buffer) is cleared at the end of each such rotation and a separate count of full rotations of the code disk is kept, then the digital resolution will also improve by a factor of p. Specifically, from Equation 5.4 we get

$$\Delta\theta_d = \frac{180°}{p2^{r-1}} = \frac{360°}{p2^r}. \tag{5.11}$$

Example 5.2
By using high-precision techniques to imprint window tracks on the code disk, it is possible to attain a window density of 500 windows/cm of diameter. Consider a 3000-window disk. Suppose that step-up gearing is used to improve resolution and the gear ratio is 10. If the word size of the output register is 16 bits, examine the displacement resolution of this device for the two cases where the register size corresponds to (1) a full rotation of the object and (2) a full rotation of the code disk.

Solution
First, consider the case in which gearing is not present. With quadrature signals, the physical resolution is

$$\Delta\theta_p = \frac{360°}{4 \times 3000} = 0.03°.$$

For a range of measurement given by $\pm 180°$, a 16-bit output provides a digital resolution of

$$\Delta\theta_d = \frac{180°}{2^{15}} = 0.005°.$$

Hence, in the absence of gearing, the overall displacement resolution is 0.03°.

Next, consider a geared encoder with gear ratio of 10, and neglect gear backlash. The physical resolution improves to 0.003°. However, in Case 1, the digital resolution remains unchanged at best. Hence, the overall displacement resolution improves to 0.005° as a result of gearing. In Case 2, the digital resolution improves to 0.0005°. Hence, the overall displacement resolution becomes 0.003°.

∎

In summary, the displacement resolution of an incremental encoder depends on the following factors:

1. Number of windows on the code track (or disk diameter)
2. Gear ratio
3. Word size of the measurement buffer.

Example 5.3

A positioning table uses a backlash-free high-precision lead screw of lead 2 cm/rev, which is driven by a servo motor with a built-in optical encoder for feedback control. If the required positioning accuracy is ± 10 μm, determine the number of windows required in the encoder track. In addition, what is the minimum bit size required for the digital data register/buffer of the encoder count?

Solution

The required accuracy is ± 10 μm. To achieve this accuracy, the required resolution for a linear displacement sensor is ± 5 μm. Lead of the lead screw is 2 cm/rev. To achieve the required resolution, the number of pulses per encoder revolution is

$$\frac{2 \times 10^{-2}\,\text{m}}{5 \times 10^{-6}\,\text{m}} = 4000 \text{ pulses/rev.}$$

Assuming that quadrature signals are available (with a resolution improvement of 4), the required number of windows in the encoder track is 1000. Percentage of physical resolution is $(1/4000) \times 100\% = 0.025\%$. Consider a buffer size of r bits, including a sign bit. Then, we need

$$2^{r-1} = 4000$$

or $r = 13$ bits.

5.3.3.4 *Interpolation*

The output resolution of an encoder can be further enhanced by interpolation. This is accomplished by adding equally spaced pulses in between every pair of pulses generated by the encoder circuit. These auxiliary pulses are not true measurements, and they can be interpreted as a linear interpolation scheme between true pulses. One method of accomplishing this interpolation is by using the two pick-off signals that are generated by the encoder (quadrature signals). These signals are nearly sinusoidal (or triangular) before shaping (say, by level detection). They can be filtered to obtain two sine signals that are 90° out of phase (i.e., a sine signal and a cosine signal). By weighted combination of these two signals, a series of sine signals can be generated such that each signal lags the preceding signal by any integer fraction of 360°. By level detection or edge detection (of rising and falling edges), these sine signals can be converted into square wave signals. Then, by logical combination of the square waves, an integer number of pulses can be generated within each encoder cycle. These are the interpolation pulses that are added to improve the encoder resolution. In practice, about 20 interpolation pulses can be added between two adjacent main pulses.

5.3.4 Velocity Measurement

Two methods are available for determining velocities using an incremental encoder: (1) pulse-counting method and (2) pulse-timing method. In the first method, the pulse count over a fixed time period (the successive time period at which the data buffer is read) is used to calculate the angular velocity. For a given period of data reading, there is a lower speed limit below which this method is not very accurate. To compute the angular velocity ω using this method, suppose that the count during a time period T is n pulses. Hence, the average time for one pulse cycle (i.e., window-to-window pitch angle) is T/n.

If there are N windows on the disk, assuming that quadrature signals are not used, the angle moved during one pulse period is $2\pi/N$ radians. Hence,

$$\text{Speed } \omega = \frac{2\pi/N}{T/n} = \frac{2\pi n}{NT}. \tag{5.12}$$

If quadrature signals are used, replace N by $4N$ in Equation 5.12.

In the second method, the time for one encoder pulse cycle (i.e., window-to-window pitch angle) is measured using a high-frequency clock signal. This method is particularly suitable for accurately measuring low speeds. In this method, suppose that the clock frequency is f Hz. If m cycles of the clock signal are counted during an encoder pulse period (i.e., window pitch, which is the interval between two adjacent windows, assuming that quadrature signals are not used), the time for that encoder cycle (i.e., the time to rotate through one encoder pitch) is given by m/f. With a total of N windows on the track, the angle of rotation during this period is $2\pi/N$ radians as before. Hence,

$$\text{Speed } \omega = \frac{2\pi/N}{m/f} = \frac{2\pi f}{Nm}. \tag{5.13}$$

If quadrature signals are used, replace N by $4N$ in Equation 5.13.

Note that a single incremental encoder can serve as both position sensor and speed sensor. Hence, a position loop and a speed loop in a control system can be closed using a single encoder, without having to use a conventional (analog) speed sensor such as a tachometer (see Chapter 4). The speed resolution of the encoder (which depends on the method of speed computation—pulse counting or pulse timing) can be chosen to meet the accuracy requirements for the speed control loop. A further advantage of using an encoder rather than a conventional (analog) motion sensor is that an analog-to-digital converter (ADC) would be unnecessary. For example, the pulses generated by the encoder may be used as interrupts for the control computer. These interrupts are then directly counted (by an up/down counter or indexer) or timed (by a clock in the data acquisition system) within the control computer, thereby providing position and velocity readings.

5.3.4.1 Velocity Resolution

The velocity resolution of an incremental encoder depends on the method that is employed to determine velocity. As both the pulse-counting method and the pulse-timing method are based on counting, the velocity resolution is given by the change in angular velocity that corresponds to a change (increment or decrement) in the count by one.

For the pulse-counting method, it is clear from Equation 5.12 that a unity change in the count n corresponds to a speed change of

$$\Delta\omega_c = \frac{2\pi}{NT}, \tag{5.14}$$

where N is the number of windows in the code track and T is the time period over which a pulse count is read. Equation 5.14 gives the velocity resolution by this method. Note that the engineering value (in rad/s) of this resolution is independent of the angular velocity itself, but when expressed as percentage of the speed, the resolution becomes better (smaller) at higher speeds. Note further from Equation 5.14 that the resolution improves with the number of windows and the count reading (sampling) period. Under transient conditions, the accuracy of a velocity reading decreases with increasing T (because, according to Shannon's sampling theorem—see Chapter 3—the sampling frequency has to be at least double the highest frequency of interest in the velocity signal). Hence, the sampling period should not be increased indiscriminately. As usual, if quadrature signals are used, N in Equation 5.14 has to be replaced by $4N$ (i.e., the resolution improves by $\times 4$).

In the pulse-timing method, the velocity resolution is given by (see Equation 5.13)

$$\Delta \omega_t = \frac{2\pi f}{Nm} - \frac{2\pi f}{N(m+1)} = \frac{2\pi f}{Nm(m+1)}, \tag{5.15}*$$

where f is the clock frequency. For large m, $(m + 1)$ can be approximated by m. Then, by substituting Equation 5.13 in Equation 5.15*, we get

$$\Delta \omega_t \simeq \frac{2\pi f}{Nm^2} = \frac{N\omega^2}{2\pi f}. \tag{5.15}$$

Note that in this case, the resolution degrades quadratically with speed. Also, the resolution degrades with the speed even when it is considered as a fraction of the measured speed:

$$\frac{\Delta \omega_t}{\omega} = \frac{N\omega}{2\pi f}. \tag{5.16}$$

This observation confirms the previous suggestion that the pulse-timing method is appropriate for low speeds. For a given speed and clock frequency, the resolution further degrades with increasing N. This is true because when N is increased the pulse period shortens and hence the number of clock cycles per pulse period also decreases. The resolution can be improved, however, by increasing the clock frequency.

Example 5.4
An incremental encoder with 500 windows in its track is used for speed measurement. Suppose that

    a. In the pulse-counting method, the count (in the buffer) is read at the rate of 10 Hz
    b. In the pulse-timing method, a clock of frequency 10 MHz is used.

Determine the percentage resolution for each of these two methods when measuring a speed of

    i. 1 rev/s
    ii. 100 rev/s.

Solution
Assume that quadrature signals are not used

    i. Speed $= 1$ rev/s

       With 500 windows, we have 500 pulses/s

       a. Pulse-counting method

          Counting period $= \dfrac{1}{10 \text{ Hz}} = 0.1$ s

Pulse count (in 0.1 s) $= 500 \times 0.1 = 50$

Percentage resolution $= \dfrac{1}{50} \times 100\% = 2\%$

b. Pulse-timing method

At 500 pulses/s,

Pulse period $= 1/500$ s $= 2 \times 10^{-3}$ s

With a 10 MHz clock,

Clock count $= 10 \times 10^{6} \times 2 \times 10^{-3} = 20 \times 10^{3}$

Percentage resolution $= \dfrac{1}{20 \times 10^{3}} \times 100\% = 0.005\%$

ii. Speed $= 100$ rev/s

With 500 windows, we have 50,000 pulses/s

a. Pulse-counting method

Pulse count (in 0.1 s) $= 50{,}000 \times 0.1 = 5000$

Percentage resolution $= \dfrac{1}{5000} \times 100\% = 0.02\%$

b. Pulse-timing method

At 50,000 pulses/s,

Pulse period $= \dfrac{1}{50{,}000}$ s $= 20 \times 10^{-6}$ s

With a 10 MHz clock,

Clock count $= 10 \times 10^{6} \times 20 \times 10^{-6} = 200$

Percentage resolution $= \dfrac{1}{200} \times 100\% = 0.5\%$

The results are summarized in Table 5.1.

Results given in Table 5.1 confirm that in the pulse-counting method the resolution improves with speed, and hence it is more suitable for measuring high speeds. Furthermore, in the pulse-timing method the resolution degrades with speed, and hence it is more suitable for measuring low speeds.

5.3.4.2 Step-Up Gearing

Consider an incremental encoder that has an N windows per track and is connected to a rotating shaft through a gear unit with step-up gear ratio p. Formulas for computing angular velocity of the shaft by (1) pulse-counting method and (2) pulse-timing method can be easily determined by using Equation 5.12 and Equation 5.13. Specifically, the angle of rotation of the shaft corresponding to window spacing (pitch) of the encoder disk now

TABLE 5.1

Comparison of Speed Resolution from an Incremental Encoder

Speed (rev/s)	Percentage Resolution	
	Pulse-Counting Method (%)	Pulse-Timing Method (%)
1.0	2	0.005
100.0	0.02	0.5

is $2\pi/(pN)$. Hence, the corresponding formulas for speed can be obtained by replacing N by pN in Equation 5.12 and Equation 5.13. We have

$$\text{For pulse-count method: } \omega = \frac{2\pi n}{pNT}. \tag{5.17}$$

$$\text{For pulse-time method: } \omega = \frac{2\pi f}{pNm}. \tag{5.18}$$

Note that these relations can also be obtained simply by dividing the encoder disk speed by the gear ratio, which gives the object speed.

As before, the speed resolution is given by the change in speed corresponding to a unity change in the count. Hence,

$$\text{For the pulse-count method: } \Delta\omega_c = \frac{2\pi(n+1)}{pNT} - \frac{2\pi n}{pNT} = \frac{2\pi}{pNT}. \tag{5.19}$$

It follows that in the pulse-count method, step-up gearing causes an improvement in the resolution.

For the pulse-time method:

$$\Delta\omega_t = \frac{2\pi f}{pNm} - \frac{2\pi f}{pN(m+1)} = \frac{2\pi f}{pNm(m+1)} \cong \frac{pN}{2\pi f}\omega^2. \tag{5.20}$$

Note that in the pulse-time approach, for a given speed, the resolution degrades with increasing p.

In summary, the speed resolution of an incremental encoder depends on the following factors:

1. Number of windows N
2. Count reading (sampling) period T
3. Clock frequency f
4. Speed ω
5. Gear ratio

In particular, gearing up has a detrimental effect on the speed resolution in the pulse-timing method, but it has a favorable effect in the pulse-counting method.

5.3.5 Data Acquisition Hardware

A method for interfacing an incremental encoder to a digital processor (digital controller) is shown schematically in Figure 5.7. In practice, a suitable interface card (e.g., servo card, encoder card, etc.) in the control computer will possess the necessary functional capabilities indicated in the figure. The pulse signals from the encoder are fed into an up/down counter, which has circuitry to detect pulses (e.g., by rising-edge detection, falling-edge detection, or by level detection) and logic circuitry to determine the direction of motion (i.e., sign of the reading) and to code the count. A pulse in one direction (say,

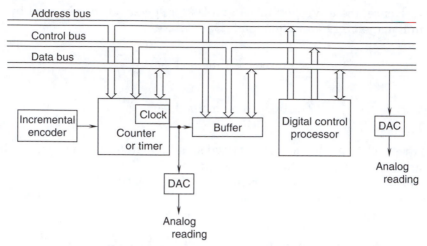

FIGURE 5.7
Computer interface for an incremental encoder.

clockwise) will increase the count by one (an upcount), and a pulse in the opposite direction will decrease the count by one (a downcount). The coded count may be directly read by the host computer, through its input/output (I/O) card without the need for an ADC. The count is transferred to a latch buffer so that the measurement is read from the buffer rather than from the counter itself. This arrangement provides an efficient means of data acquisition because the counting process can continue without interruption while the computer reads the count from the latch buffer. The digital processor (computer) identifies various components in the measurement system using addresses, and this information is communicated to the individual components through the address bus. The start, end, and nature of an action (e.g., data read, clear the counter, clear the buffer) are communicated to various devices by the computer through its control bus. The computer can command an action to a component in one direction of the bus, and the component can respond with a message (e.g., job completed) in the opposite direction. The data (e.g., the count) are transmitted through the data bus. While the computer reads (samples) data from the buffer, the control signals guarantee that no data are transferred to that buffer from the counter. It is clear that data acquisition consists of handshake operations between the main processor of the computer and auxiliary components. More than one encoder may be addressed, controlled, and read by the same three buses of the computer. The buses are conductors; for example, multicore cables carrying signals in parallel logic. Communication in serial logic is done as well, but is slower.

An incremental optical encoder generates two pulse signals one quarter of a pitch out of phase with the other. The internal electronics of the encoder may be powered by a 5 V DC supply. The two pulse signals determine the direction of rotation of the motor by one of various means (e.g., sign of the phase difference and timing of the consecutive rising edges, as described before). The encoder pulse count is stored in a buffer within the controller and is read at fixed intervals (say, 5 ms). The net count gives the position, and the difference in count at a fixed time increment gives the speed.

While measuring the displacement (position) of an object using an incremental encoder, the counter may be continuously monitored as an analog signal through a digital-to-analog converter (DAC in Figure 5.7). On the other hand, the pulse count is read by the computer only at finite time intervals. Since a cumulative count is required in displacement measurement, the buffer is not cleared in this case once the count is read by the computer.

In velocity measurement by the pulse-counting method, the buffer is read at fixed time intervals of T, which is also the counting-cycle time. The counter is cleared every time a count is transferred to the buffer, so that a new count can begin. With this method, a new reading is available at every sampling instant.

In the pulse-timing method of velocity computation, the counter is actually a timer. The encoder cycle is timed using a clock (internal or external), and the count is passed on to the buffer. The counter is then cleared and the next timing cycle is started. The buffer is periodically read by the computer. With this method, a new reading is available at every encoder cycle. Note that under transient velocities, the encoder-cycle time is variable and is not directly related to the data sampling period. In the pulse-timing method, it is desirable to make the sampling period slightly smaller than the encoder-cycle time, so that no count is missed by the processor.

More efficient use of the digital processor may be achieved by using an interrupt routine. With this method, the counter (or buffer) sends an interrupt request to the processor when a new count is ready. The processor then temporarily suspends the current operation and reads in the new data. Note that in this case the processor does not continuously wait for a reading.

5.4 Absolute Optical Encoders

An absolute encoder directly generates a coded digital word to represent each discrete angular position (sector) of its code disk. This is accomplished by producing a set of pulse signals (data channels) equal in number to the word size (number of bits) of the reading. Unlike with an incremental encoder, no pulse counting is involved. An absolute encoder may use various techniques (e.g., optical, sliding contact, magnetic saturation, and proximity sensor) to generate the sensor signal, as described before for an incremental encoder. The optical method, which uses a code disk with transparent and opaque regions and pairs of light sources and photosensors, is the most common technique.

A simplified code pattern on the disk of an absolute encoder, which uses the direct binary code, is shown in Figure 5.8a. The number of tracks (n) in this case is 4, but in practice n is in the order of 14, and may even be as high as 22. The disk is divided into 2^n sectors. Each partitioned area of the matrix thus formed corresponds to a bit of data. For example, a transparent area will correspond to binary 1 and an opaque area to binary 0. Each track has a pick-off sensor similar to that used in incremental encoders. The set of n pick-off sensors is arranged along a radial line and facing the tracks on one side of the disk. A light source (e.g., LED) illuminates the other side of the disk. As the disk rotates, the bank of pick-off sensors generates pulse signals, which are sent to n parallel data channels (or pins). At a given instant, the particular combination of signal levels in the data channels will provide a coded data word that uniquely determines the position of the disk at that time.

5.4.1 Gray Coding

There is a data interpretation problem associated with the straight binary code in absolute encoders. Note in Table 5.2 that with the straight binary code, the transition

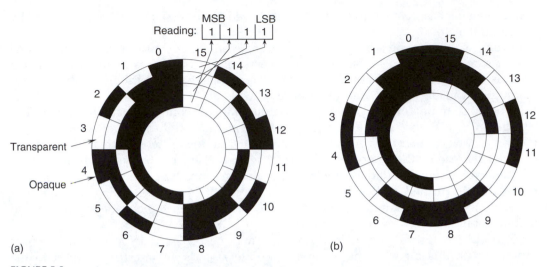

FIGURE 5.8
Illustration of the code pattern of an absolute encoder disk: (a) Binary code; (b) a gray code.

from one sector to an adjacent sector may require more than one switching of bits in the binary data. For example, the transition from 0011 to 0100 or from 1011 to 1100 requires three bit switchings, and the transition from 0111 to 1000 or from 1111 to 0000 requires four bit switchings. If the pick-off sensors are not properly aligned along a radius of the encoder disk, or if the manufacturing error tolerances for imprinting the code pattern on the disk were high, or if environmental effects have resulted in large irregularities in the sector matrix, then bit switching from one reading to the next will not take place simultaneously. This results in ambiguous readings during the transition period. For example, in changing from 0011 to 0100, if the LSB switches first, the reading becomes 0010. In decimal form, this incorrectly indicates that the rotation was from angle 3 to angle 2, whereas, it was actually a rotation from angle 3 to angle 4. Such ambiguities in data interpretation can be avoided by using a gray code, as shown in Figure 5.8b for this example. The coded representation of the sectors is given in Table 5.2. Note that in the case of gray code, each adjacent transition involves only one bit switching.

For an absolute encoder, a gray code is not essential for removing the ambiguity in bit switchings of binary code. For example, for a given absolute reading, the two adjacent absolute readings are automatically known. A reading can be checked against these two valid possibilities (or a single possibility if the direction of rotation is known) to see whether the reading is correct. Another approach is to introduce a delay (e.g., Schmitt trigger) to reading of the output. In this manner a reading will be taken only after all the bit switchings have taken place, thereby eliminating the possibility of an intermediate ambiguous reading.

5.4.1.1 Code Conversion Logic

A disadvantage of using a gray code is that it requires additional logic to convert the gray-coded number to the corresponding binary number. This logic may be provided in

TABLE 5.2

Sector Coding for a 4-bit Absolute Encoder

Sector Number	Straight Binary Code (MSB → LSB)	A Gray Code (MSB → LSB)
0	0 0 0 0	0 0 0 0
1	0 0 0 1	0 0 0 1
2	0 0 1 0	0 0 1 1
3	0 0 1 1	0 0 1 0
4	0 1 0 0	0 1 1 0
5	0 1 0 1	0 1 1 1
6	0 1 1 0	0 1 0 1
7	0 1 1 1	0 1 0 0
8	1 0 0 0	1 1 0 0
9	1 0 0 1	1 1 0 1
10	1 0 1 0	1 1 1 1
11	1 0 1 1	1 1 1 0
12	1 1 0 0	1 0 1 0
13	1 1 0 1	1 0 1 1
14	1 1 1 0	1 0 0 1
15	1 1 1 1	1 0 0 0

hardware or software. In particular, an Exclusive-Or gate can implement the necessary logic, as given by

$$B_{n-1} = G_{n-1}$$
$$B_{k-1} = B_k \oplus G_{k-1} \quad k = n-1, \ldots, 1. \tag{5.21}$$

This converts an n-bit gray-coded word $[G_{n-1}G_{n-2}, \ldots, G_0]$ into an n-bit binary-coded word $[B_{n-1}B_{n-2}, \ldots, B_0]$, where the subscript $n-1$ denotes the most significant bit (MSB) and 0 denotes the LSB. For small word sizes, the code may be given as a look-up table (see Table 5.2). Note that the gray code is not unique. Other gray codes that provide single bit switching between adjacent numbers can be developed.

5.4.2 Resolution

The resolution of an absolute encoder is limited by the word size of the output data. Specifically, the displacement (position) resolution is given by the sector angle, which is also the angular separation between adjacent transparent and opaque regions on the outermost track of the code disk; thus,

$$\Delta\theta = \frac{360°}{2^n}. \tag{5.22}$$

Where n is the number of tracks on the disk (which is equal to the number of bits in the digital reading). In Figure 5.8a, the word size of the data is 4 bits. This can represent decimal numbers from 0 to 15, as given by the 16 sectors of the disk. In each sector, the outermost element is the LSB and the innermost element is the MSB. The direct binary representation of the disk sectors (angular positions) is given in Table 5.2. The angular resolution for this simplified example is $(360/2^4)°$ or 22.5°. If $n = 14$, the angular resolution improves to $(360/2^{14})°$, or 0.022°. If $n = 22$, the resolution further improves to 0.000086°.

Step-up gear mechanisms can also be employed to improve encoder resolution. However, this has the same disadvantages as mentioned under incremental encoders (e.g., backlash, added weight and loading, and increased cost). Furthermore, when a gear is included, the absolute nature of a reading will be limited to a fraction of rotation of the main shaft; specifically, 360°/gear ratio. If a count of the total rotations of the gear shaft (encoder disk) is maintained, this will not present a problem.

An ingenious method of improving the resolution of an absolute encoder is available through the generation auxiliary pulses in between the bit switchings of the coded word. This requires an auxiliary track (usually placed as the outermost track) with a sufficiently finer pitch than the LSB track and some means of direction sensing (e.g., two pick-off sensors placed a quarter-pitch apart, to generate quadrature signals). This is equivalent to having an incremental encoder of finer resolution and an absolute encoder in a single integral unit. Knowing the reading of the absolute encoder (from its coded output, as usual) and the direction of motion (from the quadrature signal), it is possible to determine the angle corresponding to the successive incremental pulses (from the finer track) until the next absolute-word reading. Of course, if a data failure occurs in between the absolute readings, the additional accuracy (and resolution) provided by the incremental pulses will be lost.

5.4.3   Velocity Measurement

An absolute encoder can be used for angular velocity measurement as well. For this, either the pulse-timing method or the angle-measurement method may be used. With the first method, the interval between two consecutive readings is strobed (or timed) using a high-frequency strobe (clock) signal, as in the case of an incremental encoder. Typical strobing frequency is 1 MHz. The start and stop of strobing are triggered by the coded data from the encoder. The clock cycles are counted by a counter, as in the case of an incremental encoder, and the count is reset (cleared) after each counting cycle. The angular speed can be computed using these data, as discussed earlier for an incremental encoder. With the second method, the change in angle is measured from one absolute angle reading to the next, and the angular speed is computed as the ratio (angle change)/ (sampling period).

5.4.4   Advantages and Drawbacks

The main advantage of an absolute encoder is its ability to provide absolute angle readings (with a full 360° rotation). Hence, if a reading is missed, it will not affect the next reading. Specifically, the digital output uniquely corresponds to a physical rotation of the code disk, and hence a particular reading is not dependent on the accuracy of a previous reading. This provides immunity to data failure. A missed pulse (or a data failure of some sort) in an incremental encoder would carry an error into the subsequent readings until the counter is cleared.

An incremental encoder has to be powered throughout the operation of the device. Thus, a power failure can introduce an error unless the reading is reinitialized (or calibrated). An absolute encoder has the advantage that it needs to be powered and monitored only when a reading is taken.

Because the code matrix on the disk is more complex in an absolute encoder and because more light sensors are required, an absolute encoder can be nearly twice as expensive as an incremental encoder. Also, since the resolution depends on the number of tracks present, it is more costly to obtain finer resolutions. An absolute encoder does not require digital counters and buffers, however, unless resolution enhancement is effected using an auxiliary track or pulse timing is used for velocity calculation.

5.5 Encoder Error

Errors in shaft encoder readings can come from several factors. The primary sources of these errors are as follows:

1. Quantization error (due to digital word size limitations)
2. Assembly error (eccentricity of rotation, etc.)
3. Coupling error (gear backlash, belt slippage, loose fit, etc.)
4. Structural limitations (disk deformation and shaft deformation due to loading)
5. Manufacturing tolerances (errors from inaccurately imprinted code patterns, inexact positioning of the pick-off sensors, limitations and irregularities in signal generation and sensing hardware, etc.)
6. Ambient effects (vibration, temperature, light noise, humidity, dirt, smoke, etc.)

These factors can result in inexact readings of displacement and velocity and erroneous detection of the direction of motion.

One form of error in an encoder reading is hysteresis. For a given position of the moving object, if the encoder reading depends on the direction of motion, the measurement has a hysteresis error. In that case, if the object rotates from position A to position B and back to position A, for example, the initial and the final readings of the encoder will not match. The causes of hysteresis include backlash in gear couplings, loose fits, mechanical deformation in the code disk and shaft, delays in electronic circuitry and components (electrical time constants, nonlinearities, etc.), and noisy pulse signals that make the detection of pulses (say, by level detection or edge detection) less accurate.

The raw pulse signal from an optical encoder is somewhat irregular and does not consist of perfect pulses, primarily because of the variation (more or less triangular) of the intensity of light received by the optical sensor as the code disk moves through a window and because of noise in the signal generation circuitry, including the noise created by imperfect light sources and photosensors. Noisy pulses have imperfect edges. As a result, pulse detection through edge detection can result in errors such as multiple triggering for the same edge of a pulse. This can be avoided by including a Schmitt trigger (a logic circuit with electronic hysteresis) in the edge-detection circuit, so that slight irregularities in the pulse edges will not cause erroneous triggering, provided that the noise level is within the hysteresis band of the trigger. A disadvantage of this method, however, is that hysteresis will be present even when the encoder itself is perfect. Virtually noise-free pulses can be generated if two photosensors are used to detect adjacent transparent and opaque areas on a track simultaneously, and a separate circuit (a comparator) is used to create a pulse that depends on the sign of the voltage difference of the two sensor signals. This method of pulse shaping has been described earlier, with reference to Figure 5.5.

5.5.1 Eccentricity Error

Eccentricity (denoted by e) of an encoder is defined as the distance between the center of rotation C of the code disk and the geometric center G of the circular code track. Nonzero eccentricity causes a measurement error known as the eccentricity error. The primary contributions to eccentricity are

1. Shaft eccentricity (e_s)
2. Assembly eccentricity (e_a)
3. Track eccentricity (e_t)
4. Radial play (e_p)

Shaft eccentricity results if the rotating shaft on which the code disk is mounted is imperfect, so that its axis of rotation does not coincide with its geometric axis. Assembly eccentricity is caused if the code disk is improperly mounted on the shaft such that the center of the code disk does not fall on the shaft axis. Track eccentricity comes from irregularities in the imprinting process of the code track, so that the center of the track circle does not coincide with the nominal geometric center of the disk. Radial play is caused by any looseness in the assembly in the radial direction. All four of these parameters are random variables. Let their mean values be μ_s, μ_a, μ_t, and μ_p, and the standard deviations be σ_s, σ_a, σ_t, and σ_p, respectively. A very conservative upper bound for the mean value of the overall eccentricity is given by the sum of the individual mean values, each value considered positive. A more reasonable estimate is provided by the root-mean-square (rms) value, as given by

$$\mu = \sqrt{\mu_s^2 + \mu_a^2 + \mu_t^2 + \mu_p^2}. \tag{5.23}$$

Furthermore, assuming that the individual eccentricities are independent random variables, the standard deviation of the overall eccentricity is given by

$$\sigma = \sqrt{\sigma_s^2 + \sigma_a^2 + \sigma_t^2 + \sigma_p^2}. \tag{5.24}$$

Knowing the mean value μ and the standard deviation σ of the overall eccentricity, it is possible to obtain a reasonable estimate for the maximum eccentricity that can occur. It is reasonable to assume that the eccentricity has a Gaussian (or normal) distribution, as shown in Figure 5.9. The probability that the eccentricity lies between two given values

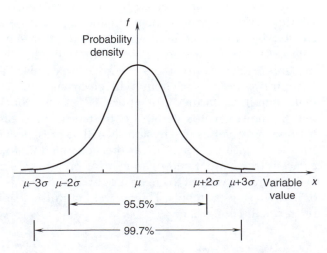

FIGURE 5.9
Gaussian (normal) probability density function.

is obtained by the area under the probability density curve within these two values (points) on the *x*-axis. In particular, for the normal distribution, the probability that the eccentricity lies within $\mu - 2\sigma$ and $\mu + 2\sigma$ is 95.5%, and the probability that the eccentricity falls within $\mu - 3\sigma$ and $\mu + 3\sigma$ is 99.7%. We can say, for example, that at a confidence level of 99.7%, the net eccentricity will not exceed $\mu + 3\sigma$.

Example 5.5
The mean values and the standard deviations of the four primary contributions to eccentricity in a shaft encoder (in millimeters) are as follows:

Shaft eccentricity = (0.1, 0.01)
Assembly eccentricity = (0.2, 0.05)
Track eccentricity = (0.05, 0.001)
Radial play = (0.1, 0.02).

Estimate the overall eccentricity at a confidence level of 96%.

Solution
The mean value of the overall eccentricity may be estimated as the rms value of the individual means, as given by Equation 5.23; thus,

$$\mu = \sqrt{0.1^2 + 0.2^2 + 0.05^2 + 0.1^2} = 0.25 \text{ mm}.$$

Using Equation 5.24, the standard deviation of the overall eccentricity is estimated as

$$\sigma = \sqrt{0.01^2 + 0.05^2 + 0.001^2 + 0.02^2} = 0.055 \text{ mm}.$$

Now, assuming a Gaussian distribution, an estimate for the overall eccentricity at a confidence level of 96% is given by

$$\hat{e} = 0.25 + 2 \times 0.055 = 0.36 \text{ mm}.$$

Once the overall eccentricity is estimated in the foregoing manner, the corresponding measurement error can be determined. Suppose that the true angle of rotation is θ and the corresponding measurement is θ_m. The eccentricity error is given by

$$\Delta\theta = \theta_\mathrm{m} - \theta. \tag{5.25}$$

Figure 5.10 presents the maximum error, which can be shown to exist when the line of eccentricity (*CG*) is symmetrically located within the angle of rotation. For this configuration, the sine rule for triangles gives

$$\frac{\sin(\Delta\theta/2)}{e} = \frac{\sin(\theta/2)}{r},$$

where *r* denotes the code track radius, which for most practical purposes can be taken as the disk radius. Hence, the eccentricity error is given by

$$\Delta\theta = 2\sin^{-1}\left(\frac{e}{r}\sin\frac{\theta}{2}\right). \tag{5.26}$$

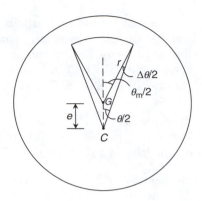

FIGURE 5.10
Nomenclature for eccentricity error. (C = center of rotation, G = geometric center of the code track.)

It is intuitively clear that the eccentricity error should not enter measurements of complete revolutions, and this can be verified by substituting $\theta = 2\pi$ in Equation 5.26. We have $\Delta\theta = 0$. For multiple revolutions, the eccentricity error is periodic with period 2π.

For small angles, the sine of an angle is approximately equal to the angle itself, in radians. Hence, for small $\Delta\theta$, the eccentricity error may be expressed as

$$\Delta\theta = \frac{2e}{r}\sin\frac{\theta}{2}. \tag{5.27}$$

Furthermore, for small angles of rotation, the fractional eccentricity error is given by

$$\frac{\Delta\theta}{\theta} = \frac{e}{r}, \tag{5.28}$$

which is, in fact, the worst-case fractional error. As the angle of rotation increases, the fractional error decreases (as shown in Figure 5.11), reaching the zero value for a full revolution. From the point of view of gross error, the worst value occurs when $\theta = \pi$, which corresponds to half a revolution. From Equation 5.26, it is clear that the maximum gross error due to eccentricity is given by

$$\Delta\theta_{max} = 2\sin^{-1}\frac{e}{r}. \tag{5.29}$$

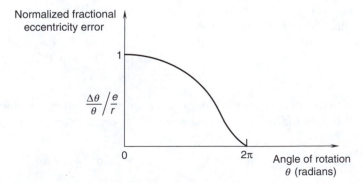

FIGURE 5.11
Fractional eccentricity error variation of an encoder with respect to the angle of rotation.

If this value is less than half the resolution of the encoder, the eccentricity error becomes inconsequential. For all practical purposes, since e is much less than r, we may use the following expression for the maximum eccentricity error:

$$\Delta\theta_{max} = \frac{2e}{r}. \tag{5.30}$$

Example 5.6
Suppose that in Example 5.5, the radius of the code disk is 5 cm. Estimate the maximum error due to eccentricity. If each track has 1000 windows, determine whether the eccentricity error is significant.

Solution
With the given level of confidence, we have calculated the overall eccentricity to be 0.36 mm. Now, from Equation 5.29 or Equation 5.30, the maximum angular error is given by

$$\Delta\theta_{max} = \frac{2 \times 0.36}{50} = 0.014 \text{ rad} = 0.83°.$$

Assuming that quadrature signals are used to improve the encoder resolution, we have

$$\text{Resolution} = \frac{360°}{4 \times 1000} = 0.09°.$$

Note that the maximum error due to eccentricity is more than 10 times the encoder resolution. Hence, eccentricity will significantly affect the accuracy of the encoder.
∎

Eccentricity of an incremental encoder also affects the phase angle between the quadrature signals if a single-track and two pick-off sensors (with circumferential offset) are used. This error can be reduced using the two-track arrangement, with the two sensors positioned along a radial line, so that eccentricity equally affects the two outputs.

5.6 Miscellaneous Digital Transducers

Now several other types of digital transducers which are useful in control engineering practice are described. In particular, digital rectilinear transducers are described. These are useful in many applications. Typical applications include x–y positioning tables, machine tools, valve actuators, read-write heads in hard disk drive (HDD) and other data storage systems, and robotic manipulators (e.g., at prismatic joints) and robot hands. The principles used in angular motion transducers described so far in Chapter 4 and Chapter 5 can be used in measuring rectilinear motions as well. Techniques of signal acquisition, interpretation, conditioning, and so on, may have similarities with the devices described so far, with those presented now.

5.6.1 Digital Resolvers

Digital resolvers, or mutual induction encoders, operate somewhat like analog resolvers, using the principle of mutual induction. They are commercially known as *Inductosyns*.

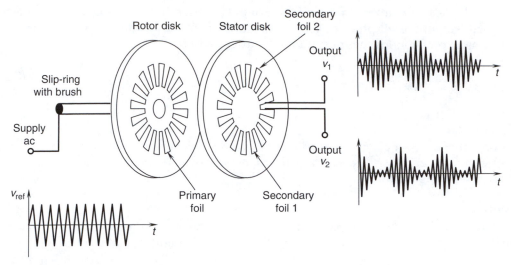

FIGURE 5.12
Schematic diagram of a digital resolver.

A digital resolver has two disks facing each other (but not in contact), one (the stator) stationary and the other (the rotor) coupled to the rotating object whose motion is measured. The rotor has a fine electric conductor foil imprinted on it, as schematically shown in Figure 5.12. The printed pattern is pulse shaped, closely spaced, and connected to a high-frequency AC supply (carrier) of voltage v_{ref}. The stator disk has two separate printed patterns that are identical to the rotor pattern, but one pattern on the stator is shifted by a quarter-pitch from the other pattern (Note: pitch = spacing between two successive crests of the foil). The primary voltage in the rotor circuit induces voltages in the two secondary (stator) foils at the same frequency; that is, the rotor and the stator are inductively coupled. These induced voltages are quadrature signals (i.e., 90° out of phase). As the rotor turns, the level of the induced voltage changes, depending on the relative position of the foil patterns on the two disks. When the foil pulse patterns coincide, the induced voltage is a maximum (positive or negative), and when the rotor foil pattern has a half-pitch offset from the stator foil pattern, the induced voltage in the adjacent segments cancel each other, producing a zero output. The output (induced) voltages v_1 and v_2 in the two foils of the stator have a carrier component at the supply frequency and a modulating component corresponding to the rotation of the disk. The latter (modulating component) can be extracted through demodulation (see Chapter 2 and Chapter 4) and converted into a proper pulse signal using pulse-shaping circuitry, as for an incremental encoder. When the rotating speed is constant, the two modulating components are periodic and nearly sinusoidal, with a phase shift of 90° (i.e., in quadrature). When the speed is not constant, the pulse width will vary with time.

As in the case of an incremental encoder, angular displacement is determined by counting the pulses. Furthermore, angular velocity is determined either by counting the pulses over a fixed time period (counter sampling period) or by timing the pulses. The direction of rotation is determined by the phase difference in the two modulating (output) signals. (In one direction, the phase shift is 90°; in the other direction, it is −90°.) Very fine resolutions (e.g., 0.0005°) may be obtained from a digital resolver, and it is usually not necessary to use step-up gearing or other techniques to improve the resolution. These transducers are usually more expensive than optical encoders. The use of a slip ring and brush to supply the carrier signal may be viewed as a disadvantage.

Consider the conventional resolver discussed in Chapter 4. Its outputs may be converted into digital form using appropriate hardware. Strictly speaking, such a device cannot be classified as a digital resolver.

5.6.2 Digital Tachometers

A pulse-generating transducer whose pulse train is synchronized with a mechanical motion may be treated as a digital transducer for motion measurement. In particular, pulse counting may be used for displacement measurement, and the pulse rate (or pulse timing) may be used for velocity measurement. As studied in Chapter 4, tachometers are devices for measuring angular velocities. According to this terminology, a shaft encoder (particularly, an incremental optical encoder) can be considered as a digital tachometer. According to popular terminology, however, a digital tachometer is a device that employs a toothed wheel to measure angular velocities.

A schematic diagram of a digital tachometer is shown in Figure 5.13. This is a magnetic induction, pulse tachometer of the variable-reluctance type. The teeth on the wheel are made of a ferromagnetic material. The two magnetic-induction (and variable-reluctance) proximity probes are placed radially facing the teeth, at quarter-pitch apart (pitch = tooth-to-tooth spacing). When the toothed wheel rotates, the two probes generate output signals that are 90° out of phase (i.e., quadrature signals). One signal leads the other in one direction of rotation and lags the other in the opposite direction of rotation. In this manner, a directional reading (i.e., velocity rather than speed) is obtained. The speed is computed either by counting the pulses over a sampling period or by timing the pulse width, as in the case of an incremental encoder.

Alternative types of digital tachometers use eddy current proximity probes or capacitive proximity probes (see Chapter 4). In the case of an eddy current tachometer, the teeth of the pulsing wheel are made of (or plated with) electricity-conducting material. The probe consists of an active coil connected to an AC bridge circuit excited by a radio-frequency (i.e., in the range 1–100 MHz) signal. The resulting magnetic field at radio frequency is modulated by the tooth-passing action. The bridge output may be demodulated and shaped to generate the pulse signal. In the case of a capacitive tachometer, the toothed wheel forms one plate of the capacitor; the other plate is the probe and is kept stationary. As the wheel turns, the gap width of the capacitor fluctuates. If the capacitor is excited by an AC voltage of high frequency (typically 1 MHz), a nearly pulse-modulated signal at that carrier frequency is obtained. This can be detected through a bridge circuit

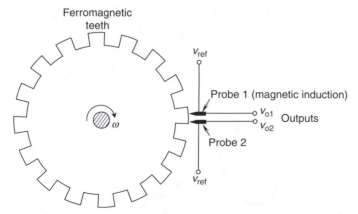

FIGURE 5.13
Schematic diagram of a pulse tachometer.

as before but using a capacitance bridge rather than an inductance bridge. In particular, by demodulating the output signal, the modulating signal can be extracted, which can be shaped to generate the pulse signal. The pulse signal generated in this manner is used in the angular velocity computation.

The advantages of digital (pulse) tachometers over optical encoders include simplicity, robustness, immunity to environmental effects and other common fouling mechanisms (except magnetic effects), and low cost. Both are noncontacting devices. The disadvantages of a pulse tachometer include poor resolution—determined by the number of teeth and size (bigger and heavier than optical encoders)—mechanical errors due to loading, hysteresis (i.e., output is not symmetric and depends on the direction of motion), and manufacturing irregularities. Note that mechanical loading will not be a factor if the toothed wheel already exists as an integral part of the original system that is sensed. The resolution (digital resolution) depends on the word size used for data acquisition.

5.6.3 Hall-Effect Sensors

Consider a semiconductor element subject to a DC voltage v_{ref}. If a magnetic field is applied perpendicular to the direction of this voltage, a voltage v_o will be generated in the third orthogonal direction within the semiconductor element. This is known as the Hall effect (observed by E.H. Hall in 1879). A schematic representation of a Hall-effect sensor is shown in Figure 5.14.

A Hall-effect sensor may be used for motion sensing in many ways; for example, as an analog proximity sensor, a limit switch (digital), or a shaft encoder. Because the output voltage v_o increases as the distance from the magnetic source to the semiconductor element decreases, the output signal v_o can be used as a measure of proximity. This is the principle behind an analog proximity sensor. Alternatively, a certain threshold level of the output voltage v_o can be used to generate a binary output, which represents the presence/absence of an object. This principle is used in a digital limit switch. The use of a toothed ferromagnetic wheel (as for a digital tachometer) to alter the magnetic flux will result in a shaft encoder. The sensitivity of a practical sensor element is of the order of 10 V/T (Note: $1T = 1$ tesla, where tesla is the unit of magnetic flux density, and is equal to 1 weber per square meter). For a Hall-effect device, the temperature coefficient of resistance is positive

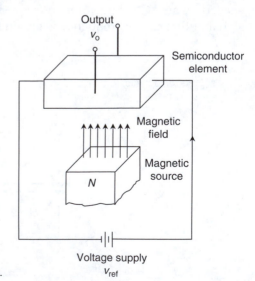

FIGURE 5.14
Schematic representation of a Hall-effect sensor.

and that of sensitivity is negative. In view of these properties, autocompensation for temperature may be achieved (as for a semiconductor strain gage—see Chapter 4).

The longitudinal arrangement of a proximity sensor, in which the moving element approaches head-on toward the sensor, is not suitable when there is a danger of overshooting the target, since it will damage the sensor. A more desirable configuration is the lateral arrangement, in which the moving member slides by the sensing face of the sensor. The sensitivity will be lower, however, with this lateral arrangement. The relationship between the output voltage v_o and the distance x of a Hall-effect sensor, measured from the moving member, is nonlinear. Linear Hall-effect sensors use calibration to linearize their output.

A practical arrangement for a motion sensor based on the Hall-effect would be to have the semiconductor element and the magnetic source fixed relative to one another in a single package. As a ferromagnetic member is moved into the air gap between the magnetic source and the semiconductor element, the flux linkage varies. The output voltage v_o changes accordingly. This arrangement is suitable for both an analog proximity sensor and a limit switch. By using a toothed ferromagnetic wheel as in Figure 5.15 to change v_o and then by shaping the resulting signal, it is possible to generate a pulse train in proportion to the wheel rotation. This provides a shaft encoder or a digital tachometer. Apart from the familiar applications of motion sensing, Hall-effect sensors are used for electronic commutation of brushless DC motors (see Chapter 7) where the field circuit of the motor is appropriately switched depending on the angular position of the rotor with respect to the stator.

Hall-effect motion transducers are rugged devices and have many advantages. They are not affected by rate effects (specifically, the generated voltage is not affected by the rate of change of the magnetic field). In addition, their performance is not severely affected by common environmental factors, except magnetic fields. They are noncontacting sensors with associated advantages as mentioned before. Some hysteresis will be present, but it is not a serious drawback in digital transducers. Miniature Hall-effect devices (mm scale) are available.

5.6.4 Linear Encoders

In rectilinear encoders (popularly called linear encoders, where "linear" does not imply linearity but refers to rectilinear motion), rectangular flat plates that move rectilinearly, instead of rotating disks, are used with the same types of signal generation and interpretation mechanisms as for shaft (rotary) encoders. A transparent plate with a series of opaque lines arranged in parallel in the transverse direction forms the stationary plate

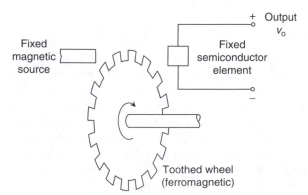

FIGURE 5.15
A Hall-effect shaft encoder or digital tachometer.

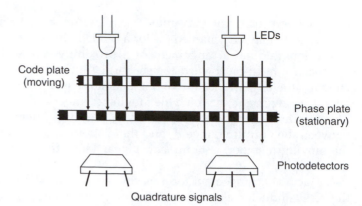

FIGURE 5.16
A rectilinear optical encoder.

(grating plate or phase plate) of the transducer. This is called the mask plate. A second transparent plate, with an identical set of ruled lines, forms the moving plate (or the code plate). The lines on both plates are evenly spaced, and the line width is equal to the spacing between adjacent lines. A light source is placed on the moving plate side, and light transmitted through the common area of the two plates is detected on the other side using one or more photosensors. When the lines on the two plates coincide, the maximum amount of light will pass through the common area of the two plates. When the lines on one plate fall on the transparent spaces of the other plate, virtually no light will pass through the plates. Accordingly, as one plate moves relative to the other, a pulse train is generated by the photosensor, and it can be used to determine rectilinear displacement and velocity, as in the case of an incremental encoder.

A suitable arrangement is shown in Figure 5.16. The code plate is attached to the moving object whose rectilinear motion is to be measured. An LED light source and a phototransistor light sensor are used to detect the motion pulses, which can be interpreted just like the way it is done for a rotary encoder. The phase plate is used, as with a shaft encoder, to enhance the intensity and the discrimination of the detected signal. Two tracks of windows in quadrature (i.e., quarter-pitch offset) would be needed to determine the direction of motion, as shown in Figure 5.16. Another track of windows at half-pitch offset with the main track (not shown in Figure 5.16) may be used as well on the phase plate, to further enhance the discrimination of the detected pulses. Specifically, when the sensor at the main track reads a high intensity (i.e., when the windows on the code plate and the phase plate are aligned) the sensor at the track that is half pitch away will read a low intensity (because the corresponding windows of the phase plate are blocked by the solid regions of the code plate).

5.6.5   Moiré Fringe Displacement Sensors

Suppose that a piece of transparent fabric is placed on another. If one piece is moved or deformed with respect to the other, we will notice various designs of light and dark patterns (lines) in motion. Dark lines of this type are called moiré fringes. In fact, the French term moiré refers to a silk-like fabric, which produces moiré fringe patterns. An example of a moiré fringe pattern is shown in Figure 5.17. Consider the rectilinear encoder, which was described earlier. When the window slits of one plate overlap with the window slits of the other plate, we get an alternating light and dark pattern. This is a special case of moiré fringes. A moiré device of this type may be used to measure rigid-body movements of one plate of the sensor with respect to the other.

FIGURE 5.17
A moiré fringe pattern.

Application of the moiré fringe technique is not limited to sensing rectilinear motions. This technology can be used to sense angular motions (rotations) and more generally, distributed deformations (e.g., elastic deformations) of one plate with respect to the other. Consider two plates with gratings (optical lines) of identical pitch (spacing) p. Suppose that initially the gratings of the two plates exactly coincide. Now, if one plate is deformed in the direction of the grating lines, the transmission of light through the two plates will not be altered. However, if a plate is deformed in the perpendicular direction to the grating lines, then the window width of that plate will be deformed accordingly. In this case, depending on the nature of the plate deformation, some transparent lines of one plate will be completely covered by the opaque lines of the other plate, and some other transparent lines of the first plate will have coinciding transparent lines on the second plate. Thus, the observed image will have dark lines (moiré fringes) corresponding to the regions with clear/opaque overlaps of the two plates and bright lines corresponding to the regions with clear/clear overlaps of the two plates. The resulting moiré fringe pattern will provide the deformation pattern of one plate with respect to the other. Such two-dimensional fringe patterns can be detected and observed by arrays of optical sensors using charge-coupled-device (CCD) elements and by photographic means. In particular, since the presence of a fringe is a binary information, binary optical sensing techniques (as for optical encoders) and digital imaging techniques may be used with these transducers. Accordingly, these devices may be classified as digital transducers. With the moiré fringe technique, very small resolutions (e.g., 0.002 mm) can be realized because finer line spacing (in conjunction with wider light sensors) can be used.

To further understand and analyze the fundamentals of moiré fringe technology, consider two grating plates with identical line pitch (spacing between the windows) p. Suppose that one plate is kept stationary. This is the plate of master gratings (or reference gratings or main gratings). The other plate, which is the plate containing index gratings or model gratings, is placed over the fixed plate and rotated so that the index gratings form an angle α with the master gratings, as shown in Figure 5.18. The lines shown are in fact the opaque regions, which are identical in size and spacing to the windows in between the opaque regions. A uniform light source is placed on one side of the overlapping pair of plates and the light transmitted through them is observed on the other side. Dark bands called moiré fringes are seen as a result, as in Figure 5.18.

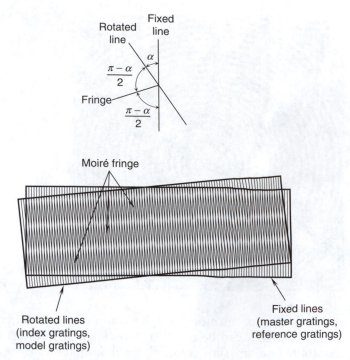

FIGURE 5.18
Formation of moiré fringes.

Rotated lines
(index gratings,
model gratings)

Fixed lines
(master gratings,
reference gratings)

A moiré fringe corresponds to the line joining a series of points of intersection of the opaque lines of the two plates because no light can pass through such points. This is further shown in Figure 5.19. Note that in the present arrangement, the line pitch of the two plates is identical and equal to p. A fringe line formed is shown as the broken line in Figure 5.19. Since the line pattern in the two plates is identical, by symmetry of the arrangement, the fringe line should bisect the obtuse angle $(\pi - \alpha)$ formed by the intersecting opaque lines. In other words, a fringe line makes an angle of $(\pi - \alpha)/2$ with the fixed gratings. Furthermore, the vertical separation (or the separation in the direction of the fixed gratings) of the moiré fringes is seen to be $p/\tan \alpha$.

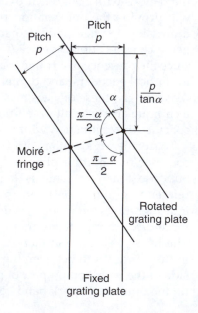

FIGURE 5.19
The orientation of moiré fringes.

In summary then, the rotation of the index plate with respect to the reference plate can be measured by sensing the orientation of the fringe lines with respect to the fixed (master or reference) gratings. Furthermore, the period of the fringe lines in the direction of the reference gratings is $p/\tan \alpha$, and when the index plate is moved rectilinearly by a distance of one grating pitch, the fringes also shift vertically by its period of $p/\tan \alpha$ (see Figure 5.19). It is clear then that the rectilinear displacement of the index plate can be measured by sensing the fringe spacing. In a two-dimensional pattern of moiré fringes, these facts can be used as local information to sense full-field motions and deformations.

Example 5.7

Suppose that each plate of a moiré fringe deformation sensor has a line pitch of 0.01 mm. A tensile load is applied to one plate in the direction perpendicular to the lines. Five moiré fringes are observed in 10 cm of the moiré image under tension. What is the tensile strain in the plate?

Solution

There is one moiré fringe in every $10/5 = 2$ cm of the plate. Hence, extension of a 2 cm portion of the plate $= 0.01$ mm, and

$$\text{tensile strain} = \frac{0.01\,\text{mm}}{2 \times 10\,\text{mm}} = 0.0005 \ \varepsilon = 500\,\mu\varepsilon$$

In this example, we have assumed that the strain distribution (or deformation) of the plate is uniform. Under nonuniform strain distributions the observed moiré fringe pattern generally will not be parallel straight lines but rather complex shapes.

5.6.6 Cable Extension Sensors

In many applications, rectilinear motion is produced from a rotary motion (say, of a motor) through a suitable transmission device, such as rack and pinion or lead screw and nut (see Chapter 8). In these cases, rectilinear motion can be determined by measuring the associated rotary motion, assuming that errors due to backlash, flexibility, and so forth, in the transmission device can be neglected. Another way to measure rectilinear motions using a rotary sensor is to use a modified sensor that has the capability to convert a rectilinear motion into a rotary motion within the sensor itself. An example would be the cable extension method of sensing rectilinear motions. This method is particularly suitable for measuring motions that have large excursions. The cable extension method uses an angular motion sensor with a spool rigidly coupled to the rotating part of the sensor (e.g., the encoder disk) and a cable that wraps around the spool. The other end of the cable is attached to the object whose rectilinear motion is to be sensed. The housing of the rotary sensor is firmly mounted on a stationary platform, so that the cable can extend in the direction of motion. When the object moves, the cable extends, causing the spool to rotate. This angular motion is measured by the rotary sensor. With proper calibration, this device can give rectilinear measurements directly. As the object moves toward the sensor, the cable has to retract without slack. This is guaranteed by using a device such as a spring motor to wind the cable back. The disadvantages of the cable extension method include mechanical loading of the moving object, time delay in measurements, and errors caused by the cable, including irregularities, slack, and tensile deformation.

5.6.7 Binary Transducers

Digital binary transducers are two-state sensors. The information provided by such a device takes only two states (on/off, present/absent, go/no-go, etc.); it can be represented by one bit. For example, limit switches are sensors used in detecting whether an object has reached its limit (or destination) of mechanical motion and are useful in sensing presence or absence and in object counting. In this sense, a limit switch is considered a digital transducer. Additional logic is needed if the direction of contact is also required. Limit switches are available for both rectilinear and angular motions. The limit of a movement can be detected by mechanical means using a simple contact mechanism to close a circuit or trigger a pulse. Although a purely mechanical device consisting of linkages, gears, ratchet wheels and pawls, and so forth can serve as a limit switch, electronic and solid-state switches are usually preferred for such reasons as accuracy, durability, a low activating force (practically zero) requirement, low cost, and small size. Any proximity sensor could serve as the sensing element of a limit switch, to detect the presence of an object. The proximity sensor signal is then used in a desired manner; for example, to activate a counter, a mechanical switch, or a relay circuit or simply as an input to a digital controller. A microswitch is a solid-state switch that can be used as a limit switch. Microswitches are commonly used in counting operations; for example, to keep a count of completed products in a factory warehouse.

There are many types of binary transducers that are applicable in detection and counting of objects. They include

1. Electromechanical switches
2. Photoelectric devices
3. Magnetic (Hall-effect, eddy current) devices
4. Capacitive devices
5. Ultrasonic devices

An electromechanical switch is a mechanically activated electric switch. The contact with an arriving object turns on the switch, thereby completing a circuit and providing an electrical signal. This signal provides the present state of the object. When the object is removed, the contact is lost and the switch is turned off. This corresponds to the absent state.

In the other four types of binary transducers listed here, a signal (light beam, magnetic field, electric field, or ultrasonic wave) is generated by a source (emitter) and is received by a receiver. A passing object interrupts the signal. This event can be detected by usual means, using the signal received at the receiver. In particular, the signal level, a rising edge, or a falling edge may be used to detect the event. The following three arrangements of the emitter–receiver pair are common:

1. Through (opposed) configuration
2. Reflective (reflex) configuration
3. Diffuse (proximity, interceptive) configuration

In the through configuration (Figure 5.20a), the receiver is placed directly facing the emitter. In the reflective configuration, the emitter–source pair is located in a single package. The emitted signal is reflected by a reflector that is placed facing the emitter–receiver package (Figure 5.20b). In the diffuse configuration as well, the emitter–reflector pair is in a single package. In this case, a conventional proximity sensor can serve the

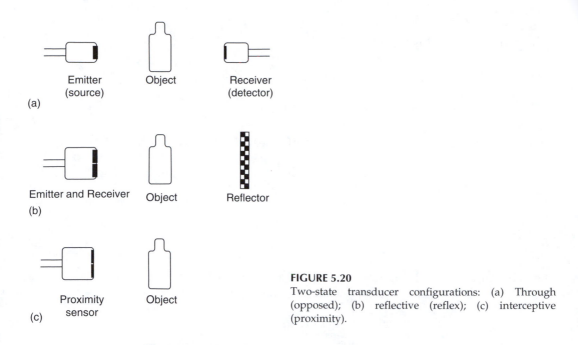

FIGURE 5.20
Two-state transducer configurations: (a) Through (opposed); (b) reflective (reflex); (c) interceptive (proximity).

purpose of detecting the presence of an object (Figure 5.20c) by using the signal diffused from the intercepting object. When the photoelectric method is used, an LED may serve as the emitter and a phototransistor may serve as the receiver. Infrared LEDs are preferred emitters for phototransistors because their peak spectral responses match.

Many factors govern the performance of a digital transducer for object detection. They include

1. Sensing range (operating distance between the sensor and the object)
2. Response time
3. Sensitivity
4. Linearity
5. Size and shape of the object
6. Material of the object (e.g., color, reflectance, permeability, and permittivity)
7. Orientation and alignment (optical axis, reflector, and object)
8. Ambient conditions (light, dust, moisture, magnetic field, etc.)
9. Signal conditioning considerations (modulation, demodulation, shaping, etc.)
10. Reliability, robustness, and design life

Example 5.8
The response time of a binary transducer for object counting is the fastest (shortest) time that the transducer needs to detect an absent-to-present condition or a present-to-absent condition and generate the counting signal (say, a pulse). Consider the counting process of packages on a conveyor. Suppose that, typically, packages of length 20 cm are placed on the conveyor at 15 cm spacing. A transducer of response time 10 ms is used for counting the packages. Estimate the allowable maximum operating speed of the conveyor.

Solution

If the conveyor speed is v cm/ms, then

$$\text{Package-present time} = \frac{20.0}{v} \text{ ms,}$$

$$\text{Package-absent time} = \frac{15.0}{v} \text{ ms.}$$

We must have a transducer response time of at least $15.0/v$ ms. Hence,

$$10.0 \le \frac{15.0}{v}$$

or

$$v \le 1.5 \, \text{cm/ms.}$$

The maximum allowable operating speed is 1.5 cm/ms or 15.0 m/s. This corresponds to a counting rate of $1.5/(20.0 + 15.0)$ packages/ms or nearly 43 packages/s.

Problems

5.1 Identify active transducers among the following types of shaft encoders and justify your claims. Also, discuss the relative merits and drawbacks of the four types of encoders.

   a. Optical encoders

   b. Sliding contact encoders

   c. Magnetic encoders

   d. Proximity sensor encoders

5.2 Consider the two quadrature pulse signals (say, A and B) from an incremental encoder. Using sketches of these signals, show that in one direction of rotation, signal B is at a high level during the up-transition of signal A, and in the opposite direction of rotation, signal B is at a low level during the up-transition of signal A. Note that the direction of motion can be determined in this manner, by using level detection of one signal during the up-transition of the other signal.

5.3 Explain why the speed resolution of a shaft encoder depends on the speed itself. What are some of the other factors that affect speed resolution? The speed of a DC motor was increased from 50 to 500 rpm. How would the speed resolution change if the speed were measured using an incremental encoder,

   a. By the pulse-counting method?

   b. By the pulse-timing method?

5.4 Describe methods of improving the displacement resolution and the velocity resolution in an encoder. An incremental encoder disk has 5000 windows. The word size of the output data is 12 bits. What is the angular (displacement) resolution of the device? Assume that quadrature signals are available but that no interpolation is used.

5.5 An incremental optical encoder that has N windows per track is connected to a shaft through a gear system with gear ratio p. Derive formulas for calculating angular velocity of the shaft by the

a. Pulse-counting method

b. Pulse-timing method

What is the speed resolution in each case? What effect does step-up gearing have on the speed resolution?

5.6 What is hysteresis in an optical encoder? List several causes of hysteresis and discuss ways to minimize hysteresis.

5.7 An optical encoder has n windows per centimeter diameter (in each track). What is the eccentricity tolerance e below which readings are not affected by eccentricity error?

5.8 Show that in the single-track, two-sensor design of an incremental encoder, the phase angle error (in quadrature signals) due to eccentricity is inversely proportional to the second power of the radius of the code disk for a given window density. Suggest a way to reduce this error.

5.9 Consider an encoder with 1000 windows in its track and capable of providing quadrature signals. What is the displacement resolution $\Delta\theta_r$ in radians? Obtain a value for the nondimensional eccentricity e/r below which the eccentricity error has no effect on the sensor reading. For this limiting value, what is $\Delta\theta_r/(e/r)$? Typically, the values for this parameter, as given by encoder manufacturers, range from 3 to 6. Note: $e =$ track eccentricity, $r =$ track radius.

5.10 What is the main advantage of using a gray code instead of straight binary code in an encoder? Give a table corresponding to a gray code different from that given in Table 5.2 for a 4-bit absolute encoder. What is the code pattern on the encoder disk in this case?

5.11 Discuss construction features and operation of an optical encoder for measuring rectilinear displacements and velocities.

5.12 A particular type of multiplexer can handle 96 sensors. Each sensor generates a pulse signal with variable pulse width. The multiplexer scans the incoming pulse sequences, one at a time, and passes the information onto a control computer.

a. What is the main objective of using a multiplexer?

b. What type of sensors could be used with this multiplexer?

5.13 A centrifuge is a device that is used to separate components in a mixture. In an industrial centrifugation process, the mixture to be separated is placed in the centrifuge and rotated at high speed. The centrifugal force on a particle depends on the mass, radial location, and the angular speed of the particle. This force is responsible for separating the particles in the mixture.

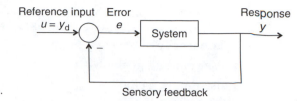

FIGURE P5.14
A feedback control loop.

Angular motion and the temperature of the container are the two key variables that have to be controlled in a centrifuge. In particular, a specific centrifugation curve is used, which consists of an acceleration segment, a constant-speed segment, and a braking (deceleration) segment, and this corresponds to a trapezoidal speed profile. An optical encoder may be used as the sensor for microprocessor-based speed control in the centrifuge. Discuss whether an absolute encoder is preferred for this purpose. Give the advantages and possible drawbacks of using an optical encoder in this application.

5.14 Suppose that a feedback control system (Figure P5.14) is expected to provide an accuracy within $\pm \Delta y$ for a response variable y. Explain why the sensor that measures y should have a resolution of $\pm(\Delta y/2)$ or better for this accuracy to be possible. An x–y table has a travel of 2 m. The feedback control system is expected to provide an accuracy of ± 1 mm. An optical encoder is used to measure the position for feedback in each direction (x and y). What is the minimum bit size that is required for each encoder output buffer? If the motion sensor used is an absolute encoder, how many tracks and how many sectors should be present on the encoder disk?

5.15 Encoders that can provide 50,000 counts/turn with ± 1 count accuracy are commercially available. What is the resolution of such an encoder? Describe the physical construction of an encoder that has this resolution.

5.16 The pulses generated by the coding disk of an incremental optical encoder are approximately triangular (actually, upward shifted sinusoidal) in shape. Explain the reason for this. Describe a method for converting these triangular (or shifted sinusoidal) pulses into sharp rectangular pulses.

5.17 Explain how the resolution of a shaft encoder could be improved by pulse interpolation. Specifically, consider the arrangement shown in Figure P5.17. When the masking windows are completely covered by the opaque regions of the moving disk, no light is received by the photosensor. The peak level of light is received when the windows of the moving disk coincide with the windows of the masking disk. The variation in the light intensity from the minimum level to the peak level is approximately linear (generating a triangular pulse), but more accurately sinusoidal and may be given by

$$v = v_0 \left(1 - \cos \frac{2\pi\theta}{\Delta\theta}\right),$$

where θ denotes the angular position of the encoder window with respect to the masking window, as shown, and $\Delta\theta$ is the window pitch angle. Note that, in the sense of rectangular pulses, the pulse corresponds to the motion in the interval

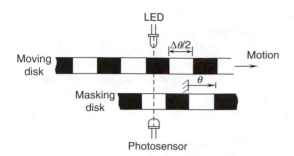

FIGURE P5.17

An encoder with a masking disk.

$\Delta\theta/4 \leq \theta \leq 3\Delta\theta/4$. By using this sinusoidal approximation for a pulse, as given previously, show that one can improve the resolution of an encoder indefinitely simply by measuring the shape of each pulse at clock cycle intervals using a high-frequency clock signal.

5.18 A Schmitt trigger is a semiconductor device that can function as a level detector or a switching element, with hysteresis. The presence of hysteresis can be used, for example, to eliminate chattering during switching caused by noise in the switching signal. In an optical encoder, a noisy signal detected by the photosensor may be converted into a clean signal of rectangular pulses by this means. The input/output characteristic of a Schmitt trigger is shown in Figure P5.18a. If the input signal is as shown in Figure P5.18b, determine the output signal.

5.19 Displacement sensing and speed sensing are essential in a position servo. If a digital controller is employed to generate the servo signal, one option would be to use an analog displacement sensor and an analog speed sensor, along with ADCs to

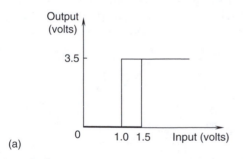

(a)

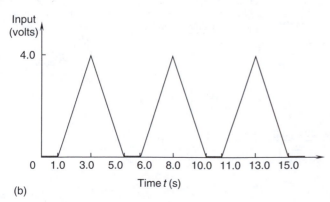

(b)

FIGURE P5.18

(a) The input/output characteristic of a Schmitt trigger; (b) a triangular input signal.

produce the necessary digital feedback signals. Alternatively, an incremental encoder may be used to provide both displacement and speed feedbacks. In this case,
ADCs are not needed. Encoder pulses will provide interrupts to the digital controller. Displacement is obtained by counting the interrupts. The speed is obtained
by timing the interrupts. In some applications, analog speed signals are needed.
Explain how an incremental encoder and a frequency-to-voltage converter (FVC)
may be used to generate an analog speed signal.

5.20 Compare and contrast an optical incremental encoder against a potentiometer, by
giving advantages and disadvantages, for an application involving the sensing of a
rotatory motion.

A schematic diagram for the servo control loop of one joint of a robotic manipulator is given in Figure P5.20.

The motion command for each joint of the robot is generated by the robot
controller, in accordance with the required trajectory. An optical incremental encoder is used for both position and velocity feedback in each servo loop. Note that
for a six-degrees-of-freedom robot there will be six such servo loops. Describe the
function of each hardware component shown in the figure and explain the operation of the servo loop.

After several months of operation, the motor of one joint of the robot was found
to be faulty. An enthusiastic engineer quickly replaced the motor with an identical
one without realizing that the encoder of the new motor was different. In particular,
the original encoder generated 200 pulses/rev, whereas the new encoder generated
720 pulses/rev. When the robot was operated the engineer noticed an erratic and
unstable behavior at the repaired joint. Discuss reasons for this malfunction and
suggest a way to correct the situation.

5.21 a. A position sensor is used in a microprocessor-based feedback control system for
accurately moving the cutter blades of an automated meat-cutting machine. The
machine is an integral part of the production line of a meat processing plant.
What are the primary considerations in selecting the position sensor for this

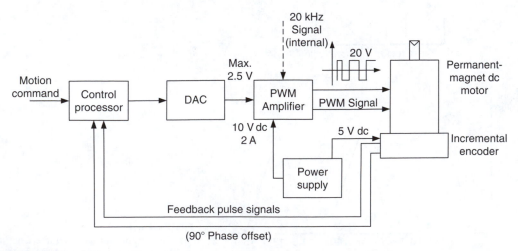

FIGURE P5.20
A servo loop of a robot.

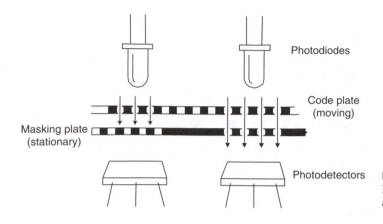

Photodiodes

Code plate
(moving)

Masking plate
(stationary)

Photodetectors

FIGURE P5.21
Photodiode-detector arrangement of
a linear optical encoder.

application? Discuss advantages and disadvantages of using an optical encoder
in comparison to an linear variable transformer (LVDT) (see Chapter 4) in this
context.

b. Figure P5.21 illustrates one arrangement of the optical components in a linear
incremental encoder.

The moving code plate has uniformly spaced windows as usual, and the fixed
masking plate has two groups of identical windows, one above each of the two
photodetectors. These two groups of fixed windows are positioned in half-pitch out
of phase so that when one detector receives light from its source directly through
the aligned windows of the two plates, the other detector has the light from its
source virtually obstructed by the masking plate.

Explain the purpose of the two sets of photodiode-detector units, giving a schematic diagram of the necessary electronics. Can the direction of motion be determined with the arrangement shown in Figure P5.21? If so, explain how this could be
done. If not, describe a suitable arrangement for detecting the direction of motion.

5.22   a. What features and advantages of a digital transducer will distinguish it from a
purely analog sensor?

b. Consider a linear incremental encoder that is used to measure rectilinear positions and speeds. The moving element is a nonmagnetic plate containing a series
of identically magnetized areas uniformly distributed along its length. The pick-
off transponder is a mutual-induction-type proximity sensor (i.e., a transformer)
consisting of a toroidal core with a primary winding and a secondary winding.
A schematic diagram of the encoder is shown in Figure P5.22. The primary
excitation v_{ref} is a high-frequency sine wave.

Explain the operation of this position encoder, clearly indicating what types of
signal conditioning would be needed to obtain a pure pulse signal. Also, sketch the
output v_o of the proximity sensor as the code plate moves very slowly. Which
position of the code plate does a high value of the pulse signal represent and which
position does a low value represent?

Suppose that the pulse period timing method is used to measure speed (v) using
this encoder. The pitch distance of the magnetic spots on the plate is p, as shown in
Figure P5.22. If the clock frequency of the pulse period timer is f, give an expression
for the speed v in terms of the clock cycle count m.

Show that the speed resolution Δv for this method may be approximated by

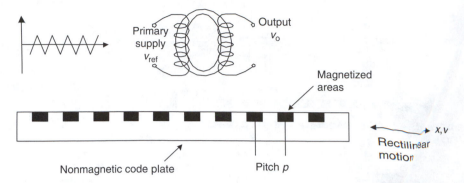

FIGURE P5.22
A linear incremental encoder of the magnetic induction type.

$$\Delta v = \frac{v^2}{pf}.$$

It follows that the dynamic range $v/\Delta v = pf/v$.

If the clock frequency is 20 MHz, the code pitch is 0.1 mm, and the required dynamic range is 100 (i.e., 40 dB), what is the maximum speed in m/s that can be measured by this method?

5.23 What is a Hall-effect tachometer? Discuss the advantages and disadvantages of a Hall-effect motion sensor in comparison with an optical motion sensor (e.g., an optical encoder).

5.24 Discuss the advantages of solid-state limit switches over mechanical limit switches. Solid-state limit switches are used in many applications, particularly in the aircraft and aerospace industries. One such application is in landing gear control, to detect up, down, and locked conditions of the landing gear. High reliability is of utmost importance in such applications. Mean time between failure (MTBF) of over 100,000 h is possible with solid-state limit switches. Using your engineering judgment, give an MTBF value for a mechanical limit switch.

5.25 Mechanical force switches are used in applications where only a force limit, rather than a continuous force signal, has to be detected. Examples include detecting closure force (torque) in valve closing, detecting fit in parts assembly, automated clamping devices, robotic grippers and hands, overload protection devices in process/machine monitoring, and product filling in containers by weight. Expensive and sophisticated force sensors are not needed in such applications because a continuous history of a force signal is not needed. Moreover, they are robust and reliable, and can safely operate in hazardous environments. Using a sketch, describe the construction of a simple spring-loaded force switch.

5.26 Consider the following three types of photoelectric object counters (or object detectors or limit switches):

1. Through (opposed) type

2. Reflective (reflex) type

3. Diffuse (proximity, interceptive) type

Classify these devices into long-range (up to several meters), intermediate range (up to 1 m), and short-range (up to fraction of a meter) detection.

5.27 A brand of autofocusing camera uses a microprocessor-based feedback control system consisting of a charge-coupled device (CCD) imaging system, a microprocessor, a drive motor, and an optical encoder. The purpose of the control system is to focus the camera automatically based on the image of the subject as sensed by a matrix of CCDs (a set of metal oxide semiconductor field-effect transistors or MOSFETs). The light rays from the subject that pass through the lens will fall onto the CCD matrix. This will generate a matrix (image frame) of charge signals, which are shifted one at a time, row by row, into an output buffer (or frame grabber) and passed on to the microprocessor after conditioning the resulting video signal. The CCD image obtained by sampling the video signal is analyzed by the microprocessor to determine whether the camera is focused properly. If not, the lens is moved by the motor so as to achieve focusing. Draw a schematic diagram for the autofocusing control system and explain the function of each component in the control system, including the encoder.

5.28 Measuring devices with frequency outputs may be considered as digital transducers. Justify this statement.

6

Stepper Motors

The actuator is the device that mechanically drives a dynamic system. Proper selection of actuators and their drive systems for a particular application is of utmost importance in the instrumentation and design of control systems. There is another ("mechatronic") perspective to the significance of actuators in the field of control systems. A typical actuator contains mechanical components like rotors, shafts, cylinders, coils, bearings, and seals, while the control and drive systems are primarily electronic in nature. Integrated design, manufacture, and operation of these two categories of components are crucial to efficient operation of an actuator. This is essentially a mechatronic problem.

Stepper motors are a popular type of actuators. Unlike continuous-drive actuators (see Chapter 7), stepper motors are driven in fixed angular steps (increments). Each step of the rotation is the response of the motor rotor to an input pulse (or a digital command). In this manner, the stepwise rotation of the rotor can be synchronized with pulses in a command pulse train, assuming of course that no steps are missed, thereby making the motor respond faithfully to the input signal (pulse sequence) in an open-loop manner. From this perspective, it is reasonable to treat stepper motors as digital actuators. Nevertheless, like a conventional continuous-drive motor, a stepper motor is also an electromagnetic actuator, in that it converts electromagnetic energy into mechanical energy to perform mechanical work. This chapter studies stepper motors, which are incremental-drive actuators. Chapter 7 discusses continuous-drive actuators.

6.1 Principle of Operation

The terms stepper motor, stepping motor, and step motor are synonymous and are often used interchangeably. Actuators that can be classified as stepper motors have been in use for more than 60 years, but only after the incorporation of solid-state circuitry and logic devices in their drive systems have stepper motors emerged as cost-effective alternatives for dc servomotors in high-speed motion-control applications. Many kinds of actuators fall into the stepper motor category, but only those that are widely used in the industry are discussed in this chapter. Note that even if the mechanism by which the incremental motion is generated differs from one type of stepper motor to the other, the same control techniques can be used in the associated control systems, making a general treatment of stepper motors possible, at least from the control point of view.

There are three basic types of stepper motors:

1. Variable-reluctance (VR) stepper motors, which have soft-iron rotors.

2. Permanent-magnet (PM) stepper motors, which have magnetized rotors.

3. Hybrid (HB) stepper motors, which have two stacks of rotor teeth forming the two poles of a permanent magnet located along the rotor axis.

FIGURE 6.1
A commercial two-stack stepper motor. (From Danaher Motion. With permission.)

The VR stepper motors and PM stepper motors operate in a more or less similar manner. Hybrid stepper motors possess characteristics of both VR steppers and PM steppers. A disadvantage of VR stepper motors is that since the rotor is not magnetized, the holding torque is practically zero when the stator windings are not energized (power off). Hence, there is no capability to hold a mechanical load at a given position under power-off conditions, unless mechanical brakes are employed. A photograph of the internal components of a two-stack stepping motor is given in Figure 6.1.

6.1.1 Permanent-Magnet (PM) Stepper Motor

To explain the operation of a permanent-magnet (PM) stepper motor, consider the simple schematic diagram shown in Figure 6.2. The stator has two sets of windings (i.e., two phases) placed at 90°. This arrangement has four salient poles in the stator, each pole geometrically separated by a 90° angle from the adjacent one. The rotor is a two-pole permanent magnet. Each phase can take one of the three states 1, 0, and −1, which are defined as follows:

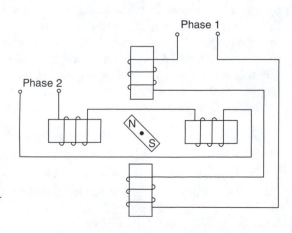

FIGURE 6.2
Schematic diagram of a two-phase permanent-magnet stepper motor.

1. State 1: current in a specified direction
2. State −1: current in the opposite direction
3. State 0: no current.

Note: As −1 is the complement state of 1, in some literature the notation 1′ is used to denote the state −1.

By switching the currents in the two phases in an appropriate sequence, either a clockwise (CW) rotation or a counterclockwise (CCW) rotation can be generated. The CW rotation sequence is shown in Figure 6.3. Note that ϕ_i denotes the state of the *i*th phase. The step angle for this motor is 45°. At the end of each step, the rotor assumes the minimum reluctance position that corresponds to the particular magnetic polarity pattern in the stator. Reluctance measures the magnetic resistance in a flux path. This is a stable

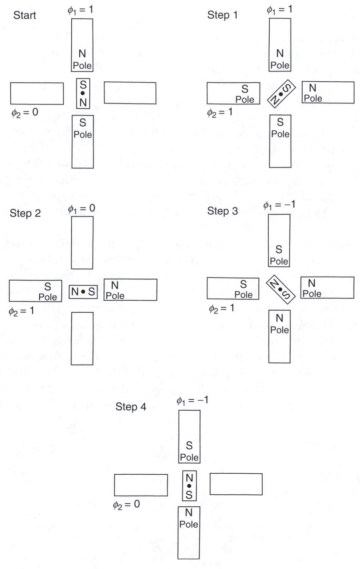

FIGURE 6.3
Stepping sequence (half stepping) in a two-phase PM stepper motor for clockwise rotation.

TABLE 6.1

Stepping Sequence (Half Stepping) for a Two-Phase PM Stepper Motor
with Two Rotor Poles

	Clockwise Rotation		
Step Number	ϕ_1	ϕ_2	
1	1	1	↑ CCW rotation
2	0	1	
3	−1	1	
4	−1	0	
5	−1	−1	
6	0	−1	
7	1	−1	
8	1	0	

equilibrium configuration and is known as the detent position for that step. When the
stator currents (phases) are switched for the next step, the minimum reluctance position
changes (rotates by the step angle) and the rotor assumes the corresponding stable
equilibrium position and the rotor turns through a single step (45° in this example).
Table 6.1 gives the stepping sequences necessary for a complete clockwise rotation.
Note that a separate pair of columns is not actually necessary to give the states for the
CCW rotation; they are simply given by the CW rotation states themselves, but tracked in
the opposite direction (bottom to top).

The switching sequence given in Table 6.1 corresponds to half stepping, which has a step
angle of 45°. Full stepping for the stator–rotor arrangement shown in Figure 6.2 corresponds to a step angle of 90°. In this case, only one phase is energized at a time. For half
stepping, both phases have to be energized simultaneously in alternate steps, as is clear
from Table 6.1.

Typically, the phase activation (switching) sequence is triggered by the pulses of an input
pulse sequence. The switching logic (which determines the states of the phases for a given
step) may be digitally generated using appropriate logic circuitry or by a simple table
lookup procedure with just eight pairs of entries given in Table 6.1. The clockwise stepping
sequence is generated by reading the table in the top-to-bottom direction, and the CCW
stepping sequence is generated by reading the same table in the opposite direction. A still
more compact representation of switching cycles is also available. Note that in one complete rotation of the rotor, the state of each phase sweeps through one complete cycle of
the switching sequence (shown in Figure 6.4a) in the clockwise direction. For clockwise
rotation of the motor, the state of phase 2 (ϕ_2) *lags* the state of phase 1 (ϕ_1) by two steps
(Figure 6.4b). For CCW rotation, ϕ_2 leads ϕ_1 by two steps (Figure 6.4c). Hence, instead of

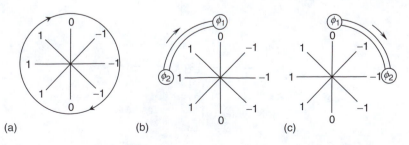

(a)   (b)   (c)

FIGURE 6.4
(a) Half-step switching states. (b) Switching logic for clockwise rotation. (c) Switching logic for CCW rotation.

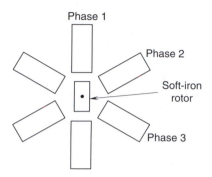

FIGURE 6.5
Schematic diagram of a three-phase variable-reluctance stepper motor.

eight pairs of numbers, just eight numbers with a delay operation would suffice to generate the phase-switching logic. Although the commands that generate the switching sequence for a phase winding could be supplied by a digital computer such as a personal computer (a software approach), it is customary to generate it through hardware logic in a device called a translator or an indexer. This approach is more effective because the switching logic for a stepper motor is fixed, as noted in the foregoing discussion. We shall say more about the translator in a later section, when dealing with motor drive electronics.

6.1.2 Variable-Reluctance (VR) Stepper Motor

Now consider the variable-reluctance (VR) stepper motor shown schematically in Figure 6.5. The rotor is a nonmagnetized soft-iron bar. If only two phases are used in the stator, there will be ambiguity regarding the direction of rotation (in full stepping). At least three phases would be needed for this two-pole rotor geometry (in full stepping, as will be clear later), as shown in Figure 6.5. The full-stepping sequence for clockwise rotation is shown in Figure 6.6. The step angle is 60°. Only one phase is energized at a time in order to execute full stepping. With VR stepping motors, the direction of the current (the polarity of a stator pole pair) is not reversed in the full-stepping sequence; only the states 1 and 0 (on and off) are used for each phase. In the case of half stepping, however, two phases have to be energized simultaneously during some steps. Furthermore, current reversals are needed in half stepping, thus requiring more elaborate switching circuitry. The advantage, however, is that the step angle would be halved to 30°, thereby providing improved motion resolution. When two phases are activated simultaneously, the minimum reluctance position is halfway between the corresponding pole pairs (i.e., 30° from the detent position that is obtained when only one of the two phases is energized), which enables half stepping. It follows that, depending on the energizing sequence of the phases, either full stepping or half stepping would be possible. As will be discussed later, microstepping provides much smaller step angles. This is achieved by changing the phase currents by small increments (rather than just on, off, and reversal) so that the detent (equilibrium) position of the rotor shifts in correspondingly small angular increments.

6.1.3 Polarity Reversal

One common feature in any stepper motor is that the stator of the motor contains several pairs of field windings that can be switched on to produce electromagnetic pole pairs (N and S). These pole pairs effectively pull the motor rotor in sequence so as to generate the torque for motor rotation. The polarities of a stator pole may have to be reversed in

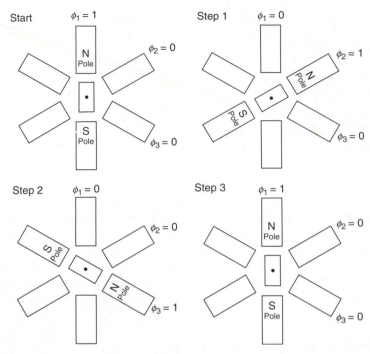

FIGURE 6.6
Full-stepping sequence for the three-phase VR stepper motor (step angle $= 60°$).

some types of stepper motors in order to carry out a stepping sequence. The polarity of a stator pole can be reversed in two ways:

1. There is only one set of windings for a group of stator poles (i.e., the case of unifilar windings). Polarity of the poles is reversed by reversing the direction of current in the winding.

2. There are two sets of windings for a group of stator poles (i.e., the case of bifilar windings), only one of which is energized at a time. One set of windings, when energized, produces one polarity for this group of poles, and the other set of windings produces the opposite polarity.

It should be clear that the drive circuitry for unifilar (i.e., single-file or single-coil) windings is somewhat complex because current reversal (i.e., bipolar) circuitry is needed. Specific-ally, a bipolar drive system is needed for a motor with unifilar windings in order to reverse the polarities of the poles (when needed). With bifilar (i.e., double-file or two-coil) wind-ings, a relatively simpler on or off switching mechanism is adequate for reversing the polarity of a stator pole because one coil gives one polarity and the other coil gives the opposite polarity, and hence current reversal is not required. It follows that a unipolar drive system is adequate for a bifilar-wound motor. Of course, a more complex (and costly) bipolar drive system may be used with a bifilar motor as well (but not necessary). Bipolar winding simply means a winding that has the capability to reverse its polarity.

For a given torque rating, as twice the number of windings as in the unifilar-wound case would be required in bifilar-wound motors, where at least half the windings are inactive at a given time. This increases the motor size for a given torque rating and will increase the friction at the bearings, thereby reducing the starting torque. Furthermore, as all the copper (i.e., windings) of a stator pole is utilized in the unifilar case, the motor

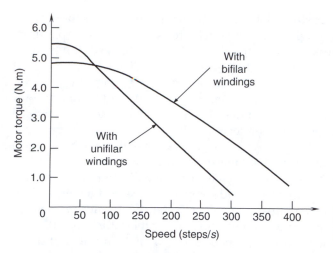

FIGURE 6.7
The effect of bifilar windings on motor torque.

torque tends to be higher. At high speeds, current reversal occurs at a higher frequency in unifilar windings. Consequently, the levels of induced voltages by self-induction and mutual induction [back electromotive force (e.m.f.)] can be significant, resulting in a degradation of the available torque from the motor. For this reason, at high speeds (i.e., at high stepping rates), the effective (dynamic) torque is typically larger for bifilar stepper motors than for their unifilar counterparts, for the same level of drive power (see Figure 6.7). At very low stepping rates, however, dissipation (friction) effects will dominate induced-voltage effects, a drawback with bifilar-wound motors. Furthermore, all the copper in a stator pole is utilized in a unifilar motor. As a result, unifilar windings provide better toque characteristics at low stepping rates, as shown in Figure 6.7.

The motor size can be reduced to some extent by decreasing the wire diameter, which results in increased resistance for a given length of wire. This decreases the current level (and torque) for a given voltage, which is a disadvantage. Increased resistance, however, means decreased electrical time constant (L/R) of the motor, which results in an improved (fast but less oscillatory) single-step response.

6.2   Stepper Motor Classification

As any actuator that generates stepwise motion can be considered a stepper motor, it is difficult to classify all such devices into a small number of useful categories. For example, toothed devices such as harmonic drives (a class of flexible-gear drives—see Chapter 8) and pawl-and-ratchet-wheel drives, which produce intermittent motions through purely mechanical means, are also classified as stepper motors. Of primary interest in today's engineering applications, however, are actuators that generate stepwise motion directly by electromagnetic forces in response to pulse (or digital) inputs. Even for these electromagnetic incremental actuators, however, no standardized classification is available.

Most classifications of stepper motors are based on the nature of the motor rotor. One such classification considers the magnetic character of the rotor. Specifically, a variable-reluctance (VR) stepper motor has a soft-iron rotor, whereas a permanent-magnet (PM) stepper motor has a magnetized rotor. The two types of motors operate in a somewhat similar manner. Specifically, the stator magnetic field (polarity) is stepped so as to change the minimum reluctance (or detent) position of the rotor in increments. Hence, both types of motors undergo similar changes in reluctance (magnetic resistance) during operation. A disadvantage of VR stepper motors is that as the rotor is not magnetized, the holding torque is practically zero when the stator windings are not energized (i.e., power-off conditions). Hence, it is not capable to hold the mechanical load at a given position under power-off conditions, unless mechanical brakes are employed. A hybrid stepper motor possesses characteristics of both VR steppers and PM steppers. The rotor of a hybrid stepper motor consists of two rotor segments connected by a shaft. Each rotor segment is a toothed wheel and is called a stack. The two rotor stacks form the two poles of a permanent magnet located along the rotor axis. Hence, an entire stack of rotor teeth is magnetized to be a single pole (which is different from the case of a PM stepper where the rotor has multiple poles). The rotor polarity of a hybrid stepper can be provided either by a permanent magnet, or by an electromagnet using a coil activated by a unidirectional dc source and placed on the stator to generate a magnetic field along the rotor axis.

Another practical classification that is used in this book is based on the number of stacks of teeth (or rotor segments) present on the rotor shaft. In particular, a hybrid stepper motor has two stacks of teeth. Further subclassifications are possible, depending on the tooth pitch (angle between adjacent teeth) of the stator and the tooth pitch of the rotor. In a single-stack stepper motor, the rotor tooth pitch and the stator tooth pitch generally have to be unequal so that not all teeth in the stator are ever aligned with the rotor teeth at any instant. It is the misaligned teeth that exert the magnetic pull, generating the driving torque. In each motion increment, the rotor turns to the minimum reluctance (stable equilibrium) position corresponding to that particular polarity distribution of the stator.

In multiple-stack stepper motors, operation is possible even when the rotor tooth pitch is equal to the stator tooth pitch, provided that at least one stack of rotor teeth is rotationally shifted (misaligned) from the other stacks by a fraction of the rotor tooth pitch. In this design, it is this interstack misalignment that generates the drive torque for each motion step. It is obvious that unequal-pitch multiple stack steppers are also a practical possibility. In this design, each rotor stack operates as a separate single-stack stepper motor. The stepper motor classifications described thus far are summarized in Figure 6.8.

Next we describe some geometric, mechanical, design, and operational aspects of single-stack and multiple-stack stepper motors. One point to remember is that some form of geometric/magnetic misalignment of teeth is necessary in both types of motors. A motion step is obtained by simply redistributing (i.e., switching) the polarities of the stator, thereby changing the minimum reluctance detent position of the rotor. Once a stable equilibrium position is reached by the rotor, the stator polarities are switched again

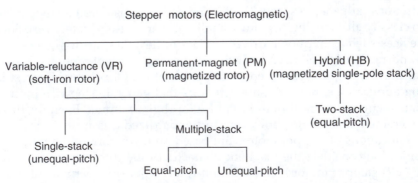

FIGURE 6.8
Classifications of stepper motors.

to produce a new detent position, and so on. In descriptive examples, it is more convenient to use variable-reluctance stepper motors. However, the principles can be extended in a straightforward manner to cover permanent-magnet and hybrid stepper motors as well.

6.2.1 Single-Stack Stepper Motors

To establish somewhat general geometric relationships for a single-stack variable-reluctance stepper motor, consider Figure 6.9. The motor has three phases of winding ($p = 3$) in the stator, and there are eight teeth in the soft-iron rotor ($n_r = 8$). The three phases are numbered 1, 2, and 3. Each phase represents a group of four stator poles wound together, and the total number of stator poles (n_s) is 12. When phase 1 is energized, one pair of diametrically opposite poles becomes N (north) poles and the other pair in that phase (located at 90° from the first pair) becomes S (south) poles. Furthermore, a geometrically orthogonal set of four teeth on the rotor will align themselves perfectly with these four stator poles. This is the minimum reluctance, stable equilibrium configuration (detent position) for the rotor under the given activation state of the stator (i.e., phase 1 is on and the other two phases are off). Observe, however, that there is a misalignment of 15° between the remaining rotor teeth and the nearest stator poles.

If the *pitch angle*, defined as the angle between two adjacent teeth, is denoted by θ (in degrees) and the number of teeth is denoted by n, we have

$$\text{Stator pitch } \theta_s = \frac{360°}{n_s}.$$

$$\text{Rotor pitch } \theta_r = \frac{360°}{n_r}.$$

For one-phase-on excitation, the step angle $\Delta\theta$, which should be equal to the smallest misalignment between a stator pole and an adjacent rotor tooth in any stable equilibrium state, is given by

$$\Delta\theta = \theta_r - r\theta_s, \quad (\text{for } \theta_r > \theta_s), \tag{6.1a}$$

$$\Delta\theta = \theta_s - r\theta_r, \quad (\text{for } \theta_r < \theta_s), \tag{6.1b}$$

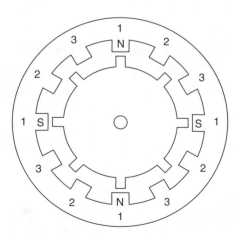

FIGURE 6.9
Three-phase single-stack VR stepper motor with 12 stator poles (teeth) and 8 rotor teeth.

where r is the largest positive integer such that $\Delta\theta$ is positive (i.e., the largest feasible r such that a misalignment in rotor and stator teeth occurs). It is clear that for the arrangement shown in Figure 6.9, $\theta_r = 360°/8 = 45°$, $\theta_s = 360°/12 = 30°$, and hence, $\Delta\theta = 45° - 30° = 15°$, as stated earlier.

Now, if phase 1 is turned off and phase 2 is turned on, the rotor will turn $15°$ in the counter-clockwise (CCW) direction to its new minimum reluctance position. If phase 3 is energized instead of phase 2, the rotor would turn $15°$ clockwise (CW). It should be clear that half this step size ($7.5°$) is also possible with the motor shown in Figure 6.9. Suppose, for example, that phase 1 is on, as before. Next suppose that phase 2 is energized while phase 1 is on, so that two like poles are in adjacent locations. As the equivalent field of the two adjoining like poles is halfway between the two poles, two rotor teeth will orient symmetrically about this pair of poles, which is the corresponding minimum-reluctance position. It is clear that this corresponds to a rotation of $7.5°$ from the previous detent position, in the CCW direction. For executing the next half step (in the CCW direction), phase 1 is turned off while phase 2 is left on. Thus, in summary, the full-stepping sequence for CCW rotation is 1-2-3-1; for CW rotation, it is 1-3-2-1. The half-stepping sequence for CCW rotation is 1-12-2-23-3-31-1; for CW rotation, it is 1-31-3-23-2-12-1.

Returning to full stepping, note that as each switching of phases corresponds to a rotation of $\Delta\theta$ and there are p number of phases, the angle of rotation for a complete switching cycle of p switches is $p\Delta\theta$. In a switching cycle, the stator polarity distribution returns to the distribution that it had in the beginning. Hence, in one switching cycle (p switches), the rotor should assume a configuration exactly like what it had in the beginning of the cycle. That is, the rotor should turn through a complete pitch angle of θ_r. Hence, the following relationship exists for the one-phase-on case

$$\theta_r = p\Delta\theta. \tag{6.2}$$

Substituting this in Equation 6.1a, we have

$$\theta_r = r\theta_s + \frac{\theta_r}{p}, \quad (\text{for } \theta_r > \theta_s) \tag{6.3a}$$

and similarly with Equation 6.1b we have

$$\theta_s = r\theta_r + \frac{\theta_r}{p}, \quad (\text{for } \theta_r < \theta_s), \tag{6.3b}$$

where θ_r is the rotor tooth pitch angle, θ_s is the stator tooth pitch angle, p is the number of phases in the stator, and r is the largest feasible positive integer.

Now, by definition of the pitch angle, Equation 6.3a gives

$$\frac{360°}{n_r} = \frac{r \times 360°}{n_s} + \frac{360°}{pn_r}$$

or

$$n_s = rn_r + \frac{n_s}{p}, \quad (\text{for } n_s > n_r), \tag{6.4a}$$

and similarly from Equation 6.3b,

$$n_r = rn_s + \frac{n_s}{p}, \quad \text{(for } n_s < n_r\text{)}, \tag{6.4b}$$

where n_r is the number of rotor teeth, n_s is the number of stator teeth, and r is the largest feasible positive integer.

Finally, the number of steps per revolution is

$$n = \frac{360°}{\Delta\theta}. \tag{6.5}$$

Example 6.1

Consider the stepper motor shown in Figure 6.9. The number of stator poles $n_s = 12$ and the number of phases $p = 3$. Assuming that $n_r < n_s$ (which is the case in Figure 6.9), substitute in Equation 6.4a:

$$12 = rn_r + \frac{12}{3}$$

or

$$rn_r = 8. \tag{i}$$

Now, $r = 1$ gives $n_r = 8$. This is the feasible case shown in Figure 6.9. Then the rotor pitch $\theta_r = 360°/8 = 45°$, and the step angle can be calculated from Equation 6.2 as

$$\Delta\theta = \frac{45°}{3} = 15°.$$

Note that this is the full-step angle, as observed earlier. Furthermore, the stator pitch is $\theta_s = 360°/12 = 30°$, which further confirms that the step angle is $\theta_r - \theta_s = 45° - 30° = 15°$. In the result (Equation i) above, $r = 2$ and $r = 4$ are not feasible solutions because, then all the rotor teeth will be fully aligned with stator poles, and there will not be a misalignment to enable stepping.

Example 6.2

Consider the motor arrangement shown in Figure 6.5. Here, $\theta_r = 180°$ and $\theta_s = 360°/6 = 60°$. Now from Equation 6.1a, $\Delta\theta = 180° - r \times 60°$. The largest feasible r in this case is 2, which gives $\Delta\theta = 180° - 2 \times 60° = 60°$. Furthermore, from Equation 6.2, we have $\Delta\theta = 180°/3 = 60°$.

Example 6.3

Consider a stepper motor with $n_r = 5$, $n_s = 2$, and $p = 2$. A schematic representation is given in Figure 6.10. In this case, $\theta_r = 360°/5 = 72°$ and $\theta_s = 360°/2 = 180°$. From Equation 6.1b, $\Delta\theta = 180° - r \times 72°$. Here, the largest feasible value for r is 2, which corresponds to a step angle of $\Delta\theta = 180° - 2 \times 72° = 36°$. This is further confirmed by Equation 6.2, which gives $\Delta\theta = 72°/2 = 36°$.

Note that this particular arrangement is feasible for a PM stepper but not for a VR stepper. The reason is simple., In a detent position (equilibrium position), the rotor

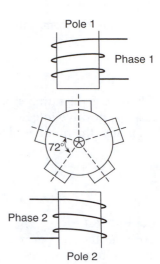

FIGURE 6.10
A two-phase two-pole stepper with a five-tooth rotor.

tooth will align itself with the stator pole (say, phase 1) that is on, and two other rotor teeth will orient symmetrically between the remaining stator pole (phase 2)that is off. Next, when phase 1 is turned off and phase 2 is turned on (for executing a full step), unless the two rotor teeth that are symmetric with phase 2 have opposite polarities, an equal force will be exerted on them by this stator pole, trying to move the rotor in opposite directions, thereby forming an unstable equilibrium position (or, ambiguity in the direction of rotation).

6.2.2 Toothed-Pole Construction

The foregoing analysis indicates that the step angle can be reduced by increasing the number of poles in the stator and the number of teeth in the rotor. Obviously, there are practical limitations to the number of poles (windings) that can be incorporated in a stepper motor. A common solution to this problem is to use toothed poles in the stator, as shown in Figure 6.11a. The toothed construction of the stator and the rotor not only improves the motion resolution (step angle) but also enhances the concentration of the magnetic field, which generates the motor torque. Also, the torque and motion characteristics become smother (smaller ripples and less jitter) as a result of the distributed tooth construction.

In the case shown in Figure 6.11a, the stator teeth are equally spaced but the pitch (angular spacing) is not identical to the pitch of the rotor teeth. In the toothed-stator construction, n_s represents the number of teeth rather than the number of poles in the stator. The number of rotor teeth has to be increased in proportion. Note that in full stepping (e.g., one phase on), after p number of switchings (steps), where p is the number of phases, the adjacent rotor tooth will take the previous position of a particular rotor tooth. It follows that the rotor rotates through θ_r (the tooth pitch of the rotor) in p steps. Thus, the relationship $\Delta\theta = \theta_r/p$ (Equation 6.2) still holds. But Equation 6.1 has to be modified to accommodate toothed poles. Toothed-stator construction can provide very small step angles—0.72° for example, or more commonly, 1.8°.

The equations for the step angle given so far in this chapter assume that the number of stator poles is identical to the number of stator teeth. In particular, Equation 6.1, Equation 6.3, and Equation 6.4 are obtained using this assumption. These equations have to be modified when there are many teeth on each stator pole. Generalization of the step

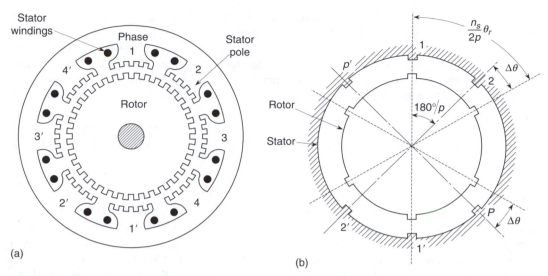

FIGURE 6.11
A possible toothed-pole construction for a stepper motor. (a) An eight-pole, four-phase motor. (b) Schematic diagram for generalizing the step angle equation.

angle equations for the case of toothed-pole construction can be made by referring to Figure 6.11b. The center tooth of each pole and the rotor tooth closest to that stator tooth are shown. This is one of the possible geometries for a single-stack toothed construction. In this case, the rotor tooth pitch θ_r is not equal to the stator tooth pitch θ_s. Another possibility for a single-stack toothed construction (which is perhaps preferable from the practical point of view) will be described later. There, θ_r and θ_s are identical, but when all the stator teeth that are wound to one of the phases are aligned with the rotor teeth, all the stator teeth wound to another phase will have a constant misalignment with the rotor teeth in their immediate neighborhood. Unlike that case, in the construction used in the present case, $\theta_r \neq \theta_s$; hence, only one tooth in a stator pole can be completely aligned with a rotor tooth.

Consider the case of $\theta_r > \theta_s$ (i.e., $n_r < n_s$). As $(\theta_r - \theta_s)$ is the offset between the rotor tooth pitch and the stator tooth pitch, and as there are $n_s/(mp)$ rotor teeth in the sector made by two adjacent stator poles, we see that the total tooth offset (step angle) $\Delta\theta$ at the second pole is given by

$$\Delta\theta = \frac{n_s}{mp}(\theta_r - \theta_s), \quad \text{for } \theta_r > \theta_s, \tag{6.6}$$

where p is the number of phases and m is the number of stator poles per phase.

Now, noting that $\Delta\theta = \theta_r/p$ is true even for the toothed construction and by substituting this in Equation 6.6 we get

$$n_s = n_r + m \text{ for } n_r < n_s \tag{6.7}$$

or in general (including the case $n_r > n_s$) we have

$$n_s = n_r \pm m. \tag{6.8}$$

A further generalization is possible if $\theta_r > 2\theta_s$ or $\theta_r < 2\theta_s$, as in the nontoothed case, by introducing an integer r.

Also, we recall that when the stator teeth are interpreted as stator poles, $n_s = mp$, and then Equation 6.6 reduces to Equation 6.1a, as expected, for $r = 1$. For the toothed-pole construction, n_s is several times the value of mp. In this case, Equation 6.6 should be used instead of Equation 6.1a. In general, p has to be replaced by n_s/m in converting an equation for a nontoothed-pole construction to the corresponding equation for a toothed-pole construction. For example, then, Equation 6.4a becomes Equation 6.7, with $r = 1$.

Finally, we observe from Figure 6.11b that the switching sequence $1\text{-}2\text{-}3\text{-}\cdots\text{-}p$ produces CCW rotations, and the switching sequence $1\text{-}p\text{-}(p\text{-}1)\text{-}\cdots\text{-}2$ produces clockwise rotations.

Example 6.4

Consider a simple design example for a single-stack VR stepper. Suppose that the number of steps per revolution, which is a functional requirement, is specified as $n = 200$. This corresponds to a step angle of $\Delta\theta = 360°/200 = 1.8°$. Assume full stepping. Design restrictions, such as size and the number of poles in the stator, govern t_s, the number of stator teeth per pole. Let us use the typical value of 6 teeth/pole. Also assume that there are two poles wound to the same stator phase. We are interested in designing a motor to meet these requirements.

Solution

First, we will derive some useful relationships. Suppose that there are m poles per phase. Hence, there are mp poles in the stator (Note: $n_s = mpt_s$.). Then, assuming $n_r < n_s$, Equation 6.7: $n_s = n_r + m$ would apply. Dividing this equation by mp, we get

$$t_s = \frac{n_r}{mp} + \frac{1}{p}. \tag{i}$$

Now

$$n_r = \frac{360°}{\theta_r} = \frac{360°}{p\Delta\theta} \quad \text{(from Equation 6.2)}$$

or

$$n_r = \frac{n}{p}. \tag{6.9}$$

Substituting this in Equation i we get

$$t_s = \frac{n}{mp^2} + \frac{1}{p}. \tag{6.10a}$$

Now, as $1/p$ is less than 1 for a stepper motor and t_s is greater than 1 for the toothed-pole construction, an approximation for Equation 6.10a can be given by

$$t_s = \frac{n}{mp^2}, \tag{6.10b}$$

where t_s is the number of teeth per stator pole, m is the number of stator poles per phase, p is the number of phases, and n is the number of steps per revolution.

In the present example, $t_s \approx 6$, $m = 2$, and $n = 200$. Hence, from Equation 6.10b, we have

$$6 \sim \frac{200}{2 \times p^2},$$

which gives $p \approx 4$. Note that p has to be an integer. Now, using Equation 6.10a, we get two possible designs for $p = 4$. First, with the specified values $n = 200$ and $m = 2$, we get $t_s = 6.5$, which is slightly larger than the required value of 6. Alternatively, with the specified $t_s = 6$ and $m = 2$, we get $n = 184$, which is slightly smaller than the specified value of 200. Either of these two designs would be acceptable. The second design gives a slightly larger step angle (Note that $\Delta\theta = 360°/n = 1.96°$ for the second design and $\Delta\theta = 360°/200 = 1.8°$ for the first design.). Summarizing the two designs, we have the following results:

For design 1:

Number of phases, $p = 4$
Number of stator poles $= 8$
Number of teeth per pole $= 6.5$
Number of steps per revolution $= 200$
Step angle $= 1.8°$
Number of rotor teeth $= 50$ (from Equation 6.9)
Number of stator teeth $= 52$

For design 2:

Number of phases, $p = 4$
Number of stator poles $= 8$
Number of teeth per pole $= 6$
Number of steps per revolution $= 184$
Step angle $= 1.96°$
Number of rotor teeth $= 46$ (from Equation 6.9)
Number of stator teeth $= 48$

Note: The number of teeth per stator pole (t_s) does not have to be an integer (see design 1). As there are interpolar gaps around the stator, it is possible to construct a motor with an integer number of actual stator teeth, even when t_s and n_s are not integers.

6.2.3 Another Toothed Construction

In the single-stack toothed construction just presented, we have $\theta_r \neq \theta_s$. An alternative design possibility exists, where $\theta_r = \theta_s$, but in this case the stator poles are located around the rotor such that when the stator teeth corresponding to one of the phases are fully aligned with the rotor teeth, the stator teeth in another phase will have a constant offset with the neighboring rotor teeth, thereby providing the misalignment that is needed for stepping. The torque magnitude of this construction is perhaps better because of this uniform tooth offset per phase, but torque ripples (jitter) would also be stronger (a disadvantage) because of sudden and more prominent changes in magnetic reluctance from pole to pole, during phase switching.

To obtain some relations that govern this construction, suppose that Figure 6.11a represents a stepper motor of this type. When the stator teeth in Pole 1 (and Pole 1') are perfectly aligned with the rotor teeth, the stator teeth in Pole 2 (and Pole 2′) will have an offset of $\Delta\theta$ with the neighboring rotor teeth. This offset is in fact the step angle, in full stepping. This offset can be either in the CCW direction (as in Figure 6.11a) or in the clockwise direction. As the pole pitch is given by $360°/pm$, we must have

$$\frac{1}{\theta_r}\left[\frac{360°}{pm}\pm\Delta\theta\right]=r, \tag{6.11}$$

where r is the integer number of rotor teeth contained within the angular sector $360°/(pm)\pm\Delta\theta$. Also, $\Delta\theta$ is the step angle (full stepping), θ_r is the rotor tooth pitch, p is the number of phases, and m is the number of stator poles per phase.

It should be clear that within two consecutive poles wound to the same phase, there are n_r/m rotor teeth. As p switchings of magnitude $\Delta\theta$ each will result in a total rotation of θ_r (Equation 6.2) by substituting this along with $n_r = 360°/\theta_r$ in Equation 6.11, and simplifying, we get:

$$n_r \pm m = pmr, \tag{6.12}$$

where n_r is the number of rotor teeth.

Example 6.5

Consider the full-stepping operation of a single-stack, equal-pitch stepper motor whose design is governed by Equation 6.12. Discuss the possibility of constructing a four-phase motor of this type that has 50 rotor teeth (i.e., step angle = 1.8°). Obtain a suitable design for a four-phase motor that uses eight stator poles. Specifically, determine the number of rotor teeth (n_r), the step angle $\Delta\theta$, the number of steps per revolution (n), and the number of teeth per stator pole (t_s).

Solution

First, with $n_r = 50$ and $p = 4$, Equation 6.12 becomes

$$50 \pm m = 4mr$$

or

$$m = \frac{50}{(4r \mp 1)}. \tag{i}$$

Note that m and r should be natural numbers (i.e., positive integers). As, such smallest value for r is 1, we see from Equation i that the largest value for m is 16. It can be easily verified that only two solutions are valid in this range; namely, $r = 1$ and $m = 10$; $r = 6$ and $m = 2$, both corresponding to the +sign in the denominator of Equation i. The first solution is not very practical. Notably, in this case, the number of poles $= 10 \times 4 = 40$, and hence the pole pitch is $360°/40 = 9°$. As each stator pole will occupy nearly this angle, it cannot have more than one tooth of pitch 7.2° (the rotor tooth pitch $= 360°/50 = 7.2°$). The second solution is more practical. In this case, the number of poles $= 2 \times 4 = 8$ and the pole pitch is $360°/8 = 45°$. Each stator pole will occupy nearly this angle, and with a tooth of pitch 7.2°, a pole can have six full teeth.

Next, consider $p = 4$ and $m = 2$ (i.e., a four-phase motor with eight stator poles, as specified in the example). Then Equation 6.12 becomes

$$n_r = 8r \mp 2. \tag{ii}$$

Hence, one possible design that is close to the previously mentioned case of $n_r = 50$ is realized with $r = 6$ (giving $n_r = 50$, for the +sign) and $r = 7$ (giving $n_r = 54$, for the −sign). Consider the latter case, where $n_r = 54$. The corresponding tooth pitch (for both rotor and stator) is

$$\theta_r = \theta_s = \frac{360°}{54} \simeq 6.67°.$$

The step angle (for full stepping) is

$$\Delta\theta = \frac{\theta_r}{p} = \frac{6.67°}{4} \simeq 1.67°.$$

The number of steps per revolution is

$$n = \frac{360°}{\Delta\theta} = pn_r = 4 \times 54 = 216.$$

$$\text{Pole pitch} = \frac{360°}{mp} = \frac{360°}{8} = 45°.$$

The maximum number of teeth that could be occupied within a pole pitch is

$$\frac{45°}{\theta_s} = \frac{45°}{6.67°} = 6.75.$$

Hence, the maximum possible number of full teeth per pole is

$$t_s = 6.$$

In a practical motor, there can be an interpolar gap of nearly half the pole angle. In that case, a suitable number for t_s would be the integer value of

$$\frac{1}{\theta_s} \frac{360°}{(8+4)}.$$

This gives

$$t_s = 4.$$

Summarizing, we have the following design parameters:

Number of phases, $p = 4$
Number of stator poles, $mp = 8$
Number of teeth per pole, $t_s =$ maximum 6 (typically 4)
Number of steps per revolution (full stepping), $n = 216$
Step angle, $\Delta\theta = 1.67°$
Number of rotor teeth, $n_r = 54$
Tooth pitch (both rotor and stator) $\approx 6.67°$

6.2.4 Microstepping

We have seen how full stepping or half stepping can be achieved simply by using an appropriate switching scheme. For example, half stepping occurs when phase switchings

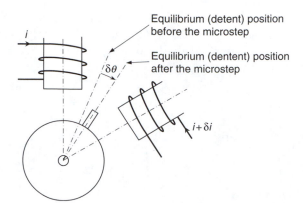

FIGURE 6.12
The principle of microstepping.

alternate between one-phase-on and two-phase-on states. Full stepping occurs when either one-phase-on switching or two-phase-on switching is used exclusively for every step. In both these cases, the current level (or state) of a phase is either 0 (off) or 1 (on). Rather than using just two current levels (the binary case), it is possible to use several levels of phase current between these two extremes, thereby achieving much smaller step angles. This is the principle behind microstepping.

Microstepping is achieved by properly changing the phase currents in small steps, instead of switching them on and off (as in the case of full stepping and half stepping). The principle behind this can be understood by considering two identical stator poles (wound with identical windings), as shown in Figure 6.12. When the currents through the windings are identical (in magnitude and direction), the resultant magnetic field will lie symmetrically between the two poles. If the current in one pole is decreased while the other current is kept unchanged, the resultant magnetic field will move closer to the pole with the larger current. As the detent position (equilibrium position) depends on the position of the resultant magnetic field, it follows that very small step angles can be achieved simply by controlling (varying the relative magnitudes and directions of) the phase currents.

Step angles of 1/125 of a full step or smaller could be obtained through microstepping. For example, 10,000 steps/revolution may be achieved. Note that the step size in a sequence of microsteps is not identical. This is because stepping is done through microsteps of the phase current, which (and the magnetic field generated by it) has a nonlinear relation with the step angle.

Motor drive units with the microstepping capability are more costly, but microstepping provides the advantages of accurate motion capabilities, including finer resolution, overshoot suppression, and smoother operation (reduced jitter and less noise), even in the neighborhood of resonance in the motor–load combination. A disadvantage is that, usually there is a reduction in the motor torque as a result of microstepping.

6.2.5 Multiple-Stack Stepper Motors

For illustration purposes, consider the longitudinal view of a three-stack stepper motor shown schematically in Figure 6.13. In this example, there are three identical stacks of teeth mounted on the same rotor shaft. There is a separate stator segment surrounding each rotor stack. One straightforward approach to designing a multiple-stack stepper motor would be to treat it as a cascaded set of identical single-stack steppers with common phase windings for all the stator segments. Then the number of phases of the motor is fixed, regardless of the number of stacks used. Such a design is simply

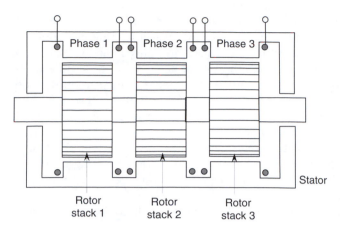

a single-stack stepper with a longer rotor and a correspondingly longer stator, thereby generating a higher torque (proportional to the length of the motor, for a given winding density and a phase current). What is considered here is not such a trivial design, but somewhat complex designs where the phase windings of a stack can operate (on, off, reversal) independently of another stack.

Both equal-pitch construction ($\theta_r = \theta_s$) and unequal-pitch construction ($\theta_r > \theta_s$ or $\theta_r < \theta_s$) are possible in multiple-stack steppers. An advantage of the unequal-pitch construction is that smaller step angles are possible than with an equal-pitch construction of the same size (i.e., same diameter and number of stacks). But the switching sequence is somewhat more complex for unequal-pitch, multiple-stack stepper motors. In particular, each stack has more than one phase and they can operate independently of the phases of another stack. First, we will examine the equal-pitch, multiple-stack construction. The operation of an unequal-pitch, multiple-stack motor should follow directly from the analysis of the single-stack case as given before. Subsequently, a hybrid stepper will be described. A hybrid stepper has a two-stack rotor, but an entire stack is magnetized with a single polarity and the two stacks have opposite polarities.

6.2.5.1 Equal-Pitch Multiple-Stack Stepper

For each rotor stack, there is a toothed stator segment around it, whose pitch angle is identical to that of the rotor ($\theta_s = \theta_r$). A stator segment may appear to be similar to that of an equal-pitch single-stack stepper (discussed previously), but this is not the case. Each stator segment is wound to a single phase, thus the entire segment can be energized (polarized) or deenergized (depolarized) simultaneously. It follows that, in the equal pitch case,

$$p = s, \tag{6.13}$$

where p is the number of phases and s is the number of rotor stacks.

The misalignment that is necessary to produce the motor torque may be introduced in one of two ways:

1. The teeth in the stator segments are perfectly aligned, but the teeth in the rotor stacks are misaligned consecutively by $1/s \times$ pitch angle.
2. The teeth in the rotor stacks are perfectly aligned, but the teeth in the stator segments are misaligned consecutively by $1/s \times$ pitch angle.

Now consider the three-stack case. Suppose that phase 1 is energized. Then the teeth in the rotor stack 1 will align perfectly with the stator teeth in phase 1 (segment 1). But the teeth in the rotor stack 2 will be shifted from the stator teeth in phase 2 (segment 2) by a one-third–pitch angle in one direction, and the teeth in rotor stack 3 will be shifted from the stator teeth in phase 3 (segment 3) by a two-thirds–pitch angle in the same direction (or a one-third–pitch angle in the opposite direction). It follows that if phase 1 is now deenergized and phase 2 is energized, the rotor will turn through one-third pitch in one direction. If, instead, phase 3 is turned on after phase 1, the rotor will turn through one-third pitch in the opposite direction. Clearly, the step angle (for full stepping) is a one-third–pitch angle for the three-stack, three-phase construction. The switching sequence 1-2-3-1 will turn the rotor in one direction, and the switching sequence 1-3-2-1 will turn the rotor in the opposite direction.

In general, for a stepper motor with s stacks of teeth on the rotor shaft, the full-stepping step angle is given by

$$\Delta\theta = \frac{\theta}{s}, \tag{6.14a}$$

where $\theta = \theta_r = \theta_s =$ tooth pitch angle. In view of Equation 6.13, we have

$$\Delta\theta = \frac{\theta}{p}. \tag{6.14b}$$

Note that the step angle can be decreased by increasing the number of stacks of rotor teeth. Increased number of stacks also means more phase windings with associated increase in the magnetic field and the motor torque. However, the length of the motor shaft increases with the number of stacks, and can result in flexural (shaft bending) vibration problems (particularly whirling of the shaft), air gap contact problems, large bearing loads, wear and tear, and increased noise. As in the case of a single-stack stepper, half stepping can be accomplished by energizing two phases at a time. Hence, in the three-stack stepper, for one direction, the half-stepping sequence is 1-12-2-23-3-31-1; in the opposite direction, it is 1-13-3-32-2-21-1.

6.2.5.2 Unequal-Pitch Multiple-Stack Stepper

Unequal-pitch multiple-stack stepper motors are also of practical interest. Very fine angular resolutions (step angles) can be achieved by this design without compromising the length of the motor. In an unequal-pitch stepper motor, each stator segment has more than one phase (p number of phases), just like in a single-stack unequal-pitch stepper. Rather than a simple cascading, however, the phases of different stacks are not wound together and can be switched on and off independently. In this manner yet finer small angles are realized, together with an added benefit of increased torque provided by the multistack design.

For a single-stack nontoothed-pole stepper, we have seen that the step angle is equal to $\theta_r - \theta_s$. In a multistack stepper, this misalignment is further subdivided into s equal steps using the interstack misalignment. Hence, the overall step angle for an unequal-pitch, multiple-stack stepper motor with nontoothed poles is given by

$$\Delta\theta = \frac{\theta_r - \theta_s}{s} \quad (\text{for } \theta_r > \theta_s). \tag{6.15}$$

For a toothed-pole multiple-stack stepper motor, we have

$$\Delta\theta = \frac{n_s(\theta_r - \theta_s)}{mps},\tag{6.16}$$

where m is the number of stator poles per phase. Alternatively, using Equation 6.2, we have

$$\Delta\theta = \frac{\theta_r}{ps}\tag{6.17}$$

for both toothed-pole and nontoothed-pole motors, where p is the number of phases in each stator segment, and s is the number of rotor stacks.

6.2.6 Hybrid Stepper Motor

Hybrid steppers are arguably the most common variety of stepping motors in engineering applications. A hybrid stepper motor has two stacks of rotor teeth on its shaft. The two rotor stacks are magnetized to have opposite polarities, as shown in Figure 6.14. There are two stator segments surrounding the two rotor stacks. Both rotor and stator have teeth and their pitch angles are equal. Each stator segment is wound to a single phase, and accordingly, the number of phases is two. It follows that a hybrid stepper is similar in mechanical design and stator winding to a two-stack equal-pitch VR stepper. There are some dissimilarities, however. First, the rotor stacks are magnetized. Second, the interstack misalignment is 1/4 of a tooth pitch (see Figure 6.15).

A full cycle of the switching sequence for the two phases is given by [0 1], [−1 0], [0 −1], [1 0], [0 1] for one direction of rotation. In fact, this sequence produces a downward movement (CW rotation, looking from the left end) in the arrangement shown in Figure 6.15, starting from the state of [0 1] shown in the figure (i.e., phase 1 off and phase 2 on with N polarity). For the opposite direction, the sequence is simply reversed; thus, [0 1], [1 0], [0 −1], [−1 0], [0 1]. Clearly, the step angle is given by

$$\Delta\theta = \frac{\theta}{4},\tag{6.18}$$

where $\theta = \theta_r = \theta_s =$ tooth pitch angle.

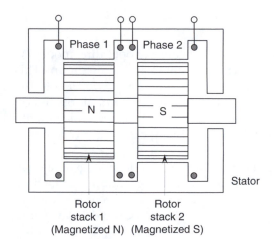

FIGURE 6.14

A hybrid stepper motor.

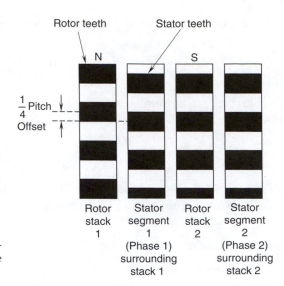

FIGURE 6.15
Rotor stack misalignment (1/4 pitch) in a hybrid stepper motor (schematically shows the state where phase 1 is off and phase 2 is on with N polarity).

Just like in the case of a PM stepper motor, a hybrid stepper has the advantage providing a holding torque (detent torque) even under power-off conditions. Furthermore, a hybrid stepper can provide very small step angles, high stepping rates, and generally good torque–speed characteristics.

Example 6.6
The half-stepping sequence for the motor represented in Figure 6.14 and Figure 6.15 may be determined quite conveniently. Starting from the state [0, 1] as before, if phase 1 is turned on to state −1 without turning off phase 2, then phase 1 will oppose the pull of phase 2, resulting in a detent position halfway between the full-stepping detent position. Next, if phase 2 is turned off while keeping phase 1 in −1, the remaining half step of the original full step will be completed. In this manner, the half-stepping sequence for CW rotation is obtained as: [0, 1], [−1, 1], [−1, 0], [−1, −1], [0, −1], [1, −1], [1, 0], [1, 1], [0, 1]. For CCW rotation, this sequence is simply reversed. Note that, as expected, in half stepping, both phases remain on during every other half step.

6.3 Driver and Controller

In principle, the stepper motor is an open-loop actuator. In its normal operating mode, the stepwise rotation of the motor is synchronized with the command pulse train. This justifies the term digital synchronous motor, which is sometimes used to denote the stepper motor. As a result of stepwise (incremental) synchronous operation, open-loop operation is adequate, at least in theory. An exception to this may result under highly transient conditions, exceeding rated torque, when pulse missing can be a problem. We will address this situation in Section 6.7.1.

A stepper motor needs a control computer or at least a hardware indexer to generate the pulse commands and a driver to interpret the commands and correspondingly generate proper currents for the phase windings of the motor. This basic arrangement is shown in Figure 6.16a. For feedback control, the response of the motor has to be sensed (say, using

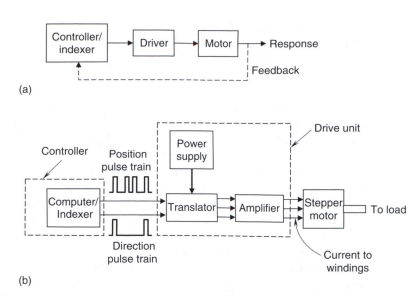

FIGURE 6.16
(a) The basic control system of a stepper motor. (b) The basic components of a driver.

an optical encoder; see Chapter 5) and fed back into the controller (see the broken-line path in Figure 6.16a) for making the necessary corrections to the pulse command, when an error is present. We will return to the subject of control in Section 6.7. The basic components of the driver for a stepper motor are identified in Figure 6.16b. A driver typically consists of a logic circuit called translator to interpret the command pulses and switch the appropriate analog circuits to generate the phase currents. As sufficiently high current levels are needed for the phase windings, depending on the motor capacity, the drive system includes amplifiers powered by a power supply.

The command pulses are generated either by a control computer (a desktop computer or a microprocessor), which is the software approach, or by a variable-frequency oscillator (or, an indexer), which is the basic hardware approach. For bidirectional motion, two pulse trains are necessary—the position-pulse train and the direction-pulse train, which are determined by the required motion trajectory. The position pulses identify the exact times at which angular steps should be initiated. The direction pulses identify the instants at which the direction of rotation should be reversed. Only a position pulse train is needed for unidirectional operation. Generation of the position pulse train for steady-state operation at a constant speed is relatively a simple task. In this case, a single command identifying the stepping rate (pulse rate), corresponding to the specified speed, would suffice. The logic circuitry within the translator will latch onto a constant-frequency oscillator, with the frequency determined by the required speed (stepping rate), and continuously cycle the switching sequence at this frequency. This is a hardware approach to open-loop control of a stepping motor. For steady-state operation, the stepping rate can be set by manually adjusting the knob of a potentiometer connected to the translator. For simple motions (e.g., speeding up from rest and subsequently stopping after reaching a certain angular position), the commands that generate the pulse train (commands to the oscillator) can be set manually. Under the more complex and transient operating conditions that are present when following intricate motion trajectories (e.g., in trajectory-following robots), however, a computer-based (or microprocessor based) generation of the pulse commands, using programmed logic, would be necessary. This is a software approach, which is usually slower than the hardware approach. Sophisticated feedback control schemes can be implemented as well through such a computer-based controller.

The translator module has logic circuitry to interpret a pulse train and translate it into the corresponding switching sequence for stator field windings (on or off or reverse state for each phase of the stator). The translator also has solid-state switching circuitry (using gates, latches, triggers, etc.) to direct the field currents to the appropriate phase windings according to the required switching state. A packaged system typically includes both indexer (or controller) functions and driver functions. As a minimum, it possesses the capability to generate command pulses at a steady rate, thus assuming the role of the pulse generator (or indexer) as well as the functions of translator and switching amplifier. The stepping rate or direction may be changed manually using knobs or through the user interface.

The translator may not have the capability to keep track of the number of steps taken by the motor (i.e., a step counter). A packaged device that has all these capabilities, including pulse generation, the standard translator functions, and drive amplifiers, is termed a preset indexer. It usually consists of an oscillator, digital microcircuitry (integrated-circuit chips) for counting and for various control functions, a translator, and drive circuitry in a single package. The required angle of rotation, stepping rate, and direction are pre-set according to a specified operating requirement. With a more sophisticated programmable indexer, these settings can be programmed through computer commands from a standard interface. An external pulse source is not needed in this case. A programmable indexer—consisting of a microprocessor and microelectronic circuitry for the control of position and speed and for other programmable functions, memory, a pulse source (an oscillator), a translator, drive amplifiers with switching circuitry, and a power supply—represents a programmable controller for a stepping motor. A programmable indexer can be programmed using a personal computer or a hand-held programmer (provided with the indexer) through a standard interface (e.g., RS232 serial interface). Control signals within the translator are in the order of 10 mA, whereas the phase windings of a stepper motor require large currents on the order of several amperes. Control signals from the translator have to be properly amplified and directed to the motor windings by means of switching amplifiers for activating the required phase sequence.

Power to operate the translator (for logic circuitry, switching circuitry, etc.) and to operate phase excitation amplifiers comes from a dc power supply (typically 24 V dc). A regulated (i.e., voltage maintained constant irrespective of the load) power supply is preferred. A packaged unit that consists of the translator (or indexer), the switching amplifiers, and the power supply, is what is normally termed a motor-drive system. The leads of the output amplifiers of the drive system carry currents to the phase windings on the stator (and to the rotor magnetizing coils located on the stator in the case of an electromagnetic rotor) of the stepping motor. The load may be connected to the motor shaft directly or through some form of mechanical coupling device (e.g., harmonic drive, tooth-timing belt drive, hydraulic amplifier, rack, and pinion; see Chapter 8).

6.3.1 Driver Hardware

The driver hardware consists of the following basic components:

1. Digital (logic) hardware to interpret the information carried by the stepping pulse signal and the direction pulse signal (i.e., step instants and the direction of motion) and provide appropriate signals to the switches (switching transistors) that actuate the phase windings. This is the translator unit of the drive hardware.

2. The drive circuit for phase windings with switching transistors to actuate the phases (on, off, reverse in the unifilar case; on, off in the bifilar case).

3. Power supply to power the phase windings.

These three components are commercially available as a single package, to operate a corresponding class of stepper motors. As there is considerable heat generation in a drive module, an integrated heat sink (or some means of heat removal) is needed as well. Consider the drive hardware for a two-phase stepper motor. The phases are denoted by A and B. A schematic representation of the drive system, which is commercially available as a single package, is shown in Figure 6.17. What is indicated is a unipolar drive (no current reversal in a phase winding). As a result, a stepper motor with bifilar windings (two coil segments for each phase) has to be used. The motor has five leads, one of which is the motor common or ground (G) and the other four are the terminals of the two bifilar coil segments (A^+, A^-, B^+, B^-).

In the drive module there are several pins, some of which are connected to the motor controller or computer (driver inputs) and some are connected to the motor leads (driver outputs). There are other pins, which correspond to the dc power supply, common ground, various control signals, etc. The pin denoted by STEP (or PULSE) receives the stepping pulse signal (from the motor controller). This corresponds to the required stepping sequence of the motor. A transition from a low level to a high level (or rising edge) of a pulse will cause the motor to move by one step. The direction in which the motor moves is determined by the state of the pin denoted by CW/$\overline{\text{CCW}}$. A logical high state at this pin (or open connection) will generate switching logic for the motor to move in the clockwise direction, and logical low (or logic common) will generate switching logic for the motor to move in the CCW direction. The pin denoted by HALF/$\overline{\text{FULL}}$ determines whether the half stepping or full stepping is carried out. Specifically, a logical low at this pin will generate the switching logic for full stepping, and the logical high will generate switching logic for half stepping. The pin denoted by RESET receives the signal for initialization of a stepping sequence. There are several other pins, which are not necessary for the present discussion. The translator interprets the logical states at the STEP, HALF/$\overline{\text{FULL}}$, and CW/$\overline{\text{CCW}}$ pins and generates the proper logic to activate the switches

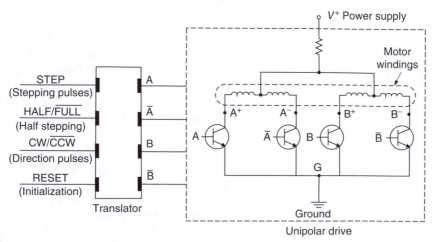

FIGURE 6.17
Basic drive hardware for a two-phase bifilar-wound stepper motor.

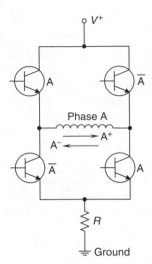

FIGURE 6.18
A bipolar drive for a single phase of a stepper motor (unifilar wound).

in the unipolar drive. Specifically, four active logic signals are generated corresponding to A (phase A on), $\overline{A}$ (phase A reversed), B (phase B on), and $\overline{B}$ (phase B reversed). These logic signals activate the four switches in the bipolar drive, thereby sending current through the corresponding winding segments or leads (A^+, A^-, B^+, B^-) of the motor.

The logic hardware is commonly available as compact chips in the monolithic form. If the motor is unifilar wound (for a two-phase stepper there should be three leads—a ground wire and two power leads for the two phases), a bipolar drive will be necessary in order to change the direction of the current in a phase winding. A schematic representation of a bipolar drive for a single phase of a stepper is shown in Figure 6.18. Note that when the two transistors marked A are on, the current flows in one direction through the phase winding, and when the two transistors marked $\overline{A}$ are on, the current flows in the opposite direction through the same phase winding. What is shown is an H-bridge circuit.

6.3.2 Motor Time Constant

As the torque generated by a stepper motor is proportional to the phase current, it is desirable for a phase winding to reach its maximum current level as quickly as possible when it is switched on. Unfortunately, as a result of self-induction, the current in the energized phase does not build up instantaneously when switched on. As the stepping rate increases, the time period that is available for each step decreases. Consequently, a phase may be turned off before reaching its desired current level in order to turn on the next phase, thereby degrading the generated torque. This behavior is illustrated in Figure 6.19.

One way to increase the current level reached by a phase winding would be to simply increase the supply voltage as the stepping rate increases. Another approach would be to use a chopper circuit (a switching circuit) to switch on and off at high frequency, a supply voltage that is several times higher than the rated voltage of a phase winding. Specifically, a sensing element (typically, a resistor) in the drive circuit detects the current level and when the desired level is reached, the voltage supply is turned off. When the current level goes below the rated level, the supply is turned on again. The required switching rate (chopping rate) is governed by the electrical time constant of the motor.

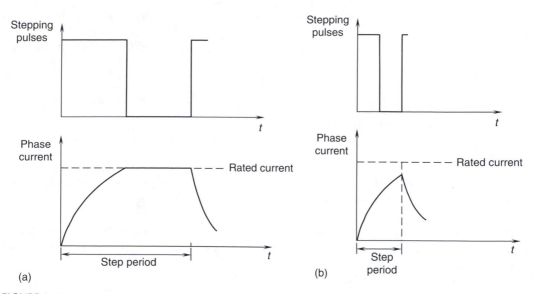

FIGURE 6.19

Torque degradation at higher stepping rates due to inductance. (a) Low stepping rate. (b) High stepping rate.

The electrical time constant of a stepper motor is given by

$$\tau_e = \frac{L}{R}, \tag{6.19}$$

where L is the inductance of the energized phase winding and R is the resistance of the energized circuit, including winding resistance. It is well known that the current buildup is given by

$$i = \frac{v}{R} \exp\left(1 - t/\tau_e\right), \tag{6.20}$$

where v is the supply voltage. The larger the electrical time constant, the slower the current buildup. The driving torque of the motor decreases due to the lower phase current. Also, because of self-induction, the current does not die out instantaneously when the phase is switched off. The instantaneous voltages caused by self-induction can be high, and they can damage the translator and other circuitry. The torque characteristics of a stepper motor can be improved (particularly at high stepping rates) and the harmful effects of induced voltages can be reduced by decreasing the electrical time constant. A convenient way to accomplish this is by increasing the resistance R. But we want this increase in R to be effective only during the transient periods (at the instants of switch-on and switch-off). During the steady period, we like to have a smaller R, which will give a larger current (and magnetic field), producing a higher torque, and furthermore lower power dissipation (and associated mechanical and thermal problems) and reduction of efficiency. This can be accomplished by using a diode and a resistor ΔR, connected in parallel with the phase winding, as shown in Figure 6.20. In this case, the current will loop through R and ΔR, as shown, during the switch-on and switch-off periods, thereby decreasing the electrical time constant to

$$\tau_e = \frac{L}{R + \Delta R}. \tag{6.21}$$

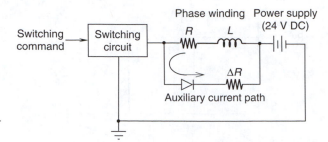

FIGURE 6.20
A diode circuit in a motor driver for decreasing the electrical time constant.

During steady conditions, however, no current flows through ΔR, as desired. Such circuits to improve the torque performance of stepping motors are commonly integrated into the motor drive hardware.

The electrical time constant is much smaller than the mechanical time constant of a motor. Hence, increasing R is not a very effective way of increasing damping in a stepper motor.

6.4   Torque Motion Characteristics

It is useful to examine the response of a stepper motor to a single-pulse input before studying the behavior under general stepping conditions. Ideally, when a single pulse is applied, the rotor should instantaneously turn through one step angle ($\Delta\theta$) and stop at that detent position (stable equilibrium position). Unfortunately, the actual single-pulse response is somewhat different from this ideal behavior. In particular, often, the rotor will oscillate about the detent position before settling down. These oscillations result primarily from the interaction of motor load inertia (the combined inertia of rotor, load, etc.) with drive torque, and not necessarily due to shaft flexibility. This behavior can be explained using Figure 6.21.

Assume single-phase energization (i.e., only one phase is energized at a time). When a pulse is applied to the translator at C, the corresponding stator phase is energized. This generates a torque (due to magnetic attraction), causing the rotor to turn toward the

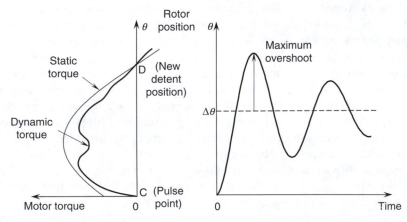

FIGURE 6.21
Single-pulse response and the corresponding single-phase torque of a stepper motor.

corresponding minimum reluctance position (detent position D). The static torque curve (broken line in Figure 6.21) represents the torque applied on the rotor from the energized phase, as a function of the rotor position θ, under ideal conditions (when dynamic effects are neglected). Under normal operating conditions, however, there will be induced voltages due to self-induction and mutual induction. Hence, a finite time is needed for the current to build up in the windings once a phase is switched on. Furthermore, there will be eddy currents generated in the rotor. These effects cause the magnetic field to deviate from the static conditions as the rotor moves at a finite speed, thereby making the dynamic torque curve different from the static torque curve, as shown in Figure 6.21. The true dynamic torque is somewhat unpredictable because of its dependence on many time-varying factors (rotor speed, rotor position, current level, etc.). The static torque curve is normally adequate to explain many characteristics of a stepper motor, including the oscillations in the single-pulse response.

It is important to note that the static torque is positive at the switching point, but is generally not maximum at that point. To explain this further, consider the three-phase VR stepper motor (with nontoothed poles) shown in Figure 6.5. The step angle $\Delta\theta$ for this arrangement is $60°$, and the full-step switching sequence for clockwise rotation is 1-2-3-1. Suppose that phase 1 is energized. The corresponding detent position is denoted by D in Figure 6.22a. The static torque curve for this phase is shown in Figure 6.22b, with the positive angle measured clockwise from the detent position D. Suppose that we turn the rotor CCW from this stable equilibrium position, using an external rotating mechanism (e.g., by hand). At position C, which is the previous detent position where phase 1 would have been energized under normal operation, there is a positive torque that tries to turn the rotor to its present detent position D. At position B, the static torque is zero, because the force from the N pole of phase 1 exactly balances that from the S pole. This point, however, is an unstable equilibrium position; a slight push in either direction will move the rotor in that direction. Position A, which is located at a rotor tooth pitch ($\theta_r = 180°$) from position D, is also a stable equilibrium position. The maximum static torque occurs at position M, which is located approximately halfway between positions B and D (at an angle $\theta_r/4 = 45°$ from the detent position). This maximum static torque is also known as the holding torque because it is the maximum resisting torque an energized motor can exert if we try to turn the rotor away from the corresponding detent position. The torque at the normal switching position (C) is less than the maximum value.

For a simplified analysis, the static torque curve is approximated to be sinusoidal. In this case, with phase 1 excited, and with the remaining phases inactive, the static torque distribution T_1 can be expressed as

$$T_1 = -T_{max} \sin n_r\theta, \tag{6.22}$$

where θ is the angular position in radians, measured from the current detent position (with phase 1 excited), n_r is the number of teeth on the rotor, and T_{max} is the maximum static torque (or, holding torque).

Equation 6.22 can be verified by referring to Figure 6.22, where $n_r = 2$. Note that Equation 6.22 is valid irrespective of whether the stator poles are toothed or not, even though the example considered in Figure 6.22 has nontoothed stator poles.

Returning to the single-pulse response shown in Figure 6.21, note that starting from rest at C, the rotor will have a positive velocity at the detent position D. Its kinetic energy (or momentum) will take it beyond the detent position. This is the first overshoot. As the same phase is still on, the torque will be negative beyond the detent position; static torque always attracts the rotor to the detent position, which is a stable equilibrium position. The rotor will decelerate because of this negative torque and will attain zero velocity at

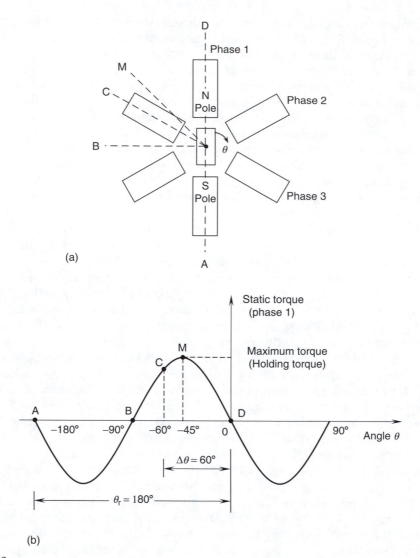

FIGURE 6.22
Static torque distribution of the VR stepper motor in Figure 6.5. (a) Schematic diagram. (b) Static torque curve for phase 1.

the point of maximum overshoot. Then, the rotor will be accelerated back toward the detent position and carried past this position by the kinetic energy, and so on. This oscillatory motion would continue forever with full amplitude ($\Delta\theta$) if there were no energy dissipation. In reality, however, there are numerous damping mechanisms—such as mechanical dissipation (frictional damping) and electrical dissipation (resistive damping through eddy currents and other induced voltages)—in the stepper motor, which will gradually slow down the rotor, as shown in Figure 6.21. Dissipated energy will appear primarily as thermal energy (temperature increase). For some stepper motors, the maximum overshoot could be as much as 80% of the step angle. Such high-amplitude oscillations with slow decay rate are clearly undesirable in most practical applications. Adequate damping should be provided by mechanical means (e.g., attaching mechanical dampers), electrical means (e.g., by further eddy current dissipation in the rotor or by using extra turns in the field windings), or by electronic means (electronic switching or multiple-phase energization) in order to suppress these oscillations. The first two techniques are wasteful, while the third approach requires switching control. The single-pulse response is often modeled using a simple oscillator transfer function.

Now, we will examine the stepper motor response when a sequence of pulses is applied to the motor under normal operating conditions. If the pulses are sufficiently spaced—typically, more than the settling time T_s of the motor (Note: $T_s \sim 4 \times$ motor time constant)—then the rotor will come to rest at the end of each step before starting the next step. This is known as single stepping. In this case, the overall response is equivalent to a cascaded sequence of single-pulse responses; the motor will faithfully follow the command pulses in synchronism. In many practical applications, however, fast responses and reasonably continuous motor speeds (stepping rates) are desired. These objectives can be met, to some extent, by decreasing the motor settling time through increased dissipation (mechanical and electrical damping). This, beyond a certain optimal level of damping, could result in undesirable effects, such as excessive heat generation, reduced output torque, and sluggish response. Electronic damping, explained in Section 6.5.2, can eliminate these problems.

As there are practical limitations to achieving very small settling times, faster operation of a stepper motor would require switching before the rotor settles down in each step. Of particular interest under high-speed operating conditions is *slewing motion*, where the motor operates at steady state in synchronism at a constant pulse rate called the slew rate. It is not necessary for the phase switching (i.e., pulse commands) to occur when the rotor is at the detent position of the old phase, but switchings (pulses) should occur in a uniform manner. As the motor moves in harmony, practically at a constant speed, the torque required for slewing is smaller than that required for transient operation (accelerating and decelerating conditions). Specifically, at a constant speed there is no inertial torque, and as a result, a higher speed can be maintained for a given level of motor torque. But, as stepper motors generate heat in their windings, it is not desirable to operate them at high speeds for long periods.

A typical displacement time curve under slewing is shown in Figure 6.23. The slew rate is given by

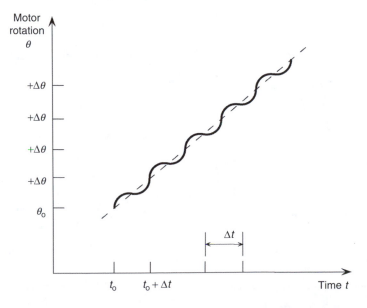

FIGURE 6.23
Typical slewing response of a stepping motor.

$$R_s = \frac{1}{\Delta t} \text{steps/s,} \tag{6.23}$$

where Δt denotes the time between successive pulses under slewing conditions. Note that Δt could be significantly smaller than the motor settling time, T_s. Some periodic oscillation (or hunting) is possible under slewing conditions, as seen in Figure 6.23. This is generally unavoidable, but its amplitude can be reduced by increasing damping. The slew rate depends as well on the external load connected to the motor. In particular, motor damping, bearing friction, and torque rating set an upper limit to the slew rate.

To attain slewing conditions, the stepper motor has to be accelerated from a low speed by ramping. This is accomplished by applying a sequence of pulses with a continuously increasing pulse rate $R(t)$. Strictly speaking, ramping represents a linear (straight line) increase of the pulse rate, as given by

$$R(t) = R_o + \frac{(R_s - R_o)t}{n \, \Delta t}, \tag{6.24}$$

where R_o is the starting pulse rate (typically zero), R_s is the final pulse rate (slew rate), and n is the total number of pulses.

If exponential ramping is used, the pulse rate is given by

$$R(t) = R_s - (R_s - R_o)e^{-t/\tau}. \tag{6.25}$$

If the time constant τ of the ramp is equal to $n\Delta t/4$, a pulse rate of 0.98, R_s is reached in a total of n pulses (Note: $e^{-4} = 0.02$.). In practice, the pulse rate is often increased beyond the slew rate, in a time interval shorter than what is specified for acceleration, and then decelerated to the slew rate by pulse subtraction at the end. In this manner, the slew rate is reached more quickly. In general, during upramping (acceleration), the rotor angle trails the pulse command, and during downramping (deceleration), the rotor angle leads the pulse command. These conditions are illustrated in Figure 6.24. The ramping rate cannot be chosen arbitrarily, and is limited by the torque–speed characteristics of the motor. If ramping rates

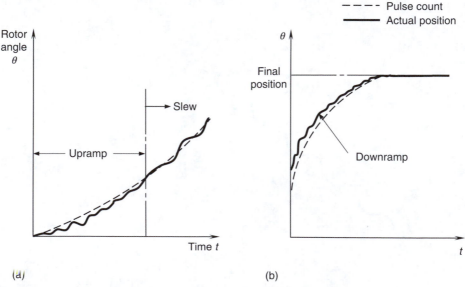

FIGURE 6.24

Ramping response. (a) Accelerating motion. (b) Decelerating motion.

beyond the capability of the particular motor are attempted, it is possible that the response will go completely out of synchronism and the motor will stall.

In transient operation of stepper motors, nonuniform stepping sequences might be necessary, depending on the complexity of the motion trajectory and the required accuracy. Consider, for example, the three-step drive sequence shown in Figure 6.25. The first pulse is applied at A when the motor is at rest. The resulting positive torque (curve 1) of the energized phase will accelerate the motor, causing an overshoot beyond the detent position (see broken line). The second pulse is applied at B, the point of intersection of the torque curves 1 and 2, which is before the detent position. This switches the torque to curve 2, which is the torque generated by the newly energized phase. Fast acceleration is possible in this manner because the torque is kept positive up to the second detent position. Note that the average torque is maximum when switching is done at the point of intersection of successive torque curves. The resulting torque produces a larger overshoot beyond the second detent position. As the rotor moves beyond the second detent position, the torque becomes negative, and the motor begins to slow down. The third pulse is applied at C when the rotor is close to the required final position. Note that the corresponding torque (curve 3) is relatively small, because the rotor is near its final (third) detent position. As a result of this and in view of the previous negative torque, the overshoot from the final detent position is relatively small, as desired. The rotor then quickly settles down to the final position, as there exists some damping or friction in the motor and its bearings.

Drive sequences can be designed in this manner to produce virtually any desired motion in a stepper motor. The motor controller is programmed to generate the appropriate pulse train in order to achieve the required phase switchings for a specified motion. Such drive sequences are useful also in compensating for missed pulses and in electronic damping. These two topics will be discussed later in the chapter.

Note that in order to simplify the discussion and illustration, we have used static torque curves in Figure 6.25. This assumes instant buildup of current in the energized phase and instant decay of current in the deenergized phase, thus neglecting all induced voltages and eddy currents. In reality, however, the switching torques will not be generated

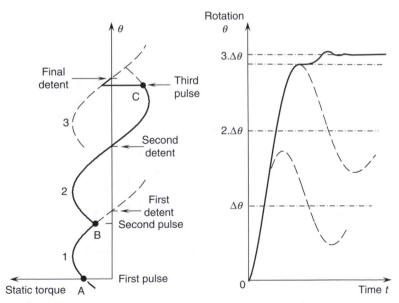

FIGURE 6.25
Torque–response diagram for a three-step drive sequence.

instantaneously. The horizontal lines with sharp ends to represent the switching torques, as shown in Figure 6.25, are simply approximations, and in practice the entire torque curve will be somewhat irregular. These dynamic torque curves should be used for accurate switching control in sophisticated practical applications.

6.4.1 Static Position Error

If a stepper motor does not support a static load (e.g., spring-like torsional element), the equilibrium position under power-on conditions would correspond to the zero-torque (detent) point of the energized phase. If there is a static load T_L, however, the equilibrium position would be shifted to $-\theta_e$, as shown in Figure 6.26. The offset angle θ_e is called the static position error.

Assuming that the static torque curve is sinusoidal, we can obtain an expression for θ_e. First, note that the static torque curve for each phase is periodic with period $p.\Delta\theta$ (equal to the rotor pitch θ_r), where p is the number of phases and $\Delta\theta$ is the step angle. As an example, this relationship is shown for the three-phase case in Figure 6.27. Accordingly, the static torque curve may be expressed as

$$T = -T_{\max}\sin\left(\frac{2\pi\theta}{p\Delta\theta}\right), \tag{6.26}$$

where $T_{\max}$ denotes the maximum torque. Equation 6.26 can be directly obtained by substituting Equation 6.2 in Equation 6.22. Note that under standard switching conditions, Equation 6.26 governs for $-\Delta\theta \leq \theta \leq 0$. With reference to Figure 6.26, the static position error is given by

$$T_L = -T_{\max}\sin\left[\frac{2\pi(-\theta_e)}{p.\Delta\theta}\right]$$

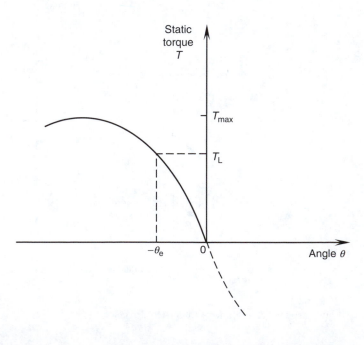

FIGURE 6.26
Representation of the static position error.

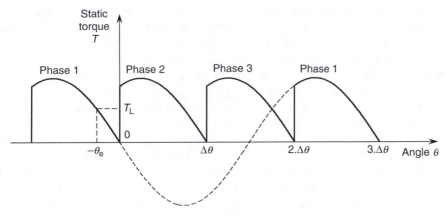

FIGURE 6.27
Periodicity of the single-phase static torque distribution (a three-phase example).

or

$$\theta_e = \frac{p.\Delta\theta}{2\pi} \sin^{-1}\left(\frac{T_L}{T_{max}}\right). \tag{6.27}$$

If n denotes the number of steps per revolution, Equation 6.27 may be expressed as

$$\theta_e = \frac{p}{n} \sin^{-1}\left(\frac{T_L}{T_{max}}\right). \tag{6.28}$$

It is intuitively clear that the static position error decreases with the number of steps per revolution.

Example 6.7
Consider a three-phase stepping motor with 72 steps/revolution. If the static load torque is 10% of the maximum static torque of the motor, determine the static position error.

Solution
In this problem,

$$\frac{T_L}{T_{max}} = 0.1, \quad p = 3, \quad n = 72.$$

Now, using Equation 6.28, we have

$$\theta_e = \frac{3}{72} \sin^{-1} 0.1 = 0.0042 \text{ rad} = 0.24°.$$

Note that this is less than 5% of the step angle.

6.5 Damping of Stepper Motors

Lightly damped oscillations in stepper motors are undesirable in applications that require single-step motions or accurate trajectory under transient conditions. Also, in slewing motions (where the stepping rate is constant), high-amplitude oscillations can result if the resonant frequency of the motor shaft–load combination coincides with the stepping frequency. Damping has the advantages of suppressing overshoots, increasing the decay rate of oscillations (i.e., shorter settling time) and decreasing the amplitude of oscillations under resonant conditions. Unfortunately, heavy damping has drawbacks, such as sluggish response (longer rise time, peak time, or delay), large time constants, thermal problems, wear, and reduction of the net output torque. On the average, however, the advantages of damping outweigh the disadvantages, in stepper motor applications.

Several techniques are employed to damp stepper motors. Most straightforward are the conventional techniques of damping, which use mechanical and electrical energy dissipation. Usually, mechanical damping is provided by a torsional damper attached to the motor shaft. Methods of electrical damping include eddy current dissipation in the rotor, the use of magnetic hysteresis and saturation effects, and increased resistive dissipation by adding extra windings to the motor stator. For example, solid-rotor construction has higher hysteresis losses due to magnetic saturation than laminated-rotor construction has. These direct techniques of damping have undesirable side effects, such as excessive heat generation, reduction of the net output torque of the motor, and decreased speed of response. Electronic damping methods have been developed to overcome such shortcomings. These methods are nondissipative, and are based on employing properly designed switching schemes for phase energization so as to inhibit overshoots in the final stage of response. A general drawback of electronic damping is that the associated switching sequences are complex (irregular) and depend on the nature of a particular motion trajectory. A rather sophisticated controller may be necessary as a result. The level of damping achieved by electronic damping is highly sensitive to the time sequence of the switching scheme. Accordingly, a high level of intelligence concerning the actual response of the motor is required to effectively use electronic damping methods. Note, also, that in the design stage, damping in a stepper motor can be improved or optimized by judicious choice of values for motor parameters (e.g., resistance of the windings, rotor size, material properties of the rotor, and air gap width).

6.5.1 Mechanical Damping

A convenient, practical method for damping of stepper motors is to connect an inertia element to the motor shaft through an energy dissipation medium, such as a viscous fluid (e.g., silicone) or a solid friction surface (e.g., brake lining). A common example for the first type of torsional dampers is the Houdaille damper (or viscous torsional damper) and the second type (which depends on Coulomb-type friction) it is the Lanchester damper.

The effectiveness of torsional dampers on stepper motors can be examined using a linear dynamic model for the single-step oscillations. From Figure 6.28a, it is evident that in the neighborhood of the detent position, the static torque due to the energized phase is approximately linear, and this torque acts as an electromagnetic spring. In this region, the torque can be expressed by

$$T = -K_m\theta, \tag{6.29}$$

where θ is the angle of rotation measured from the detent position and K_m is the torque constant (or magnetic stiffness or torque gradient) of the motor.

Damping forces also come from such sources as bearing friction, resistive dissipation in windings, eddy current dissipation in the rotor, and magnetic hysteresis. If the combined

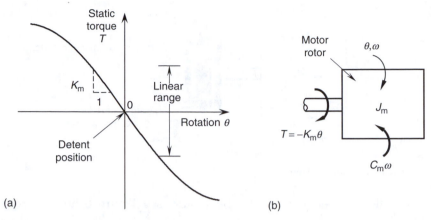

FIGURE 6.28
Model for single-step oscillations of a stepper motor. (a) Linear torque approximation. (b) Rotor free-body diagram.

contribution from these internal dissipation mechanisms is represented by a single damping constant C_m, the equation of motion for the rotor near its detent position (equilibrium position) can be written as

$$J_m \frac{d\omega}{dt} = -C_m\omega - K_m\theta, \tag{6.30}$$

where J_m is the overall inertia of the rotor and

$$\omega = \frac{d\theta}{dt} \text{ is motor speed.}$$

Note that for a motor with an external load, the load inertia has to be included in J_m. Equation 6.30 is expressed in terms of θ; thus,

$$J_m\ddot{\theta} + C_m\dot{\theta} + K_m\theta = 0. \tag{6.31}$$

The solution of this second-order ordinary differential equation is obtained using the maximum overshoot point as the initial state:

$$\dot{\theta}(0) = 0 \quad \text{and} \quad \theta(0) = \alpha\Delta\theta.$$

The constant α represents the fractional overshoot. Its magnitude can be as high as 0.8. The undamped natural frequency of single-step oscillations is given by

$$\omega_n = \sqrt{\frac{K_m}{J_m}}, \tag{6.32}$$

and the damping ratio is given by

$$\zeta = \frac{C_m}{2\sqrt{K_mJ_m}}. \tag{6.33}$$

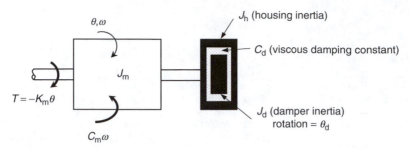

FIGURE 6.29
A stepper motor with a Houdaille damper.

With a Houdaille damper attached to the motor (see Figure 6.29), the equations of motion are

$$(J_m + J_h)\ddot{\theta} = -C_m\dot{\theta} - K_m\theta - C_d(\dot{\theta} - \dot{\theta}_d),$$

(6.34)

$$J_d\ddot{\theta}_d = C_d(\dot{\theta} - \dot{\theta}_d),$$

(6.35)

where θ_d is the angle of rotation of the damper inertia, J_d is the moment of inertia of the damper, and J_h is the moment of inertia of the damper housing. It is assumed that the damper housing is rigidly attached to the motor shaft.

In Figure 6.30, a typical response of a mechanically damped stepper motor is compared with the response when the external damper is disconnected. Observe the much faster decay when the external damper is present. One disadvantage of this method of damping, however, is that it always adds inertia to the motor (note the J_h term in Equation 6.34). This reduces the natural frequency of the motor (Equation 6.32) and, hence, decreases the speed of response (or bandwidth). Other disadvantages include reduction of the effective torque, wear and tear of the moving elements, and increased heat generation, which may require special cooling means.

A Lanchester damper is similar to a Houdaille damper, except that the former depends on nonlinear (Coulomb) friction instead of viscous damping. Hence, a stepper motor with

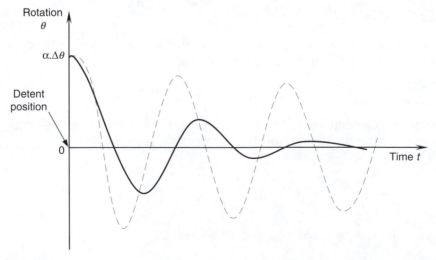

FIGURE 6.30
Typical single-step response of a stepper motor with a Houdaille damper (solid line: with damper; broken line: without damper).

a Lanchester damper can be analyzed in a manner similar to what was presented for a Houdaille damper, but the equations of motion are nonlinear now, because the frictional torque is of Coulomb type. Coulomb frictional torque has a constant magnitude for a given reaction force but acts opposite to the direction of relative motion between the rotor (and damper housing) and the damper inertia element. The reaction force on the friction lining can be adjusted using spring-loaded bolts, thereby changing the frictional torque. There are two limiting states of operation: (1) if the reaction force is very small, the motor is virtually uncoupled (disengaged) from the damper and (2) if the reaction force is very large, the damper inertia will be rigidly attached to the damper housing, thus moving as a single unit. In either case, there is very little dissipation. Maximum energy dissipation takes place under some intermediate condition. For constant-speed operation, by adjusting the reaction force, the damper inertia element can be made to rotate at the same speed as the rotor, thereby eliminating dissipation and torque loss under these steady conditions in which damping is usually not needed. This is an advantage of friction dampers.

6.5.2 Electronic Damping

Damping of stepper motor response by electronic switching control is an attractive method of overshoot suppression for several reasons. For instance, it is not an energy dissipating method. In that sense, it is actually an electronic control technique rather than a damping technique. By properly timing the switching sequence, virtually a zero overshoot response can be realized. Another advantage is that the reduction in net output torque is insignificant in this case in comparison with the torque losses in direct (mechanical) damping methods. A majority of electronic damping techniques depend on a two-step procedure that is straightforward in principle:

1. Decelerate the final-step response of the motor so as to avoid large overshoots from the final detent position.
2. Energize the final phase (i.e., apply the last pulse) when the motor response is very close to the final detent position (i.e., when the torque is very small).

It is possible to come up with many switching schemes that conform to these two steps. Generally, such schemes differ only in the manner in which response deceleration is brought about (in step 1 listed above). Three common methods of response deceleration are

1. The pulse turn-off method: Turn off the motor (all phases) for a short time.
2. The pulse reversal method: Apply a pulse in the opposite direction (i.e., energize the reverse phase) for a short time.
3. The pulse delay method: Maintain the present phase beyond its detent position for a short time.

These three types of switching schemes can be explained using the static torque response curves in Figure 6.31 through Figure 6.33. In all three figures, the static torque curve corresponding to the last pulse (i.e., last energized phase) is denoted by 2. The static curve corresponding to the next-to-last pulse is denoted by 1.

In the pulse turn-off method (Figure 6.31), the last pulse is applied at A, as usual. This energizes phase 2, turning off phase 1. The rotor accelerates toward its final detent position because of the positive torque that is present. At point B, which is sufficiently close to the final detent position, phase 2 is shut off. From B to C, all phases of the motor are inactive, and the static torque is zero. The motor decelerates during this interval,

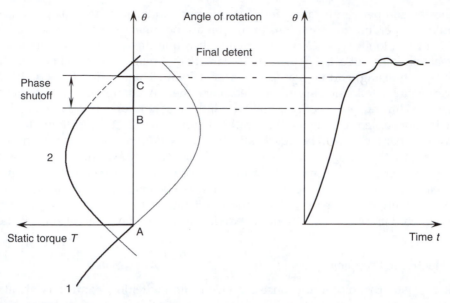

FIGURE 6.31
The pulse turn-off method of electronic damping.

giving a peak response that is very close to, but below, the final detent position. At point C, the last phase (phase 2) is energized again. As the corresponding static torque is very small (in comparison with the maximum torque) but positive, the motor will accelerate slowly (assume a purely inertial load) to the final detent position. By properly choosing the points B and C, the overshoot can be made sufficiently small. This choice requires knowledge of the actual response of the motor. The amount of final overshoot can be very sensitive to the timing of the switching points B and C. Furthermore, the actual response θ will depend on mechanical damping and other load characteristics as well.

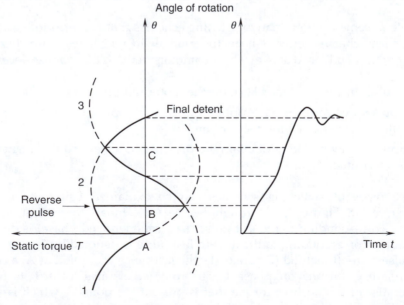

FIGURE 6.32
The pulse reversal method of electronic damping.

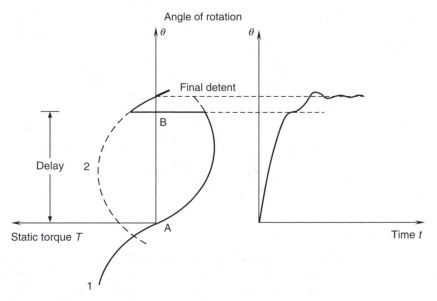

FIGURE 6.33
The pulse delay method of electronic damping. Damping of stepper motors.

The pulse reversal method is illustrated in Figure 6.32. The static torque curve corresponding to the second pulse before last is denoted by 3. As usual, the last phase (phase 2) is energized at A. The motor will accelerate toward the final detent position. At point B (located at less than half the step angle from A), phase 2 is shut off and phase 3 is turned on (Note: The forward pulse sequence is 1–2–3–1, and the reverse pulse sequence is 1–3–2–1.). The corresponding static torque is negative over some duration (Note: For a three-phase stepper motor, this torque is usually negative up to the halfway point of the step angle and positive thereafter.). Consequently, the motor will decelerate first and then accelerate (assume a purely inertial load); the overall decelerating effect is not as strong as in the previous method (Note: If faster deceleration is desired, phase 1 should be energized, instead of phase 3, at B.). At point C, the static torque of phase 3 becomes equal to that of phase 2. To avoid large overshoots, phase 3 is turned off at point C, and the last phase (phase 2) is energized again. This will drive the motor to its final detent position.

In the pulse delay method (Figure 6.33), the last phase is not energized at the detent position of the previous step (point A). Instead, phase 1 is continued on beyond this point. The resulting negative torque will decelerate the response. If intentional damping is not employed, the overshoot beyond A could be as high as 80% (Note: In the absence of any damping, 100% overshoot is possible.). When the overshoot peak is reached at B, the last phase is energized. As the static torque of phase 2 is relatively small at this point and will reach zero at the final detent position, the acceleration of the motor is slow. Hence, the final overshoot is maintained within a relatively small value. It is interesting to note that if 100% overshoot is obtained with phase 1 energized, the final overshoot becomes zero in this method, thus producing ideal results.

In all these techniques of electronic damping, the actual response depends on many factors, particularly the dynamic behavior of the load. Hence, the switching points cannot be exactly prespecified unless the true response is known ahead of time (through tests, simulations, etc.). In general, accurate switching may require measurement of the actual response and use of that information in real time to apply the switching pulses. Note that in Figure 6.31 through Figure 6.33 static torque curves are used to explain electronic damping. In practice, however, currents in the phase windings neither decay nor build up instantaneously, following a pulse command. Induced voltages, eddy currents, and magnetic hysteresis effects are primarily responsible for this behavior. These factors, in addition to external loads, can complicate the nature of dynamic torque and, hence, the true response of a stepping motor. This can make an accurate preplanning of switching points rather difficult in electronic damping. In the foregoing discussion, we have assumed that the mechanical damping (including bearing friction) of the motor is negligible and that the load connected to the motor is a pure inertia. In practice, the net torque available to drive the combined rotor–load inertia is smaller than the electromagnetic torque generated at the rotor. Hence, in practice, the accelerations obtained are not quite as high as what Figure 6.31 through Figure 6.33 suggest. Nevertheless, the general characteristics of motor response will be the same as those shown in these figures.

6.5.3 Multiple Phase Energization

A popular and relatively simple method that may be classified under electronic damping is multiple-phase energization. With this method, two phases are excited simultaneously (e.g., 13–21–32–13). One is the standard stepping phase and the other is the damping phase. The damping phase provides a deceleration effect. Specifically, the damping phase corresponds to rotation in the reverse direction, but it is energized at a fraction of the stepping voltage (rated voltage), together with the stepping phase (which is energized with the full voltage). As noted earlier in the chapter, the step angle remains unchanged when more than one phase is energized simultaneously (as long as the number of phases activated at a time is the same, which is the case here). It has been observed that this switching sequence provides a better response (less overshoot) than the single-phase energization method (e.g., 1–2–3–1), particularly for single-stack stepper motors. The damping phase (which is the reverse phase) provides a negative torque, and it not only reduces the overshoot but also the speed of response. Increased magnetic hysteresis and saturation effects of the ferromagnetic materials in the motor, as well as higher energy dissipation through eddy currents when two phases are energized simultaneously, are other factors that enhance damping and reduce the speed of response, in simultaneous multiphase energization. Another factor is that multiple-phase excitation results in wider overlaps of magnetic flux between switchings, giving smoother torque transitions. Note, however, that there can be excessive heat generation with this method. This may be reduced, to some extent, by further reducing the voltage of the damping phase, typically to half the normal rated voltage.

6.6 Stepping Motor Models

In the preceding sections, we have discussed variable-reluctance (VR) stepper motors, which have nonmagnetized soft-iron rotors, and permanent-magnet (PM) stepper motors, which have magnetized rotors. As noted, hybrid (HB) stepper motors are a special type of PM stepping motors. Specifically, a hybrid motor has two rotor stacks, which are magnetized to have opposite polarities (one rotor stack is the N pole and the other is the S pole). Also, there is a tooth misalignment between the two rotor stacks. As usual, stepping is achieved by switching the phase excitations.

In the analysis of stepper motors under steady operation at low speeds, we usually do not need to differentiate between VR motors, PM motors, and HB motors. But under transient conditions, the torque characteristics of the three types of motors can differ considerably. In particular, the torque in PM and HB motors varies somewhat linearly with the magnitude of the phase current (as rotor field is provided by permanent magnets), whereas the torque in a VR motor varies nearly quadratically with the phase current (as the stator field links with the rotor, which does not have its own magnetic field).

6.6.1 A Simplified Model

Under steady-state operation of a stepper motor at low speeds, the motor (magnetic) torque can be approximated by a sinusoidal function, as given by Equation 6.22 or Equation 6.26. Hence, the simplest model for any type of stepping motor (VR, PM, or HB) is the torque source given by

$$T = -T_{\max} \sin n_r \theta \tag{6.36}$$

or equivalently,

$$T = -T_{\max} \sin\left(\frac{2\pi\theta}{p\cdot\Delta\theta}\right), \tag{6.37}$$

where $T_{\max}$ is the maximum torque during a step (holding torque), $\Delta\theta$ is the step angle, n_r is the number of rotor teeth, and p is the number of phases.

Note that θ is the angular position of the rotor measured from the detent position of the presently excited phase, as indicated in Figure 6.34a. Hence, $\theta = -\Delta\theta = -\theta_r/p$ at the

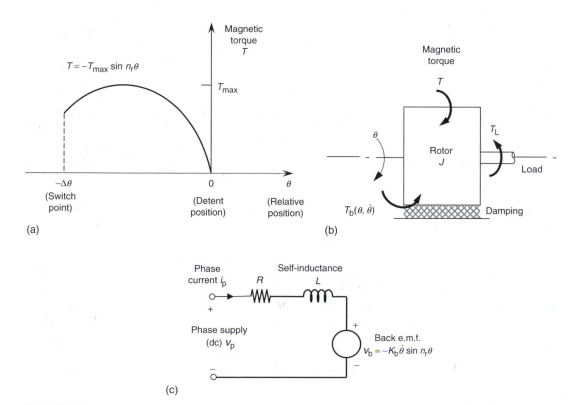

FIGURE 6.34
Stepper motor models. (a) Torque source model. (b) Mechanical model. (c) Equivalent circuit for an improved model.

previous detent position, where the present phase is switched on, and $\theta = 0$ at the approaching detent position. The coordinate frame of θ is then shifted again to a new origin ($+\Delta\theta$) when the next phase is excited at the approaching detent position of the conventional method of switching. Hence, θ gives the relative position of the rotor during each step. The absolute position is obtained by adding θ to the absolute rotor angle at the approaching detent position.

The motor model becomes complete with the mechanical dynamic equation for the rotor. With reference to Figure 6.34b, Newton's second law gives

$$T - T_L - T_b(\theta, \dot\theta) = J\ddot\theta, \tag{6.38}$$

where T_L is the resisting torque (reaction) on the motor by the driven mechanical load (i.e., load torque), T_b $(\theta, \dot\theta)$ is the dissipative resisting torque (viscous damping torque, frictional torque, etc.) on the motor, and J is the motor–rotor inertia.

Note that T_L will depend on the nature of the external load. Furthermore, T_b $(\theta, \dot\theta)$ will depend on the nature of damping. If viscous damping is assumed, T_b may be taken as proportional to $\dot\theta$. On the other hand, if Coulomb friction is assumed, the magnitude of T_b is taken to be constant, and the sign of T_b is the sign of $\dot\theta$. In the case of general dissipation (e.g., a combination of viscous, Coulomb, and structural damping), T_b is a nonlinear function of both θ and $\dot\theta$. Note that the torque source model may be used for all three (VR, PM, and HB) types of stepping motors.

6.6.2 An Improved Model

Under high-speed and transient operation of a stepper motor, many of the quantities that were assumed to be constant in the torque source model will vary with time as well as rotor position. In particular, for a given supply voltage v_p to a phase winding, the associated phase current i_p will not be constant. Also, inductance L in the phase circuit will vary with the rotor position. Furthermore, a voltage v_b (a back e.m.f.) will be induced in the phase circuit because of the changes in magnetic flux resulting from the speed of rotation of the rotor (in all three, VR, PM, and HB motors). It follows that an improved dynamic model is needed to represent the behavior of a stepper motor under high-speed and transient conditions. Such a model is described now. Instead of using rigorous derivations, motor equations are obtained from an equivalent circuit using qualitative considerations.

As magnetic flux linkage of the phase windings changes as a result of variations in the phase current, a voltage is induced in the phase windings. Hence, a self-inductance (L) should be included in the circuit. Although a mutual inductance should also be included to account for voltages induced in a phase winding as a result of current variations in the other phase windings, this voltage is usually smaller than the self-induced voltage. Hence, in the present model we neglect mutual inductance. Furthermore, flux linkage of the phase windings changes as a result of the motion of the rotor. This induces a voltage v_b (termed a back e.m.f.) in the phase windings. This voltage is present irrespective of whether the rotor is a VR type, a PM type, or an HB type. Also, phase windings will have a finite resistance R. It follows that an approximate equivalent circuit (neglecting mutual induction, in particular) for one phase of a stepper motor can be represented as in Figure 6.34c. The phase circuit equation is

$$v_p = Ri_p + L\frac{di_p}{dt} + v_b, \tag{6.39}$$

where v_p is the phase supply voltage (dc), i_p is the phase current, v_b is the back e.m.f. due to rotor motion, R is the resistance in the phase winding, and L is the self-inductance of the phase winding.

The back e.m.f. is proportional to the rotor speed $\dot{\theta}$ and it will also vary with the rotor position θ. The variation with position is periodic with period θ_r. Hence, using only the fundamental term in a Fourier series expansion, we have

$$v_b = -k_b \dot{\theta} \sin n_r\theta, \tag{6.40}$$

where $\dot{\theta}$ is the rotor speed, θ is the rotor position (as defined in Figure 6.34a), n_r is the number of rotor teeth, and k_b is the back e.m.f. constant. As θ is negative in a conventional step (which is from $\theta = -\Delta\theta$ to $\theta = 0$), we note that v_b is positive for positive $\dot{\theta}$.

Self-inductance L also varies with the rotor position θ. This variation is periodic with period θ_r. Now, retaining only the constant and the fundamental terms in a Fourier series expansion, we have

$$L = L_o + L_a \cos n_r\theta, \tag{6.41}$$

where L_o and L_a are appropriate constants and angle θ is as defined in Figure 6.34a.

Equation 6.39 through Equation 6.41 are valid for all three types of stepper motors (VR, PM, and HB). The torque equation will depend on the type of stepper motor, however.

6.6.2.1   Torque Equation for PM and HB Motors

In a permanent-magnet (PM) and hybrid stepper motor, magnetic flux is generated by both the phase current i_p and the magnetized rotor. The flux from the magnetic rotor is constant, but its linkage with the phase windings will be modulated by the rotor position θ. Hence, retaining only the fundamental term in a Fourier series expansion, we have

$$T = -k_m i_p \sin n_r\theta, \tag{6.42}$$

where i_p is the phase current and k_m is the torque constant for the PM or HB motor.

6.6.2.2   Torque Equation for VR Motors

In a variable-reluctance (VR) stepper motor, the rotor is not magnetized; hence, there is no magnetic flux generation from the rotor. The flux generated by the phase current i_p is linked with the phase windings. The flux linkage is coupled with the motor rotor and as a result it is modulated by the motion of the VR motor rotor. Hence, retaining only the fundamental term in a Fourier series expansion, the torque equation for a VR stepper motor may be expressed as

$$T = -k_r i_p^2 \sin n_r\theta, \tag{6.43}$$

where k_r is the torque constant for the VR motor. Note that torque T depends on the phase current i_p in a quadratic manner in the VR stepper motor. This makes a VR motor more nonlinear than a PM motor or an HB motor.

In summary, to compute the torque T at a given rotor position, we first have to solve the differential equation given by Equation 6.39 through Equation 6.41 for known values of the rotor position θ and the rotor speed $\dot{\theta}$ and for a given (constant) phase supply voltage v_p. Initially, as a phase is switched on, the phase current is zero. The model parameters R, L_o, L_a, and k_b are assumed to be known (either experimentally or from the manufacturer's data sheet). Then torque is computed using Equation 6.42 for a PM or HB stepper motor or using Equation 6.43 for a VR stepper motor. Again, the torque constant (k_m or k_r) is assumed to be known. The simulation of the model then can be completed by using this torque in the mechanical dynamic Equation 6.38 to determine the rotor position θ and the rotor speed $\dot{\theta}$.

6.7   Control of Stepper Motors

Open-loop operation is adequate for many applications of stepper motors, particularly at low speeds and in steady-state operation. The main shortcoming of open-loop control is that the actual response of the motor is not measured; consequently, it is not known whether a significant error is present, for example due to missed pulses.

6.7.1   Pulse Missing

There are two main reasons for pulse missing:

1. Particularly under variable-speed conditions, if the successive pulses are received at a high frequency (high stepping rate), the phase translator might not respond to a received pulse, and the corresponding phase would not be energized before the next pulse arrives. This may occur, for instance, due to a malfunction in the translator or the drive circuit.

2. Because of a malfunction in the pulse source, a pulse might not actually be generated, even when the motor is operating at well below its rated capacity (low-torque, low-speed, and low-transient conditions). Extra (erroneous) pulses can be generated as well by a faulty pulse source or drive circuitry.

If a pulse is missed by the motor, the response has to catch up somehow (e.g., by a subsequent overshoot in motion), or else an erratic behavior may result, causing the rotor to oscillate and probably stall eventually. Under relatively favorable conditions, particularly with small step angles, if a single pulse is missed, the motor will decelerate so that a complete cycle of pulses is missed; then it will lock in again with the input pulse sequence. In this case, the motor will trail the correct trajectory by a rotor tooth pitch angle (θ_r). Here, pulses equal in number to the total phases (p) of the motor (Note: $\theta_r = p\Delta\theta$.) are missed. In this manner, it is also possible to lose accuracy by an integer multiple of θ_r because of a single missed pulse. Under adverse conditions, however, pulse missing can lead to a highly nonsynchronous response or even complete stalling of the motor.

In summary, the missing (or dropping) of a pulse can be interpreted in two ways. First, a pulse can be lost between the pulse generator (e.g., a command computer or controller) and the translator. In this case, the logic sequence within the translator that energizes motor phases will remain intact. The next pulse to arrive at the translator will be interpreted as the lost pulse and will energize the phase corresponding to the lost pulse. As a result, a time delay is introduced to the command (pulse) sequence. The second interpretation of a missed pulse is that the pulse actually reached the translator, but the corresponding motor phase was not energized because of some hardware problem in the translator or other drive circuits. In this case, the next pulse reaching the translator will not energize the phase corresponding to the missed pulse but will energize the phase corresponding to the received pulse. This interpretation is termed missing of phase activation.

In both interpretations of pulse missing, the motor will decelerate because of the negative torque from the phase that was not switched off. Depending on the timing of subsequent pulses, a negative torque can continue to exist in the motor, thereby eventually stalling the motor. Motor deceleration due to pulse missing can be explained using the static torque approximation, as shown in Figure 6.35. Consider a three-phase motor with one-phase-on excitation (i.e., only one phase is excited at a given time). Suppose that under normal operating conditions, the motor runs at a constant speed and phase activation is brought about at points A, B, C, D, etc. in Figure 6.35, using a pulse sequence sent into the translator. These points are equally spaced (with the horizontal axis as the angle of rotation θ, not time t) because of constant speed operation. The torque generated by the motor under normal operation, without pulse missing, is shown as a solid line in Figure 6.35. Note that phase 1 is excited at point A, phase 2 is excited at point B, phase 3 is excited at point C, and so on. Now let us examine the two cases of pulse missing.

In the first case (Figure 6.35a), a pulse is missed at B. Phase 1 continues to be active, providing a negative torque. This slows down the motor. The next pulse is received when the rotor is at position B' (not C) because of rotor deceleration (note that pulses are sent at equal time intervals for constant-speed operation). At point B', phase 2 (not phase 3) is excited in this case, because the translator interprets the present pulse as the pulse that

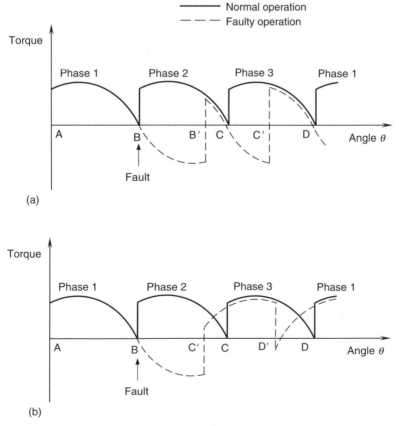

FIGURE 6.35
Motor deceleration due to pulse missing: (a) Case of a missed pulse; (b) Case of a missed phase activation.

was lost. The next pulse is received at C′, and so on. The resulting torque is shown by the broken line in Figure 6.35a. As this torque could be significantly less than the torque in the absence of missed pulses—depending on the locations of points B′, C′, and so on—the motor might decelerate continuously and finally stall.

In the second case (Figure 6.35b), the pulse at B fails to energize phase 2. This decelerates the motor because of the negative torque generated by the existing phase 1. The next pulse is received at point C′ (not C) because the motor has slowed down. This pulse excites phase 3 (not phase 2, unlike in the previous case), because the translator assumes that phase 2 has been excited by the previous pulse. The subsequent pulse arrives at point D′ (not D) because of the slowed speed of the motor. The corresponding motor torque is shown by the broken line in Figure 6.35b. In this case as well, the net torque can be much smaller than what is required to maintain the normal operating speed, and the motor may stall. To avoid this situation, pulse missing should be detected by response sensing (e.g., using an optical encoder; see Chapter 5), and proper corrective action taken by modifying the future switching sequence in order to accelerate the motor back into the desired trajectory. In other words, feedback control is required.

6.7.2 Feedback Control

Feedback control is used to compensate for motion errors in stepper motors. A block diagram for a typical closed-loop control system is shown in Figure 6.36. This should be compared with Figure 6.16a. The noted improvement in the feedback control scheme is that the actual response of the stepper motor is sensed and compared with the desired response; if an error is detected, the pulse train to the drive system is modified appropriately to reduce the error. Typically, an optical incremental encoder (see Chapter 5) is employed as the motion transducer. This device provides two pulse trains that are in phase quadrature (or, alternatively, a position pulse sequence and a direction change pulse may be provided), giving both the magnitude and the direction of rotation of the stepper motor. The encoder pitch angle should be made equal to the step angle of the motor for ease of comparison and error detection. When feedback control is employed, the resulting closed-loop system can operate near the rated capacity (torque, speed, acceleration, etc.) of the stepper motor, perhaps exceeding these ratings at times but without introducing excessive error and stability problems (e.g., hunting).

A simple closed-loop device that does not utilize sophisticated control logic is the feedback encoder–driven stepper motor. In this case, the drive pulses, except for the very first pulse, are generated by a feedback encoder itself, which is mounted on the motor shaft. This mechanism is particularly useful for operations requiring steady acceleration and deceleration under possible overload conditions, when there is likelihood of a

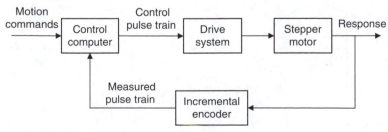

FIGURE 6.36
Feedback control of a stepper motor.

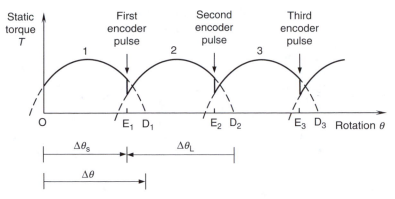

FIGURE 6.37
Operation of a feedback encoder–driven stepper motor.

pulse missing. The principle of operation of a feedback encoder–driven stepper motor may be explained using Figure 6.37. The starting pulse is generated externally at the initial detent position O. This will energize phase 1 and drive the rotor toward the corresponding detent position D_1. The encoder disc is positioned such that the first pulse from the encoder is generated at E_1. This pulse is automatically fed back as the second pulse is input to the motor (translator). This pulse will energize phase 2 and drive the rotor toward the corresponding detent position D_2. During this step, the second pulse from the encoder is generated at E_2, which is automatically fed back as the third pulse is input to the motor, energizing phase 3 and driving the motor toward the detent position D_3, and so on. Note that phase switching occurs (because of an encoder pulse) every time the rotor has turned through a fixed angle $\Delta\theta_s$, from the previous detent position. This angle is termed the switching angle. The encoder pulse leads the corresponding detent position by an angle $\Delta\theta_L$. This angle is termed the lead angle. Note from Figure 6.37 that

$$\Delta\theta_s + \Delta\theta_L = 2\Delta\theta, \tag{6.44}$$

where $\Delta\theta$ denotes the step angle.

For the switching angle position (or lead angle position) shown in Figure 6.37, the static torque on the rotor is positive throughout the motion. As a result, the motor will accelerate steadily until the motor torque exactly balances damping torque, other speed-dependent resistive torques, and load torque. The resulting final steady-state condition corresponds to the maximum speed of operation for a feedback encoder–driven stepper motor. This maximum speed usually decreases as the switching angle is increased beyond the point of intersection of two adjacent torque curves. For example, if $\Delta\theta_s$ is increased beyond $\Delta\theta$, there is a negative static torque from the present phase (before switching), which tends to somewhat decelerate the motor. But the combined effect of the before-switching torque and the after-switching torque is to produce an overall increase in speed until the speed limit is reached. This is generally true, provided that the lead angle $\Delta\theta_L$ is positive (the positive direction, is as indicated by the arrowhead in Figure 6.37). The lead angle may be adjusted either by physically moving the signal pick-off point on the encoder disc or by introducing a time delay into the feedback path of the encoder signal. The former method is less practical, however.

Steady decelerations can be achieved using feedback encoder–driven stepper motors if negative lead angles are employed. In this case, switching to a particular phase occurs when the rotor has actually passed the detent position for that phase. The

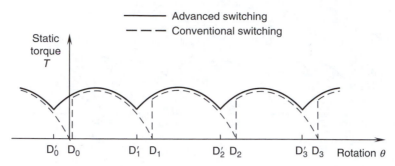

FIGURE 6.38
The effect of advancing the switching pulses.

resulting negative torque will steadily decelerate the rotor, eventually bringing it to a halt. Negative lead angles may be obtained by simply adding a time delay into the feedback path. Alternatively, the same effect (negative torque) can be generated by blanking out (using a blanking gate) the first two pulses generated by the encoder and using the third pulse to energize the phase that would be energized by the first pulse for an accelerating operation.

The feedback encoder–driven stepper motor is just a simple form of closed-loop control. Its application is normally limited to steadily accelerating (upramping), steadily decelerating (downramping), and steady-state (constant speed or slewing) operations. More sophisticated feedback control systems require point-by-point comparison of the encoder pulse train with the desired pulse train and injection of extra pulses or extraction (blanking out) of existing pulses at proper instants so as to reduce the error. A commercial version of such a feedback controller uses a count-and-compare card. More complex applications of closed-loop control include switching control for electronic damping (see Figure 6.31 through Figure 6.33), transient drive sequencing (see Figure 6.25), and dynamic torque control.

6.7.3 Torque Control through Switching

Under standard operating conditions for a stepper motor, phase switching (by a pulse) occurs at the present detent position. It is easy to see from the static torque diagram in Figure 6.38, however, that a higher average torque is possible by advancing the switching time to the point of intersection of the two adjacent torque curves (before and after switching). In the figure, the standard switching points are denoted as D_0, D_1, D_2, and so on, and the advanced switching points as D'_0, D'_1, D'_2, and so forth. Note that in the case of advanced switching, the static torque always remains greater than the common torque value at the point of intersection. This confirms what is intuitively clear; motor torque can be controlled by adjusting the switching point. The resulting actual magnitude of torque, however, will depend on the dynamic conditions that exist. For low speeds, the dynamic torque may be approximated by the static torque curve, making the analysis simpler. As the speed increases, the deviation from the static curve becomes more pronounced, for reasons that were mentioned earlier.

Example 6.8
Suppose that the switching point is advanced beyond the zero-torque point of the switched phase, as shown in Figure 6.39. The switching points are denoted by D'_0, D'_1, D'_2, and so on. Although the static torque curve takes negative values in some regions under this advanced switching sequence, the dynamic torque stays positive at all times.

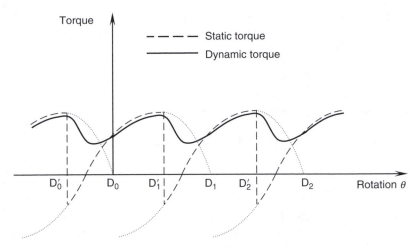

FIGURE 6.39
Dynamic torque at high speeds.

The main reason for this is that a finite time is needed for the current in the turned-off phase to decay completely because of induced voltages and eddy current effects.

6.7.4 Model-Based Feedback Control

The improved motor model, as presented before, is useful in computer simulation of stepper motors; for example, for performance evaluation. Such a model is also useful in model-based feedback control of stepper motors where the model provides a relationship between the motor torque and the motion variables θ and $\dot{\theta}$. From the model then, we can determine the required phase-switching points in order to generate a desired motor torque (to drive the load). Actual values of θ and $\dot{\theta}$ (e.g., as measured using an incremental optical encoder) are used in model-based computations.

A simple feedback control strategy for a stepper motor is outlined now. Initially, when the motor is at rest, the phase current $i_p = 0$. Also, $\theta = -\Delta\theta$ and $\dot{\theta} = 0$. As the phase is switched on to drive the motor, the motor Equation 6.39 is integrated in real time, using a suitable integration algorithm and an appropriate time step. Simultaneously, the desired position is compared with the actual (measured) position of the load. If the two are sufficiently close, no phase-switching action is taken. But suppose that the actual position lags behind the desired position. Then we compute the present motor torque using the model: Equation 6.40, Equation 6.41, and Equation 6.42 or Equation 6.43, and repeat the computations, assuming (hypothetically) that the excitation is switched to one of the two adjoining phases. As we need to accelerate the motor, we should actually switch to the phase that provides a torque larger than the present torque. If the actual position leads the desired position, however, we need to decelerate the motor. In this case, we switch to the phase that provides a torque smaller than the present torque or we turn off all the phases. The time taken by the phase current to build up to its full value is approximately equal to 4τ, where τ is the electrical time constant for each phase, as approximated by $\tau = L_o/R$. Hence, when a phase is hypothetically switched, numerical integration has to be performed for a time period of 4τ before the torques are compared. It follows that the performance of this control approach will depend on the operating speed of the motor, the computational efficiency of the integration algorithm, and the available computing power. At high speeds, less time is available for control computations. Ironically, it is at high speeds that control problems are severe, and sophisticated control techniques are needed; hence, hardware implementations of switching are desired. For better control, phase switching has to be based on the motor speed as well as the motor position.

6.8   Stepper Motor Selection and Applications

Earlier in the chapter we have discussed design problem that addressed the selection of geometric parameters (number of stator poles, number of teeth per pole, number of rotor teeth, etc.) for a stepper motor. Selection of a stepper motor for a specific application cannot be made on the basis of geometric parameters alone, however. Torque and speed considerations are often more crucial in the selection process. For example, a faster speed of response is possible if a motor with a larger torque-to-inertia ratio is used. In this context, it is useful to review some terminology related to torque characteristics of a stepper motor.

6.8.1   Torque Characteristics and Terminology

The torque that can be provided to a load by a stepper motor depends on several factors. For example, the motor torque at constant speed is not the same as that when the motor passes through that speed (i.e., under acceleration, deceleration, or general transient conditions). In particular, at constant speed there is no inertial torque. Also, the torque losses due to magnetic induction are lower at constant stepping rates in comparison with variable stepping rates. It follows that the available torque is larger under steady (constant speed) conditions. Another factor of influence is the magnitude of the speed. At low speeds (i.e., when the step period is considerably larger than the electrical time constant), the time taken for the phase current to build up or turn off is insignificant compared with the step time. Then, the phase current waveform can be assumed rectangular. At high stepping rates, the induction effects dominate, and as a result a phase may not reach its rated current within the duration of a step. As a result, the generated torque will be degraded. Furthermore, as the power provided by the power supply is limited, the torque × speed product of the motor is limited as well. Consequently, as the motor speed increases, the available torque must decrease in general. These two are the primary reasons for the characteristic shape of a speed–torque curve of a stepper motor where the peak torque occurs at a very low (typically zero) speed, and as the speed increases the available torque decreases. Eventually, at a particular limiting speed (known as the no-load speed), the available torque becomes zero.

The characteristic shape of the speed–torque curve of a stepper motor is shown in Figure 6.40. Some terminology is given as well. What is given may be interpreted as experimental data measured under steady operating conditions (and averaged and interpolated). The given torque is called the pull-out torque and the corresponding speed is the pull-out speed. In industry, this curve is known as the pull-out curve.

Holding torque is the maximum static torque (see Equation 6.36, for instance), and is different from the maximum (pull out) torque defined in Figure 6.40. In particular, the holding torque can be about 40% greater than the maximum pull-out torque, which is typically equal to the starting torque (or stand-still torque). Furthermore, the static torque becomes higher if the motor has more than one stator pole per phase and if all these poles are excited at a time. The residual torque is the maximum static torque that is present when the motor phases are not energized. This torque is practically zero for a VR

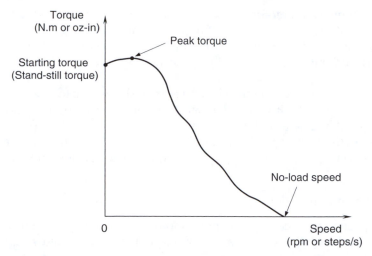

FIGURE 6.40
The speed–torque characteristics of a stepper motor.

motor, but is not negligible for a PM motor or an HB motor. In some industrial literature, detent torque takes the same meaning as the residual torque. In this context, detent torque is defined as the torque ripple that is present under power-off conditions. A more appropriate definition for detent torque is the static torque at the present detention position (equilibrium position) of the motor, when the next phase is energized. According to this definition, detent toque is equal to $T_{max} \sin 2\pi/p$, where T_{max} is the holding torque and p is the number of phases.

Some further definitions of speed–torque characteristics of a stepper motor are given in Figure 6.41. The pull-out curve or the slew curve here takes the same meaning as that given in Figure 6.40. Another curve known as the start–stop curve or pull-in curve is given as well.

As noted before, the pull-out curve (or slew curve) gives the speed at which the motor can run under steady (constant speed) conditions, under rated current and using appropriate drive circuitry. But, the motor is unable to steadily accelerate to the slew speed, starting from rest and applying a pulse sequence at constant rate corresponding to the slew speed. Instead, it should be accelerated first up to the pull-in speed by applying a pulse sequence corresponding to this speed. After reaching the start–stop region (pull-in

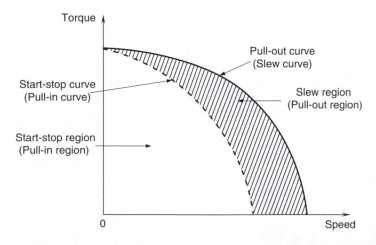

FIGURE 6.41
Further speed–torque characteristics and terminology.

region) in this manner, the motor can be accelerated to the pull-out speed (or to a speed lower than this, within the slew region). Similarly, when stopping the motor from a slew speed, it should be first decelerated (by downramping) to a speed in the start–stop region (pull-in region), and only when this region is reached satisfactorily, the stepping sequence should be turned off.

As the drive system determines the current and the switching sequence of the motor phases and the rate at which the switching pulses are applied, it directly affects the speed–torque curve of a motor. Accordingly, what is given in a product data sheet should be interpreted as the speed–torque curve of the particular motor when used with a specified drive system and a matching power supply, and for operation at rated values.

6.8.2 Stepper Motor Selection

The effort required in selecting a stepper motor for a particular application can be reduced if the selection is done in a systematic manner. The following steps provide some guidelines for the selection process:

Step 1: List the main requirements for the particular application, according to the conditions and specifications for the application. These include operational requirements such as speed, acceleration, and required accuracy and resolution, and load character-istics, such as size, inertia, fundamental natural frequencies, and resistance torques.

Step 2: Compute the operating torque and stepping rate requirements for the particular application. Newton's second law is the basic equation employed in this step. Specifically, the required torque rating is given by

$$T = T_R + J_{eq}\frac{\omega_{max}}{\Delta t}, \tag{6.45}$$

where T_R is the net resistance torque on the motor, J_{eq} is the equivalent moment of inertia (including rotor, load, gearing, dampers, etc.), ω_{max} is the maximum operating speed, and Δt is the time taken to accelerate the load to the maximum speed, starting from rest.

Step 3: Using the torque vs. stepping rate curves (i.e., pull-out curves) for a group of commercially available stepper motors, select a suitable stepper motor. The torque and speed requirements determined in Step 2 and the accuracy and resolution requirements specified in Step 1 should be used in this step.

Step 4: If a stepper motor that meets the requirements is not available, modify the basic design. This may be accomplished by changing the speed and torque requirements by adding devices such as gear systems (e.g., harmonic drive; see Chapter 8) and amplifiers (e.g., hydraulic amplifiers).

Step 5: Select a drive system that is compatible with the motor and that meets the operational requirements in Step 1.

Motors and appropriate drive systems are prescribed in product manuals and catalogs available from the vendors. For relatively simple applications, a manually controlled preset indexer or an open-loop system consisting of a pulse source (oscillator) and a translator could be used to generate the pulse signal to the translator in the drive unit. For more complex transient tasks, a software controller (a microprocessor or a personal computer) or a customized hardware controller may be used to generate the desired pulse command in open-loop operation. Further, sophistication may be incorporated by using digital-signal-processor–based closed-loop control with encoder feedback, for tasks that require very high accuracy under transient conditions and for operation near the rated capacity of the motor.

The single most useful piece of information in selecting a stepper motor is the torque vs. stepping rate curve (i.e., the pull-out curve). Other parameters that are valuable in the selection process include:

1. The step angle or the number of steps per revolution
2. The static holding torque (maximum static torque of motor when powered at rated voltage)
3. The maximum slew rate (maximum steady-state stepping rate possible at rated load)
4. The motor torque at the required slew rate (pull-out torque, available from the pull-out curve)
5. The maximum ramping slope (maximum acceleration and deceleration possible at rated load)
6. The motor time constants (no-load electrical time constant and mechanical time constant)
7. The motor natural frequency (without an external load and near detent position)
8. The motor size (dimensions of: poles, stator and rotor teeth, air gap and housing; weight, rotor moment of inertia)
9. The power supply ratings (voltage, current, and power)

There are many parameters that determine the ratings of a stepper motor. For example, the static holding torque increases with the number of poles per phase that are energized, decreases with the air gap width and tooth width, and increases with the rotor diameter and stack length. Furthermore, the minimum allowable air gap width should exceed the combined maximum lateral (flexural) deflection of the rotor shaft caused by thermal deformations and the flexural loading, such as magnetic pull and static and dynamic mechanical loads. In this respect, the flexural stiffness of the shaft, the bearing characteristics, and the thermal expansion characteristics of the entire assembly become important. Field winding parameters (diameter, length, resistivity, etc.) are chosen by giving due consideration to the required torque, power, electrical time constant, heat generation rate, and motor dimensions. Note that a majority of these are design parameters that cannot be modified in a cost-effective manner during the motor selection stage.

6.8.2.1 Positioning (x–y) Tables

A common application of stepper motors is in positioning tables (see Figure 6.42a). A two-axis (x–y) table requires two stepper motors of nearly equal capacity. The values of the following parameters are assumed to be known:

- Maximum positioning resolution (displacement/step)
- Maximum operating velocity to be attained in less than a specified time
- Weight of the x–y table
- Maximum resistance force (primarily friction) against table motion.

A schematic diagram of the mechanical arrangement for one of the two axes of the table is shown in Figure 6.42b. A lead screw (see Chapter 8) is used to convert the rotary motion of the motor into rectilinear motion. Free-body diagrams for the motor rotor and the table are shown in Figure 6.43.

Now, we will derive a somewhat generalized relation for this type of application. The equations of motion (from Newton's second law) are

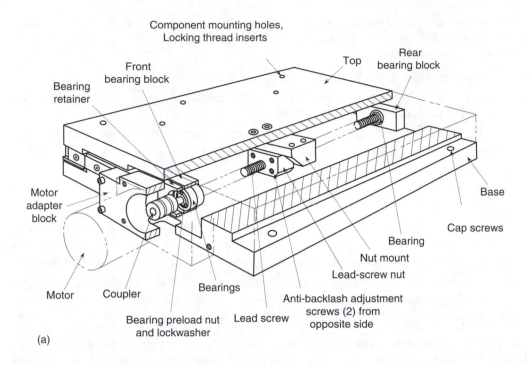

(a)

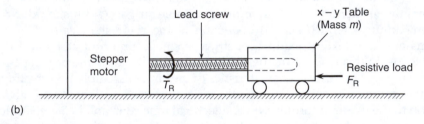

(b)

FIGURE 6.42
(a) A single axis of a positioning table. (b) An equivalent model.

For the rotor,

$$T - T_R = J\alpha \tag{6.46}$$

For the table,

$$F - F_R = ma, \tag{6.47}$$

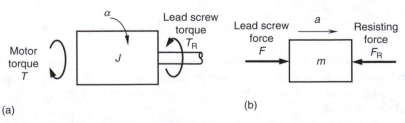

(a)   (b)

FIGURE 6.43
Free-body diagrams. (a) Motor rotor. (b) Table.

where T is the motor torque, T_R is the resistance torque from the lead screw, J is the equivalent moment of inertia of the rotor, α is the angular acceleration of the rotor, F is the driving force from the lead screw, F_R is the external resistance force on the table, m is the equivalent mass of the table, and a is the acceleration of the table.

Assuming a rigid lead screw without backlash, the compatibility condition is written as

$$a = r\alpha, \tag{6.48}$$

where r denotes the transmission ratio (rectilinear motion/angular motion) of the lead screw. The load transmission equation for the lead screw is

$$F = \frac{e}{r}T_R, \tag{6.49}$$

where e denotes the fractional efficiency of the lead screw. Finally, Equation 6.46 through Equation 6.49 can be combined to give

$$T = \left(J + \frac{mr^2}{e}\right)\frac{a}{r} + \frac{r}{e}F_R. \tag{6.50}$$

Example 6.9

A schematic diagram of an industrial conveyor unit is shown in Figure 6.44. In this application, the conveyor moves intermittently at a fixed rate, thereby indexing the objects on the conveyor through a fixed distance d in each time period T. A triangular speed profile is used for each motion interval, having an acceleration and a deceleration that are equal in magnitude (see Figure 6.45). The conveyor is driven by a stepper motor. A gear unit with step-down speed ratio $p:1$, where $p > 1$, may be used if necessary, as shown in Figure 6.45.

a. Explain why the equivalent moment of inertia, J_e, at the motor shaft, for the overall system, is given by

$$J_e = J_m + J_{g1} + \frac{1}{p^2}(J_{g2} + J_d + J_s) + \frac{r^2}{p^2}(m_c + m_L),$$

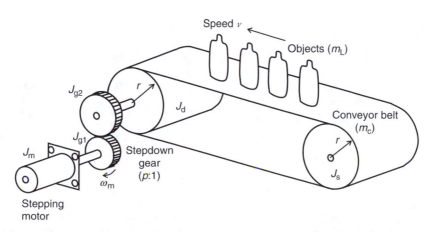

FIGURE 6.44
Conveyor unit with intermittent motion.

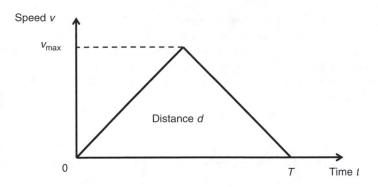

FIGURE 6.45
Speed profile for a motion period of the conveyor.

where J_m, J_{g1}, J_{g2}, J_d, and J_s are the moments of inertia of the motor rotor, drive gear, driven gear, drive cylinder of the conveyor, and the driven cylinder of the conveyor, respectively; m_c and m_L are the overall masses of the conveyor belt and the moved objects (load), respectively; and r is the radius of each of the two conveyor cylinders.

b. Four models of stepping motor are available for the application. Their specifications are given in Table 6.2 and the corresponding performance curves are given in Figure 6.46. The following values are known for the system: $d = 10\,\text{cm}$, $T = 0.2\,\text{s}, r = 10\,\text{cm}, m_c = 5\,\text{kg}, m_L = 5\,\text{kg}, J_d = J_s = 2.0 \times 10^{-3}\,\text{kg.m}^2$. Also two gear units with $p = 2$ and 3 are available, and for each unit $J_{g1} = 50 \times 10^{-6}$ kg.m^2 and $J_{g2} = 200 \times 10^{-6}\,\text{kg.m}^2$.

Indicating all calculations and procedures, select a suitable motor unit for this application. You must not use a gear unit unless it is necessary to have one with

TABLE 6.2

Stepper Motor Data

Model		Stepping Motor Specifications			
		50SM	101SM	310SM	1010SM
NEMA motor frame size		23	23	34	42
Full step angle	Degrees		1.8		
Accuracy	Percent		± 3 (noncumulative)		
Holding torque	oz.in	38	90	370	1050
	N.m	0.27	0.64	2.61	7.42
Detent torque	oz.in	6	18	25	20
	N.m	0.04	0.13	0.18	0.14
Rated phase current	Amps	1	5	6	8.6
Rotor inertia	oz.in.s²	1.66×10^{-3}	5×10^{-3}	26.5×10^{-3}	114×10^{-3}
	kg.m²	11.8×10^{-6}	35×10^{-6}	187×10^{-6}	805×10^{-6}
Maximum radial load	lb	15	15	35	40
	N	67	67	156	178
Maximum thrust load	lb	25	25	60	125
	N	111	11	267	556
Weight	lb	1.4	2.8	7.8	20
	kg	0.6	1.3	3.5	9.1
Operating temperature	°C		−55 to +50		
Storage temperature	°C		−55 to +130		

Source: From Aerotech Inc. With permission.

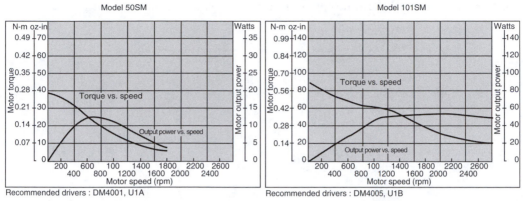

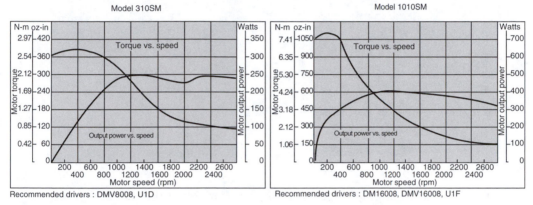

FIGURE 6.46

Stepper motor performance curves. (From Aerotech Inc. With permission.)

the available motors. What is the positioning resolution of the conveyor (rectilinear) for the final system?

Note: Assume an overall system efficiency of 80% regardless of whether a gear unit is used.

Solution

a. Angular speed of the motor and drive gear $= \omega_m$.

Angular speed of the driven gear and conveyor cylinders $= \dfrac{\omega_m}{p}$.

Rectilinear speed of the conveyor and objects, $v = \dfrac{r\omega_m}{p}$.

Kinetic energy of the overall system

$$= \frac{1}{2}(J_m + J_{g1})\omega_m^2 + \frac{1}{2}(J_{g2} + J_d + J_s)\left(\frac{\omega_p}{p}\right)^2 + \frac{1}{2}(m_c + m_L)\left(\frac{r\omega_m}{p}\right)^2,$$

$$= \frac{1}{2}\left[J_m + J_{g1} + \frac{1}{p^2}(J_{g2} + J_d + J_s) + \frac{r^2}{p^2}(m_c + m_L)\right]\omega_m^2,$$

$$= \frac{1}{2}J_e\omega_m^2.$$

Hence, the equivalent moment of inertia as felt at the motor rotor is

$$J_e = J_m + J_{g1} + \frac{1}{p^2}(J_{g2} + J_d + J_s) + \frac{r^2}{p^2}(m_c + m_L).$$

b. From the triangular speed profile we have

$$d = \frac{1}{2}v_{max}T.$$

Substituting numerical values,

$$0.1 = \frac{1}{2}v_{max}0.2$$

or

$$v_{max} = 1.0 \text{ m/s}.$$

The acceleration or deceleration of the system

$$a = \frac{v_{max}}{T/2} = \frac{1.0}{0.2/2} \text{ m/s}^2 = 10.0 \text{ m/s}^2.$$

Corresponding angular acceleration or deceleration of the motor

$$\alpha = \frac{pa}{r}.$$

With an efficiency of η, the motor torque T_m that is needed to accelerate or decelerate the system is given by

$$\eta T_m = J_e\alpha = J_e\frac{pa}{r} = \left[J_m + J_{g1} + \frac{1}{p^2}(J_{g2} + J_d + J_s) + \frac{r^2}{p^2}(m_c + m_L)\right]\frac{pa}{r}.$$

Maximum speed of the motor

$$\omega_{max} = \frac{pv_{max}}{r}.$$

Without gears, we have

$$\eta T_m = [J_m + J_d + J_s + r^2(m_c + m_L)]\frac{a}{r}$$

and

$$\omega_{max} = \frac{v_{max}}{r}.$$

Now, we substitute numerical values.

TABLE 6.3

Data for Selecting a Motor Without a Gear Unit

Motor Model	Available Torque at ω_{max} (N.m)	Motor–Rotor Inertia ($\times 10^{-6}$ kg.m^2)	Required Torque (N.m)
50 SM	0.26	11.8	13.0
101 SM	0.60	35.0	13.0
310 SM	2.58	187.0	13.0
1010 SM	7.41	805.0	13.1

Case 1: Without gears

For an efficiency value $\eta = 0.8$ (i.e., 80% efficient), we have

$$0.8T_m = [J_m + 2 \times 10^{-3} + 2 \times 10^{-3} + 0.1^2(5+5)]\frac{10}{0.1} \text{ N.m}$$

or

$$T_m = 125.0[J_m + 0.104] \text{ N.m}$$

and

$$\omega_{max} = \frac{1.0}{0.1} \text{ rad/s} = 10 \times \frac{60}{2\pi} \text{ rpm} = 95.5 \text{ rpm}.$$

The operating speed range is 0–95.5 rpm. Note that the torque at 95.5 rpm is less than the starting torque for the first two motor models, and not so for the second two models (see the speed–torque curves in Figure 6.46). We must use the weakest point (i.e., lowest torque) in the operating speed range, in the motor selection process. Allowing for this requirement, Table 6.3 is formed for the four motor models.

It is seen that without a gear unit, the available motors cannot meet the system requirements.

Case 2: With gears

Note: Usually the system efficiency drops when a gear unit is introduced. In the present exercise, we use the same efficiency for reasons of simplicity.

TABLE 6.4

Data for Selecting a Motor With a Gear Unit

Motor Model	Available Torque at ω_{max} (N.m)	Motor–Rotor Inertia ($\times 10^{-6}$ kg.m^2)	Required Torque (N.m)
50 SM	0.25	11.8	6.53
101 SM	0.58	35.0	6.53
310 SM	2.63	187.0	6.57
1010 SM	7.41	805.0	6.73

With an efficiency of 80%, we have $\eta = 0.8$. Then,

$$0.8T_m = \left[J_m + 50 \times 10^{-6} + \frac{1}{p^2}(200 \times 10^{-6} + 2 \times 10^{-3} + 2 \times 10^{-3}) + \frac{0.1^2}{p^2}(5+5) \right] p \times \frac{10}{0.1} \text{ N.m}$$

and

$$\omega_{max} = \frac{1.0\, p}{0.1} \text{ rad/s} = 10\, p \times \frac{60}{2\pi} \text{ rpm,}$$

or

$$T_m = 125.0 \left[J_m + 50 \times 10^{-6} + \frac{1}{p^2} \times 104.2 \times 10^{-3} \right] p \text{ N.m}$$

and

$$\omega_{max} = 95.5p \text{ rpm.}$$

For the case of $p = 2$, we have $\omega_{max} = 191.0$ rpm. Table 6.4 is formed for the present case.

It is seen that with a gear of speed ratio $p = 2$, motor model 1010 SM satisfies the requirement.

With full stepping, step angle of the rotor $= 1.8°$. Corresponding step in the conveyor motion is the positioning resolution. With $p = 2$ and $r = 0.1$ m, the positioning resolution is

$$\frac{1.8°}{2} \times \frac{\pi}{180°} \times 0.1 = 1.57 \times 10^{-3} \text{ m.}$$

6.8.3 Stepper Motor Applications

More than one type of actuator may be suitable for a given application. In the present discussion, we indicate situations where stepper motor is a suitable choice as an actuator. It does not, however, rule out the use of other types of actuators for the same application (see Chapter 7).

Stepper motors are particularly suitable for positioning, ramping (constant acceleration and deceleration), and slewing (constant speed) applications at relatively low speeds. Typically they are suitable for short and repetitive motions at speeds less than 2000 rpm. They are not the best choice for servoing or trajectory following applications, because of jitter and step (pulse) missing problems (dc and ac servomotors are better for such applications; see Chapter 7). Encoder feedback will make the situation better, but at a higher cost and controller complexity. Generally, however, stepper motor provides a low-cost option in a variety of applications.

The stepper motor is a low-speed actuator that may be used in applications that require torques as high as 15 N.m (2121 oz.in.). For heavy-duty applications, torque amplification may be necessary. One way to accomplish this is by using a hydraulic actuator in cascade with the motor. The hydraulic valve (typically a rectilinear spool valve as described in Chapter 7), which controls the hydraulic actuator (typically a piston–cylinder device), may be driven by a stepper motor through suitable gearing for speed reduction as well as for rotary–rectilinear motion conversion. Torque amplification by an order of magnitude is possible with such an arrangement. Of course, the time constant will increase and operating bandwidth will decrease because of the sluggishness of hydraulic components. Also, a certain amount of backlash will be introduced by the gear system. Feedback control will be necessary to reduce the position error, which is usually present in open-loop hydraulic actuators.

Stepper motors are incremental actuators. As such, they are ideally suited for digital control applications. High-precision open-loop operation is possible as well, provided that the operating conditions are well within the motor capacity. Early applications of stepper motor were limited to low-speed, low-torque drives. With rapid developments in solid-state drives and microprocessor-based pulse generators and controllers, however, reasonably high-speed operation under transient conditions at high torques and closed-loop control has become feasible. As brushes are not used in stepper motors, there is no danger of spark generation. Hence, they are suitable in hazardous environments. But, heat generation and associated thermal problems can be significant at high speeds.

There are numerous applications of stepper motors. For example, stepper motor is particularly suitable in printing applications (including graphic printers, plotters, and electronic typewriters) because the print characters are changed in steps and the printed lines (or paper feed) are also advanced in steps. Stepper motors are commonly used in x–y tables. In automated manufacturing applications, stepper motors are found as joint actuators and end effector (gripper) actuators of robotic manipulators, parts assembly and inspection systems, and as drive units in programmable dies, parts-positioning tables, and tool holders of machine tools (milling machines, lathes, etc.). In automotive applications, pulse windshield wipers, power window drives, power seat mechanisms, automatic carburetor control, process control applications, valve actuators, and parts-handling systems use stepper motors. Other applications of stepper motors include source and object positioning in medical and metallurgical radiography, lens drives in autofocus cameras, camera movement in computer vision systems, and paper feed mechanisms in photocopying machines.

The advantages of stepper motors include the following:

1. Position error is noncumulative. A high accuracy of motion is possible, even under open-loop control.
2. The cost is relatively low. Furthermore, considerable savings in sensor (measuring system) and controller costs are possible when the open-loop mode is used.
3. Because of the incremental nature of command and motion, stepper motors are easily adoptable to digital control applications.
4. No serious stability problems exist, even under open-loop control.
5. Torque capacity and power requirements can be optimized and the response can be controlled by electronic switching.
6. Brushless construction has obvious advantages (see Chapter 7).

The disadvantages of stepper motors include the following:

1. They are low-speed actuators. The torque capacity is typically less than 15 N.m, which may be low compared to what is available from torque motors.
2. They have limited speed (limited by torque capacity and by pulse-missing problems due to faulty switching systems and drive circuitry).
3. They have high vibration levels due to stepwise motion.
4. Large errors and oscillations can result when a pulse is missed under open-loop control.
5. Thermal problems can be significant when operating at high speeds.

In most applications, the merits of stepper motors outweigh the drawbacks.

Problems

6.1 Consider the two-phase PM stepper motor shown in Figure 6.2. Show that in full stepping, the sequence of states of the two phases is given by Table P6.1. What is the step angle in this case?

6.2 Consider the variable-reluctance stepper motor shown schematically in Figure 6.5. The rotor is a nonmagnetized soft-iron bar. The motor has a two-pole rotor and a three-phase stator. Using a schematic diagram show the half-stepping sequence for a full clockwise rotation of this motor. What is the step angle? Indicate an advantage and a disadvantage of half stepping over full stepping.

6.3 Consider a stepper motor with three rotor teeth ($n_r = 3$), two rotor stator poles ($n_s = 2$), and two phases ($p = 2$). What is the step angle for this motor in full stepping? Is this a VR motor or a PM motor? Explain.

6.4 Explain why a two-phase variable-reluctance stepper motor is not a physical reality, in full stepping. A single-stack variable-reluctance stepper motor with nontoothed poles has n_r teeth in the rotor, n_s poles in the stator, and p phases of winding. Show that

$$n_r = \left(r + \frac{1}{p}\right)n_s,$$

where r is the largest positive integer (natural number), such that $n_r > r n_s$.

6.5 For a single-stack stepper motor that has toothed poles, for the case $\theta_s > \theta_r$, show that,

$$\Delta\theta = \frac{n_s}{mp}(\theta_s - r\theta_r),$$

$$\theta_s = r\theta_r + \frac{m\theta_r}{n_s},$$

$$n_r = r n_s + m,$$

where $\Delta\theta$ is the step angle, θ_r is the rotor tooth pitch, θ_s is the stator tooth pitch, n_r is the number of teeth in the rotor, n_s is the number of teeth in the stator, p is the number of phases, m is the number of stator poles per phase, and r is the largest integer such that $\theta_s - r\theta_r > 0$. Assume that the stator teeth are uniformly distributed around the rotor. Derive the corresponding equations for the case $\theta_s < \theta_r$.

TABLE P6.1

Stepping Sequence (Full Stepping) for a Two-Phase PM Stepper Motor with Two Rotor Poles

State of ϕ_1		State of ϕ_2	
1	CW ↑	0	↓ CCW
0		1	
−1		0	
0		−1	

6.6 For a stepper motor with m stator poles per phase, show that the number of teeth in a stator pole is given by

$$t_s = \frac{n}{mp^2} - \frac{1}{p},$$

where n denotes the number of steps per revolution, for the case $n_r > n_s$ (Hint: This relation is the counterpart of Equation 6.10a). Pick suitable parameters for a four-phase, eight-pole motor, using this relation, if the step angle is required to be 1.8°. Can the same step be obtained using a three-phase stepper motor?

6.7 Consider the single-stack, three-phase VR stepper motor shown in Figure 6.9 ($n_r = 8$ and $n_s = 12$). For this arrangement, compare the following phase-switching sequences:

a. 1-2-3-1

b. 1-12-2-23-3-31-1

c. 12-23-31-12

What is the step angle, and how would you reverse the direction of rotation in each case?

6.8 Describe the principle of operation of a single-stack VR stepper motor that has toothed poles in the stator. Assume that the stator teeth are uniformly distributed around the rotor. If the motor has 5 teeth/pole and 2 pole pairs/phase and provides 500 full steps/revolution, determine the number of phases in the stator. Also determine the number of stator poles, the step angle, and the number of teeth in the rotor.

6.9 Figure P6.9 shows a schematic diagram of a stepper motor. What type of stepper is this? Describe the operation of this motor. In particular, discuss whether four separate phases are needed or whether the phases of the opposite stator poles may be connected together, giving a two-phase stepper. What is the step angle of the motor

a. In full stepping?

b. In half stepping?

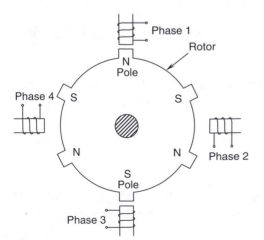

FIGURE P6.9
Schematic diagram of a stepper motor.

6.10 So far, in the problems on toothed single-stack stepper motors, we have assumed that $\theta_r \neq \theta_s$. Now consider the case of $\theta_r = \theta_s$. In a single-stack stepper motor of this type, the stator–rotor tooth misalignment that is necessary to generate the driving torque is achieved by offsetting the entire group of teeth on a stator pole (not just the central tooth of the pole) by the step angle $\Delta\theta$ with respect to the teeth on the adjacent stator pole. The governing equations are Equation 6.11 and Equation 6.12. There are two possibilities, as given by the $+$ sign and the $-$ sign in these equations. The $+$ sign governs the case in which the offset is generated by reducing the pole pitch. The $-$ sign governs the case where the offset of $\Delta\theta$ is realized by increasing the pole pitch. Show that in this latter case it is possible to design a four-phase motor that has 50 rotor teeth. Obtain appropriate values for tooth pitch (θ_r and θ_s), full-stepping step angle $\Delta\theta$, number of steps per revolution (n), number of poles per phase (m), and number of stator teeth per pole (t_s) for this design.

6.11 The stepper motor shown in Figure 6.11a uses the balanced pole arrangement. Specifically, all the poles wound to the same phase are uniformly distributed around the rotor. In Figure 6.11, there are 2 poles/phase. Hence, the two poles connected to the same phase are placed at diametrically opposite locations. In general, in the case of m poles per phase, the poles connected to the same phase would be located at angular intervals of $360°/m$. What are the advantages of this balanced pole arrangement?

6.12 In connection with the phase windings of a stepper motor, explain the following terms:

a. Unifilar (or monofilar) winding

b. Bifilar winding

c. Bipolar winding.

Discuss why the torque characteristics of a bifilar-wound motor are better than those of a unifilar-wound motor at high stepping rates.

6.13 For a multiple-stack variable-reluctance stepper motor whose rotor tooth pitch angle is not equal to the stator tooth pitch angle (i.e., $\theta_r \neq \theta_s$), show that the step angle may be expressed by

$$\Delta\theta = \frac{\theta_r}{ps},$$

where p is the number of phases in each stator segment, and s is the number of stacks of rotor teeth on the shaft.

6.14 Describe the principle of operation of a multiple-stack VR stepper motor that has toothed poles in each stator stack. Show that if $\theta_r < \theta_s$, the step angle of this type of motor is given by

$$\Delta\theta = \frac{n_s}{mps}(\theta_s - \theta_r),$$

where θ_r is the rotor tooth pitch, θ_s is the stator tooth pitch, n_s is the number of teeth in the stator, p is the number of phases, m is the number of poles per phase, and s is the number of stacks.

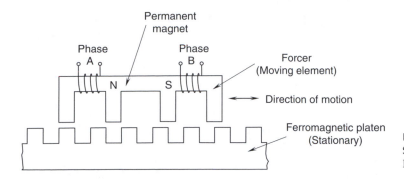

FIGURE P6.16
Schematic representation of a linear hybrid stepping motor.

Assume that the stator teeth are uniformly distributed around the rotor and that the phases of different stator segments are independent (i.e., can be activated independently).

6.15 The torque of a stepping motor can be increased by increasing its diameter, for a given coil density (the number of turns per unit area) of the stator poles, and current rating. Alternatively, the motor torque can be increased by introducing multiple stacks (resulting in a longer motor) for a given diameter, coil density, and current rating. Giving reasons indicate which design is generally preferred.

6.16 The principle of operation of a (hybrid) linear stepper motor is indicated in the schematic diagram of Figure P6.16. The toothed platen is a stationary member made of ferromagnetic material, which is not magnetized. The moving member is termed the forcer, which has four groups of teeth (only 1 tooth/group is shown in the figure, for convenience). A permanent magnet has its N pole located at the first two groups of teeth and the S pole located at the next two groups of teeth, as shown. Accordingly, the first two groups are magnetized to take the N polarity and the next two groups take the S polarity. The motor has two phases, denoted by A and B. Phase A is wound between the first two groups of teeth and phase B is wound between the second two groups of teeth of the forcer, as shown. In this manner, when phase A is energized, it will create an electromagnet with opposite polarities located at the first two groups of teeth. Hence, one of these first two groups of teeth will have its magnetic polarity reinforced, whereas the other group will have its polarity neutralized. Similarly, phase B, when energized, will strengthen one of the next two groups of teeth while neutralizing the other group. The teeth in the four groups of the forcer have quadrature offsets as follows. Second group has an offset of 1/2 tooth pitch with respect to the first group. The third group of teeth has an offset of 1/4 tooth pitch with respect to the first group in one direction, and the fourth group has an offset of 1/4 pitch with respect to the first group in the opposite direction (Hence, the fourth group has an offset of 1/2 pitch with respect to the third group of teeth.). The phase windings are bipolar (i.e., the current in a coil can be reversed).

a. Describe the full-stepping cycle of this motor, for motion to the right and for motion to the left.

b. Give the half-stepping cycle of this motor, for motion to the right and for motion to the left.

6.17 When a phase winding of a stepper motor is switched on, ideally the current in the winding should instantly reach the full value (hence providing the full magnetic

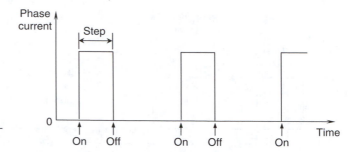

FIGURE P6.17
Ideal phase current waveform for a step-per motor.

field instantly). Similarly, when a phase is switched off, its current should become zero immediately. It follows that the ideal shape of phase current history is a rectangular pulse sequence, as shown in Figure P6.17. In actual motors, however, the current curves deviate from the ideal rectangular shape, primarily because of the magnetic induction in the phase windings. Using sketches indicate how the phase current waveform would deviate from this ideal shape under the following conditions:

a. Very slow stepping

b. Very fast stepping at a constant stepping rate

c. Very fast stepping at a variable (transient) stepping rate.

A stepper motor has a phase inductance of 10 mH and a phase resistance of 5 Ω. What is the electrical time constant of each phase in a stepper motor? Estimate the stepping rate below which magnetic induction effects can be neglected so that the phase current waveform is almost a rectangular pulse sequence.

6.18 Consider a stepper motor that has 2 poles/phase. The pole windings in each phase may be connected either in parallel or in series, as shown in Figure P6.18. In each case, determine the required ratings for phase power supply (rated current, rated voltage, rated power) in terms of current i and resistance R, as indicated in Figure P6.18a. Note that the power rating should be the same for both cases, as is intuitively clear.

6.19 Some industrial applications of stepper motors call for very high stepping rates under variable load (variable motor torque) conditions. As motor torque depends directly on the current in the phase windings (typically 5 A/phase), one method of obtaining a variable-torque drive is to use an adjustable resistor in the drive circuit. An alternative method is to use a chopper drive. Switching transistors, diodes, or thyristors are used in a chopper circuit to periodically bypass (chop) the current through a phase winding. The chopped current passes through a free-wheeling diode back to the power supply. The chopping interval and chopping frequency are

FIGURE P6.18
Pole windings in a phase of a stepper motor that has 2 poles/phase. (a) Parallel connection. (b) Series connection.

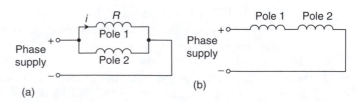

adjustable. Discuss the advantages of chopper drives compared to the resistance drive method.

6.20 Define and compare the following pairs of terms in the context of electromagnetic stepper motors:

a. Pulses and steps

b. Step angle and resolution

c. Residual torque and static holding torque

d. Translator and drive system

e. PM stepper motor and VR stepper motor

f. Single-stack stepper and multiple-stack stepper

g. Stator poles and stator phases

h. Pulse rate and slew rate.

6.21 Compare the VR stepper motor with the PM stepper motor with respect to the following considerations:

a. Torque capacity for a given motor size

b. Holding torque

c. Complexity of switching circuitry

d. Step size

e. Rotor inertia.

The hybrid stepper motor possesses characteristics of both the VR and the PM types of stepper motors. Consider a typical construction of a hybrid stepper motor, as shown schematically in Figure P6.21. The rotor has two stacks of teeth made of ferromagnetic material, joined together by a permanent magnet which assigns opposite polarities to the two rotor stacks. The tooth pitch is the same for both stacks, but the two stacks have a tooth misalignment of half a tooth pitch ($\theta_r/2$). The stator may consist of a common tooth stack for both rotor stacks (as in the figure), or it may consist of two separate tooth stack segments that are in complete alignment with each other, one surrounding each rotor stack. The number of teeth in the stator are not equal to the number of teeth in each rotor stack. The stator is made up of several toothed poles that are equally spaced around the rotor. Half the poles are

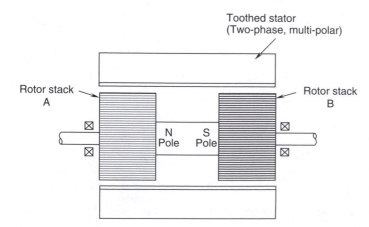

FIGURE P6.21
Schematic diagram of a hybrid stepper motor.

connected to one phase and the other half are connected to the second phase. The current in each phase may be turned on and off or reversed using switching amplifiers. The switching sequence for rotation in one direction (say, CW) would be A^+, B^+, A^-, B^-; for rotation in the opposite direction (CCW), it would be A^+, B^-, A^-, B^+, where A and B denote the two phases and the superscripts + and − denote the direction of current in each phase. This may also be denoted by [1, 0], [0, 1], [−1, 0], [0, −1] for CW rotation and [1, 0], [0, −1], [−1, 0], [0, 1] for CCW rotation.

Consider a motor that has 18 teeth in each rotor stack and 8 poles in the stator, with 2 teeth/stator pole. The stator poles are wound to the two phases as follows: Two radially opposite poles are wound to the same phase, with identical polarity. The two radially opposite poles that are at 90° from this pair of poles are also wound to the same phase, but with the field in the opposite direction (i.e., opposite polarity) to the previous pair.

a. Using suitable sketches of the rotor and stator configurations at the two stacks, describe the operation of this hybrid stepper motor.

b. What is the step size of the motor?

6.22 A Lanchester damper is attached to a stepper motor. A sketch is shown in Figure P6.22. Write equations to describe the single-step response of the motor about the detent position. Assume that the flywheel of the damper is not locked onto its housing at any time. Let T_d denote the magnitude of the frictional torque of the damper. Give appropriate initial conditions. Using a computer simulation, plot the motor response, with and without the damper, for the following parameter values:

Rotor + load inertia, $J_m = 4.0 \times 10^{-3}$ N.m^2

Damper housing inertia, $J_h = 0.2 \times 10^{-3}$ N.m^2

Damper flywheel inertia, $J_d = 1.0 \times 10^{-3}$ N.m^2

Maximum overshoot, $\theta(0) = 1.0°$

Static torque constant (torque gradient), $K_m = 114.6$ N.m/rad

Damping constant of the motor when the Lanchester damper is disconnected, $C_m = 0.08$ N.m/rad/s

Magnitude of the frictional torque, $T_d = 80.0$ N.m.

Estimate the resonant frequency of the motor using the given parameter values, and verify it using the simulation results.

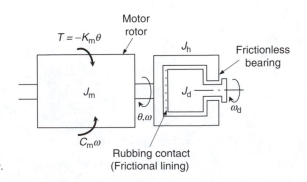

FIGURE P6.22
A Lanchester damper attached to a stepper motor.

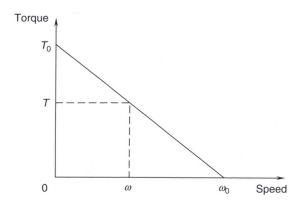

FIGURE P6.26
An approximate speed–torque curve for a stepper motor.

6.23 Compare and contrast the three electronic damping methods illustrated in Figure 6.31 through Figure 6.33. In particular, address the issue of effectiveness in relation to the speed of response and the level of final overshoot.

6.24 In the pulse reversal method of electronic damping, suppose that phase 1 is energized, instead of phase 3, at point B in Figure 6.32. Sketch the corresponding static torque curve and the motor response. Compare this new method of electronic damping with the pulse reversal method illustrated in Figure 6.32.

6.25 A relatively convenient method of electronic damping uses simultaneous multi-phase energization, where more than one phase are energized simultaneously and some of the simultaneous phases are excited with a fraction of the normal operating (rated) voltage. A simultaneous two-phase energization technique has been suggested for a three-phase, single-stack stepper motor. If the standard sequence of switching of the phases for forward motion is given by 1-2-3-1, what is the corresponding simultaneous two-phase energization sequence?

6.26 The torque vs. speed curve of a stepper motor is approximated by a straight line, as shown in Figure P6.26. The following two parameters are given: T_o is the torque at zero speed (starting torque or stand-still torque), and ω_o is the speed at zero torque (no-load speed).

Suppose that the load resistance is approximated by a rotary viscous damper with damping constant b. Assuming that the motor directly drives the load; without any speed reducers, determine the steady-state speed of the load and the corresponding drive torque of the stepper motor.

6.27 The speed–torque curve of a stepper motor is shown in Figure P6.27. Explain the shape, particularly the two dips, of this curve.

Suppose that with one phase on, the torque of a stepper motor in the neighborhood of the detent position of the rotor is given by the linear relationship

$$T = -K_m\theta,$$

where θ is the rotor displacement measured from the detent position and K_m is the motor torque constant (or magnetic stiffness or torque gradient). The motor is

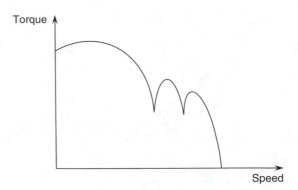

FIGURE P6.27
Typical speed–torque curve of a stepper motor.

directly coupled to an inertial load. The combined moment of inertia of the motor rotor and the inertial load is $J = 0.01$ kg.m^2. If $K_m = 628.3$ N.m/rad, at what stepping rates would you expect dips in the speed–torque curve of the motor–load combination?

6.28 The torque source model may be used to represent all three VR, PM, and HB types of stepper motors at low speeds and under steady operating conditions. What assumptions are made in this model?

A stepper motor has an inertial load rigidly connected to its rotor. The equivalent moment of inertia of rotor and load is $J = 5.0 \times 10^{-3}$ kg.m^2. The equivalent viscous damping constant is $b = 0.5$ N.m/rad/s. The number of phases $p = 4$, and the number of rotor teeth $n_r = 50$. Assume full stepping (step angle $= 1.8°$). The mechanical model for the motor is

$$T = b\dot{\bar{\theta}} + J\ddot{\bar{\theta}},$$

where $\bar{\theta}$ is the absolute position of the rotor.

a. Assuming a torque source model with $T_{max} = 100$ N.m, simulate and plot the motor response $\bar{\theta}$ as a function of t for the first 10 steps, starting from rest. Assume that in open-loop control, switching is always at the detent position of the present step. You should pay particular attention to the position coordinate, because $\bar{\theta}$ is the absolute position from the starting point and θ is the relative position measured from the approaching detent position of the current step. Plot the response on the phase plane (with speed $\dot{\bar{\theta}}$ as the vertical axis and position $\bar{\theta}$ as the horizontal axis).

b. Repeat part (a) for the first 150 steps of motion. Check whether a steady state (speed) is reached or whether there is an unstable response.

c. Consider the improved PM motor model with torque due to one excited phase given by Equation 6.39 through Equation 6.42, with $R = 2.0\ \Omega$, $L_o = 10.0$ mH, $L_a = 2.0$ mH, $k_b = 0.05$ V/rad/s, $v_p = 20.0$ V, and $k_m = 10.0$ N. m/A. Starting from rest and switching at each detent position, simulate the motor response for the first 10 steps. Plot θ vs. t to the same scale as in part (a). Also, plot the response on the phase plane to the same scale as in part (a). Note that at each switching point, the initial condition of the phase current i_p is zero. For example, simulation may be done by picking about 100 integration steps for each motor step. In each integration step, first for known θ and $\dot{\theta}$, integrate Equation 6.39 along with Equation 6.40 and Equation 6.41 to determine i_p. Substitute this in Equation 6.42

to compute torque T for the integration step. Then, use this torque and integrate the mechanical equation to determine $\bar{\theta}$ and $\dot{\bar{\theta}}$. Repeat this for the subsequent integration steps. After the detent position is reached, repeat the integration steps for the new phase, with zero initial value for current, but using $\bar{\theta}$ and $\dot{\bar{\theta}}$, as computed before, as the initial values for position and speed. Note that $\theta = \bar{\theta}$.

d. Repeat part (c) for the first 150 motor steps. Plot the curves to the same scale as in part (b).

e. Repeat parts (c) and (d), this time assuming a VR motor with torque given by Equation 6.43 and $k_r = 1.0$ N.m/A². The rest of the model is the same as for the PM motor.

f. Suppose that the fifth pulse did not reach the translator. Simulate the open-loop response of the three motor models during the first 10 steps of motion. Plot the response of all three motor models (torque source, improved PM, and improved VR) to the same scale as before. Give both the time history response and the phase plane trajectory for each model.

g. Suppose that the fifth pulse was generated and translated but the corresponding phase was not activated. Repeat part (f) under these conditions.

h. If the rotor position is measured, the motor can be accelerated back to the desired response by properly choosing the switching point. The switching point for maximum average torque is the point of intersection of the two adjacent torque curves, not the detent point. Simulate the response under a feedback control scheme of this type to compensate for the missed pulse in parts (f) and (g). Plot the controlled responses to the same scale as for the earlier results. Each simulation should be done for all three motor models and the results should be presented as a time history as well as a phase plane trajectory. Also, both pulse losing and phase losing should be simulated in each case. Explain how the motor response would change if the mechanical dissipation were modeled by Coulomb friction rather than by viscous damping.

6.29 A stepper motor with rotor inertia, J_m, drives a free (i.e., no-load) gear train, as shown in Figure P6.29. The gear train has two meshed gear wheels. The gear wheel attached to the motor shaft has inertia J_1 and the other gear wheel has inertia J_2. The gear train steps down the motor speed by the ratio 1:r ($r < 1$). One phase of the motor is energized, and once the steady state is reached, the gear system is turned (rotated) slightly from the corresponding detent position and released.

a. Explain why the system will oscillate about the detent position.

b. What is the natural frequency of oscillation (neglecting electrical and mechanical dissipations) in radians per second?

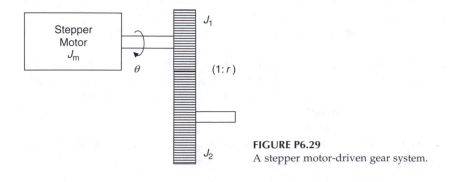

FIGURE P6.29
A stepper motor-driven gear system.

c. What is the significance of this frequency in a control system that uses a stepper motor as the actuator?

(Hint: Static torque for the stepper motor may be taken as

$$T = -T_{max} \sin\left(\frac{2\pi\theta}{p\Delta\theta}\right)$$

with the usual notation.)

6.30 Using the sinusoidal approximation for static torque in a three-phase variable-reluctance stepper motor, the torques T_1, T_2, and T_3 due to the three phases (1, 2, and 3) activated separately, may be expressed as

$$T_1 = -T_{max} \sin n_r\theta,$$

$$T_2 = -T_{max} \sin\left(n_r\theta - \frac{2\pi}{3}\right),$$

$$T_3 = -T_{max} \sin\left(n_r\theta - \frac{4\pi}{3}\right),$$

where θ is the angular position of the rotor measured from the detent position of phase 1, and n_r is the number of rotor teeth.

Using the trigonometric identity,

$$\sin A + \sin B = 2\sin\left(\frac{A+B}{2}\right)\cos\left(\frac{A-B}{2}\right)$$

shows that

$$T_1 + T_2 = -T_{max} \sin\left(n_r\theta - \frac{\pi}{3}\right),$$

$$T_2 + T_3 = -T_{max} \sin\left(n_r\theta - \pi\right),$$

$$T_3 + T_1 = -T_{max} \sin\left(n_r\theta - \frac{5\pi}{3}\right),$$

Using these expressions, show that the step angle for the switching sequence 1–2–3 is $\theta_r/3$ and the step angle for the switching sequence 1-12-2-23-3-31 is $\theta_r/6$. Determine the step angle for the two-phase-on switching sequence 12-23-31.

6.31 A stepper motor misses a pulse during slewing (high-speed stepping at a constant rate in steady state). Using a displacement vs. time curve, explain how a logic controller may compensate for this error by injecting a special switching sequence.

6.32 Briefly discuss the operation of a microprocessor-controlled stepper motor. How would it differ from the standard setup in which a hardware indexer is employed? Compare and contrast table lookup, programmed stepping, and hardware stepping methods for stepper motor translation.

6.33 Using a static torque diagram, indicate the locations of the first two encoder pulses for a feedback encoder–driven stepper motor for steady deceleration.

6.34 Suppose that the torque produced by a stepping motor when one of the phases is energized can be approximated by a sinusoidal function with amplitude T_{max}. Show that with the advanced switching sequence shown in Figure 6.38, for a three-phase stepper motor ($p = 3$), the average torque generated is approximately $0.827T_{max}$. What is the average torque generated with conventional switching?

6.35 A lectern (or podium) in an auditorium is designed to adjust its height automatically, depending on the height of the speaker. An ultrasonic gage measures the height of the speaker and sends a command to the logic hardware controller of a stepper motor, which adjusts the lectern vertically through a rack-and-pinion drive. The dead load of the moving parts is supported by a bellow device. A schematic diagram of this arrangement is shown in Figure P6.35. The following design requirements have been specified: Time to adjust a maximum stroke of 1 m = 5 s. Mass of the lectern = 50 kg. Maximum resistance to vertical motion = 5 kg. Displacement resolution = 0.5 cm/step.

Select a suitable stepper motor system for this application. You may use the ratings of the four commercial stepper motors, as given in Table 6.2 and Figure 6.46.

6.36 In connection with a stepper motor, explain the terms:
- Stand still or stalling torque
- Residual torque
- Holding torque.

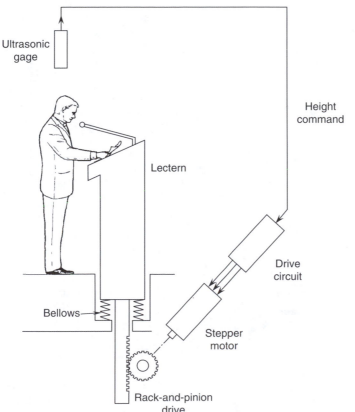

FIGURE P6.35
An automated lectern.

Figure P6.36a shows an automated salmon heading system. The fish moves horizontally along a conveyor toward a doubly inclined rotary cutter. The cutter mechanism generates a symmetric V-cut near the gill region of the fish, thereby improving the overall product recovery. Before a fish enters the cutter blades it passes over a positioning platform. The vertical position of the platform is automatically adjusted using a lead screw and nut arrangement, which is driven by a stepper motor in the open-loop mode. Specifically, the thickness of a fish is measured using an ultrasound sensor and is transmitted to the drive system of the stepper motor. The drive system commands the stepper motor to adjust the vertical position of the platform according to this measurement so that the fish will enter the cutter blade pair in a symmetrical orientation. The cutter blades are continuously driven by two ac motors. The positioning trajectory of the drive system of the stepper motor is always triangular, starting from rest, uniformly accelerating to a desired speed during the first half of the positioning time and then uniformly decelerating to rest during the second half.

The throughput of the machine is 2 fish/s. Even though this would make available the full period of 500 ms for thickness sensing and transmission, it will only provide a fraction of the time for positioning of the platform. More specifically, the platform cannot be positioned for the next fish until the present fish completely leaves the platform, and the positioning has to be completed before the fish enters the cutter.

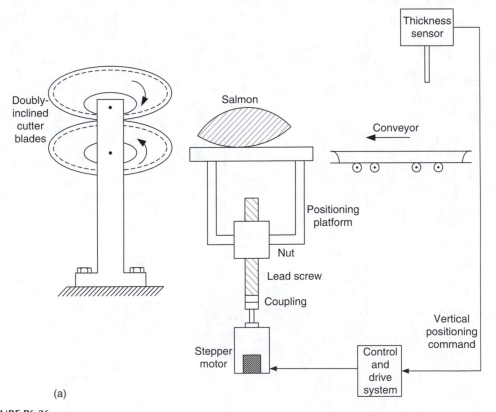

(a)

FIGURE P6.36
(a) An automated fish cutting system.

(*continued*)

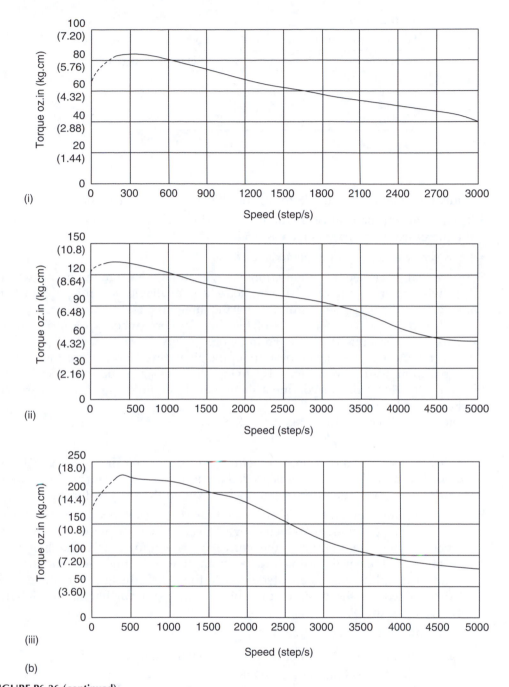

FIGURE P6.36 (continued)
(b) Speed–torque characteristics of the motors: (i) Model 1; (ii) Model 2; (iii) Model 3.

For this reason, the time available for positioning of the platform is specified as 200 ms. Primary specifications for the positioning system are as follows:

- Positioning resolution of 0.1 mm/step.
- Maximum positioning range of 2 cm with the positioning time not exceeding 200 ms, while following a triangular speed trajectory.

TABLE P6.36

Some Useful Data for the Available Motors

Stepper Motor	Step Angle	Rotor Inertia (kg.cm^2)
Model 1	1.8°	0.23
Model 2	1.8°	0.67
Model 3	1.8°	1.23

- Equivalent mass of a fish, platform, and the lead-screw nut = 10 kg.
- The equivalent moment of inertia of the lead screw and coupling (excluding the motor–rotor inertia) is given as 2.5 kg.cm^2.
- Lead screw efficiency may be taken as 80%.

Suppose that three stepper motors (models 1, 2, and 3) of a reputed manufacturer along with their respective drive systems are available for this application. Table P6.36 provides some useful data for the three motors.

The speed–torque characteristics of the three motors (when appropriate drive systems are incorporated) are shown in Figure P6.36b.

Select the most appropriate motor out of the given three models, for the particular application. Justify your choice by giving all necessary equations and calculations in detail. In particular you must show that all required specifications are met by the selected motor.

Note: g = 981 cm/s^2 and 1 kg.cm = 13.9 oz.in.

6.37 a. In theory, a stepper motor does not require a feedback sensor for its control. But, in practice, a feedback encoder is needed for accurate control, particularly under transient and dynamic loading conditions. Explain the reasons for this.

b. A material transfer unit in an automated factory is sketched in Figure P6.37. The unit has a conveyor, which moves objects on to a platform. When an object reaches the platform, the conveyor is stopped and the height of the object is measured using a laser triangulation unit. Then, the stepper motor of the platform is activated to raise the object through a distance that is determined on the basis of the object height, for further manipulation/ processing of the object.

The following parameters are given:

Mass of the heaviest object that is raised = 3.0 lb (1.36 kg)

Mass of the platform and nut = 3.0 lb

Inertia of the lead screw and coupling = 0.001 oz.in.s^2 (0.07 kg.m^2)

Maximum travel of the platform = 1.0 in (2.54 cm)

Positioning time = 200 ms

Assume a four-pitch lead screw of 80% efficiency. Also, neglect any external resistance to the vertical motion of the object, apart from gravity.

Out of the four choices of stepper motor that are given in Table 6.2 and Figure 6.46, which one would you pick to drive the platform? Justify your selection by giving all the computational details of the approach.

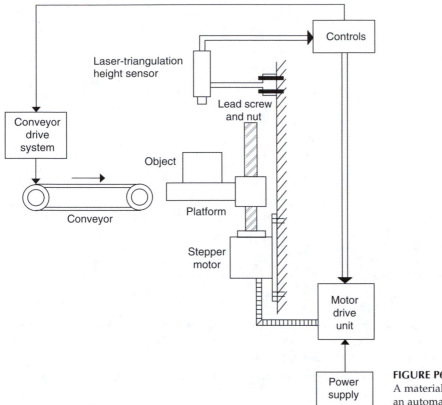

FIGURE P6.37
A material transfer unit in an automated factory.

6.38 a. What parameters or features determine the step angle of a stepper motor? What is microstepping? Briefly explain how microstepping is achieved.

b. A stepper motor–driven positioning platform is schematically shown in Figure P6.38.

Suppose that the maximum travel of the platform is L and this is accomplished in a time period of Δt. A trapezoidal velocity profile is used with a region of constant speed V in between an initial region of constant acceleration from rest and a final region of constant deceleration to rest, in a symmetric manner.

i. Show that the acceleration is given by

$$a = \frac{V^2}{V.\Delta t - L}.$$

The platform is moved using a mechanism of light, inextensible cable, and a pulley, which is directly (without gears) driven by a stepper motor. The platform moves on a pair of vertical guideways that use linear bearings and, for design purposes, the associated frictional resistance to platform motion may be neglected. The frictional torque at the bearings of the pulley is not negligible, however. Suppose that

$$\frac{\text{Frictional torque of the pulley}}{\text{Load torque on the pulley from the cable}} = e.$$

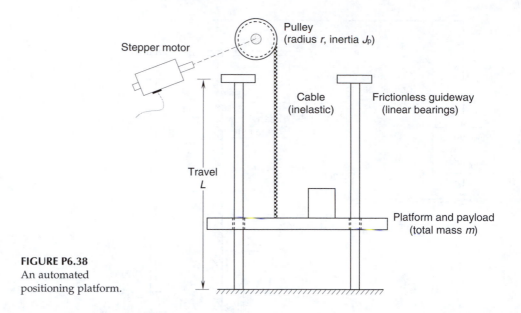

FIGURE P6.38
An automated
positioning platform.

Also, the following parameters are known: J_p is the moment of inertia of the pulley about the axis of rotation, r is the radius of the pulley, and m is the equivalent mass of the platform and its payload.

ii. Show that the maximum operating torque that is required from the stepper motor is given by

$$T = \left[J_m + J_p + (1 + e)mr^2\right]\frac{a}{r} + (1 + e)rmg,$$

where J_m is the moment of inertia of the motor rotor.

iii. Suppose that $V = 8.0\,\text{m/s}, L = 1.0\,\text{m}, \Delta t = 1.0\,\text{s}, m = 1.0\,\text{kg}, J_p = 3.0 \times 10^{-4}\,\text{kg.m}^2$, $r = 0.1$ m, and $e = 0.1$.

Four models of stepper motor are available, and their specifications given in Table 6.2 and Figure 6.46. Select the most appropriate motor (with the corresponding drive system) for this application. Clearly indicate all your computations and justify your choice.

iv. What is the position resolution of the platform, as determined by the chosen motor?

6.39 a. Consider a stepper motor of moment of inertia J_m, which drives a purely inertial load of moment of inertia J_L, through a gearbox of speed reduction r:1, as shown in Figure P6.39a.

Note that $\omega_L = \omega_m/r$,

where ω_m is the motor speed and ω_L is the load speed.

i. Show that the motor torque T_m may be expressed as

$$T_m = \left(rJ_m + \frac{J_L}{er}\right)\dot{\omega}_L,$$

where e is the gear efficiency.

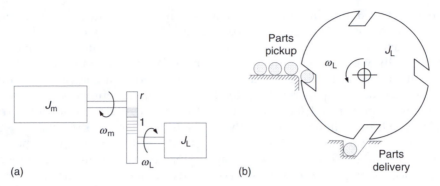

FIGURE P6.39
(a) Stepper motor driving an inertial load. (b) A parts transfer mechanism: an example of inertial load.

    ii. For optimal conditions of load acceleration express the required gear ratio r in terms of J_L, J_m, and e.

b. An example of a rotary load that is driven by a stepper motor is shown in Figure P6.39b. Here, in each quarter revolution of the load rotor, a part is transferred from the pickup position to the delivery position. The equivalent moment of inertia of the rotor, which carries a part, is denoted by J_L.

    Suppose that $J_L = 12.0 \times 10^{-3}$ kg.m^2. The required rate of parts transfer is 7 parts/s. A stepper motor is used to drive the load. A gearbox may be employed as well. Four motor models are available and their parameters are given in Table P6.39.

    The speed–torque characteristics of the motors are given in Figure 6.46. Assume that the step angle of each motor is 1.8°. The gearbox efficiency may be taken as 0.8.

    i. Prepare a table giving the optimal gear ratio, the operating speed of motor, the available torque, and the required torque, for each of the four models of motor, assuming that a gearbox with optimal gear ratio is employed in each case. On this basis, which motor would you choose for the present application?

    ii. Now consider the motor chosen in i. Suppose that three gearboxes of speed reduction 5, 8, and 10 may be available to you. Is a gearbox required in the present application, with the chosen motor? If so, which gearbox would you choose? Make your decision by computing the available torque and the required torque (with the motor chosen in i), for the four values of r given by 1, 5, 8, and 10.

    iii. What is the positioning resolution of the parts transfer system? What factors can affect this value?

TABLE P6.39

Motor Parameter Values

Motor Model	Motor Inertia, J_m ($\times 10^{-6}$ kg.m^2)
50SM	11.8
101SM	35
310SM	187
1010SM	805

6.40 Piezoelectric stepper motors are actuators that convert vibrations in a piezoelectric element (say, PZT) generated by an ac voltage (reverse piezoelectric effect) into rotary motion. Step angles in the order of $0.001°$ can be obtained by this method. In one design, as the piezoelectric PZT rings vibrate due to an applied ac voltage, radial bending vibrations are produced in a conical aluminum disc. These vibrations impart twisting (torsional) vibrations onto a beam element. The twisting motion is subsequently converted into a rotary motion of a frictional disc, which is frictionally coupled with the top surface of the beam. Essentially, because of the twisting motion, the two top edges of the beam push the frictional disc tangentially in a stepwise manner. This forms the output member of the piezoelectric stepper motor. List several advantages and disadvantages of this motor. Describe an application in which a miniature stepper motor of this type could be used.

7

Continuous-Drive Actuators

An actuator is a device that mechanically drives a control system. There are many classifications of actuators. Those that directly operate a process (load, plant) are termed process actuators. Joint motors in a robotic manipulator are good examples of process actuators. In process control applications in particular, actuators are often used to operate controller components (final control elements), such as servovalves, as well. Actuators in this category are termed control actuators. Actuators that automatically use response error signals from a process in feedback to correct the time-varying behavior of the process (i.e., to drive the process according to a desired response trajectory) are termed servo actuators. In particular, the motors that use position, speed, and perhaps load torque measurements and armature current or field current in feedback, to drive a load according to a specified motion trajectory, are termed servomotors.

One broad classification of actuators separates them into two types: incremental-drive actuators and continuous-drive actuators. Stepper motors, which are driven in fixed angular steps, represent the class of incremental-drive actuators. They can be considered as digital actuators, which are pulse-driven devices. Each pulse received at the driver of a digital actuator causes the actuator to move by a predetermined, fixed increment of displacement. Stepper motors were studied in Chapter 6. Most actuators used in control applications are continuous-drive devices. Examples are direct current (dc) servo motors, induction motors, hydraulic and pneumatic motors, and piston-cylinder drives (rams). Microactuators are actuators that are able to generate very small (microscale) actuating forces or torques and motions. In general, they can be neither developed nor analyzed as scaled-down versions of regular actuators. Separate and more innovative procedures of design, construction, and analysis are necessary for microactuators. Micromachined, millimeter-size micromotors with submicron accuracy are useful in modern information storage systems. Distributed or multilayer actuators constructed using piezoelectric, electrostrictive, magnetostrictive, or photostrictive materials are used in advanced and complex applications such as adaptive structures.

In the early days of analog control, servo actuators were exclusively continuous-drive devices. Since the control signals in this early generation of (analog) control systems generally were not discrete pulses, the use of pulse-driven incremental actuators was not feasible in those systems. DC servomotors and servovalve-driven hydraulic and pneumatic actuators were the most widely used types of actuators in industrial control systems, particularly because digital control was not available. Furthermore, the control of alternating current (ac) actuators was a difficult task at that time. Now, ac motors are also widely used as servomotors, employing modern methods of phase voltage control and frequency control through microelectronic-drive systems and using field feedback compensation through digital signal processing (DSP) chips. It is interesting to note that actuator control using pulse signals is no longer limited to digital actuators. Pulse-width modulated (PWM) signals through PWM amplifiers (rather than linear amplifiers) are increasingly used to drive continuous-drive actuators such as dc servomotors and hydraulic servos. Furthermore, it should be pointed out that electronic-switching commutation in dc motors is quite similar to the method of phase switching used for driving stepper motors.

For an actuator, requirements of size, torque or force, speed, power, stroke, motion resolution, repeatability, duty cycle, and operating bandwidth can differ significantly, depending on the particular application and the specific function of the actuator within the control system. Furthermore, the capabilities of an actuator will be affected by its drive system. Although the cost of sensors and transducers is a deciding factor in low-power applications and in situations where precision, accuracy, and resolution are of primary importance, the cost of actuators can become crucial in moderate to high power control applications. It follows that the proper design and selection of actuators can have a significant economical impact in many applications of industrial control. The applications of actuators are immense, spanning over industrial, manufacturing, transportation, medical, instrumentation, and household appliance fields.

This chapter discusses the principles of operation, mathematical modeling, analysis, characteristics, performance evaluation, methods of control, and sizing or selection of the more common types of continuous-drive actuators used in control applications. In particular, dc motors, ac induction motors, ac synchronous motors, and hydraulic and pneumatic actuators are considered. Fluidic devices are introduced.

7.1 DC Motors

The dc motor converts dc electrical energy into rotational mechanical energy. A major part of the torque generated in the rotor (armature) of the motor is available to drive an external load. The dc motor is probably the earliest form of electric motor. Because of features such as high torque, speed controllability over a wide range, portability, well-behaved speed–torque characteristics, easier and accurate modeling, and adaptability to various types of control methods, dc motors are still widely used in numerous control applications including robotic manipulators, transport mechanisms, disk drives, positioning tables, machine tools, and servovalve actuators.

The principle of operation of a dc motor is illustrated in Figure 7.1. Consider an electric conductor placed in a steady magnetic field at right angles to the direction of the field.

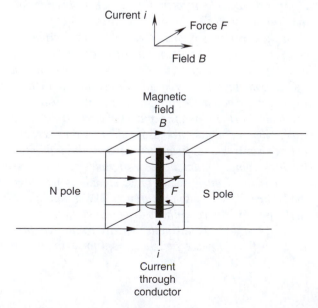

FIGURE 7.1
Operating principle of a dc motor.

Flux density B is assumed constant. If a dc current is passed through the conductor, the magnetic flux due to the current loops around the conductor, as shown in the figure. Consider a plane through the conductor, parallel to the direction of flux of the magnet. On one side of this plane, the current flux and the field flux are additive; on the opposite side, the two magnetic fluxes oppose each other. As a result, an imbalance magnetic force F is generated on the conductor, normal to the plane. This force is given by (Lorentz's law)

$$F = Bil, \tag{7.1}$$

where B is the flux density of the original field, i is the current through the conductor, and l is the length of the conductor.

Note that if the field flux is not perpendicular to the length of the conductor, it can be resolved into a perpendicular component that generates the force and to a parallel component that has no effect. The active components of i, B, and F are mutually perpendicular and form a right-hand triad, as shown in Figure 7.1. Alternatively, in the vector representation of these three quantities, the vector F can be interpreted as the cross product of the vectors i and B. Specifically, $F = i \times B$.

If the conductor is free to move, the generated force moves it at some velocity v in the direction of the force. As a result of this motion in the magnetic field B, a voltage is induced in the conductor. This is known as the back electromotive force or back e.m.f., and is given by

$$v_b = Blv. \tag{7.2}$$

According to Lenz's law, the flux due to the back e.m.f. v_b opposes the flux due to the original current through the conductor, thereby trying to stop the motion. This is the cause of electrical damping in motors, which is discussed later. Equation 7.1 determines the armature torque (motor torque) and Equation 7.2 governs the motor speed.

7.1.1 Rotor and Stator

A dc motor has a rotating element called rotor or armature. The rotor shaft is supported on two bearings in the motor housing. The rotor has many closely spaced slots on its periphery. These slots carry the rotor windings, as shown in Figure 7.2a. Assuming the field flux is in the radial direction of the rotor, the force generated in each conductor will be in the tangential direction, thereby generating a torque (force × radius), which drives the rotor. The rotor is typically a laminated cylinder made from a ferromagnetic material. A ferromagnetic core helps concentrate the magnetic flux toward the rotor. The lamination reduces the problem of magnetic hysteresis and limits the generation of eddy currents and associated dissipation (energy loss by heat generation) within the ferromagnetic material. More advanced dc motors use powdered-iron-core rotors rather than the laminated–iron-core variety, thereby further restricting the generation and conduction or dissipation of eddy currents and reducing various nonlinearities such as hysteresis. The rotor windings (armature windings) are powered by the supply voltage v_a.

The fixed magnetic field, which interacts with the rotor coil and generates the motor torque, is provided by a set of fixed magnetic poles around the rotor. These poles form the stator of the motor. The stator may consist of two opposing poles of a permanent magnet (PM). In industrial dc motors, however, the field flux is usually generated not by a permanent magnet but electrically in the stator windings, by an electromagnet, as schematically shown in Figure 7.2a. Stator poles are constructed from ferromagnetic

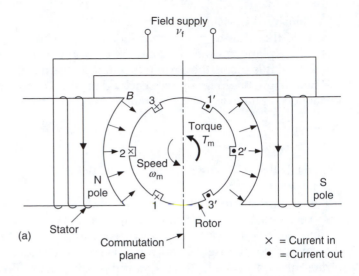

(a)

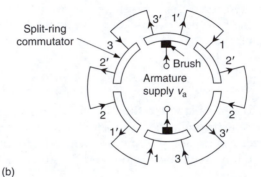

FIGURE 7.2
(a) Schematic diagram of a dc motor.
(b) Commutator wiring.   (b)

sheets (i.e., a laminated construction). The stator windings are powered by supply voltage v_f, as shown in Figure 7.2a. Furthermore, note that in Figure 7.2a, the net stator magnetic field is perpendicular to the net rotor magnetic field, which is along the commutation plane. The resulting forces that attempt to pull the rotor field toward the stator field may be interpreted as the cause of the motor torque, which is maximum when the two fields are at right angles.

The rotor in a conventional dc motor is called the *armature* (voltage supply to the armature windings is denoted by v_a). This nomenclature is particularly suitable for electric generators because the windings within which the useful voltage is induced (generated) are termed armature windings. According to this nomenclature, armature windings of an ac machine are located in the stator, not in the rotor. Stator windings in a conventional dc motor are termed field windings. In an electric generator, the armature moves relative to the magnetic field of the field windings, generating the useful voltage output. In synchronous ac machines, the field windings are the rotor windings. A dc motor may have more than two stator poles and far more conductor slots than what is shown in Figure 7.2a. This enables the stator to provide a more uniform and radial magnetic field. For example, some rotors carry more than 100 conductor slots.

7.1.2 Commutation

A plane known as the "commutation plane" symmetrically divides two adjacent stator poles of opposite polarity. In the two-pole stator shown in Figure 7.2a, the commutation plane is at right angles to the common axis of the two stator poles, which is the direction of the stator magnetic field. It is noted that on one side of the plane, the field is directed toward the rotor, whereas on the other side, the field is directed away from the rotor. Accordingly, when a rotor conductor rotates from one side of the plane to the other side, the direction of the generated torque will be reversed at the commutator plane, pushing the rotor in the opposite direction. Such a scenario is not useful since the average torque will be zero in that case.

In order to maintain the direction of torque in each conductor group (one group is numbered 1, 2, and 3 and the other group is numbered $1'$, $2'$, and $3'$ in Figure 7.2a), the direction of current in a conductor has to change as the conductor crosses the commutation plane. Physically, this may be accomplished by using a split-ring and brush commutator, shown schematically in Figure 7.2b, which is explained now. The armature voltage is applied to the rotor windings through a pair of stationary conducting blocks made of graphite (i.e., conducting soft carbon), which maintain sliding contact with the split ring. These contacts are called brushes because historically, they were made of bristles of copper wire in the form of a brush. The graphite contacts are cheaper, more durable primarily due to reduced sparking (arcing) problems, and provide more contact area (and hence, less electrical contact resistance). In addition, the contact friction is lower. The split-ring segments, equal in number to the conductor slot pairs (or loops) in the rotor, are electrically insulated from one another, but the adjacent segments are connected by the armature windings in each opposite pair of rotor slots, as shown in Figure 7.2b. For the rotor position shown in Figure 7.2, note that when the split ring rotates in the counterclockwise direction through 30°, the current paths in conductors 1 and $1'$ reverse but the remaining current paths are unchanged, thus achieving the required commutation. Mechanically, this is possible because the split ring is rigidly mounted on the rotor shaft, as shown in Figure 7.3.

7.1.3 Static Torque Characteristics

Let us examine the nature of the static torque generated by a dc motor. For static torque we assume that the motor speed is low so that the dynamic effects need not be explicitly included in the discussion. Consider a two-pole permanent magnet stator and a planar coil that is free to rotate about the motor axis, as shown in Figure 7.4a. The coil (rotor, armature) is energized by current i_a as shown. The flux density vector of the stator magnetic filed is B and the unit vector normal to the plane of the coil is n. The angle

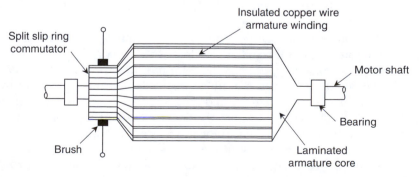

FIGURE 7.3
Physical construction of the rotor of a dc motor.

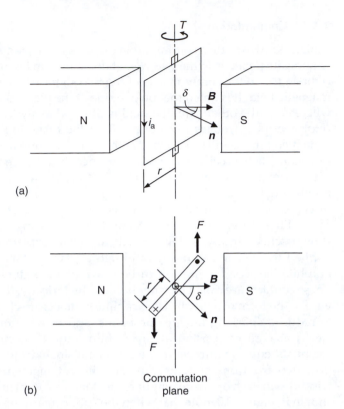

FIGURE 7.4
(a) Torque generated in a planar rotor.
(b) Nomenclature.   (b)

between **B** and **n** is δ, which is known as the torque angle. It should be clear from Figure 7.4b that the torque T generated in the rotor is given by

$$T = F \times 2r \sin \delta$$

which, in view of Equation 7.1, becomes

$$T = Bi_a l \times 2r \sin \delta$$

or

$$T = Ai_a B \sin \delta, \tag{7.3}$$

where l is the axial length of the rotor, r is the radius of the rotor, and A is the face area of the planar rotor.

Suppose that the rotor rotation starts by coinciding with the commutation plane, where $\delta = 0$ or π, and the rotor rotates through an angle of 2π. The corresponding torque profile is shown in Figure 7.5a. Next suppose that the rotor has three planar coil segments placed at 60° apart, and denoted by 1, 2, and 3, as in Figure 7.2. Note from Figure 7.2b that current switching occurs at every 60° rotation, and in a given instant two coil segments are energized. Figure 7.5b shows the torque profile of each coil segment and the overall torque profile due to the three-segment rotor in Figure 7.2. Note that the torque profile has improved (i.e., larger torque magnitude and smaller variation) as a result of the multiple coil segments, with shorter commutation angles. The torque profile can be further improved by incorporating still more coil segments, with correspondingly shorter commutation angles, but the design of the split-ring and brush arrangement becomes

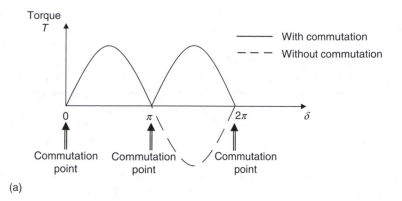

(a)

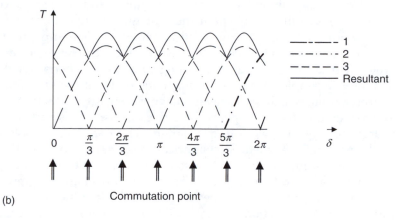

(b)

FIGURE 7.5
(a) Torque profile from a coil segment due to commutation. (b) Resultant torque from a rotor with three-coil segments.

more challenging then. Hence, there is a design limitation to achieving uniform torque profiles in a dc motor. It should be clear from Figure 7.2a that if the stator field can be made radial, then B is always perpendicular to n and hence sin δ becomes equal to 1. In that case, the torque profile is uniform, under ideal conditions.

7.1.4 Brushless DC Motors

There are several shortcomings of the slip-ring and brush mechanisms, which are used for current transmission through moving members. Even with the advances from the historical copper brushes to modern graphite contacts, many disadvantages remain, including rapid wear out, mechanical loading, wear and heating due to sliding friction, contact bounce, excessive noise, and electrical sparks (arcing) with the associated dangers in hazardous (e.g., chemical) environments, problems of oxidation, problems in applications that require wash down (e.g., in food processing), and voltage ripples at switching points. Conventional remedies to these problems—such as the use of improved brush designs and modified brush positions to reduce arcing—are inadequate in sophisticated applications. In addition, the required maintenance (to replace brushes and resurface the split-ring commutator) can be rather costly. Electronic communication, as used in brushless dc motors, is able to overcome these problems.

Brushless dc motors have permanent-magnet rotors. Since in them the polarities of the rotor cannot be switched as the rotor crosses a commutation plane, commutation is accomplished by electronically switching the current in the stator winding segments. Note that this is the reverse of what is done in brushed commutation, where the stator polarities are fixed and the rotor polarities are switched when crossing a commutation plane. The stator windings of a brushless dc motor can be considered the armature windings whereas for a brushed dc motor, rotor is the armature. In concept, brushless dc motors are somewhat similar to permanent magnet stepper motors (see Chapter 6) and to some types of ac motors. By definition, a dc motor should use a dc supply to power the motor. The torque–speed characteristics of dc motor are different from those of stepper motor or ac motor. Furthermore, permanent-magnet motors are less nonlinear than the electromagnet motors because the field strength generated by a permanent magnet is rather constant and independent of the current through a coil. This is true whether the permanent magnet is in the stator (i.e., a brushed motor) or in the rotor (i.e., a brushless dc motor or a PM stepper motor).

Figure 7.6 schematically shows a brushless dc motor and associated commutation circuitry. The rotor is a multiple-pole permanent magnet. Conventional ferrite magnets and alnico or ceramic magnets are economical but their field-strength/mass ratio is relatively low compared with more costly rare earth magnets. Hence, for a given torque rating, the rotor inertia can be reduced by using rare earth material for the rotor of a brushless dc motor. Examples of rare earth magnetic material are samarium cobalt and neodymium–iron–boron, which can generate magnetic energy levels that are more than 10 times those for ceramic–ferrite magnets, for a given mass. This is particularly desirable when high torque is required, as in torque motors. The popular two-pole rotor design consists of a diametrically magnetized cylindrical magnet, as shown in Figure 7.6. The stator windings are distributed around the stator in segments of winding groups. Each winding segment has a separate supply lead. Figure 7.6 shows a four-segment stator. Two diametrically opposite segments are connected together so that they carry current simultaneously but in opposite directions. Commutation is accomplished by energizing each pair of diametrically opposite segments sequentially, at time instants determined by the rotor position. This commutation could be achieved through mechanical means, using

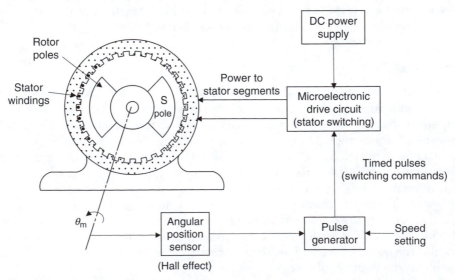

FIGURE 7.6
A brushless dc motor system.

a multiple contact switch driven by the motor itself. Such a mechanism would defeat the purpose, however, because it has most of the drawbacks of regular commutation using split rings and brushes. Modern brushless motors use microelectronic switching for commutation.

7.1.4.1 Constant-Speed Operation

For constant-speed operation, open-loop switching may be used. In this case, speed setting is provided as the input to a timing pulse generator. It generates a pulse sequence starting at zero pulse rate and increasing (ramping) to the final rate, which corresponds to the speed setting. Each pulse causes the driver circuit, which has proper switching circuitry, to energize a pair of stator segments. In this manner, the input pulse signal activates the stator segments sequentially, thereby generating a stator field, which rotates at a speed that is determined by the pulse rate. This rotating magnetic field would accelerate the rotor to its final speed. A separate command (or a separate pulse signal) is needed to reverse the direction of rotation, which is accomplished by reversing the switching sequence.

7.1.4.2 Transient Operation

Under transient motions of a brushless dc motor, it is necessary to know the actual position of the rotor for accurate switching of the stator field circuitry. An angular position sensor (e.g., a shaft encoder or more commonly a Hall effect sensor; see Chapter 6) is used for this purpose, as shown in Figure 7.6. By switching the stator segments at the proper instants, it is possible to maximize the motor torque. To explain this further, consider a brushless dc motor that has two rotor poles and four stator winding segments. Let us number the stator segments as in Figure 7.7a and also define the rotor angle θ_m as shown. The typical shape of the static torque curve of the motor when segment 1 is energized (with segment 1' automatically energized in the opposite direction) is shown in Figure 7.7b, as a function of θ_m. When segment 2 is energized, the torque distribution would be identical, but shifted to the right through 90°. Similarly, if segment 1' is energized in the positive direction (with segment 1 energized in the opposite direction), the corresponding torque distribution would be shifted to the right by an additional 90°, and so on. The superposition of these individual torque curves is shown in Figure 7.7c. It should be clear that to maximize the motor torque, switching has to be done at the points of intersection of the torque curves corresponding to the adjacent stator segments (as for stepper motors, see Chapter 7). These switching points are indicated as A, B, C, and D in Figure 7.7c. Under transient motions, position measurement would be required to determine these switching positions accurately. An effective solution is to mount Hall effect sensors at switching points (which are fixed) around the stator. The voltage pulse generated by each of these sensors, when a rotor pole passes the sensor, is used to switch the appropriate field windings.

Note from Figure 7.7c that positive average torque is possible even if the switching positions are shifted from these ideal locations by less than 90° to either side. It follows that the motor torque can be controlled by adjusting the switching locations with respect to the actual position of the rotor. The smoothness and magnitude of the motor torque, the accuracy of operation, and motor controllability can be improved by increasing the number of winding segments in the stator. This, however, increases the number of power lines and the complexity of the commutation circuitry. The commutation electronics for modern brushless dc motors is available as a single integrated circuit (IC) instead of discrete circuits using transistor switches.

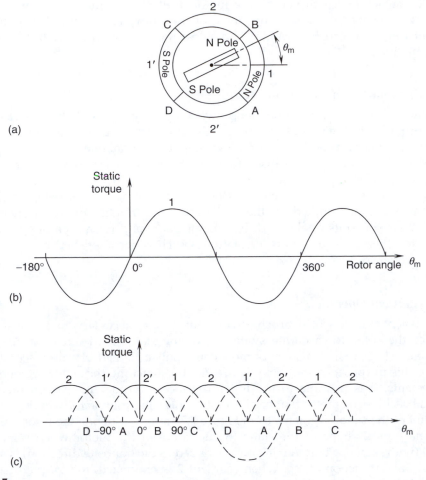

FIGURE 7.7
(a) A brushless dc motor. (b) Static torque curve with no switching (one-stator segment energized). (c) Switching sequence for maximum average torque.

Advantages of brushless dc motors primarily result from the disadvantages of using split rings and brushes for commutation, as noted before. Primary among them are the high efficiency, low mass for a specified torque rating, low maintenance, longer life, improved safety, and quieter operation. The drawbacks include the additional cost due to sensing and switching hardware. Two-state on/off switching generates torque ripples due to induction effects. This problem can be reduced by using transient (gradual) switching or shaped (e.g., ramp, sinusoidal) drive signals. Brushless dc motors with neodymium–iron–boron rotors can generate high torques (over 30 N.m). Motors in the continuous operating torque range of 0.5 N.m (75 oz. in.) to 30 N.m (270 lb. in.) are commercially available and used in general-purpose applications as well as in servo systems. For example, motors in the power range 0.0.1 to 5 hp, operating at speeds up to 7200 rpm, are available. Fractional horsepower applications include optical scanners, computer disk drives, instrumentation applications, surgical drills, and other medical devices. Medium-to-high power applications include robots, positioning devices, power blowers, industrial refrigerators, heating-ventilation-and-air-conditioning (HVAC) systems, and positive displacement pump drives. Many of these applications are for constant-speed operations, where ac motors are equally suitable. There are variable speed servo applications (e.g., robotics and inspection devices) and high acceleration applications (e.g., spinners and centrifuges) for which dc motors are preferred over ac motors.

7.1.5 Torque Motors

Conventionally, torque motors are high-torque dc motors with permanent magnet stators. These actuators characteristically possess a linear (straight line) torque–speed relationship, primarily because of their high-strength permanent-magnet stators, which provide a fairly constant and uniform magnetic field. The magnet should have high flux density per unit volume of the magnet material, yielding a high torque/mass ratio for a torque motor. Furthermore, coercivity (resistance to demagnetization) should be high and the cost has to be moderate. Rare earth materials (e.g., samarium cobalt, $SmCo_5$) possess most of these desirable characteristics, although their cost could be high. Conventional and low-cost ferrite magnets and alnico (aluminum–nickel–cobalt) or ceramic magnets provide a relatively low torque/mass ratio. Hence, rare earth magnets are widely used in torque motor and servomotor applications. As a comparison, a typical rare earth motor may produce a peak torque of over 27 N.m, with a torque/mass ratio of over 6 N.m/kg, whereas an alnico motor of identical dimensions and mass may produce a peak torque of about half the value (less than 15 N.m, with a torque/mass ratio of about 3.4 N.m/kg).

When operating at high torques (e.g., a thousand or more newton-meters; Note: 1 N.m = 0.74 lb. ft), the motor speeds have to be quite low for a given level of power. One straightforward way to increase the output torque of a motor (with a corresponding reduction in speed) is to employ a gear system (typically using worm gears) with high gear reduction. Gear drives introduce undesirable effects such as backlash, additional inertia loading, higher friction, increased noise, lower efficiency, and additional maintenance. Backlash in gears would be unacceptable in high-precision applications. Frictional loss of torque, wear problems, and the need for lubrication must also be considered. Furthermore, the mass of the gear system reduces the overall torque/mass ratio and the useful bandwidth of the actuator. For these reasons, torque motors are particularly suitable for high-precision, direct-drive applications (e.g., direct-drive robot arms) that require high-torque drives without having to use speed reducers and gears. Torque motors are usually more expensive than the conventional types of dc motors. This is not a major drawback, however, because torque motors are often custom-made and are supplied as units that can be directly integrated with the process (load) within a common housing. For example, the stator might be integrated with one link of a robot arm and the rotor with the next link, thus forming a common joint in a direct-drive robot. Torque motors are widely used as valve actuators in hydraulic servo valves, where large torques and very small displacements are required.

Brushless torque motors have permanent magnet rotors and wound stators, with electronic commutation. Consider a brushless dc motor with electronic commutation. The output torque can be increased by increasing the number of magnetic poles. Since direct increase of the magnetic poles has serious physical limitations, a toothed construction, as in stepping motors, could be employed for this purpose. Torque motors of this type have toothed ferromagnetic stators with field windings on them. Their rotors are similar in construction to those of variable-reluctance (VR) stepping motors. A harmonic drive is a special type of gear reducer that provides very large speed reductions (e.g., 200:1) without backlash problems. The harmonic drive is often integrated with conventional motors to provide very high torques, particularly in backlash-free servo applications. The principle of operation of a harmonic drive is discussed in Chapter 8.

7.2 DC Motor Equations

Consider a dc motor with separate windings in the stator and the rotor. Each coil has a resistance (R) and an inductance (L). When a voltage (v) is applied to the coil, a current (i) flows through the circuit, thereby generating a magnetic field. As discussed before, forces are produced in the rotor windings, and an associated torque (T_m), which turns the rotor. The rotor speed (ω_m) causes the magnetic flux linkage with the rotor coil from the stator field to change at a corresponding rate, thereby generating a voltage (back e.m.f.) in the rotor coil.

Equivalent circuits for the stator and the rotor of a conventional dc motor are shown in Figure 7.8a. Since the field flux is proportional to field current i_f, we can express the magnetic torque of the motor as

$$T_m = k i_f i_a,\tag{7.4}$$

which directly follows Equation 7.1. Next, in view of Equation 7.2, the back e.m.f. generated in the armature of the motor is given by

$$v_b = k' i_f \omega_m,\tag{7.5}$$

where i_f is the field current, i_a is the armature current, and ω_m is the angular speed of the motor, and k and k' are motor constants, which depend on factors such as the rotor dimensions, the number of turns in the armature winding, and the permeability (inverse of reluctance) of the magnetic medium. In the case of ideal electrical-to-mechanical energy conversion at the rotor (where the rotor coil links with the stator field), we have $T_m\omega_m = v_b \times i_a$ when consistent units are used (e.g., torque in newton-meters, speed in radians per second, voltage in volts, and current in amperes). Then we observe that

$$k = k'.\tag{7.6}$$

The field circuit equation is obtained by assuming that the stator magnetic field is not affected by the rotor magnetic field (i.e., the stator inductance is not affected by the rotor) and that there are no eddy current effects in the stator. Then, from Figure 7.8a,

$$v_f = R_f i_f + L_f \frac{d i_f}{dt},\tag{7.7}$$

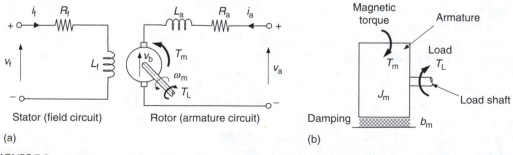

(a)

(b)

FIGURE 7.8
(a) The equivalent circuit of a conventional dc motor (separately excited). (b) Armature mechanical loading diagram.

where v_f is the supply voltage to the stator, R_f is the resistance of the field winding, and L_f is the inductance of the field winding.

The equation for the armature rotor circuit is written as (see Figure 7.8a)

$$v_a = R_a i_a + L_a \frac{di_a}{dt} + v_b, \tag{7.8}$$

where v_a is the supply voltage to the armature, R_a is the resistance of the armature winding, and L_a is the leakage inductance in the armature winding.

It should be emphasized here that the primary inductance or mutual inductance in the armature winding (due to its coupling with the stator field) is represented in the back e.m.f. term v_b. The leakage inductance, which is usually neglected, represents the fraction of the armature flux that is not linked with the stator and is not used in the generation of useful torque. This represents a self-inductance effect in the armature.

The mechanical equation of the motor is obtained by applying Newton's second law to the rotor. Assuming that the motor drives some load, which requires a load torque T_L to operate, and that the frictional resistance in the armature can be modeled by a linear viscous term, we have (see Figure 7.8b)

$$J_m \frac{d\omega_m}{dt} = T_m - T_L - b_m \omega_m, \tag{7.9}$$

where J_m is the moment of inertia of the rotor and b_m is the equivalent mechanical damping constant for the rotor.

Note that the load torque may be due, in part, to the inertia of the external load that is coupled to the motor shaft. If the coupling flexibility is neglected (i.e., a rigid shaft), the load inertia may be directly added to (i.e., lumped with) the rotor inertia after accounting for the possible existence of a speed reducer (gear, harmonic drive, etc.). In general, however, a separate set of equations is necessary to represent the dynamics of the external load.

Equation 7.4 through Equation 7.9 form the dynamic model for a dc motor. In obtaining this model, we have made several assumptions and approximations. In particular, we have either approximated or neglected the following factors:

1. Coulomb friction and associated dead-band effects
2. Magnetic hysteresis (particularly in the stator core, but in the armature as well if not a brushless motor)
3. Magnetic saturation (in both stator and the armature)
4. Eddy current effects (laminated core reduces this effect)
5. Nonlinear constitutive relations for magnetic induction (in which case inductance L is not constant)
6. Brush contact resistance, finite width contact of brushes, and other types of noise and nonlinearities in split-ring commutators
7. The effect of the rotor magnetic flux (armature flux) on the stator magnetic flux (field flux)

7.2.1 Steady-State Characteristics

In selecting a motor for a given application, its steady-state characteristics are a major determining factor. In particular, steady-state torque–speed curves are employed for this purpose. The rationale is that, if the motor is able to meet the steady-state operating requirements, with some design conservatism, it should be able to tolerate some deviations under transient conditions of short duration. In the separately excited case shown in Figure 7.8a, where the armature circuit and field circuit are excited by separate and independent voltage sources, it can be shown that the steady-state torque–speed curve is a straight line. To verify this, we set the time derivatives in Equation 7.7 and Equation 7.8 to zero, as this corresponds to steady-state conditions. It follows that i_f is constant for a fixed supply voltage v_f. By substituting Equation 7.4 and Equation 7.5 in Equation 7.8, we get

$$v_a = \frac{R_a}{k i_f} T_m + k' i_f \omega_m.$$

Under steady-state conditions in the field circuit, we have from Equation 7.7,

$$i_f = \frac{v_f}{R_f}.$$

It follows that the steady-state torque–speed characteristics of a separately excited dc motor may be expressed as

$$\omega_m + \frac{R_a R_f^2}{k k' v_f^2} T_m = \frac{R_f v_a}{k' v_f} \tag{7.10a}$$

Now, since v_a and v_f are constant supply voltages from a regulated power supply, on defining the constant parameters T_s and ω_o, Equation 7.10 can be expressed as

$$\frac{\omega_m}{\omega_o} + \frac{T_m}{T_s} = 1, \tag{7.10b}$$

where ω_o is the no-load speed (at steady state, assuming zero damping) and T_s is the stalling torque (or starting torque) of the motor.

It should be noted from Equation 7.9 that if there is no damping ($b_m = 0$), the steady-state magnetic torque (T_m) of the motor is equal to the load torque (T_L). In practice, however, there is mechanical damping on the rotor, and the load torque is smaller than the motor torque. In particular, the motor stalls at a load torque smaller than T_s. The idealized characteristic curve given by Equation 7.10 is shown in Figure 7.9.

7.2.1.1 Bearing Friction

The primary source of mechanical damping in a motor is the bearing friction. Roller bearings have low friction. But, since the balls make point contacts on the bearing sleeve, they are prone to damage due to impact and wear problems. Roller bearings provide better contact capability (line contact), but can produce roller creep and noisy operation. For ultraprecision and specialized applications, air bearings and magnetic bearings are suitable, which offer the capability of active control and very low friction. A linear viscous model is normally adequate to represent bearing damping. For more accurate analysis, sophisticated models (e.g., Sribeck model) may be incorporated.

Example 7.1

A load is driven at constant power under steady-state operating conditions, using a separately wound dc motor with constant supply voltages to the field and armature

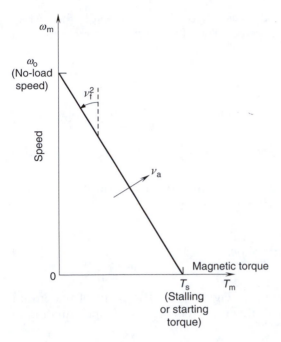

FIGURE 7.9

Steady-state speed–torque characteristics of a separately wound dc motor.

windings. Show that, in theory, two operating points are possible. Also show that one of the operating points is stable and the other one is unstable.

Solution

As shown in Figure 7.9, the steady-state characteristic curve of a dc motor with windings that are separately excited by constant voltage supplies, is a straight line. The constant-power curve for the load is a hyperbola because the product $T\omega_m$ is constant in this case. The two curves shown in Figure 7.10 intersect at point P and point Q. At point P, if there is a slight decrease in the speed of operation, the motor (magnetic) torque increases to T_{PM} and the load torque demand increases to T_{PL}. However, since $T_{PM} > T_{PL}$, the system accelerates back to point P. It follows that point P is a stable operating point. Alternatively, at point Q, if the speed drops slightly, the magnetic torque of the motor increases to T_{QM} and the load

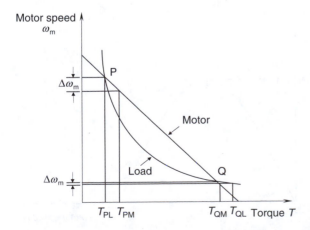

FIGURE 7.10

Operating points for a constant-power load driven by a dc motor.

torque demand increases to T_{QL}. However, in this case, $T_{QM} < T_{QL}$. As a result, the system decelerates further, subsequently stalling the system. Therefore, it can be concluded that point Q is an unstable operating point.

7.2.1.2 Output Power

The output power of a motor is given by

$$p = T_m \omega_m. \tag{7.11}$$

Equation 7.10b applies for a dc motor excited by a regulated power supply, in steady state. Substitute this in Equation 7.11, for T_m. We get the output power

$$p = T_s \left(1 - \frac{\omega_m}{\omega_o}\right) \omega_m. \tag{7.12}$$

Equation 7.12 has a quadratic shape, as shown in Figure 7.11. The point of maximum power is obtained by differentiating Equation 7.12 with respect to speed, and equating to zero; thus,

$$\frac{dp}{d\omega_m} = T_s \left(1 - \frac{\omega_m}{\omega_o}\right) - \frac{T_s}{\omega_o} \omega_m = T_s \left(1 - 2\frac{\omega_m}{\omega_o}\right) = 0.$$

It follows that the speed at which the motor provides the maximum power is given by half the no-load speed:

$$\omega_{pmax} = \frac{\omega_o}{2}. \tag{7.13}$$

From Equation 7.12, the corresponding maximum power is

$$p_{max} = \frac{1}{4} T_s \omega_o. \tag{7.14}$$

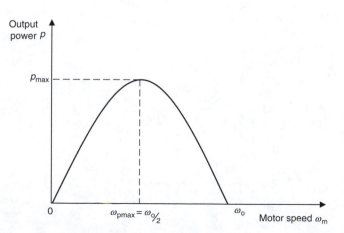

FIGURE 7.11
The output power curve of a dc motor at steady state.

7.2.1.3 *Combined Excitation of Motor Windings*

The shape of the steady-state speed–torque curve will change if a common voltage supply is used to excite both the field winding and the armature winding. Here, the two windings have to be connected together. There are three common ways the windings of the rotor and the stator are connected. They are known as

1. Shunt-wound motor
2. Series-wound motor
3. Compound-wound motor

In a shunt-wound motor, the armature windings and the field windings are connected in parallel. In the series-wound motor, they are connected in series. In the compound-wound motor, part of the field windings is connected with the armature windings in series and the other part is connected in parallel. These three connection types of the rotor and the stator windings of a dc motor are shown in Figure 7.12. Note that in a shunt-wound motor at steady state, the back e.m.f. v_b depends directly on the supply voltage. Since the

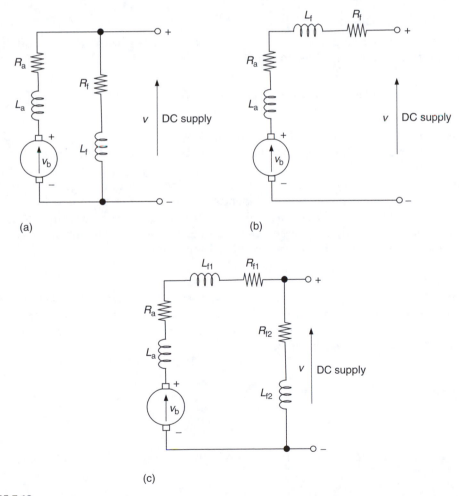

FIGURE 7.12
(a) A shunt-wound motor. (b) A series-wound motor. (c) A compound-wound motor.

TABLE 7.1

Influence of the Winding Configuration on the Steady-State Characteristics of a dc Motor

DC Motor Type	Field Coil Resistance	Speed Controllability	Starting Torque
Shunt-wound	High	Good	Average
Series-wound	Low	Poor	High
Compound-wound	Parallel high, series low	Average	Average

back e.m.f. is proportional to the speed, it follows that speed controllability is good with the shunt-wound configuration. In a series-wound motor, the relation between v_b and the supply voltage is coupled through both the armature windings and the field windings. Hence its speed controllability is relatively poor. But in this case, a relatively large current flows through both windings at low speeds of the motor (when the back e.m.f. is small), giving a higher starting torque. Also, the operation is approximately at constant power in this case. These properties are summarized in Table 7.1. Since both speed controllability and higher starting torque are desirable characteristics, compound-wound motors are used to obtain a performance in between the two extremes.

7.2.1.4 Speed Regulation

Variation in the operating speed of a motor due to changes in the external load is measured by the percentage speed regulation. Specifically,

$$\text{Percentage speed regulation} = \frac{(\omega_o - \omega_f)}{\omega_f} \times 100\%, \tag{7.15}$$

where ω_o is the no-load speed and ω_f is the full-load speed.

   This is a measure of the speed stability of a motor; the smaller the percentage speed regulation, the more stable the operating speed under varying load conditions (particularly in the presence of load disturbances). In the shunt-wound configuration, the back e.m.f., and hence the rotating speed, depends directly on the supply voltage. Consequently, the armature current and the related motor torque have virtually no effect on the speed. For this reason, the percentage speed regulation is relatively small for shunt-wound motors, resulting in improved speed stability.

Example 7.2

An automated guideway transit (AGT) vehicle uses a series-wound dc actuator in its magnetic suspension system. If the desired control bandwidth (see Chapter 3) of the active suspension (in terms of the actuator force) is 40 Hz, what is the required minimum bandwidth for the input voltage signal?

Solution

The actuating force is

$$F = k i_a i_f = k i^2, \tag{i}$$

where i denotes the common current through both windings of the actuator. Consider a harmonic component

$$v(\omega) = v_o \sin \omega t \tag{ii}$$

of the input voltage to the windings, where ω denotes the frequency of the chosen frequency component. The field current is given by

$$i(\omega) = i_o \sin(\omega t + \phi) \tag{iii}$$

at this frequency, where ϕ denotes the phase shift. Substitute Equation (iii) in Equation (i) to determine the corresponding actuating force:

$$F = ki_o^2 \sin^2(\omega t + \phi) = ki_o^2[1 - \sin(2\omega t + 2\phi)]/2.$$

It follows that there is an inherent frequency doubling in the suspension system. As a result, the required minimum bandwidth for the input voltage signal is 20 Hz.

Example 7.3
Consider the three types of winding connections for dc motors, shown in Figure 7.12. Derive equations for the steady-state torque–speed characteristics in the three cases. Sketch the corresponding characteristic curves. Using these curves, discuss the behavior of the motor in each case.

Solution
Shunt-Wound Motor
Note that at steady state, the inductances are not present in the motor equivalent circuit. For the shunt-wound dc motor (Figure 7.12a), the field current is

$$i = v/R_f = \text{constant.} \tag{i}$$

The armature current is

$$i_a = [v - v_b]/R_a. \tag{ii}$$

The back e.m.f. for a motor speed of ω_m is given by

$$v_b = k'i_f\omega_m. \tag{iii}$$

Substituting Equation (i) through Equation (iii) in the motor magnetic torque equation

$$T_m = ki_fi_a,$$

we get

$$\omega_m + \left(\frac{R_aR_f^2}{kk'v^2}\right)T_m = \frac{R_f}{k'}. \tag{7.16a}$$

Note that Equation 7.16a represents a straight line with a negative slope of magnitude $\left(\dfrac{R_aR_f^2}{kk'v^2}\right)$.

   Since this magnitude is typically small, it follows that good speed regulation (constant-speed operation and relatively low sensitivity of the speed to torque changes) can be obtained using a shunt-wound motor. The characteristic curve for the shunt-wound dc motor is shown in Figure 7.13a. The starting torque T_s is obtained by setting $\omega_m = 0$ in Equation 7.16a. The no-load speed ω_o is obtained by setting $T_m = 0$ in the same equation.

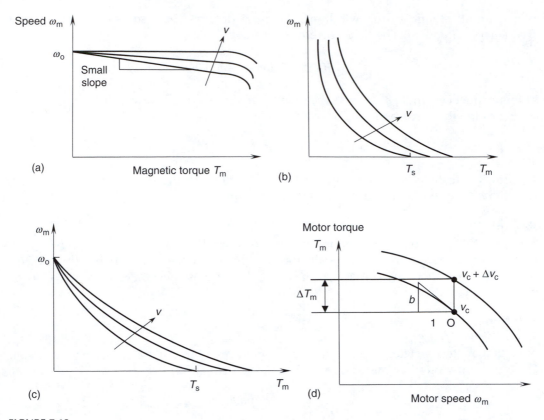

FIGURE 7.13
Torque–speed characteristic curves for dc motors. (a) Shunt-wound. (b) Series-wound. (c) Compound-wound, and (d) Linearization of the general case.

The corresponding expressions are tabulated in Table 7.2. Note that if the input voltage v is increased, the starting torque increases but the no-load speed remains unchanged, as sketched in Figure 7.13a.

Series-Wound Motor
At steady state, for the series-wound dc motor shown in Figure 7.12b, the field current is equal to the armature current; thus,

$$i_a = i_f = \frac{v - v_b}{R_a + R_f}.$$
(v)

TABLE 7.2

Comparison of dc Motor Winding Types

Winding Type	No-Load Speed ω_o	Starting Torque T_s
Shunt-wound	$\dfrac{R_f}{k'}$	$\dfrac{kv^2}{R_a R_f}$
Series-wound	∞	$\dfrac{kv^2}{(R_a + R_f)^2}$
Compound-wound	$\dfrac{R_{f2}}{k'}$	$\dfrac{kv^2}{R_a + R_{f1}}\left[\dfrac{1}{R_a + R_{f1}} + \dfrac{1}{R_{f2}}\right]$

The back e.m.f. is given by Equation (iii) as before. The motor magnetic torque is given by

$$T_m = ki_f^2. \tag{vi}$$

From these relations, we get the following equation for the steady-state speed–torque relation of a series-wound motor:

$$\omega_m = \frac{v}{k'}\sqrt{\frac{k}{T_m}} - \frac{R_a + R_f}{k'}. \tag{7.16b}$$

This equation is sketched in Figure 7.13b. Note that the starting torque, as given in Table 7.2, increases with the input voltage v. In the present case, the no-load speed is infinite. For this reason, the motor coasts at low loads. It follows that speed regulation in series-wound motors is poor. Starting torque and low-speed operation are satisfactory, however.

Compound-Wound Motor
Figure 7.12c gives the equivalent circuit for a compound-wound dc motor. Note that part of the field coil is connected in series with the rotor windings and the other part is connected in parallel with the rotor windings. Under steady-state conditions, the currents in the two parallel branches of the circuit are given by

$$i_a = i_{f1} = \frac{v - v_b}{R_a + R_{f1}}, \tag{vii}$$

$$i_{f2} = \frac{v}{R_{f2}}. \tag{viii}$$

Note that the total field current that generates the stator field is

$$i_f = i_{f1} + i_{f2} \tag{ix}$$

which, in view of Equation (vii), Equation (viii), and Equation (iii), becomes

$$i_f = v\left[\frac{1}{R_a + R_{f1}} + \frac{1}{R_{f2}}\right] - \frac{k'i_f\omega_m}{R_a + R_{f1}}.$$

Consequently,

$$i_f = v\left[\frac{1}{R_a + R_{f1}} + \frac{1}{R_{f2}}\right]\bigg/\left[1 + \frac{k'\omega_m}{R_a + R_{f1}}\right]. \tag{x}$$

The motor magnetic torque is given by

$$T_m = ki_f i_a = ki_f\frac{v - v_b}{R_a + R_{f1}} = ki_f\frac{v - k'i_f\omega_m}{R_a + R_{f1}}. \tag{xi}$$

Finally, by substituting Equation (x) in Equation (xi), we get the steady-state torque–speed relationship; thus,

$$
\begin{aligned}
T_{\mathrm{m}} &= \frac{kv^2\left(\dfrac{1}{R_{\mathrm{a}}+R_{\mathrm{f1}}}+\dfrac{1}{R_{\mathrm{f2}}}\right)\left[1-k'\omega_{\mathrm{m}}\left(\dfrac{1}{R_{\mathrm{a}}+R_{\mathrm{f1}}}+\dfrac{1}{R_{\mathrm{f2}}}\right)\Big/\left(1+\dfrac{k'\omega_{\mathrm{m}}}{R_{\mathrm{a}}+R_{\mathrm{f1}}}\right)\right]}{(R_{\mathrm{a}}+R_{\mathrm{f1}})\left(1+\dfrac{k'\omega_{\mathrm{m}}}{R_{\mathrm{a}}+R_{\mathrm{f1}}}\right)} \\[2mm]
&= \frac{kv^2\left(\dfrac{1}{R_{\mathrm{a}}+R_{\mathrm{f1}}}+\dfrac{1}{R_{\mathrm{f2}}}\right)\left(1-\dfrac{k'\omega_{\mathrm{m}}}{R_{\mathrm{f2}}}\right)}{(R_{\mathrm{a}}+R_{\mathrm{f1}})\left(1+\dfrac{k'\omega_{\mathrm{m}}}{R_{\mathrm{a}}+R_{\mathrm{f1}}}\right)^2}.
\end{aligned}
\tag{7.16c}
$$

This equation is sketched in Figure 7.13c. The expressions for the starting torque and the no-load speed are given in Table 7.2.

By comparing the foregoing results, we can conclude that good speed regulation and high starting torques are available from a shunt-wound motor, and nearly constant-power operation is possible with a series-wound motor. The compound-wound motor provides a trade-off between these two.

7.2.2 Experimental Model

We have noticed that, in general, the speed–torque characteristic of a dc motor is non-linear. A linearized dynamic model can be extracted from the speed–torque curves. One of the parameters of the model is the damping constant. First we will examine this.

7.2.2.1 *Electrical Damping Constant*

Newton's second law governs the dynamic response of a motor. In Equation 7.9, for example, b_{m} denotes the mechanical (viscous) damping constant and represents mechanical dissipation of energy. As is intuitively clear, mechanical damping torque opposes motion—hence the negative sign in the $b_{\mathrm{m}}\omega_{\mathrm{m}}$ term in Equation 7.9. Note further that the magnetic torque T_{m} of the motor is also dependent on speed ω_{m}. In particular, the back e.m.f., which is governed by ω_{m}, produces a magnetic field, which tends to oppose the motion of the motor rotor. This acts as a damper, and the corresponding damping constant is given by

$$
b_{\mathrm{e}} = -\frac{\partial T_{\mathrm{m}}}{\partial \omega_{\mathrm{m}}}.
\tag{7.17a}
$$

This parameter is termed the electrical damping constant. Caution should be exercised when experimentally measuring b_{e}. Note that in constant speed tests, the inertia torque of the rotor will be zero; there is no torque loss due to inertia. Torque measured at the motor shaft includes as well the torque reduction due to mechanical dissipation (mechanical damping) within the rotor, however. Hence the magnitude b of the slope of the speed–torque curve as obtained by steady-state tests, is equal to $b_{\mathrm{e}}+b_{\mathrm{m}}$, where b_{m} is the equivalent viscous damping constant representing mechanical dissipation at the rotor.

7.2.2.2 *Linearized Experimental Model*

To extract a linearized experimental model for a dc motor, consider the speed–torque curves shown in Figure 7.13d. For each curve, the excitation voltage v_{c} is maintained constant. This is the voltage that is used in controlling the motor, and is termed control voltage. It can be, for example, the armature voltage, the field voltage, or the voltage that excites both armature and field windings in the case of combined excitation (e.g., shunt-wound motor). One curve in Figure 7.13d is obtained at control voltage v_c and the other curve is obtained at $v_c + \Delta v_c$. Note also that a tangent can be drawn at a selected point (operating point O) of a speed–torque curve. The magnitude b of the slope, which is negative, corresponds to a damping constant, which includes both electrical and mechanical damping effects. What mechanical damping effects are included in this parameter depends entirely on the nature of mechanical damping that was present during the test (primarily bearing friction). We have the damping constant as the magnitude of the slope at the operating point:

$$b = -\frac{\partial T_m}{\partial \omega_m}\bigg|_{v_c = \text{constant}}. \tag{7.17b}$$

Next draw a vertical line through the operating point O. The torque intercept ΔT_m between the two curves can be determined in this manner. Since a vertical line is a constant speed line, we have the voltage gain:

$$k_v = \frac{\partial T_m}{\partial v_c}\bigg|_{\omega_m = \text{constant}} = \frac{\Delta T_m}{\Delta v_c}. \tag{7.18}$$

Now, using the well-known relation for total differential we have

$$\delta T_m = \frac{\partial T_m}{\partial \omega_m}\bigg|_{v_c} \delta \omega_m + \frac{\partial T_m}{\partial v_c}\bigg|_{\omega_m} \delta v_c.$$
$$= -b\delta \omega_m + k_v \delta v_c \tag{7.19}$$

Equation 7.19 is the linearized model of the motor. This may be used in conjunction with the mechanical equation of the motor rotor, for the incremental motion about the operating point:

$$J_m \frac{d\delta \omega_m}{dt} = \delta T_m - \delta T_L. \tag{7.20}$$

Note that Equation 7.20 is the incremental version of Equation 7.9 except that the overall damping constant of the motor (including mechanical damping) is included in Equation 7.19. The torque needed to drive the rotor inertia, however, is not included in Equation 7.19 because the steady-state curves are used in determining the parameters for this equation. The inertia term is explicitly present in Equation 7.20.

Example 7.4

Split-field series-wound dc motors are sometimes used as servo actuators. A motor circuit for this arrangement, under steady-state conditions, is shown in Figure 7.14. The field windings are divided into two identical parts and supplied by a differential amplifier (such as a push/pull amplifier) such that the magnetic fields in the two winding segments oppose each other. In this manner, the difference in the two input voltage signals (i.e., an error signal) is employed in driving the motor. Split-field dc motors are used in low-power applications. Determine the electrical damping constant of the motor shown in Figure 7.14.

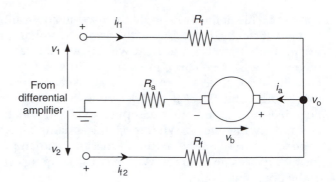

FIGURE 7.14
A split-field series-wound dc motor.

Solution
Suppose that $v_1 = \bar{v} + \Delta v/2$ and $v_2 = \bar{v} - \Delta v/2$, where $\bar{v}$ is a constant representing the average supply voltage. Hence,

$$v_1 - v_2 = \Delta v, \tag{i}$$

$$v_1 + v_2 = 2\bar{v}. \tag{ii}$$

The motor is controlled using the differential voltage Δv. In a servo actuator, this differential voltage corresponds to a feedback error signal. Using the notation shown in Figure 7.14, the field current is given by (because the magnetic fields of the two stator winding segments oppose each other)

$$i_f = i_{f1} - i_{f2}. \tag{iii}$$

The armature current is given by (see Figure 7.14)

$$i_a = i_{f1} + i_{f2}. \tag{iv}$$

Hence, the motor magnetic torque can be expressed as

$$T_m = k i_a i_f = k(i_{f1} + i_{f2})(i_{f1} - i_{f2}). \tag{v}$$

In addition, the node voltage is

$$v_o = v_1 - i_{f1}R_f = v_2 - i_{f2}R_f.$$

Using this fact along with Equation i, we get

$$i_{f1} - i_{f2} = \frac{v_1 - v_2}{R_f} = \frac{\Delta v}{R_f} \tag{vi}$$

and

$$2v_o = v_1 + v_2 - R_f(i_{f1} + i_{f2}). \tag{vii}$$

However, it is clear from Figure 7.14 along with the motor back e.m.f. equation that

$$v_o = v_b + i_a R_a = k' i_f \omega_m + i_a R_a,$$

where v_b denotes the back e.m.f. in the rotor. Hence, in view of Equation iii and Equation iv, we have

$$v_o = k'(i_{f1} - i_{f2})\omega_m + (i_{f1} + i_{f2})R_a. \tag{viii}$$

Substitute Equation viii in Equation vii to eliminate v_o. We get

$$\frac{v_1 + v_2}{2} = \frac{R_f}{2}(i_{f1} + i_{f2}) + k'(i_{f1} - i_{f2})\omega_m + R_a(i_{f1} + i_{f2}). \tag{xi}$$

Substitute Equation ii and Equation vi in Equation ix. We get,

$$\bar{v} = \frac{k'}{R_f}\Delta v\omega_m + \left(R_a + \frac{R_f}{2}\right)(i_{f1} + i_{f2}). \tag{x}$$

Substitute Equation vi in Equation v. We get,

$$T_m = \frac{k}{R_f}(i_{f1} + i_{f2})\Delta v. \tag{xi}$$

Substitute Equation xi in Equation x. We get,

$$\bar{v} = \frac{k'}{R_f}\Delta v\omega_m + \left(R_a + \frac{R_f}{2}\right)\frac{R_f}{k\Delta v}T_m$$

or

$$T_m + \frac{kk'\Delta v^2}{R_f^2(R_a + R_f/2)}\omega_m = \frac{k\bar{v}\Delta v}{R_f(R_a + R_f/2)}. \tag{7.21}$$

This is a linear relationship between T_m and ω_m. Now, according to Equation 7.17a, the electrical damping constant for a split-field series-wound dc motor is given by

$$b_e = \frac{kk'\Delta v^2}{R_f^2(R_a + R_f/2)}. \tag{7.17c}$$

Note that the damping is zero under balanced conditions ($\Delta v = 0$). But damping increases quadratically with the differential voltage Δv.

7.3 Control of DC Motors

Both speed and torque of a dc motor may have to be controlled for proper performance in a given application of a dc motor. As we have seen, by using proper winding arrangements, dc motors can be operated over a wide range of speeds and torques. Because of this adaptability, dc motors are particularly suitable as variable-drive actuators. Historically, ac motors were employed almost exclusively in constant-speed applications, but their use in variable-speed applications was greatly limited because speed control of ac motors was found to be quite difficult, by conventional means. Since variable-speed control of a dc motor is quite convenient and straightforward, dc motors have dominated in industrial control applications for many decades.

Following a specified motion trajectory is called servoing, and servomotors (or servo actuators) are used for this purpose. The vast majority of servomotors are dc motors with feedback control of motion. Servo control is essentially a motion control problem, which involves the control of position and speed. There are applications, however, that require torque control, directly or indirectly, but they usually require more sophisticated sensing and control techniques.

Control of a dc motor is accomplished by controlling either the stator field flux or the armature flux. If the armature and field windings are connected through the same circuit (see Figure 7.12), both techniques are incorporated simultaneously. Specifically, the two methods of control are

1. Armature control
2. Field control

In armature control, the field voltage in the stator circuit is kept constant and the input voltage v_a to the rotor circuit is varied in order to achieve a desired performance (i.e., to reach specified values of position, speed, torque, etc.). It is assumed that the conditions in the field are steady, and particularly the field current (or the magnetic field in the stator) is assumed constant. Since v_a directly determines the motor back e.m.f., after allowance is made for the impedance drop due to resistance and leakage inductance of the armature circuit, it follows that armature control is particularly suitable for speed manipulation over a wide range of speeds (typically, 10 dB or more). The motor torque can be kept constant simply by keeping the armature current at a constant value because the field current is virtually a constant in the case of armature control (see Equation 7.4).

In field control, the armature voltage is kept constant and the input voltage v_f to the field circuit is varied. It is assumed that the armature current (and hence the rotor magnetic field) is also maintained constant in the field control. Note further that leakage inductance in the armature circuit is relatively small, and the associated voltage drop can be neglected. From Equation 7.4, it can be seen that since i_a is kept more or less constant, the torque varies in proportion to the field current i_f. Also, since the armature voltage is kept constant, the back e.m.f. remains virtually unchanged. Hence, it follows from Equation 7.5 that the speed will be inversely proportional to i_f. This means that in field control, when the field voltage is increased the motor torque increases whereas the motor speed decreases, so that the output power remains somewhat constant. For this reason, field control is particularly' suitable for constant power drives under varying torque–speed conditions, such as those present in material winding mechanisms (e.g., winding of wire, paper, metal sheet, and so on).

7.3.1 DC Servomotors

If the system characteristics and loading conditions are very accurately known and if the system is stable, it is possible in theory, to schedule the input signal to a motor (e.g., the armature voltage in armature control or field voltage in field control) so as to obtain a desired response (e.g., a specified motion trajectory or torque) from it. Parameter variations, model uncertainties, and external disturbances can produce errors that will build up (integrate) rapidly and will display unstable behavior in this case of open-loop control or computed-input control (inverse-model control). Instability is not acceptable in control system implementations. Feedback control is used to reduce these errors and to improve the control system performance, particularly with regard to stability, robustness, accuracy, and speed of response. In feedback control systems, response variables are sensed and fed back to the driver end of the system so as to reduce the response error.

Servomotors are motors with motion feedback control, which are able to follow a specified motion trajectory. In a dc servomotor system, both angular position and speed might be measured (using shaft encoders, tachometers, resolvers, rotary-variable differential transformers (RVDTs), potentiometers, etc.; see Chapter 4 and Chapter 5) and compared with the desired position and speed. The error signal (= desired response−actual response) is conditioned and compensated using analog circuitry or is processed by a digital hardware processor or control computer, and is supplied to drive the servomotor toward the desired response. Both position feedback and velocity feedback are usually needed for accurate position control. For speed control, velocity feedback alone might be adequate, but position error can build up. On the other hand, if only position feedback is used, a large error in velocity is possible, even when the position error is small. Under certain conditions (e.g., high gains, large time delays), with position feedback alone, the control system may become marginally stable or even unstable. For this reason, dc servo systems historically employed tachometer feedback (velocity feedback) in addition to other types of feedback, primarily position feedback. In early generations of commercial servomotors, the motor and the tachometer were available as a single package, within a common housing. Modern servomotors typically have a single built-in optical encoder mounted on the motor shaft. This encoder is able to provide both position and speed measurements for servo control (see Chapter 5).

Motion control (position and speed control) implies control of motor torque as well (albeit indirectly), since it is the motor torque that causes the motion. In applications where torque itself is a primary output (e.g., metal forming operations, machining, micromanipulation, grasping, and tactile operations) and in situations where small motion errors could produce large unwanted forces (e.g., in parts assembly), direct control of motor torque would be desirable. In some applications of torque control, this is accomplished using feedback of the armature current or the field current which determine the motor torque. The motor torque (magnetic torque), however, is not equal to the load torque or the torque transmitted through the output shaft of the motor. Hence, for precise torque control, direct measurement of torque (e.g., using strain-gage, piezo-electric, or inductive sensors; see Chapter 4) would be required.

A schematic representation of an analog dc servomotor system is given in Figure 7.15. The actuator in this case is a dc motor. The sensors might include a tachometer to measure

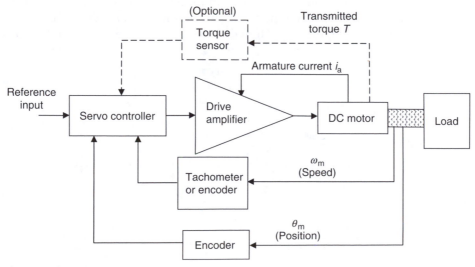

FIGURE 7.15
A dc servomotor system.

angular speed, a potentiometer to measure angular position, and a strain gage torque sensor, which is optional. More commonly, however, a single optical encoder is provided to measure both angular position and speed. The process (the system that is driven) is represented by the load block in the figure. Signal-conditioning (filters, amplifiers, etc.) and compensating (lead, lag, etc.) circuitry are represented by a single block. The power supply to the servo amplifier (and to the motor) is not shown in the figure. The motor and encoder are usually available as an integral unit, possibly with a tachometer as well, mounted on a common shaft. An additional position sensor (encoder, RVDT, potentio-meter, resolver, etc.) may be attached to the load itself since in the presence of shaft flexibility, backlash, etc., the motor motion is not identical to the load motion.

7.3.2 Armature Control

In an armature-controlled dc motor, the armature voltage v_a is used as the control input, while keeping the conditions in the field circuit constant. In particular, the field current i_f (or, the magnetic field in the stator) is assumed constant. Consequently, Equation 7.4 and Equation 7.5 can be written as

$$T_m = k_m i_a \tag{7.22}$$

$$v_b = k'_m \omega_m. \tag{7.23}$$

The parameters k_m and k'_m are termed the torque constant and the back e.m.f. constant, respectively. Note that with consistent units, $k_m = k'_m$ in the case of ideal electrical-to-mech-anical energy conversion at the motor rotor. In the Laplace domain, Equation 7.8 becomes

$$v_a - v_b = (L_a s + R_a)i_a. \tag{7.24}$$

Note that, for convenience, time domain variables (functions of t) are used to denote their Laplace transforms (functions of s). It is understood, however, that the time functions are not identical to the Laplace functions. In the Laplace domain, Equation 7.9 becomes

$$T_m - T_L = (J_m s + b_m)\omega_m, \tag{7.25}$$

where J_m and b_m denote the moment of inertia and the rotary viscous damping constant, respectively, of the motor rotor. Equation 7.22 through Equation 7.25 are represented in the block diagram form, in Figure 7.16. Here the speed ω_m is taken as the motor output. If

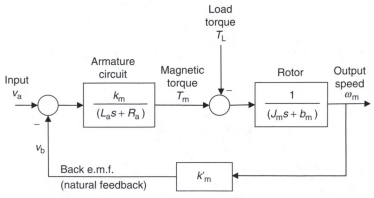

FIGURE 7.16
Open-loop block diagram for an armature-controlled dc motor.

the motor position θ_m is considered the output, it is obtained by passing ω_m through an integration block $1/s$. Note, further, that the load torque T_L, which is the useful (effective) torque transmitted to the load that is driven, is an (unknown) input to the system. Usually, T_L increases with ω_m because a larger torque is necessary to drive a load at a higher speed. If a linear (and dynamic) relationship exists between T_L and ω_m at the load, a feedback path can be completed from the output speed to the input load torque through a proper load transfer function (load block). In particular, if the load is a pure inertia that is rigidly attached to the motor shaft, then it can be simply added to the motor inertia, and the load-torque input path can be removed. The system shown in Figure 7.16 is not a feedback control system. The feedback path, which represents the back e.m.f., is a natural feedback and is characteristic of the process (dc motor); it is not an external control feedback loop.

The overall transfer relation for the system is obtained by first determining the output for each input with the other input removed, and then adding the two output components obtained in this manner, in view of the principle of superposition, which holds for a linear system. We get

$$\omega_m = \frac{k_m}{\Delta(s)} v_a - \frac{(L_a s + R_a)}{\Delta(s)} T_L, \tag{7.26}$$

where $\Delta(s)$ is the characteristic polynomial of the system, given by

$$\Delta(s) = (L_a s + R_a)(J_m s + b_m) + k_m k_m'. \tag{7.27}$$

This is a second-order polynomial in the Laplace variable s, and the system is 2nd order.

7.3.2.1 Motor Time Constants

The electrical time constant of the armature is

$$\tau_a = \frac{L_a}{R_a}, \tag{7.28}$$

which is obtained from Equation 7.8 or Equation 7.24. The mechanical response of the rotor is governed by the mechanical time constant,

$$\tau_m = \frac{J_m}{b_m}, \tag{7.29}$$

which is obtained from Equation 7.9 or Equation 7.25. Usually, τ_m is several times larger than τ_a because the leakage inductance L_a is quite small (leakage of the magnetic flux linkage is negligible for high-quality dc motors). Hence, τ_a can be neglected in comparison to τ_m for most practical purposes. In that case, the transfer functions in Equation 7.26 become first order.

Note that the characteristic polynomial is the same for both transfer functions in Equation 7.26, regardless of the input (v_a or T_L). This should be the case because $\Delta(s)$ determines the natural response of the system and the poles (eigenvalues) of the system, and it does not depend on the system input. True time constants of the motor are obtained by first solving the characteristic equation $\Delta(s) = 0$ to determine the two roots (poles or eigenvalues), and then taking the reciprocal of the magnitudes (Note: only the real part of the two roots is used for this purpose if the roots are complex). For an armature-controlled dc motor, these true time constants are not the same as τ_a and τ_m because of the presence of the coupling term $k_m k'_m$ in $\Delta(s)$ (see Equation 7.27). This also follows from the presence of the natural feedback path (back e.m.f.) in Figure 7.16.

Example 7.5

Determine an expression for the dominant time constant of an armature-controlled dc motor. What is the speed behavior (response) of the motor to a unit step input in armature voltage, in the absence of a mechanical load?

Solution

By neglecting the electrical time constant in Equation 7.27, we have the approximate characteristic polynomial

$$\Delta(s) = R_a(J_m s + b_m) + k_m k'_m.$$

This is expressed as

$$\Delta(s) = k'(\tau s + 1),$$

where τ is the overall dominant time constant of the system. It follows that the dominant time constant is given by

$$\tau = \frac{R_a J_m}{(R_a b_m + k_m k'_m)}. \tag{7.30}$$

With $T_L = 0$, the motor transfer relation is

$$\omega_m = \frac{k}{(\tau s + 1)} v_a, \tag{7.31a}$$

where the dc gain is

$$k = \frac{k_m}{(R_a b_m + k_m k'_m)}. \tag{7.32}$$

Equation 7.31a corresponds to the system input–output differential equation:

$$\omega \frac{d\omega_m}{dt} + \omega_m = k v_a. \tag{7.31b}$$

The speed response to a unit step change in v_a, with zero initial conditions, is

$$\omega_m(t) = k(1 - e^{-t/\tau}). \tag{7.33}$$

This is a nonoscillatory response. In practical situations, some oscillations will be present in the free response because, invariably, a load inertia is coupled to the motor through a shaft, which has some flexibility (i.e., not rigid).

7.3.2.2   *Motor Parameter Measurement*

The parameters k and τ are functions of the motor parameters, as clear from Equation 7.30 and Equation 7.32. These two parameters can be determined by a time-domain test, where a step input is applied to the motor drive system and the response as given by Equation 7.33 is determined using either a digital oscilloscope or a data acquisition

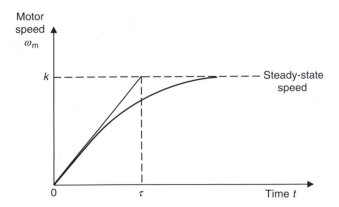

FIGURE 7.17
Open-loop step response of motor speed.

computer. The step response as given by Equation 7.33 is sketched in Figure 7.17. Note that the steady-state value of the speed is k. The slope of the response curve is obtained by differentiating Equation 7.33. Then by setting $t = 0$, we obtain the initial slope as

$$\frac{d\omega_m}{dt}(0) = \frac{k}{\tau}. \tag{7.34}$$

This line is drawn in Figure 7.17, which, according to Equation 7.34 intersects the steady-state level at time $t = \tau$. It follows that from an experimentally determined step response curve, it is possible to estimate the two parameters k and τ.

Alternatively, a frequency-domain test can be carried out by applying a sine input and measuring the speed response, for a series of frequencies (or, by applying a transient input, measuring the speed response, and computing the ratio of the Fourier transforms of the response and the input). This gives the frequency transfer function (see Equation 7.31a)

$$G(j\omega) = \frac{k}{(\tau j\omega + 1)}. \tag{7.35}$$

The Bode diagram of the frequency response may be plotted as in Figure 7.18 (i.e., the log magnitude and phase angle of the frequency-transfer function plotted against the frequency). From the Bode magnitude plot, it is seen that

$$\text{DC gain} = 20 \ \log_{10} k. \tag{7.36}$$

From either the magnitude plot or the phase plot, the corner frequency where the low-frequency asymptote (slope $= 0$ dB/decade) intersects the high-frequency asymptote (slope $= -20$ dB/decade), is given by

$$\text{Corner frequency, } \omega_c = \frac{1}{\tau}. \tag{7.37}$$

In this manner, the frequency response plot can be used to estimate the two parameters k and τ.

Example 7.6
Analytical modeling may not be feasible for some complex engineering systems, and modeling using experimental data (this is known as experimental modeling or system

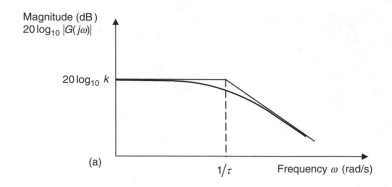

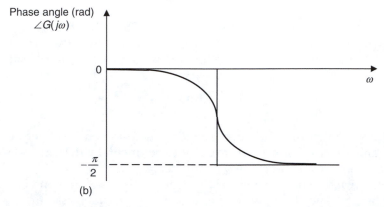

FIGURE 7.18
Open-loop frequency response of motor speed (Bode diagram).

identification) might be the only available recourse. One approach is to use the measured response to a test input, as discussed before. In another approach, experimentally determined steady-state torque–speed characteristics are used to determine (approximately) a dynamic model for a motor. To illustrate this latter approach, consider an armature-controlled dc motor. Sketch steady-state speed–load torque curves using the input voltage (armature voltage) v_a as a parameter that is constant for each curve but varies from curve to curve. Obtain an equation to represent these curves. Now consider an armature-controlled dc motor driving a load of inertia J_L, which is connected directly to the motor rotor through a shaft that has torsional stiffness k_L. The viscous damping constant at the load is b_L. Obtain the system transfer function, with the load position θ_L as the output and armature supply voltage v_a as the input.

Solution
From Equation 7.10a, the steady-state speed–torque curves for a separately excited dc motor are given by

$$T_m + \left[\frac{kk'v_f^2}{R_aR_f^2}\right]\omega_m = \left[\frac{kv_f}{R_aR_f}\right]v_a. \tag{7.38}$$

Since v_f is a constant for armature-controlled motors, we can define two new constants k_m and b_e as

$$k_m = \frac{kv_f}{R_f} \tag{7.39}$$

and

$$b_e = \frac{kk'v_f^2}{R_a R_f^2} = \frac{k_m k_m'}{R_a}. \tag{7.40}$$

Hence, Equation 7.38 becomes

$$T_m + b_e \omega_m = \frac{k_m}{R_a} v_a. \tag{7.41}$$

Note that k_m is the torque constant defined by Equation 7.22, k_m' is the back e.m.f. constant defined by Equation 7.23, and b_e is the electrical damping constant defined by Equation 7.17a. Note, however, that because of the presence of mechanical dissipation, the torque T_{Ls} supplied to the load at steady-state (constant-speed) conditions is less than the motor magnetic torque T_m. Specifically, assuming linear viscous damping,

$$T_{Ls} = T_m - b_m \omega_m. \tag{7.42}$$

It should be noted that if the motor speed is not constant, the output torque of the motor is further affected because some torque is used up in accelerating (or decelerating) the rotor inertia. Obviously, this factor does not enter into constant-speed tests. Now, by substituting Equation 7.42 in Equation 7.41, we have

$$T_{Ls} + (b_m + b_e)\omega_m = \frac{k_m}{R_a} v_a. \tag{7.43}$$

In constant-speed motor tests we measure T_{Ls} not T_m. It follows from Equation 7.43 that the steady-state speed–torque curves (characteristic curves) for an armature-controlled dc motor are parallel straight lines with a negative slope of magnitude $b_m + b_e$ These curves are sketched in Figure 7.19. Note from Equation 7.43 that the parameters

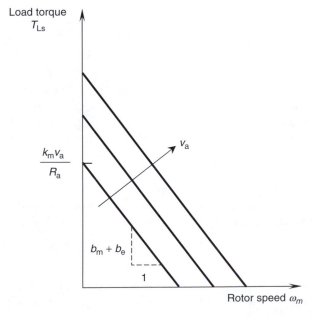

FIGURE 7.19
Steady-state speed–torque curves for an armature-controlled dc motor.

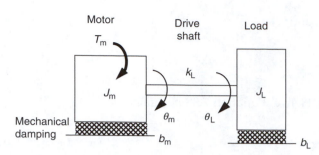

FIGURE 7.20
A motor driving an inertial load.

$b_m + b_e$ and k_m/R_a can be directly extracted from an experimentally determined characteristic curve. Once this is accomplished, Equation 7.43 is completely known and can be used for modeling the control system.

The system given in this example is shown in Figure 7.20. Suppose that θ_m denotes the motor angle of rotation. Newton's second law gives the rotor equation:

$$T_m - k_L(\theta_m - \theta_L) - b_m\dot\theta_m = J_m\ddot\theta_m \tag{i}$$

and the load equation:

$$-k_L(\theta_L - \theta_m) - b_L\dot\theta_L = J_L\ddot\theta_L. \tag{ii}$$

Substituting Equation 7.41 into Equation i and Equation ii, and taking Laplace transforms, we get

$$\frac{k_m}{R_a}v_a + k_L\theta_L = \left[J_ms^2 + (b_m + b_e)s + k_L\right]\theta_m, \tag{iii}$$

$$k_L\theta_m = (J_Ls^2 + b_Ls + k_L)\theta_L. \tag{iv}$$

As usual, we use the same symbol to denote the Laplace transforms as well as its time function. By substituting Equation iv in Equation iii and after straightforward algebraic manipulation, we obtain the system transfer function

$$\frac{\theta_L}{v_a} = \frac{k_Lk_m/R_a}{s[J_mJ_Ls^3 + \{J_L(b_m + b_e) + J_mb_L\}s^2 + \{k_L(J_L + J_m) + b_L(b_m + b_e)\}s + k_Lb_L + k_L(b_m + b_e)]}. \tag{7.44}$$

Note that k_m/R_a and $b_m + b_e$ are the experimentally determined parameters. The mechanical parameters k_L, b_L, and J_L are assumed to be known. Notice the free integrator that is present in the transfer function given by Equation 7.44. This gives a pole (eigenvalue) at the origin of the s-plane ($s = 0$). It represents the rigid-body mode of the system, implying that the load is not externally restrained by a spring.

■

We have seen that for some winding configurations, the speed–torque curve is not linear. Thus, the slope of a characteristic curve is not constant. Hence, an experimentally determined model would be valid only for an operating region in the neighborhood of the point where the slope was determined.

Example 7.7

A dc motor uses 2 hp under no-load conditions to maintain a constant speed of 600 rpm. The motor torque constant $k_m = 1$ V.s, the rotor moment of inertia $J_m = 0.1$ kg.m², and the armature circuit parameters are $R_a = 10\ \Omega$ and $L_a = 0.01$ H. Determine the electrical damping constant, the mechanical damping constant, the electrical time constant of the armature circuit, the mechanical time constant of the rotor, and the true time constants of the motor.

Solution
With consistent units, $k'_m = k_m$. Hence, from Equation 7.40, the electrical damping constant is

$$b_e = \frac{k_m^2}{R_a} = \frac{1}{10} = 0.1 \text{ N.m/rad/s}.$$

It is given that the power absorbed by the motor at no-load conditions is

$$2 \text{ hp} = 2 \times 746 \text{ W} = 1492 \text{ W}$$

and the corresponding speed is

$$\omega_m = \frac{600}{60} \times 2\pi \text{ rad/s} = 20\pi \text{ rad/s}.$$

This power is used against electrical and mechanical damping, at constant speed ω_m. Hence,

$$(b_m + b_e)\omega_m^2 = 1492$$

or

$$b_m + b_e = \frac{1492}{(20\pi)^2} = 0.38 \text{ N.m/rad/s}.$$

It follows that the mechanical damping constant is

$$b_m = 0.38 - 0.1 = 0.28 \text{ N.m/rad/s}.$$

From Equation 7.28 and Equation 7.29,

$$\tau_a = \frac{0.01}{10} = 0.001 \text{ s},$$

$$\tau_m = \frac{0.1}{0.28} = 0.36 \text{ s}.$$

Note that τ_m is several orders larger than τ_a. In view of Equation 7.27, the characteristic polynomial of the motor transfer function can be written as

$$\Delta(s) = R_a[b_m(\tau_a s + 1)(\tau_m s + 1) + b_e]. \qquad (7.45)$$

The poles (eigenvalues) are given by $\Delta(s) = 0$. Hence,

$$0.28(0.001\,s + 1)(0.36\,s + 1) + 0.1 = 0$$

or

$$s^2 + 1010\,s + 3800 = 0.$$

Solving this characteristic equation for the motor eigenvalues, we get

$$\lambda_1 = -3.8 \text{ and } \lambda_2 = -1006.$$

Note that the two poles are real and negative. This means that any disturbance in the motor speed will die out exponentially without oscillations.

The time constants are given by the reciprocals of the magnitudes of the real parts of the eigenvalues. Hence, the true time constants are

$$\tau_1 = 1/3.8 = 0.26\,\text{s},$$
$$\tau_2 = 1/1006 = 0.001\,\text{s}.$$

The smaller time constant τ_2, which derives primarily from the electrical time constant of the armature circuit, can be neglected for all practical purposes. The larger time constant τ_1 comes not only from the mechanical time constant τ_m (rotor inertia/mechanical damping constant) but also from the electrical damping constant (back e.m.f. effect) b_e. Hence, τ_1 is not equal to τ_m, even though the two values are of the same order of magnitude.

7.3.3 Field Control

In field-controlled dc motors, the armature voltage is kept constant, and the field voltage is used as the control input. It is assumed that i_a (and the rotor magnetic field) is maintained constant (Note: Leakage inductance in the armature circuit, and the associated voltage drop is negligible as well). Then, Equation 7.4 can be written as

$$T_m = k_a i_f, \tag{7.46}$$

where k_a is the electromechanical torque constant for the motor. The back e.m.f. relation and the armature circuit equation are not used in this case. Equation 7.7 and Equation 7.9 are written in the Laplace form as

$$v_f = (L_f s + R_f) i_f, \tag{7.47}$$
$$T_m - T_L = (J_m s + b_m)\omega_m. \tag{7.48}$$

Equation 7.46 through Equation 7.48 can be represented by the open-loop block diagram given in Figure 7.21.

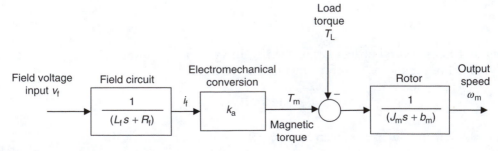

FIGURE 7.21
Open-loop block diagram for a field-controlled dc motor.

It should be mentioned that even though i_a is assumed constant, this is not strictly true. This should be clear from the armature circuit equation (Equation 7.8). In field control, it is the armature supply voltage v_a that is kept constant. Even though the leakage inductance L_a can be neglected, i_a depends as well on the back e.m.f. v_b, which changes with the motor speed as well as the field current i_f. Under these conditions, the block representing k_a in Figure 7.21 is not a constant gain, and in fact it is not linear. At least, a feedback will be needed into this block from output speed. This will also add another electrical time constant, which depends on the dynamics of the armature circuit. It will also introduce a coupling effect between the mechanical dynamics (of the rotor) and the armature circuit electronics. For the present purposes, however, we assume that k_a is a constant gain.

Now, we return to Figure 7.21. Since the system is linear, the principle of superposition holds. According to this, the overall output ω_m is equal to the sum of the individual outputs due to the two inputs v_f and T_L, taken separately. It follows that the transfer relationship is given by

$$\omega_m = \frac{k_a}{(L_f s + R_f)(J_m s + b_m)} v_f - \frac{1}{(J_m s + b_m)} T_L. \tag{7.49}$$

In this case, the electrical time constant originates from the field circuit and is given by

$$\tau_f = \frac{L_f}{R_f}. \tag{7.50}$$

The mechanical time constant τ_m of the field-controlled motor is the same as that for the armature-controlled motor, and can be defined by Equation 7.29:

$$\tau_m = \frac{J_m}{b_m}. \tag{7.29}$$

The characteristic polynomial of the open-loop field-controlled motor is

$$\Delta(s) = (L_f s + R_f)(J_m s + b_m). \tag{7.51}$$

It follows that τ_f and τ_m are the true time constants of the system, unlike in an armature-controlled motor. This is so because, in the case of field control, the mechanical dynamics are uncoupled with the electrical dynamics, which is not the case in armature control (due to the back e.m.f. natural feedback path). As in an armature-controlled dc motor, however, the electrical time constant is several times smaller and can be neglected in comparison to the mechanical time constant. Furthermore, as for an armature-controlled motor, the speed and the angular position of a field-controlled motor have to be measured and fed back for accurate motion control.

7.3.4 Feedback Control of DC Motors

Open-loop operation of a dc motor, as represented by Figure 7.16 for armature control and Figure 7.21 for field control, can lead to excessive error and even instability, particularly because of the unknown load input and also due to the integration effect when position (not speed) is the desired output (as in positioning applications). Feedback control is necessary under these circumstances.

In feedback control, the motor response (position, speed, or both) is measured using an appropriate sensor and fed back into the motor controller, which generates the control signal for the drive hardware of the motor. An optical encoder (see Chapter 5) can be used to sense both position and speed and a tachometer may be used to measure the speed alone (see Section 4.4.1). The following three types of feedback control are important:

1. Velocity feedback
2. Position plus velocity feedback
3. Position feedback with a multiterm controller

7.3.4.1 Velocity Feedback Control

Velocity feedback is particularly useful in controlling the motor speed. In velocity feedback, motor speed is sensed using a device such as a tachometer or an optical encoder, and is fed back to the controller, which compares it with the desired speed, and the error is used to correct the deviation. Additional dynamic compensation (e.g., lead or lag compensation) may be needed to improve the accuracy and the effectiveness of the controller, and can be provided using either analog circuits or digital processing. The error signal is passed through the compensator in order to improve the performance of the control system.

7.3.4.2 Position Plus Velocity Feedback Control

In position control, the motor angle θ_m is the output. In this case, the open-loop system has a free integrator, and the characteristic polynomial is $s(\tau s + 1)$. This is a marginally stable system, in view of the pole at O. For example, if a slight disturbance or model error is present, it will be integrated out, which can lead to a diverging error in the motor angle. In particular, the load torque T_L is an input to the system, and is not completely known. In control systems terminology, this is a disturbance (an unknown input), which can cause unstable behavior in the open-loop system. In view of the free integrator associated with the position output, the resulting unstable behavior cannot be corrected using velocity feedback alone. Position feedback is needed to remedy the problem. Both position and velocity feedback are needed. The feedback gains for the position and velocity signals can be chosen so as to obtain the desired response (speed of response, overshoot limit, steady-state accuracy, etc.). Block diagram of a position plus velocity feedback control system for a dc motor is shown in Figure 7.22. The motor block in this diagram is given by Figure 7.16 for an armature-controlled motor, and by Figure 7.21 for a field-controlled motor (Note: load torque input is integral in either of these two models). The drive unit (see Section 7.4) of the motor is represented by an amplifier of gain k_a. Control system design

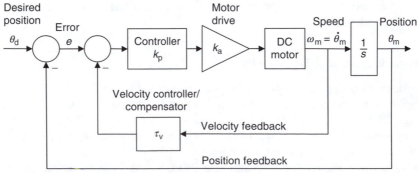

FIGURE 7.22
Position plus velocity feedback control of a dc motor.

involves selection of proper parameter values for sensors and other components in the control system, in order to satisfy performance specifications.

7.3.4.3 Position Feedback with Proportional, Integral, and Derivative Control

A popular method of controlling a dc motor is to use just position feedback, and then compensate for the error using a three-term controller having the proportional, integral, and derivative (PID) actions. A block diagram for this control system is shown in Figure 7.23.

Each term of the PID controller provides specific benefits. There are some undesirable side effects as well. In particular, proportional action improves the speed of response and reduces the steady-state error but it tends to increase the level of overshoot (i.e., system becomes less stable). Derivative action adds damping, just like velocity feedback, thereby making the system more stable (less overshoot). In doing so, it does not degrade the speed of response, however, which is a further advantage. But, the derivative action amplifies high-frequency noise and disturbances. Strictly speaking, a pure derivative action is not physically realizable using analog hardware. The integral action reduces the steady-state error (typically reduces it to zero), but it tends to degrade the system stability and the speed of response. A lead compensator provides an effect somewhat similar to the derivative action (while being physically realizable) whereas a lag compensator provides an integrator-like effect.

In the control system of a dc motor (Figure 7.22 or Figure 7.23), the desired position command may be provided by a potentiometer as a voltage signal. The measurements of position and speed are also provided as voltage signals. Specifically, in the case of an optical encoder, the pulses are detected by a digital pulse counter, and read into the digital controller (see Chapter 5). This reading has to be calibrated to be consistent with the desired position command. In the case of a tachometer, the velocity reading is generated as a voltage (see Section 7.3.4.1), which has to be calibrated then, to be consistent with the desired position signal.

It is noted that proportional plus derivative control (PPD control or PD control) with position feedback has a similar effect as position plus velocity (speed) feedback control. But, the two are not identical because the former, when placed in the forward path of the feedback loop, adds a zero to the system transfer function. That would require further considerations in the controller design, and affect the motor response. In particular, the zero modifies the sign and the ratio in which the two response components corresponding to the two poles contribute to the overall response.

Example 7.8

Consider the position and velocity control system of Figure 7.22. Suppose that the motor model is given by the transfer function $k_m/(\tau_m s + 1)$. Determine the closed-loop transfer function θ_m/θ_d. Next consider PPD control system (Figure 7.23, with the integral controller

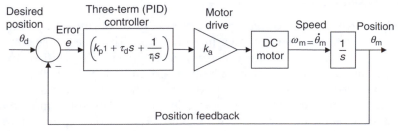

FIGURE 7.23
PID control of the position response of a dc motor.

removed) and the same motor model. What is the corresponding closed-loop transfer function θ_m/θ_d? Compare these two types of control, particularly with respect to speed of response, stability (percentage overshoot), and steady-state error.

Solution

From Figure 7.22, we can write

$$(\theta_d - \theta_m \tau_v s \, \theta_m) k_p k_a \frac{k_m}{(\tau_m s + 1)} = s \theta_m.$$

Hence,

$$\frac{\theta_m}{\theta_d} = \frac{k}{[\tau_m s^2 + (1 + k\tau_v)s + k]}, \qquad (7.52)$$

where $k = k_p k_a k_m$.

Now from Figure 7.23, with the integral control action removed, we can write

$$(\theta_d - \theta_m) k_p (1 + \tau_d s) k_a \frac{k_m}{(\tau_m s + 1)} = s \theta_m.$$

On simplification, we get

$$\frac{\theta_m}{\theta_d} = \frac{k(1 + \tau_d s)}{[\tau_m s^2 + (1 + k\tau_d)s + k]}. \qquad (7.53)$$

It is seen that the characteristic polynomials (denominators of the transfer functions) are identical, in the two cases. As a result, it is possible to place the closed-loop poles (eigenvalues) at desirable locations, in both cases.

The PPD controller introduces a zero to the transfer function, however, as seen in the numerator of Equation 7.53. This zero can have a significant effect on the transient response of the motor. In particular, the zero contributes a time-derivative term, which can be significant in the beginning (start-up conditions of the response). Hence, a larger overshoot (than for the position plus velocity control) results. But, the same derivative action causes the response to settle down quickly to the steady-state value.

The stead-state gain (or, dc gain) of both transfer functions 7.52 and 7.53 is equal to 1 (which is obtained by setting $s = 0$). It follows that the steady-state error is zero in both cases.

7.3.5 Phase-Locked Control

Phase-locked control is an effective approach to control dc motors. A block diagram of a phase-locked servo system is shown in Figure 7.24. This is a phase control method. The position command is generated according to the desired (specified) motion of the motor, using a signal generator (e.g., a voltage-controlled oscillator) or using digital means. This reference signal is in the form of a pulse train, which is quite analogous to the output signal of an incremental encoder (Chapter 5). The rotation of the motor (or load) is measured using an incremental encoder. The encoder pulse train forms the feedback signal. The reference signal and the feedback signal are supplied to a phase detector, which generates a signal representing the phase difference between the two

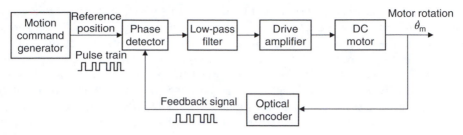

FIGURE 7.24
Schematic diagram of a phase-locked servo.

signals and possibly some unwanted high-frequency components. The unwanted components are removed using a low-pass filter, and the resulting (error) signal is supplied to the drive amplifier of the motor. The error signal drives the motor so as to obtain the desired motion. The objective is to maintain a fixed phase difference (ideally, a zero phase difference) between the reference pulse signal and the position (encoder) pulse signal. Under these conditions, the two signals are synchronized or phase-locked together. Any deviation from the locked conditions generates an error signal, which brings the motor motion back in phase with the reference command. In this manner, deviations due to external disturbances, such as load changes on the motor, are also corrected.

One method of determining the phase difference of two pulse signals is by detecting the edge transitions (Chapter 5). An alternative method is to take the product of the two signals and then low-pass filter the result. To illustrate this second method, suppose that the primary (harmonic) components of the reference pulse signal and the response pulse signal are ($u_o \sin \theta_u$) and ($y_o \sin \theta_y$), respectively, where

$$\theta_u = \omega t + \phi_u,$$
$$\theta_y = \omega t + \phi_y.$$

Note that ω is the frequency of the two pulse signals (assumed to be the same) and ϕ denotes the phase angle. The product signal is

$$p = u_o y_o \sin \theta_u \sin \theta_y$$
$$= \frac{1}{2} u_o y_o [\cos(\theta_u - \theta_y) - \cos(\theta_u + \theta_y)].$$

Consequently,

$$p = \frac{1}{2} u_o y_o \cos(\phi_u - \phi_y) - \frac{1}{2} u_o y_o \cos(2\omega t + \phi_u + \phi_y). \tag{7.54}$$

Low-pass filtering removes the high-frequency component of frequency 2ω, leaving the signal

$$e = \frac{1}{2} u_o y_o \cos(\phi_u - \phi_y). \tag{7.55}$$

This is a nonlinear function of the phase difference ($\phi_u - \phi_y$). Note that by applying a $\pi/2$ phase shift to the original two signals, we also could have determined $(1/2)u_o y_o \sin(\phi_u - \phi_y)$. In this manner, the magnitude and sign of ($\phi_u - \phi_y$) are determined. Very accurate position control can be obtained by driving this phase difference to zero. This is the objective of phase-locked control; the phase angle of the output is locked to the phase angle of the command signal. In more sophisticated phase-locked servos, the frequency differences are also detected—for example, using pulse counting—and compensated. This is analogous to the classic PPD control. It is clear that phase-locked servos are velocity control devices as well, because velocity is proportional to the pulse frequency. When the two pulse signals are synchronized, the velocity error also approaches zero, subject to the available resolution of the control system components. Typically, speed error levels of $\pm 0.002\%$ or less are possible using phase-locked servos. Additionally, the overall cost of a phase-locked servo system is usually less than that of a conventional analog servo system, because less-expensive solid-state devices replace bulky analog control circuitry.

7.4 Motor Driver

The driver of a dc motor is a hardware unit, which generates the necessary current to energize the windings of the motor. The motor torque can be controlled by controlling the current generated by the driver. By receiving feedback from a motion sensor (encoder, tachometer, etc.), the angular position and the speed of the motor can be controlled. Note that when an optical encoder is provided as a mounted sensor of a motor—a typical situation—it is not necessary to use a tachometer as well, because the encoder can generate both position and speed measurements (see Chapter 5). The drive unit primarily consists of a drive amplifier, with additional circuitry and a dc power supply. In typical applications of motion control and servoing, the drive unit is a servo amplifier with auxiliary hardware. The driver is commanded by a control input provided by a host computer (personal computer or PC) through an interface (input–output) card. A suitable arrangement is shown in Figure 7.25. Also, typically, the driver parameters (e.g., amplifier gains) are software programmable and can be set by the host computer.

The control computer receives a feedback signal of the motor motion, through the interface board, and generates a control signal, which is provided to the drive amplifier, again through the interface board. Any control scheme can be programmed (say, in C language) and implemented in the control computer. In addition to typical servo control schemes such as PID and position-plus-velocity feedback, other advanced control algorithms [e.g., optimal control techniques such as linear quadratic regulator (LQR) and

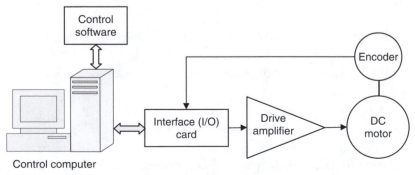

FIGURE 7.25
Components of a dc motor control system.

linear quadratic Gaussian (LQG), adaptive control techniques such as model-referenced adaptive control, switching control techniques such as sliding-mode control, nonlinear control schemes such as feedback linearization technique (FLT), and intelligent control techniques such as fuzzy logic control] may be implemented in this manner. If the computer does not have the processing power to carry out the control computations at the required speed (i.e., control bandwidth), a DSP may be incorporated into the computer. But, with modern computers, which can provide substantial computing power at low cost, DSPs are not needed in most applications.

7.4.1   Interface Card

The I/O card is a hardware module with associated driver software, based in a host computer (PC), and connected through its bus (e.g., ISA bus). It forms the input–output link between the motor and the controller. It can provide many (say, eight) analog signals to drive many (eight) motors, and hence termed a multiaxis card. It follows that the digital-to-analog conversion (DAC) capability (see Chapter 2) is built into the I/O card (e.g., 16 bit DAC including a sign bit, ± 10 V output voltage range). Similarly, the analog-to-digital conversion (ADC) function (see Chapter 2) is included in the I/O card (e.g., eight analog input channels with 16 bit ADC including a sign bit, ± 10 V output voltage range). These input channels can be used for analog sensors such as tachometers, potentiometers, and strain gages. Equally important are the encoder channels to read the pulse signals from the optical encoders mounted on the dc servomotors. Typically, the encoder input channels are equal in number to the analog output channels (and the number of axes, e.g., eight). The position pulses are read using counters (e.g., 24-bit counters), and the speed is determined by the pulse rate. The rate at which the encoder pulses are counted can be quite high (e.g., 10 MHz). In addition a number of bits (e.g., 32) of digital input and output may be available through the I/O card, for use in simple digital sensing, control, and switching functions. The principles of ADC, DAC, and other signal modification devices are discussed in Chapter 2.

7.4.2   Drive Unit

The primary hardware component of the motor drive system is the drive amplifier. In typical motion control applications, these amplifiers are called servo amplifiers. Two basic types of drive amplifiers are commercially available:

1. Linear amplifier
2. PWM amplifier

A linear amplifier generates a voltage output, which is proportional to the control input provided to it. Since the output voltage is proportioned by dissipative means (using resistor circuitry), this is a wasteful and inefficient approach. Furthermore, fans and heat sinks have to be provided to remove the generated heat, particularly in continuous operation. To understand the inefficiency associated with a linear amplifier, suppose that the operating output range of the amplifier is 0 to 20 V, and that the amplifier is powered by a 20 V power supply. Under a particular operating condition, suppose that the motor is applied 10 V and draws a current of 4 A. The power used by the motor then is 10×4 W $= 40$ W. Still, the power supply provides 20 V at 5 A, thereby consuming 100 W. This means, 60 W of power is dissipated, and the efficiency is only 40%. The efficiency can be made close to 100% using modern pulse-width modulation (PWM)

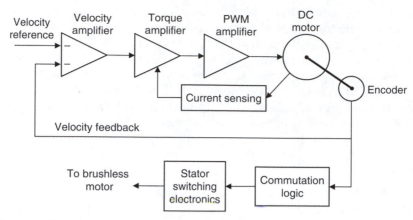

FIGURE 7.26
The main components of a PWM-drive system for a dc motor.

amplifiers, which are nondissipative devices, and depend on high-speed switching at constant voltage to control the power supplied to the motor, as discussed subsequently.

Modern servo amplifiers use pulse-width modulation (PWM) to drive servomotors efficiently under variable-speed conditions, without incurring excessive power losses. Integrated microelectronic design makes them compact, accurate, and inexpensive. The components of a typical PWM-drive system are shown in Figure 7.26. Other signal-conditioning hardware (e.g., filters) and auxiliary components such as isolation hardware, safety devices including tripping hardware, and cooling fan are not shown in the figure. In particular, note the following components, connected in series:

1. A velocity amplifier (a differential amplifier)
2. A torque amplifier
3. A PWM amplifier

The power can come from an ac line supply, which is rectified in the drive unit to provide the necessary dc power for the electronics. Alternatively, leads may be provided for an external direct-current (dc) power supply (e.g., 15 V DC). The reference velocity signal and the feedback signal (from an encoder or a tachometer) are connected to the input leads of the velocity amplifier. The resulting difference (error signal) is conditioned and amplified by the torque amplifier to generate a current corresponding to the required torque (corresponding to the driving speed). The motor current is sensed and fed back to this amplifier, to improve the torque performance of the motor. The output from the torque amplifier is used as the modulating signal to the PWM amplifier. The reference switching frequency of a PWM amplifier is high (in the order of 25 kHz). The pulse-width modulation is accomplished by varying the duty cycle of the generated pulse signal, through switching control, as explained next. The PWM signal from the amplifier (e.g., at 10 V) is used to energize the field windings of a dc motor. A brushless dc motor needs electronic commutation. This may be accomplished by using the encoder signal to determine the motor rotation and based on that, time the switching of the current through the stator windings.

7.4.3 Pulse-Width Modulation

The final control of a dc motors is accomplished by controlling the supply voltage to either the armature circuit or the field circuit. A dissipative method of achieving this involves using a variable resistor in series with the supply source to the circuit. This is a wasteful method and also has other disadvantages. Notably, the heat generated at the control resistor has to be removed promptly to avoid malfunction and damage due to high temperatures. As noted before, a linear amplifier with a variable gain is also dissipative and inefficient. A much more desirable way to control the voltage to a dc motor is by using a solid-state switch to vary the off time of a fixed voltage level, while keeping the period (or inverse frequency) of the on-time constant. Specifically, the duty cycle of a pulse signal is varied while maintaining the switching frequency constant.

Consider the voltage pulse signal shown in Figure 7.27. The following notation is used:

T = pulse period (i.e., interval between the successive on times)
T_o = on period (i.e., interval between on time to the next off time).

Then, the duty cycle is given by the percentage

$$d = \frac{T_o}{T} \times 100\%. \tag{7.56a}$$

The voltage level v_{ref} and the pulse frequency $1/T$ are kept fixed, and what is varied is T_o. In this manner, PWM is achieved by "chopping" the reference voltage over a part of the switching period so that the average voltage is varied. As discussed later, it is easy to see that, with respect to an output pulse signal, the duty cycle is given by the ratio of average output to the peak output; specifically,

$$\text{Duty cycle} = \frac{\text{Average output}}{\text{Peak output}} \times 100\%. \tag{7.56b}$$

Equation 7.56 verifies that the average level of a PWM signal is proportional to the duty cycle (or the on-time period T_o) of the signal. It follows that the output level (i.e., the average value) of a PWM signal can be varied simply by changing the signal-on time period (in the range 0 to T) or equivalently by changing the duty cycle (in the range 0% to 100%). This relationship between the average output and the duty cycle is linear. Hence, a digital or software means of generating a PWM signal would be to use a straight line from 0 to the maximum signal level, spanning the period (T) of the signal. For a given output level, the straight line segment at this height, when projected on the time axis, gives the required on-time interval (T_o).

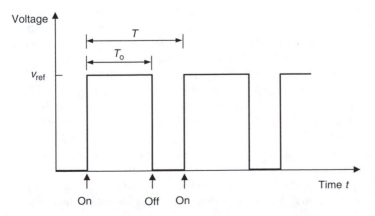

FIGURE 7.27
Duty cycle of a PWM signal.

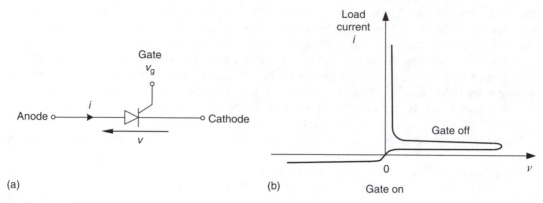

FIGURE 7.28
(a) Symbol for a thyristor. (b) Ideal characteristic curve of a thyristor.

A discrete chopper circuit for pulse-width modulation may be developed using a solid-state switch known as the thyristor. The thyristor is also known as a silicon-controlled rectifier, a solid-state–controlled rectifier, a semiconductor-controlled rectifier, or simply an SCR. It is a pellet made of four layers (pnpn) of semiconductor material (e.g., silicon with a trace of dope material). It has three terminals—the anode, the cathode, and the gate—as shown in Figure 7.28a. The anode and the cathode are connected to the circuit that carries the load current i. When the gate potential v_g is less than or equal to zero with respect to the cathode, the thyristor cannot conduct in either direction ($i = 0$). When v_g is made positive, the thyristor conducts from anode to cathode but not in the opposite direction (i.e., it acts like a basic diode). In other words, a positive firing signal (i.e., a positive trigger voltage) v_g closes (turn on) the switch. To open (turn off) the switch again, we not only have to make v_g zero (or slightly negative) with respect to the cathode, but also the load current from the anode to the cathode has to be zero (or slightly negative). This is the natural mode of operation of a thyristor. When the supply voltage is dc, it does not drop to zero; hence, the thyristor would be unable to turn itself off. In this case, a commutating circuit that can make the voltage across the thyristor slightly negative has to be employed. This is called forced commutation (as opposed to natural commutation) of a thyristor. Note that when a thyristor is conducting, it offers virtually no resistance, and the voltage drop across the thyristor can be neglected for practical purposes. An idealized voltage–current characteristic curve of a thyristor is shown in Figure 7.28b. Solid-state switching devices are lossless (or nondissipative) in nature.

A basic thyristor circuit using a dc power supply, which may be used in dc motor control, is shown in Figure 7.29a. The dc supply voltage is v_{ref}, the PWM voltage signal supplied to the armature circuit is v_a, and the back e.m.f. in the motor is v_b. The nature of these voltage signals is shown in Figure 7.29b. Since the supply voltage v_{ref} to the armature circuit is chopped in generating the PWM signal v_a, the circuit in Figure 7.29a is usually known as a chopper circuit. In armature control, the field circuit is separately excited, as shown. The armature resistance R_a is neglected to provide a qualitative explanation of the voltages appearing in various parts of the circuit; R_a should be included in a more accurate model. The inductance L_o includes the usual armature leakage inductance, self-inductance, and so forth (normally denoted by L_a), and also an external inductance that is needed to avoid large fluctuations in armature current i_a, since v_a is pulsating. Alternatively, a series-wound motor in which the field inductance L_f is connected in series with the armature may be used to increase L_o. A free-wheeling diode

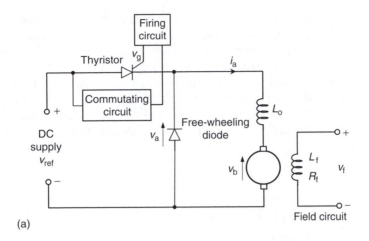

(a)

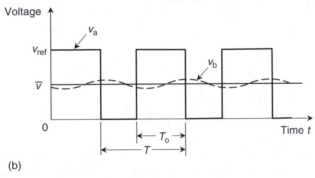

(b)

FIGURE 7.29
(a) A thyristor (SCR) control circuit (chopper) for a dc motor. (b) Circuit voltage signals.

provides a path for i_a during the off period of v_a so as to avoid large voltage buildup in the armature.

Initially, the voltage v_g applied to the gate terminal closes (turns on) the SCR, causing v_{ref} to be applied to the armature circuit. Since v_{ref} is a dc voltage, however, the SCR does not open (turn off) by itself. Hence, a commutating circuit that is capable of applying a slightly negative voltage to the anode of the SCR is needed. The commutating circuit usually consists of a capacitor that is charged to provide the required voltage, a diode, and a thyristor.

The average voltage $\bar{v}$ supplied to the armature circuit is the average of v_a. This is given by

$$\bar{v} = \frac{T_o}{T} v_{ref},\qquad(7.57)$$

where T_o is the on-time interval of the supplied voltage pulse and T is the pulse period. In other words, the reference voltage is fractioned by the duty cycle (see Equation 7.56b) to obtain the average voltage. As noted before, $\bar{v}$ can be varied either by changing T_o (called pulse-width modulation or PWM) or by changing T (pulse frequency modulation, or PFM). What is common is PWM, where the pulse frequency (or T) is kept constant. This method of pulsing control is employed in chopper drive circuits or PWM amplifiers for dc motors.

If v_a were a constant, there would not be a potential drop across the inductor L_o, and v_b would be a constant equal to v_a. It follows that the average value (i.e., the dc component) of v_b is equal to the average value of v_a, which is denoted by $\bar{v}$ in Equation 7.57. When the motor speed is properly regulated (i.e., speed fluctuations are small), and assuming that the conditions in the field circuit are steady, the back e.m.f. v_b is nearly a constant. Hence,

$$v_b \sim \bar{v}. \tag{7.58}$$

Note that the voltage across L_o is $v_a - v_b$. It follows that

$$L_o \frac{di_a}{dt} = v_a - v_b. \tag{7.59}$$

Now, in view of Equation 7.58, we can write

$$L_o \frac{di_a}{dt} = v_a - \bar{v} \tag{7.60}$$

or

$$i_a = \frac{1}{L_o} \int (v_a - \bar{v}) dt. \tag{7.61}$$

Equation 7.61 indicates that the change in the armature current is proportional to the area between the v_a curve and the $\bar{v}$ line shown in Figure 7.29b. In particular, starting from a steady-state value of i_a, the armature current rises by

$$\Delta i_a = \frac{1}{L_o} (v_{ref} - \bar{v}) T_o$$

over the time period T_o in which the thyristor is on. Substituting Equation 7.57, we get

$$\Delta i_a = \frac{v_{ref}}{L_o T} (T - T_o) T_o. \tag{7.62}$$

Then, over the time interval $(T - T_o)$ during which the thyristor is off, the armature current drops by

$$\frac{1}{L_o} \bar{v} (T - T_o),$$

which is equal to

$$\frac{v_{ref}}{L_o T} (T - T_o) T_o.$$

As a result, the armature current goes back to the initial value i_a. This cycle repeats over the subsequent pulse cycles of duration T. It follows that the fluctuation in the armature current is given by Equation 7.62. For a given T (i.e., a given pulse frequency), this amplitude is maximum when $T_o = T/2$. Hence,

$$(\Delta i_a)_{max} = \frac{T v_{ref}}{4 L_o}. \tag{7.63}$$

Note that the current fluctuations can be reduced by increasing L_o and decreasing T for a given supply voltage v_{ref}. Note, further, that since $v_{ref} - \bar{v}$ is constant, it follows from Equation 7.61 that the armature current increases or decreases linearly with time.

Example 7.9

Consider the chopper circuit given in Figure 7.29a. The chopper frequency is 200 Hz, the series inductance in the armature circuit is 50 mH, and the supply dc voltage to the chopper is 100 V. Determine the amplitude of the maximum (worst case) fluctuation in the armature current.

Solution

Since we are interested in the worst case of current fluctuations, we use Equation 7.63. Then,

$$T = \frac{1}{200}\,\text{s}, L_o = 0.05\,\text{mH}, v_{ref} = 100\,\text{V}.$$

Substituting values, we get

$$(\Delta i_a)_{max} = \frac{100}{200 \times 4 \times 0.05} 2.5\,\text{A}.$$

∎

High-power dc motors are usually driven by rectified ac supplies (single phase or three phase). In this case as well, motor control can be accomplished by using thyristor circuits. Full-wave circuits are those that use both the positive and negative parts of an ac supply voltage. A full-wave single-phase control circuit for a dc motor is shown in Figure 7.30. It uses two thyristors. For convenience, the armature resistance R_o is not shown, although it is important in the analysis of the circuit. The inductance L_o contains the usual armature leakage component L_a as well as an external inductor, which is

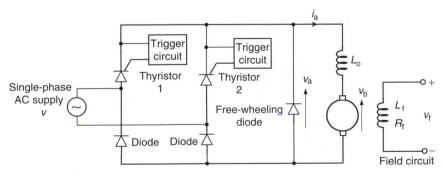

FIGURE 7.30
Full-wave single-phase control circuit for a dc motor.

connected in series with the armature to reduce surges in the armature current. Furthermore, a free-wheeling diode is used to provide a current path when the two thyristors are turned off. Two additional diodes are provided to complete the current path for each half of the supply wave period.

Various signals through the circuit are sketched in Figure 7.31. The supply voltage v is shown in Figure 7.31a. The broken line in this figure is the voltage v_a that would result if the two thyristors were replaced by diodes. Figure 7.31b shows the dc voltage v_a supplied to the armature circuit and the back e.m.f. v_b across the armature; T_o is the firing time of each thyristor from the time when the voltage supplied to a thyristor begins to build from zero. Specifically, during the positive half of the supply voltage v, thyristor 1 will be triggered after time T_o, and during the negative half of the supply voltage, thyristor 2 will be triggered after time T_o. Note that in this full-wave circuit, the negative half of v also appears as positive in v_a (across the free-wheeling diode). The back e.m.f. v_b is reasonably constant, and so is the armature current i_a. Note that the voltage across L_o is $v_a - v_b$. Hence, the armature current variation is given by the area between the two curves v_a and v_b, as is clear from Equation 7.61. As in the case of dc power supply, the motor can be

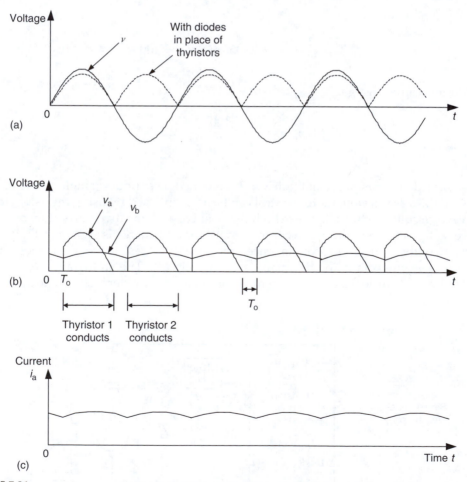

FIGURE 7.31
Voltages and armature current for the circuit in Figure 7.30. (a) Supply voltage. (b) Voltage to the armature circuit and back e.m.f. (c) Armature current.

controlled either by varying the thyristor firing time T_o for a given supply frequency or by varying the supply frequency for a given firing time. The latter (PWM) is for more common.

The two diodes in the supply part of the circuit in Figure 7.30 may be replaced by two thyristors. In that case, the two thyristors in each current path (i.e., for positive v and negative v) have to be triggered simultaneously. The operation of a three-phase control circuit for a dc motor can be analyzed by direct extension of the concepts presented here.

7.5 DC Motor Selection

DC motors, dc servomotors in particular, are suitable for applications requiring continuous operation (continuous duty) at high levels of torque and speed. Brushless permanent magnet motors with advanced magnetic material provide high torque/mass ratio, and are preferred for continuous operation at high throughput (e.g., component insertion machines in the manufacture of printed-circuit boards, portioning and packaging machines, printing machines) and high speeds (e.g., conveyors, robotic arms), in hazardous environments (where spark generation from brushes would be dangerous), and in applications that need minimal maintenance and regular wash down (e.g., in food processing applications). For applications that call for high torques and low speeds at high precision (e.g., inspection, sensing, product assembly), torque motors or regular motors with suitable speed reducers (e.g., harmonic drives, gear units using worm gears, etc.; see Chapter 8) may be employed.

A typical application involves a rotation stage, which produces rotary motion for the load. If an application requires linear (rectilinear) motions, a linear stage has to be used. One option is to use a rotary motor with a rotatory-to-linear motion transmission device such as lead screw or ball screw and nut, rack and pinion, or conveyor belt (see Chapter 8). This approach introduces some degree of nonlinearity and other errors (e.g., friction, backlash). For high-precision applications, linear motor provides a better alternative. The operating principle of a linear motor is similar to that of a rotary motor, except linearly moving armatures on linear bearing or guideways are used instead of rotors mounted on rotary bearings.

When selecting a dc motor for a particular application, a matching drive unit has to be chosen as well. Due consideration must be given to the requirements (specifications) of power, speed, accuracy, resolution, size, weight, and cost, when selecting a motor and a drive system. In fact vendor catalogs give the necessary information for motors and matching drive units, thereby making the selection far more convenient. Additionally, a suitable speed transmission device (harmonic drive, gear unit, lead screw and nut, etc.) may have to be chosen as well, depending on the application.

7.5.1 Motor Data and Specifications

Torque and speed are the two primary considerations in choosing a motor for a particular application. Speed–torque curves are available, in particular, from the manufacturer or commercial supplier. The torques given in these curves are typically the maximum torques (known as peak torques), which the motor can generate at the indicated speeds. A motor should not be operated continuously at these torques (and current levels) because of the dangers of overloading, wear, and malfunction. The peak values have to be reduced (say, by 50%) in selecting a motor to match the torque requirement for continuous operation. Alternatively, the continuous torque values as given by the manufacturer should be used in the motor selection.

Motor manufacturers' data that are usually available to users include the following:

1. Mechanical data
   (a) Peak torque (e.g., 65 N.m)
   (b) Continuous torque at zero speed or continuous stall torque (e.g., 25 N.m)
   (c) Frictional torque (e.g., 0.4 N.m)
   (d) Maximum acceleration at peak torque (e.g., 33×10^3 rad/s^2)
   (e) Maximum speed or no-load speed (e.g., 3000 rpm)
   (f) Rated speed or speed at rated load (e.g., 2400 rpm)
   (g) Rated output power (e.g., 5100 W)
   (h) Rotor moment of inertia (e.g., 0.002 kg.m^2)
   (i) Dimensions and weight (e.g., 14 cm diameter, 30 cm length, 20 kg)
   (j) Allowable axial load or thrust (e.g., 230 N)
   (k) Allowable radial load (e.g., 700 N)
   (l) Mechanical (viscous) damping constant (e.g., 0.12 N.m/krpm)
   (m) Mechanical time constant (e.g., 10 m.s)

2. Electrical data
   (a) Electrical time constant (e.g., 2 m.s)
   (b) Torque constant (e.g., 0.9 N.m/A for peak current or 1.2 N.m/A rms current)
   (c) Back e.m.f. constant (e.g., 0.95 V/rad/s for peak voltage)
   (d) Armature/field resistance and inductance (e.g., 1.0 Ω, 2 mH)
   (e) Compatible drive unit data (voltage, current, etc.)

3. General data
   (a) Brush life and motor life (e.g., 5×10^8 revolutions at maximum speed)
   (b) Operating temperature and other environmental conditions (e.g., 0 to 40°C)
   (c) Thermal resistance (e.g., 1.5°C/W)
   (d) Thermal time constant (e.g., 70 min)
   (e) Mounting configuration

Quite commonly, motors and drive systems are chosen from what is commercially available. Customized production may be required, however, in highly specialized research and development applications where the cost may not be a primary consideration. The selection process typically involves matching the engineering specifications for a given application with the data of commercially available motor systems.

7.5.2 Selection Considerations

When a specific application calls for large speed variations (e.g., speed tracking over a range of 10 dB or more), armature control is preferred. Note, however, that at low speeds (typically, half the rated speed), poor ventilation and associated temperature buildup can cause problems. At very high speeds, mechanical limitations and heating due to frictional dissipation become determining factors. For constant-speed applications, shunt-wound motors are preferred. Finer speed regulation may be achieved using a servo system with encoder or tachometer feedback or with phase-locked operation. For constant power

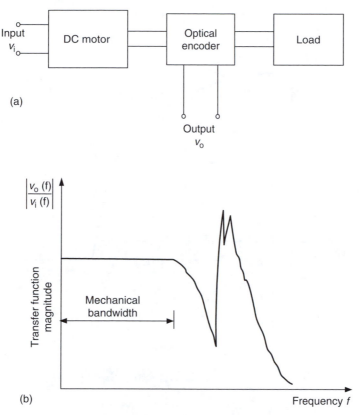

FIGURE 7.32
Determination of the mechanical bandwidth of a dc motor. (a) Test setup. (b) Test result.

applications, the series-wound or compound-wound motors are preferable over shunt-wound units. If the shortcomings of mechanical commutation and limited brush life are critical, brushless dc motors should be used.

For high-speed and transient operations of a dc motor, its mechanical time constant (or mechanical bandwidth) is an important consideration. This is limited by the moment of inertia of the rotor (armature) and the load, shaft flexibility, and the dynamics of the mounted instrumentation, such as tachometers and encoders. The mechanical bandwidth of a dc motor can be determined by simply measuring the velocity transducer signal v_o for a transient drive signal v_i and computing the ratio of their Fourier spectra. This procedure and the result are illustrated in Figure 7.32. A better way of computing this transfer function is by the cross-spectral density method. The flat region of the resulting frequency transfer function (magnitude) plot determines the mechanical bandwidth of the motor.

A simple way to establish the operating conditions of a motor is by using its torque–speed curve, as illustrated in Figure 7.33. What is normally provided by the manufacturer is the peak torque curve, which gives the maximum torque the motor (with a matching drive system) can provide at a given speed, for short periods (say, 30% duty cycle). The actual selection of a motor should be based on its continuous torque, which is the torque that the motor is able to provide continuously at a given speed, for long periods without overheating or damaging the unit. If the continuous torque curve is not provided by the manufacturer, the peak torque curve should be reduced by about 50% (or even by 70%) for matching with the specified operating requirements.

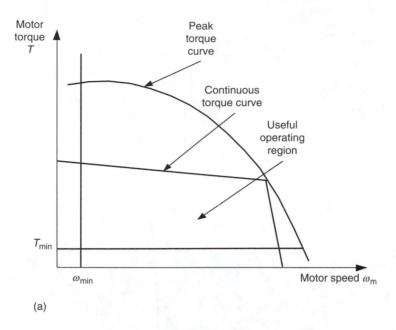

(a)

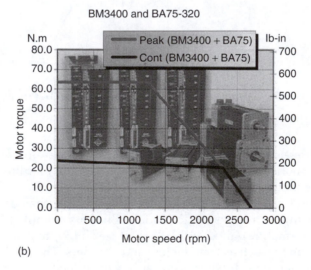

(b)

FIGURE 7.33
(a) Representation of the useful operating region for a dc motor. (b) Speed–torque characteristics of a commercial brushless dc servomotor with a matching amplifier. (From Aerotech, Inc. With permission.)

The minimum operating torque T_{min} is limited mainly by loading considerations. The minimum speed ω_{min} is determined primarily by operating temperature. These boundaries along with the continuous torque curve define the useful operating region of the particular motor (and its drive system), as indicated in Figure 7.33a. The optimal operating points are those that fall within this segment on the continuous torque–speed curve. The upper limit on speed may be imposed by taking into account transmission limitations in addition to the continuous torque–speed capability of the motor system.

7.5.3 Motor Sizing Procedure

Motor sizing is the term used to denote the procedure of matching a motor (and its drive system) to a load (demand of the specific application). The load may be given by a load curve, which is the speed–torque curve representing the torque requirements for operating the load at various speeds (see Figure 7.34). Clearly, greater torques are needed to drive a load at higher speeds. For a motor and a load, the acceptable operating range is the interval where the load curve overlaps with the operating region of the motor (segment AB in Figure 7.34). The optimal operating point is the point where the load curve intersects with the speed–torque curve of the motor (point A in Figure 7.34).

Sizing a dc motor is similar to sizing a stepper motor, as studied in Chapter 6. The same equations may be used for computing the load torque (demand). The motor characteristic (i.e., speed–torque curve) gives the available torque, as in the case of a stepper motor. The main difference is, a stepper motor is not suitable for continuous operation for long periods and at high speeds, whereas a dc motor can perform well in such situations. In this context, a dc motor can provide high torques, as given by its peak torque curve, for short periods, and reduced torques, as given by its continuous torque curve for long periods of operation. In the motor sizing procedure, then, the peak torque curve may be used for short periods of acceleration and deceleration, but the continuous torque curve (or the peak torque curve reduced by about 50%) must be used for continuous operation for long periods.

7.5.3.1 *Inertia Matching*

The motor rotor inertia (J_m) should not be very small compared with the load inertia (J_L). This is particularly critical in high-speed and highly repetitive (high-throughput) applications. Typically, for high-speed applications, the value of J_L/J_m may be in the range of 5 to 20. For low-speed applications, J_L/J_m can be as high as 100. This assumes direct-drive applications.

A gear transmission may be needed between the motor and the load in order to amplify the torque available from the motor, which also reduces the speed at which the load is driven. Then, further considerations have to be made in inertia matching. In particular, neglecting the inertial and frictional loads due to gear transmission, it can be shown that best acceleration conditions for the load are possible if (see Chapter 2, Section 2.2.3 under impedance matching of mechanical devices)

$$\frac{J_L}{J_m} = r^2, \tag{7.64}$$

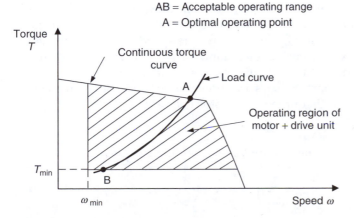

FIGURE 7.34
Sizing a motor for a given load.

where r is the step-down gear ratio (i.e., motor speed/load speed). Since J_L/r^2 is the load inertia as felt at the motor rotor, the optimal condition (Equation 7.64) is when this equivalent inertia, which moves at the same acceleration as the rotor, is equal to the rotor inertia (J_m).

7.5.3.2 Drive Amplifier Selection

Usually, the commercial motors come with matching drive systems. If this is not the case, some useful sizing computations can be done to assist the process of selecting a drive amplifier. As noted before, even though the control procedure becomes linear and convenient when linear amplifiers are used, it is desirable to use PWM amplifiers in view of their high efficiency (and associated low-thermal dissipation).

The required current and voltage ratings of the amplifier, for a given motor and a load, may be computed rather conveniently. The required motor torque is given by

$$T_m = J_m \alpha + T_L + T_f, \qquad (7.65a)$$

where α is the highest angular acceleration needed from the motor in the particular application, T_L is the worst-case load torque, and T_f is the frictional torque on the motor.

If the load is a pure inertia (J_L), Equation 7.65a becomes

$$T_m = (J_m + J_L)\alpha + T_f. \qquad (7.65b)$$

The current required to generate this torque in the motor is given by

$$i = \frac{T_m}{k_m}, \qquad (7.66)$$

where k_m is the torque constant of the motor.

The voltage (armature control) required to drive the motor is given by

$$v = k'_m \omega_m + Ri, \qquad (7.67)$$

where $k'_m = k_m$ is the back e.m.f. constant, R is the winding resistance, and ω_m is the highest operating speed of the motor in driving the load. The leakage inductance, which is small, is neglected. For a PWM amplifier, the supply voltage (from a dc power supply) is computed by dividing the voltage in Equation 7.67 by the lowest duty cycle of operation.

Example 7.10

A load of moment of inertia $J_L = 0.5$ kg.m^2 is ramped up from rest to a steady speed of 200 rpm in 0.5 s using a dc motor and a gear unit of step-down speed ratio $r = 5$. A schematic representation of the system is shown in Figure 7.35a and the speed profile of the load is shown in Figure 7.35b. The load exerts a constant resistance of $T_R = 55$ N.m throughout the operation. The efficiency of the gear unit is $e = 0.7$. Check whether the commercial brushless dc motor and its drive unit, whose characteristics are shown in Figure 7.33b, is suitable for this application. The moment of inertia of the motor rotor is $J_m = 0.002$ kg.m^2.

Solution

The load equation to compute the torque required from the motor is given by

$$T_m = \left(J_m + \frac{J_L}{er^2}\right)r\alpha + \frac{T_R}{er}, \qquad (7.68)$$

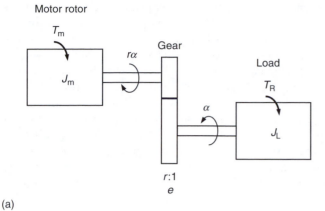

(a)

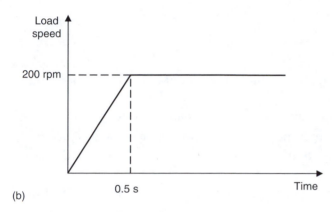

(b)   0.5 s   Time

FIGURE 7.35
(a) A load driven by a dc motor through a gear transmission. (b) Speed profile of the load.

where α is the load acceleration, and the remaining parameters are as defined in the example. The derivation of Equation 7.68 is straightforward. In particular, see the derivation of a similar equation for positioning table driven by a stepper motor (Chapter 6, Equation 6.50). From the given speed profile,

Maximum load speed $= 200$ rpm $= 20.94$ rad/s

Load acceleration $= \dfrac{20.94}{0.5}$ rad/s^2 $= 42$ rad/s^2.

Substitute the numerical values in Equation 7.68, under worst-case conditions, to compute the required torque from the motor. We have

$$T_m = \left(0.002 + \frac{0.05}{0.7 \times 5^2}\right)5 \times 42 + \frac{55.0}{0.7 \times 5}\text{N.m} = 1.02 + 15.71\,\text{N.m} = 16.73 \text{ N.m.}$$

Under worst-case conditions, at least this much of torque would be required from the motor, operating at a speed of $200 \times 5 = 1000$ rpm. Note from Figure 7.33b that the load point (1000 rpm, 16.73 N.m) is sufficiently below even the continuous torque curve of the given motor (with its drive unit). Hence this motor is adequate for the task.

7.6 Induction Motors

With the widespread availability of alternating current (ac) as an economical form of power supply for operating industrial machinery and household appliances, much attention has been given to the development of ac motors. Because of the rapid progress made in this area, ac motors have managed to replace dc motors in many industrial applications until the revival of the dc motor, particularly as a servomotor in control system applications. However, ac motors are generally more attractive than conventional dc motors, in view of their robustness, lower cost, simplicity of construction, and easier maintenance, especially in heavy duty (high-power) applications (e.g., rolling mills, presses, elevators, cranes, material handlers, and operations in paper, metal, petrochemical, cement, and other industrial plants) and in continuous constant-speed operations (e.g., conveyors, mixers, agitators, extruders, pulping machines, household and industrial appliances such as refrigerators, heating-ventilation-and-air-conditioning or HVAC devices such as pumps, compressors, and fans). Many industrial applications using ac motors may involve continuous operation throughout the day for over 6 days/week. Moreover, advances in control hardware and software and the low cost of microelectronics have led to advance controllers for ac motors, which can emulate the performance of variable-speed drives of dc motors; for example, ac servomotors that rival their dc counterpart.

Some advantages of ac motors are:

1. Cost-effectiveness
2. Convenient power source (standard power grid providing single-phase and three-phase ac supplies)
3. No commutator and brush mechanisms needed in many types of ac motors
4. Low-power dissipation, low rotor inertia, and lightweight in some designs
5. Virtually no electric spark generation or arcing (less hazardous in chemical environments)
6. Capability of accurate constant-speed operation without needing servo control (with synchronous ac motors)
7. No drift problems in ac amplifiers in drive circuits (unlike linear dc amplifiers)
8. High reliability, robustness, easy maintenance, and long life

The primary disadvantages include

1. Low starting torque (zero starting torque in synchronous motors)
2. Need of auxiliary starting devices for ac motors with zero starting torque
3. Difficulty of variable-speed control or servo control (this problem hardly exists now in view of modern solid-state and variable-frequency drives with devices having field feedback compensation)

We discuss two basic types of ac motors:

1. Induction motors (asynchronous motors)
2. Synchronous motors

7.6.1 Rotating Magnetic Field

The operation of an ac motor can be explained using the concept of a rotating magnetic field. A rotating field is generated by a set of windings uniformly distributed around a circular stator and excited by ac signals with uniform phase differences. To illustrate this, consider a standard three-phase supply. The voltage in each phase is 120° out of phase with the voltage in the next phase. The phase voltages can be represented by

$$v_1 = a \cos \omega_p t,$$

$$v_2 = a \cos\left(\omega_p t - \frac{2\pi}{3}\right),$$

$$v_3 = a \cos\left(\omega_p t - \frac{4\pi}{3}\right), \tag{7.69}$$

where ω_p is the frequency of each phase of the ac signal (i.e., the line frequency). Note that v_1 leads v_2 by $2\pi/3$ radians and v_2 leads v_3 by the same angle. Furthermore, since v_1 leads v_3 by $4\pi/3$ radians, it is correct to say that v_1 lags v_3 by $2\pi/3$ radians. In other words, v_1 leads $-v_3$ by $(\pi - 2\pi/3)$, which is equal to $\pi/3$. Now consider a group of three windings, each of which has two segments (a positive segment and a negative segment) uniformly arranged around a circle (stator), as shown in Figure 7.36, in the order v_1, $-v_3$, v_2, $-v_1$, v_3, $-v_2$. Note that each winding segment has a phase difference of $\pi/3$ (or 60°) from the adjacent segment. The physical (geometric) spacing of adjacent winding segments is also 60°. Now, consider the time interval $\Delta t = \pi/(3\omega_p)$. The status of $-v_3$ at the end of a time interval of Δt is identical to the status of v_1 in the beginning of the time interval. Similarly, the status of v_2 after a time Δt becomes that of $-v_3$ in the beginning, and so on. In other words, the voltage status (and hence the magnetic field status) of one segment becomes identical to that of the adjacent segment in a time interval Δt. This means that the

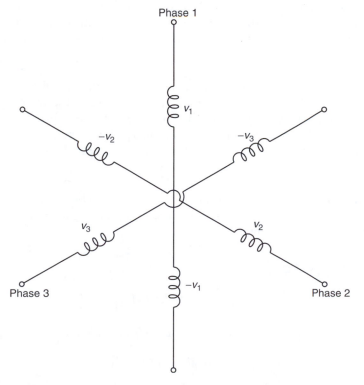

FIGURE 7.36
The generation of a rotating magnetic field using a three-phase supply and two winding sets per phase.

magnetic field generated by the winding segments appears to rotate physically around the circle (stator) at angular velocity ω_p.

It is not necessary for the three sets of three-phase windings to be distributed over the entire 360° angle of the circle. Suppose that, instead, these three sets (six segments) of windings are distributed within the first 180° of the circle, at 30° apart and a second three sets (identical to the first three sets) are distributed similarly within the remaining 180°. Then, the field would appear to rotate at half the speed ($\omega_p/2$), because in this case, Δt is the time taken for the field to rotate through 30°, not 60°. It follows that the general formula for the angular speed ω_p of the rotating magnetic field generated by a set of winding segments uniformly distributed on a stator and excited by an ac supply, is

$$\omega_f = \frac{\omega_p}{n}, \tag{7.70}$$

where ω_p is the frequency of the ac signal in each phase (i.e., line frequency) and n is the number of pairs of winding sets used per phase (i.e., number of pole pairs per phase).

Note that when $n=1$, there are two coils (positive and negative) for each phase (i.e., there are two poles per phase). Similarly, when $n=2$, there are four coils for each phase. Hence, n denotes the number of pole pairs per phase in a stator. In this manner, the speed of the rotating magnetic field can be reduced to a fraction of the line frequency simply by adding more sets of windings. These windings occupy the stator of an ac motor. The number of phases and the number of segments wound to each phase determine the angular separation of the winding segments around the stator. For example, for the three-phase, one-pole pair per phase arrangement shown in Figure 7.36, the physical separation of the winding segments is 60°. For a two-phase supply with one-pole pair per phase, the physical separation is 90°, and the separation is halved to 45° if two-pole pairs are used per phase. It is the rotating magnetic field, produced in this manner, which generates the driving torque by interacting with the rotor windings. The nature of this interaction determines whether a particular motor is an induction motor or a synchronous motor.

Example 7.11

Another way to interpret the concept of a rotating magnetic field is to consider the resultant field due to the individual magnetic fields in the stator windings. Consider a single set of three-phase windings arranged geometrically as in Figure 7.36. Suppose that the magnetic field due to phase 1 is denoted by $a \sin \omega_p t$. Show that the resultant magnetic field has an amplitude of $3a/2$ and that the field rotates at speed ω_p.

Solution

The magnetic field vectors in the three sets of windings are shown in Figure 7.37a. These can be resolved into two orthogonal components, as shown in Figure 7.37b. The component in the vertical direction (upwards) is

$$a \sin \omega_p t - a \sin\left(\omega_p t - \frac{2\pi}{3}\right) \cos\frac{\pi}{3} - a \sin\left(\omega_p t - \frac{4\pi}{3}\right) \cos\frac{\pi}{3}$$

$$= a \sin \omega_p t - \frac{a}{2}\left[\sin\left(\omega_p t - \frac{2\pi}{3}\right) + \sin\left(\omega_p t - \frac{4\pi}{3}\right)\right]$$

$$= a \sin \omega_p t - a \sin\left(\omega_p t - \pi\right)\cos\frac{\pi}{3}$$

$$= a \sin \omega_p t + \frac{a}{2}[\sin \omega_p t]$$

$$= \frac{3a}{2}\sin \omega_p t.$$

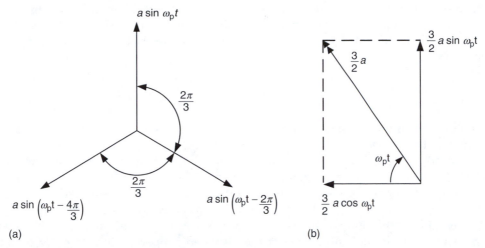

FIGURE 7.37
An alternative interpretation of a rotating magnetic field. (a) Magnetic fields of the windings. (b) Resultant magnetic field.

Note that in deriving this result, we have used the following trigonometric identities:

$$\sin A + \sin B = 2\sin\left(\frac{A+B}{2}\right)\cos\left(\frac{A-B}{2}\right)$$

and

$$\sin(A - \pi) = -\sin A.$$

The horizontal component of the magnetic fields, which is directed to the left, is

$$a\sin\left(\omega_p t - \frac{4\pi}{3}\right)\sin\frac{\pi}{3} - a\sin\left(\omega_p t - \frac{2\pi}{3}\right)\sin\frac{\pi}{3}$$

$$= \frac{\sqrt{3}}{2}a\left[\sin\left(\omega_p t - \frac{4\pi}{3}\right) - \sin\left(\omega_p t - \frac{2\pi}{3}\right)\right]$$

$$= \sqrt{3}a\cos(\omega_p t - \pi)\sin\left(-\frac{\pi}{3}\right)$$

$$= \frac{3a}{2}\cos\omega_p t.$$

Here, we have used the following trigonometric identities:

$$\sin A - \sin B = 2\cos\frac{A+B}{2}\sin\frac{A-B}{2},$$

$$\cos(A - \pi) = -\cos A,$$

$$\sin(-A) = -\sin A,$$

The resultant of the two orthogonal components is a vector of magnitude $3a/2$, making an angle $\omega_p t$ with the horizontal component, as shown in Figure 7.37b. It follows that the resultant magnetic field has a magnitude of $3a/2$ and rotates in the clockwise direction at speed ω_p rad/s.

7.6.2 Induction Motor Characteristics

The stator windings of an induction motor generate a rotating magnetic field, as explained in the previous section. The rotor windings are purely secondary windings, which are not energized by an external voltage. For this reason, no commutator-brush devices are needed in induction motors (see Figure 7.38). The core of the rotor is made of ferromagnetic laminations in order to concentrate the magnetic flux and to minimize dissipation (primarily due to eddy currents). The rotor windings are embedded in the axial direction on the outer cylindrical surface of the rotor and are interconnected in groups. The rotor windings may consist of uninsulated copper or aluminum (or any other conductor) bars (a cage rotor), which are fitted into slots in the end rings at the two ends of the rotor. These end rings complete the paths for electrical conduction through the rods. Alternatively, wire with one or more turns in each slot (a wound rotor) may be used.

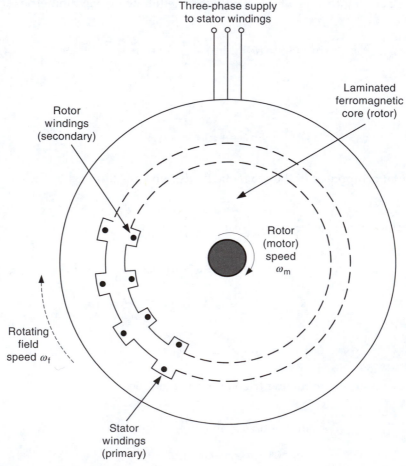

FIGURE 7.38
Schematic diagram of an induction motor.

First, consider a stationary rotor. The rotating magnetic field in the stator intercepts the rotor windings, thereby generating an induced voltage (and current) due to mutual induction or transformer action (hence the name induction motor). The resulting secondary magnetic flux from the induced current in the rotor interacts with the primary, rotating magnetic flux, thereby producing a torque in the direction of rotation of the stator field. This torque drives the rotor. As the rotor speed increases, initially the motor torque also increases (rather moderately) because of secondary interactions between the stator circuit and the rotor circuit, even though the relative speed of the rotating field with respect to the rotor decreases, which reduces the rate of change of flux linkage and hence the direct transformer action (Note: the relative speed is termed the slip rate). Quite soon, the maximum torque will be reached.

Further increase in rotor speed (i.e., a decrease in slip rate) sharply decreases the motor torque, until at synchronous speed (i.e., zero slip rate), the motor torque becomes zero. This behavior of an induction motor is illustrated by the typical characteristic curve given in Figure 7.39. From the starting torque T_s to the maximum torque (which is known as the breakdown torque) T_{max}, the motor behavior is unstable. This can be explained as follows. An incremental increase in speed causes an increase in torque, which further increases the speed. Similarly, an incremental reduction in speed brings about a reduction in torque that further reduces the speed. The portion of the curve from T_{max} to the zero torque (or, no-load or synchronous) condition represents the region of stable operation. Under normal operating conditions, an induction motor should operate in this region.

The fractional slip S for an induction motor is given by

$$S = \frac{\omega_f - \omega_m}{\omega_f} \tag{7.71}$$

Even when there is no external load, the synchronous operating condition (i.e., $S = 0$) is not achieved by an induction motor at steady state, because of the presence of frictional torque, which opposes the rotor motion. When an external torque (load torque) T_L is present, under normal operating conditions, the slip rate increases further so as to increase the motor torque to support this load torque. As is clear from Figure 7.39, in

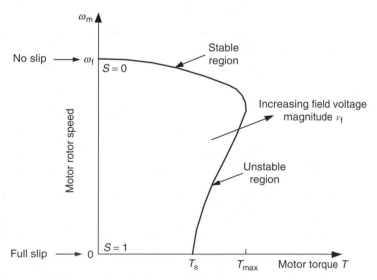

FIGURE 7.39
Torque–speed characteristic curve of an induction motor.

the stable region of the characteristic curve, the induction motor is quite insensitive to torque changes; a small change in speed would require a very large change in torque (in comparison with an equivalent dc motor). For this reason, an induction motor is relatively insensitive to load variations and can be regarded as a constant-speed machine. Note that if the rotor speed is increased beyond the synchronous speed (i.e., $S < 0$), the motor becomes a generator.

7.6.3 Torque–Speed Relationship

It is instructive to determine the torque–speed relationship for an induction motor. This relationship provides insight into possible control methods for induction motors. The equivalent circuits of the stator and the rotor for one phase of an induction motor are shown in Figure 7.40a. The circuit parameters are R_f, the stator coil resistance; L_f, the stator leakage inductance; R_c, the stator core iron loss resistance (eddy current effects, etc.); L_c, the stator core (magnetizing) inductance; L_r, the rotor leakage inductance; and R_r, the rotor coil resistance.

The magnitude of the ac supply voltage for each phase of the stator windings is v_f at the line frequency ω_p. The rotor current generated by the induced e.m.f. is i_r. After allowing for the voltage drop due to stator resistance and stator leakage inductance, the voltage that is available for mutual induction is denoted by v. This is also the induced voltage in the secondary (rotor) windings at standstill, assuming the same number of turns. This induced voltage changes linearly with slip S, because the induced voltage is proportional

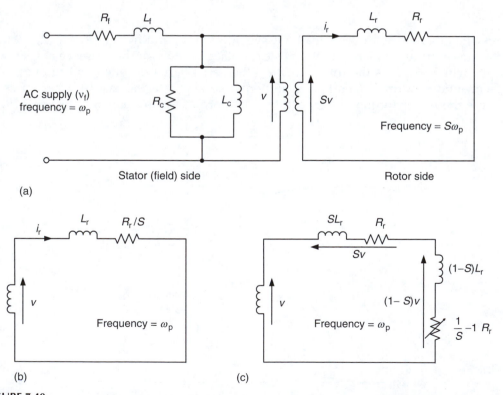

FIGURE 7.40
(a) Stator and rotor circuits for an induction motor. (b) Rotor circuit referred to the stator side. (c) Representation of available mechanical power using the rotor circuit.

to the relative velocity of the rotating field with respect to the rotor (i.e., $\omega_f - \omega_m$), as is evident from Equation 7.2. Hence, the induced voltage in the rotor windings (secondary windings) is Sv. Note, further, that at standstill (when $S = 1$), the frequency of the induced voltage in the rotor is ω_p. At synchronous speed of rotation (when $S = 0$), this frequency is zero because the magnetic field is fixed and constant relative to the rotor in this case. Now, assuming a linear variation of frequency of the induced voltage between these two extremes, we note that the frequency of the induced voltage in the rotor circuit is $S\omega_p$. These observations are indicated in Figure 7.40a.

Using the frequency domain (complex) representation for the out-of-phase currents and voltages, the rotor current i_r in the complex form is given by

$$i_r = \frac{Sv}{(R_r + jS\omega_p L_r)} = \frac{v}{(R_r/S + j\omega_p L_r)}. \tag{7.72}$$

From Equation 7.72, it is clear that the rotor circuit can be represented by a resistance R_r/S and an inductance L_r in series and excited by voltage v at frequency ω_p. This is in fact the rotor circuit referred to the stator side, as shown in Figure 7.40b. This circuit can be grouped into two parts, as shown in Figure 7.40c. The inductance SL_r and resistance R_r in series, with a voltage drop Sv, are identical to the rotor circuit in Figure 7.40a. Note that SL_r has to be used as the inductance in the new equivalent circuit segment, instead of L_r in the original rotor circuit, for the sake of circuit equivalence. The reason is simple. The new equivalent circuit operates at frequency ω_p whereas the original rotor circuit operates at frequency $S\omega_p$ (Note: impedance of an inductor is equal to the product of inductance and frequency of excitation). The second voltage drop $(1 - S)v$ in Figure 7.40c represents the back e.m.f. due to rotor–stator field interaction; it generates the capacity to drive an external load (mechanical power). Note here that the back e.m.f. governs the current in the rotor circuit and hence the generated torque. It follows that the available mechanical power, per phase, of an induction motor is given by $i_r^2(1/S - 1)R_r$. Hence,

$$T_m \omega_m = pi_r^2 \left(\frac{1}{S} - 1 \right) R_r \tag{7.73}$$

where T_m is the motor torque generated in the rotor, ω_m is the rotor speed of the motor, p is the number of supply phases, and i_r is the magnitude of the current in the rotor. The magnitude of the current in the rotor circuit is obtained from Equation 7.72; thus,

$$i_r = \frac{v}{\sqrt{R_r^2/S^2 + \omega_p^2 L_r^2}}. \tag{7.74}$$

By substituting Equation 7.74 in Equation 7.73, we get

$$T_m = pv^2 \frac{S(1 - S)}{\omega_m} \frac{R_r}{\left(R_r^2 + S^2 \omega_p^2 L_r^2 \right)} \tag{7.75}$$

From Equation 7.70 and Equation 7.71, we can express the number of pole pairs per phase of stator winding as

$$n = \frac{\omega_p}{\omega_m} (1 - S). \tag{7.76}$$

Equation 7.76 is substituted in Equation 7.75; thus,

$$T_\mathrm{m} = \frac{pnv^2 SR_\mathrm{r}}{\omega_\mathrm{p}\left(R_\mathrm{r}^2 + S^2\omega_\mathrm{p}^2 L_\mathrm{r}^2\right)}. \tag{7.77}$$

If the resistance and the leakage inductance in the stator are neglected, v is approximately equal to the stator excitation voltage v_f. This gives the torque–slip relationship:

$$T_\mathrm{m} = \frac{pnv_\mathrm{f}^2 SR_\mathrm{r}}{\omega_\mathrm{p}\left(R_\mathrm{r}^2 + S^2\omega_\mathrm{p}^2 L_\mathrm{r}^2\right)}. \tag{7.78}$$

Note that by using Equation 7.76, it is possible to express S in Equation 7.78 in terms of the rotor speed ω_m. This results in a torque–speed relationship, which gives the characteristic curve shown in Figure 7.39. Specifically, we employ the fact that the motor speed ω_m is related to slip through

$$S = \frac{\omega_\mathrm{p} - n\omega_\mathrm{m}}{\omega_\mathrm{p}}. \tag{7.79}$$

Note, further, from Equation 7.78 that the motor torque is proportional to the square of the supply voltage v_f.

Example 7.12
In the derivation of Equation 7.78, we assumed that the number of effective turns per phase in the rotor is equal to that in the stator. This assumption is generally not valid, however. Determine how the equation should be modified in the general case. Suppose that

$$r = \frac{\text{number of effective turns per phase in the rotor}}{\text{number of effective turns per phase in the stator}}.$$

Solution
At standstill ($S = 1$), the induced voltage in the rotor is rv and the induced current is i_r/r. Hence, the impedance in the rotor circuit is given by

$$Z_\mathrm{r} = \frac{rv}{i_\mathrm{r}/r} = r^2 \frac{v}{i_\mathrm{r}} \tag{7.80}$$

or

$$Z_\mathrm{r} = r^2 Z_\mathrm{req}.$$

It follows that the true rotor impedance (or resistance and inductance) simply has to be divided by r^2 to obtain the equivalent impedance. In this general case of $r \neq 1$, the resistance R_r and the inductance L_r should be replaced by $R_\mathrm{req} = R_\mathrm{r}/r^2$ and $L_\mathrm{req} = L_\mathrm{r}/r^2$, in Equation 7.78.

Example 7.13
Consider a three-phase induction motor that has one-pole pair per phase. The equivalent resistance and leakage inductance in the rotor circuit are $8\ \Omega$ and $0.06\ \mathrm{H}$, respectively. The motor supply voltage is 115 V in each phase, at a line frequency of 60 Hz. Compute the torque–speed curve for the motor.

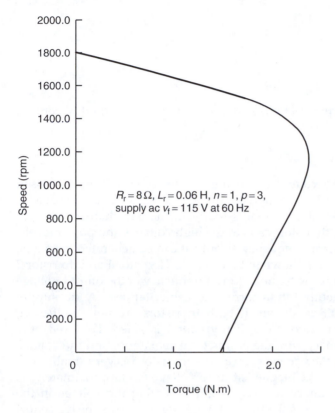

$R_r = 8\,\Omega,\ L_r = 0.06\,\text{H},\ n = 1,\ p = 3,$
supply ac $v_f = 115\,\text{V}$ at 60 Hz

FIGURE 7.41
Torque–speed curve for an induction motor.

Solution

In this example, $R_r = 8\ \Omega$, $L_r = 0.06$ H, $v_f = 115$ V, $n = 1$, $p = 3$, and $\omega_p = 60 \times 2\pi$ rad/s. Now, using Equation 7.78 along with Equation 7.79, we can compute the torque–speed curve. The result is shown in Figure 7.41.

7.7 Induction Motor Control

DC motors are widely used in servo control applications because of their simplicity and flexible speed–torque capabilities. In particular, dc motors are easy to control and they operate accurately and efficiently over a wide range of speeds. The initial cost and the maintenance cost of a dc motor, however, are generally higher than those for a comparable ac motor. AC motors are rugged and are most common in medium- to high-power applications involving fairly constant-speed operation. Of late, much effort has been invested in developing improved control methods for ac motors, and significant progress is seen in this area. The present day ac motors having advanced drive systems with frequency control and field feedback compensation can provide speed control that is comparable to the capabilities of dc servomotors (e.g., 1:20 or 26 dB range of speed variation).

Since fractional slip S determines motor speed ω_m, Equation 7.78 suggests several possibilities for controlling an induction motor. Four possible methods for induction motor control are

1. Excitation frequency control (ω_p)
2. Supply voltage control (v_f)
3. Rotor resistance control (R_r)
4. Pole changing (n)

What is given in parentheses is the parameter that is adjusted in each method of control.

7.7.1 Excitation Frequency Control

Excitation frequency control can be accomplished using a thyristor circuit. As discussed under dc motor drive system (Section 7.4.2), a thyristor (or SCR) is a semiconductor device, which possesses very effective, efficient, and nondissipative switching characteristics, at very high frequencies. Furthermore, thyristors can handle high voltages and power levels.

By using an inverter circuit, a variable-frequency ac output can be generated from a dc supply. A single-phase inverter circuit is shown in Figure 7.42. Thyristor 1 and thyristor 2 are gated by their firing circuits according to the required frequency of the output voltage v_o. The primary winding of the output transformer is center-tapped. A dc supply voltage v_{ref} is applied to the circuit as shown. If both thyristors are not conducting, the voltage across the capacitor C is zero. Now, if thyristor 1 is gated (i.e., fired), the current in the upper half of the primary winding builds to its maximum and the voltage across that half reaches v_{ref} (because the voltage drop across thyristor 1 is very small). As a result of the corresponding change in the magnetic flux, a voltage v_{ref} (approximately) is induced in the lower half of the primary winding, complementing the voltage in the upper half. Accordingly, the voltage across the primary winding (or across the capacitor) is approximately $2v_{ref}$. Now, if thyristor 2 is fired, the voltage at point A becomes v_{ref}. Since the capacitor is already charged to $2v_{ref}$, the voltage at point B becomes $3v_{ref}$. This means that a voltage of $2v_{ref}$ is applied across thyristor 1 in the nonconducting direction.

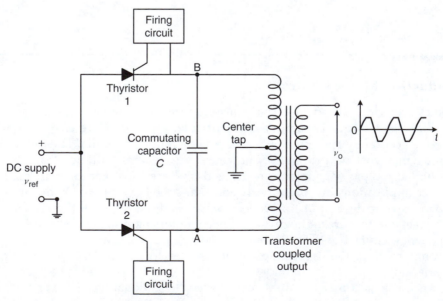

FIGURE 7.42
A single-phase inverter circuit for frequency control.

As a result, thyristor 1 will be turned off. Then, as before, a voltage $2v_{ref}$ is generated in the primary winding, but in the opposite direction, because it is thyristor 2 that is conducting now. In this manner, an approximately rectangular pulse sequence of ac voltage v_o is generated at the circuit output. The frequency of the voltage is equal to the inverse of the firing interval between the two thyristors. A three-phase inverter can be formed by triplicating the single-phase inverter and by phasing the firing times appropriately.

Modern drive units for induction motors use pulse-width modulation (PWM) and advanced microelectronic circuitry incorporating a single monolithic integrated circuit chip with more than 30,000 circuit elements, rather than discrete semiconductor elements. The block diagram in Figure 7.43a shows a frequency control system for an induction motor. A standard ac supply (three phase or single phase) is rectified and filtered to provide the dc supply to the three-phase PWM inverter circuit. This device generates a nearly sinusoidal three-phase output at a specified frequency. Firing of the switching circuitry according to the required frequency of the ac output, is commanded by a hardware controller. If the control requirements are simple, a variable-frequency

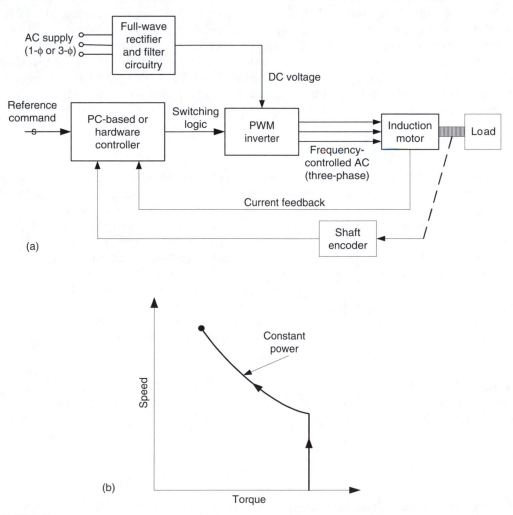

FIGURE 7.43
(a) Variable frequency control of an induction motor. (b) A typical control strategy.

oscillator or a voltage-to-frequency converter may be used instead. Alternatively, a digital computer (PC)-based controller may be used to vary the drive frequency and to adjust other control parameters in a more flexible manner, using software. The controller may use hardware logic or software to generate the switching signal (reference frequency), while taking into account external (human-operator) commands and sensor feedback signals. A variable-frequency drive for an ac motor can effectively operate in the open-loop mode. Sensor feedback may be employed, however, for more accurate performance. Feedback signals may include shaft encoder readings (motor angle) for speed control and current (stator current, rotor current in wound rotors, dc current to PWM inverter, etc.) particularly for motor torque control. A typical control strategy is shown in Figure 7.43b. In this case, the control processor provides a two-mode control scheme. In the initial mode, the torque is kept constant while accelerating the motor. In the other mode, the power is kept constant while further increasing the speed. Both modes of operation can be achieved through frequency control. Strategies of specified torque profiles (torque control) or specified speed profiles (speed control) can be implemented in a similar manner.

Programmable, microprocessor-based variable-frequency drives for ac motors are commercially available. One such drive is able to control the excitation frequency in the range 0.1 to 400 Hz with a resolution of 0.01 Hz. A three-phase ac voltage in the range 200 to 230 V or 380 to 460 V is generated by the drive, depending on the input ac voltage. AC motors with frequency control are employed in many applications, including variable-flow control of pumps, fans and blowers, industrial manipulators (robots, hoists, etc.), conveyors, elevators, process plant and factory instrumentation, and flexible operation of production machinery for flexible (variable output) production. In particular, ac motors with frequency control and sensor feedback are able to function as servomotors (i.e., ac servos).

7.7.2 Voltage Control

From Equation 7.77, it is seen that the torque of an induction motor is proportional to the square of the supply voltage. It follows that an induction motor can be controlled by varying its supply voltage. This may be done in several ways. For example, amplitude modulation of the ac supply, using a ramp generator, directly accomplishes this objective by varying the supply amplitude. Alternatively, by introducing zero-voltage regions (i.e., blanking out or "chopping") periodically (at high frequency) in the ac supply, for example, using a thyristor circuit with firing delays as in PWM, accomplishes voltage control by varying the root-mean-square (rms) value of the supply voltage. Voltage control methods are appropriate for small induction motors, but they provide poor efficiency when control over a wide speed range is required. Frequency control methods are recommended in low-power applications. An advantage of voltage control methods over frequency control methods is the lower stator copper loss.

Example 7.14
Show that the fractional slip vs. motor torque characteristic of an induction motor, at steady state, may be expressed by

$$T_m = \frac{aSv_f^2}{\left[1 + (S/S_b)^2\right]} \tag{7.81}$$

Identify (give expressions for) the parameters a and S_b. Show that S_b is the slip corresponding to the breakdown torque (maximum torque) T_{max}. Obtain an expression for T_{max}.

An induction motor with parameter values $a = 4 \times 10^{-3}$ N.m/V^2 and $S_b = 0.2$ is driven by an ac supply that has a line frequency of 60 Hz. Stator windings have two "pole pairs" per phase. Initially, the line voltage is 500 V. The motor drives a mechanical load, which can be represented by an equivalent viscous damper with damping constant $b = 0.265$ N.m/rad/s. Determine the operating point (i.e., the values of torque and speed) for the system. Suppose that the supply voltage is dropped by 50% (to 250 V) using a voltage control scheme. What is the new operating point? Is this a stable operating point? In view of your answer, comment on the use of voltage control in induction motors.

Solution
First, we note that Equation 7.78 can be expressed as Equation 7.81, with

$$a = \frac{pn}{\omega_p R_r} \tag{7.82}$$

and

$$S_b = \frac{R_r}{\omega_p L_r}. \tag{7.83}$$

The breakdown torque is the peak torque and is defined by

$$\frac{\partial T_m}{\partial \omega_m} = 0.$$

We express

$$\frac{\partial T_m}{\partial \omega_m} = \frac{\partial T_m}{\partial S} \frac{\partial S}{\partial \omega_m} = -\frac{1}{\omega_f} \frac{\partial T_m}{\partial S},$$

where we have differentiated Equation 7.71 with respect to ω_m and substituted the result. It follows that the breakdown torque is given by

$$\frac{\partial T_m}{\partial S} = 0.$$

Now, differentiate Equation 7.81 with respect to S and equate to zero. We get

$$\left[1 + \left(\frac{S}{S_b} \right)^2 \right] - S \left[2 \frac{S}{S_b^2} \right] = 0$$

or

$$1 - \left(\frac{S}{S_b} \right)^2 = 0.$$

It follows that $S = S_b$ corresponds to the breakdown torque. Substituting this in Equation 7.81, we have

$$T_{max} = \frac{1}{2} a S_b v_f^2. \tag{7.84}$$

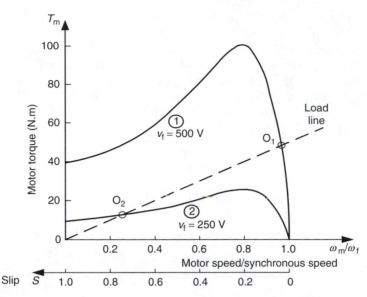

FIGURE 7.44
Speed–torque curves for induction motor voltage control.

Next, the speed–torque curve is computed using the given parameter values in Equation 7.81 and plotted as shown in Figure 7.44 for the two cases $v_f = 500$ V and $v_f = 250$ V. Note that with $S_b = 0.2$, we have, from Equation 7.84, $(T_{max})_1 = 100$ N.m and $(T_{max})_2 = 25$ N.m. These values are confirmed from the curves in Figure 7.44.

The load curve is given by

$$T_m = b\omega_m$$

or

$$T_m = b\omega_f \frac{\omega_m}{\omega_f}.$$

Now, from Equation 7.70, the synchronous speed is computed as

$$\omega_f = \frac{60 \times 2\pi}{2} \, \text{rad/s} = 188.5 \, \text{rad/s}.$$

Hence,

$$b\omega_f = 0.265 \times 188.5 = 50 \, \text{N.m.}$$

This is the slope of the load line shown in Figure 7.44. The points of intersection of the load line and the motor characteristic curve are the steady-state operating points. They are for case 1 ($v_f = 500$ V):

1. Operating torque $= 48$ N.m
2. Operating slip $= 4\%$
3. Operating speed $= 1728$ rpm

for case 2 ($v_f = 250$ V):

1. Operating torque $= 12$ N.m
2. Operating slip $= 77\%$
3. Operating speed $= 414$ rpm

Note that when the supply voltage is halved, the torque drops by a factor of 4 and the speed drops by about 76%. However what is worse is that the new operating point (O_2) is in the unstable region (i.e., from $S = S_b$ to $S = 1$) of the motor characteristic curve. It follows that large drops in supply voltage are not desirable, and as a result the efficiency of the motor can degrade significantly with voltage control.

7.7.3 Rotor Resistance Control

It can be seen from Equation 7.78 that an induction motor can be controlled by varying R_r. Since this is a dissipative technique, it is also a wasteful method. This was a commonly used method for induction motor control before the development of more efficient variable-frequency switching circuits, digital signal processing (DSP) chips, microelectronic-drive systems, and related control techniques. The rotor of an induction motor has a closed circuit (resistive and inductive), which is not connected to an external power supply, unlike in the case of a dc motor. In the wound-rotor design, windings are usually arranged and connected as polyphase groups (e.g., a delta configuration (Δ) or star configuration (Y) in three-phase motors), just like the stator windings, but without a supply voltage. The current in the rotor circuit is generated purely by magnetic induction, but it determines the torque–speed characteristics of the motor. The motor response is controlled by changing the rotor resistance. This can be accomplished by connecting a variable resistance to each phase externally through a slip-ring and brush arrangement, as schematically shown in Figure 7.45 for

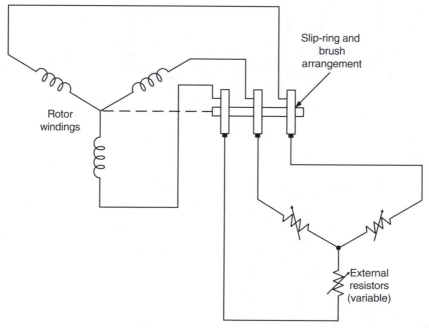

FIGURE 7.45
Rotor resistance control of an induction motor (three phase).

the three-phase star (Y) connection. Rotor resistance control has the same disadvantages as in the voltage control method. In particular, the motor efficiency drops considerably when the motor operates over a wide range of speeds. Furthermore, the energy dissipated by the control resistors results in thermal problems. Heat sinks, fans, and other cooling methods may have to be employed, particularly for continuous operation.

7.7.4 Pole-Changing Control

The number of pole pairs per phase in the stator windings (n) is a parameter in the speed–torque Equation 7.78. It follows that changing the parameter n is an alternative method for controlling an induction motor. This can be accomplished by switching the supply connections in the stator windings in some manner. The principle is illustrated in Figure 7.46. Consider the windings in one phase of the stator. With the coil currents as in Figure 7.46a, the magnetic fields in the alternate pairs of adjacent coils cancel out. When the coil currents are as in Figure 7.46b, all adjacent pairs of coils have complementary magnetic fields. As a result, the number of pole pairs per phase is doubled when the stator windings are switched from the configuration in Figure 7.46a to that in Figure 7.46b. Note that when the stator windings are switched into a certain configuration of poles, the same switching should be done simultaneously to the rotor windings as well. This results in an additional complexity for the control circuitry in the case of a wound rotor. In a squirrel-cage rotor, a separate switching mechanism is not necessary for the rotor because it automatically reacts to configure itself according to the winding configuration in the stator. For this reason, cage rotor induction motors are better suited for pole-changing control.

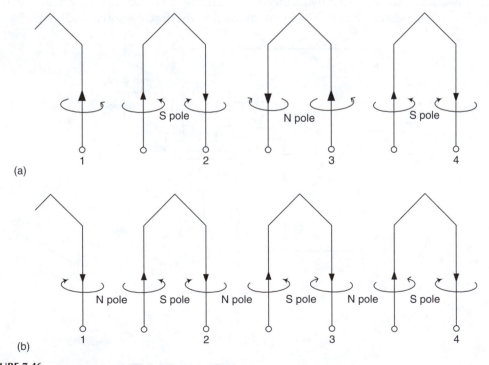

FIGURE 7.46
Pole-changing control of an induction motor. In changing from (a) to (b), the number of pole pairs per phase is doubled by reversing the currents in stator windings 1 and 3.

7.7.5 Field Feedback Control (Flux Vector Drive)

An innovative method for controlling ac motors is through field feedback (or flux vector) compensation. This approach can be explained using the equivalent circuit shown in Figure 7.40c. Note that this circuit separates the rotor-equivalent impedance into two parts—a nonproductive part with a voltage drop Sv and a torque-producing part with a voltage drop $(1-S)v$—as discussed previously. There exist magnetic field vectors (or complex numbers) that correspond to these two parts of circuit impedance. As clear from Figure 7.40c, these magnetic flux components depend on the slip S and hence the rotor speed and also on the current. In the present method of control, the magnetic field vector associated with the first part of impedance is sensed using speed measurement (from an encoder) and motor current measurement (from a current–to-voltage transducer), and compensated for (i.e., removed through feedback) in the stator circuit. As a result, only the second part of impedance (and magnetic field vector), which corresponds to the back e.m.f., remains. Hence, the ac motor behaves quite like a dc motor that has an equivalent torque-producing back e.m.f. More sophisticated schemes of control may use a model of the motor. Flux vector control has been commercially implemented in ac motors using customized digital signal processor (DSP) chips. Feedback of rotor current can further improve the performance of a flux vector drive. A flux vector drive tends to be more complex and costly than a variable-frequency drive. Need of sensory feedback introduces a further burden in this regard.

7.7.6 A Transfer-Function Model for an Induction Motor

The true dynamic behavior of an induction motor is generally nonlinear and time varying. For small variations about an operating point, however, linear relations can be written. On this basis, a transfer-function model can be established for an induction motor, as we have done for a dc motor. The procedure described in this section uses the steady-state speed–torque relationship for an induction motor to determine the transfer function model. The basic assumption here is that this steady-state relationship, if the inertia effects are modeled by some means, can represent the dynamic behavior of the motor for small changes about an operating point (steady-state) with reasonable accuracy.

Suppose that a motor rotor, which has moment of inertia J_m and mechanical damping constant b_m (mainly from the bearings) is subjected to a variation δT_m in the motor torque and an associated change $\delta \omega_m$ in the rotor speed, as shown in Figure 7.47. In general, these changes may arise from a change δT_L in the load torque, and a change δv_f in supply voltage.

Newton's second law gives

$$\delta T_m - \delta T_L = J_m \delta \dot{\omega}_m + b_m \delta \omega_m. \tag{7.85}$$

Now, use a linear steady-state relationship (motor characteristic cure) to represent the variation in motor torque as a function of the incremental change $\delta \omega_m$ in speed and a variation δv_f in the supply voltage. We get

$$\delta T_m = -b_e \delta \omega_m + k_v \delta v_f. \tag{7.86}$$

By substituting Equation 7.86 in Equation 7.85 and using the Laplace variable s, we have

$$\delta \omega_m = \frac{k_v}{[J_m s + b_m + b_e]} \delta v_f - \frac{1}{[J_m s + b_m + b_e]} \delta T_L. \tag{7.87}$$

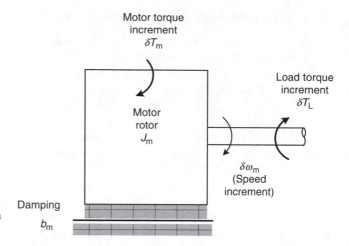

FIGURE 7.47
Incremental load model for an induction motor.

In the transfer function Equation 7.87, note that $\delta\omega_m$ is the output, δv_f is the control input, and δT_L is an unknown (disturbance) input. The motor transfer function $\delta\omega_m/\delta v_f$ is given by

$$G_m(s) = \frac{k_v}{[J_m s + b_m + b_e]}. \tag{7.88}$$

The motor time constant τ is

$$\tau = \frac{J_m}{b_m + b_e}. \tag{7.89}$$

Now it remains to identify the parameters b_e (analogous to electrical damping in a dc motor) and k_v (a voltage gain parameter, as for a dc motor). To accomplish this, we use Equation 7.81, which can be written in the form

$$T_m = k(S)v_f^2, \tag{7.90}$$

where

$$k(S) = \frac{aS}{1 + (S/S_b)^2}. \tag{7.91}$$

Now, using the well-known relation in differential calculus

$$\delta T_m = \frac{\partial T_m}{\partial \omega_m}\delta\omega_m + \frac{\partial T_m}{\partial v_f}\delta v_f$$

we have

$$b_e = -\frac{\partial T_m}{\partial \omega_m} \text{ and } k_v = \frac{\partial T_m}{\partial v_f}.$$

But

$$\frac{\partial T_m}{\partial \omega_m} = \frac{\partial T_m}{\partial S} \frac{\partial S}{\partial \omega_m} = -\frac{1}{\omega_f} \frac{\partial T_m}{\partial S}.$$

Thus,

$$b_e = \frac{1}{\omega_f} \frac{\partial T_m}{\partial S}, \tag{7.92}$$

where ω_f is the synchronous speed of the motor. By differentiating Equation 7.91 with respect to S, we have

$$\frac{\partial k}{\partial S} = a \frac{1 - (S/S_b)^2}{\left[1 + (S/S_b)^2\right]^2}. \tag{7.93}$$

Hence,

$$b_e = \frac{a v_f^2}{\omega_f} \frac{1 - (S/S_b)^2}{\left[1 + (S/S_b)^2\right]^2}. \tag{7.94}$$

Next, by differentiating Equation 7.90 with respect to v_f, we have

$$\frac{\partial T_m}{\partial v_f} = 2k(S)v_f.$$

Accordingly, we get

$$k_v = \frac{2aSv_f}{1 + (S/S_b)^2}, \tag{7.95}$$

where S_b is the fractional slip at the breakdown (maximum) torque and a is a motor torque parameter defined by Equation 7.82. If we wish to include the effects of the electrical time constant τ_e of the motor, we may include the factor $\tau_e s + 1$ in the denominator (characteristic polynomial) on the right-hand side of Equation 7.87. Since τ_e is usually an order of magnitude smaller than τ as given by Equation 7.89, no significant improvement in accuracy results through this modification. Finally, note that the constants b_e and k_v can be obtained graphically as well using experimentally determined speed–torque curves for an induction motor for several values of the line voltage v_f, using a procedure similar to what we have described for a dc motor.

Example 7.15

A two-phase induction motor can serve as an ac servomotor. The field windings are identical and are placed in the stator with a geometric separation of 90°, as shown in Figure 7.48a. One of the phases is excited by a fixed reference ac voltage $v_{ref} \cos \omega_p t$. The other phase is 90° out of phase from the reference phase; it is the control phase, with voltage amplitude v_c. The motor is controlled by varying the voltage v_c.

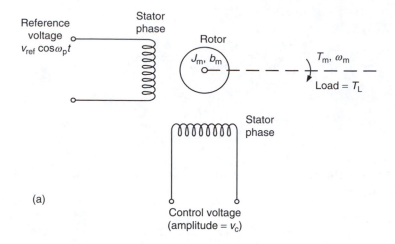

(a)

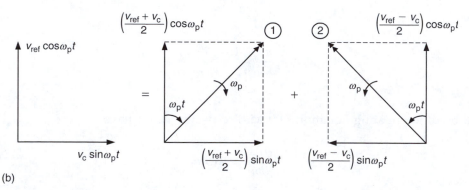

(b)

FIGURE 7.48
(a) A two-phase induction motor functioning as an ac servomotor. (b) Equivalent representation of the magnetic field vector in the stator.

1. With the usual notation, obtain an expression for the motor torque T_m in terms of the rotor speed ω_m and the input voltage v_c
2. Indicate how a transfer function model may be obtained for this ac servo
    (a) Graphically, using the characteristic curves of the motor
    (b) Analytically, using the relationship obtained in Part 1

Solution
Note that since $v_c \neq v_f$, the two phases are not balanced. Hence, the resultant magnetic field vector in this two-phase induction motor does not rotate at a constant speed ω_p; as a result, the relations derived previously cannot be applied directly. The first step, then, is to decompose the field vector into two components that rotate at constant speeds. This is accomplished in Figure 7.48b. The field component 1 is equivalent to that of an induction motor supplied with a line voltage of $(v_{ref} + v_c)/2$, and it rotates in the clockwise direction at speed ω_p. The field component 2 is equivalent to that generated with a line voltage of $(v_{ref} - v_c)/2$, and it rotates in the counterclockwise direction.

Suppose that the motor rotates in the clockwise direction at speed ω_p. The slip for the equivalent system 1 is

$$S = \frac{\omega_p - \omega_m}{\omega_p}$$

and the slip for the equivalent system 2 is

$$S' = \frac{\omega_p + \omega_m}{\omega_p} = 2 - S,$$

which are in opposite directions. Now, using the relationship for an induction motor with a balanced multiphase supply (Equation 7.90), we have

$$T_m = k(S)\left[\frac{v_{ref} + v_c}{2}\right]^2 - k(2 - S)\left[\frac{v_{ref} - v_c}{2}\right]^2. \tag{7.96}$$

In this derivation, we have assumed that the electrical and magnetic circuits are linear, so the principle of superposition holds. The function $k(S)$ is given by the standard Equation 7.91, with a and S_b defined by Equation 7.82 and Equation 7.83, respectively. In this example, there is only one-pole pair per phase ($n = 1$). Hence, the synchronous speed ω_f is equal to the line frequency ω_p. The motor speed ω_m is related to S through the usual Equation 7.87.

To obtain the transfer-function relation for operation about an operating point, we use the differential relation

$$\delta T_m = \frac{\partial T_m}{\partial \omega_m}\delta\omega_m + \frac{\partial T_m}{\partial v_c}\delta v_c$$
$$= -b_e\delta\omega_m + k_v\delta v_c.$$

As derived previously, the transfer relation is

$$\delta\omega_m = \frac{k_v}{[J_m s + b_m + b_e]}\delta v_c - \frac{1}{[J_m s + b_m + b_e]}\delta T_L,$$

where T_L denotes the load torque. It remains to be shown how to determine the parameters b_e and k_v, both graphically and analytically.

In the graphic method, we need a set of speed–torque curves for the motor, for several values of v_c in the operating range. Note that experimental measurements of motor torque contain the mechanical damping torque in the bearings. The actual electromagnetic torque of the motor is larger than the measured torque at steady state, and the difference is the frictional torque. As a result, adjustments have to be made to the measured torque curve in order to get the true speed–motor torque curves. If this is done, the parameters b_e and k_v can be determined graphically as indicated in Figure 7.49. Each curve is a constant v_c curve. Hence, the magnitude of its slope gives b_e. Note that $\partial T_m/\partial v_c$ is evaluated at constant ω_m. Hence, the parameter k_v has to be determined on a vertical line (where $\omega_m = $ constant). If two curves, one for the operating value of v_c and the other for a unit increment in v_c, are available, as shown in Figure 7.49, the value of k_v is simply the vertical separation of the two curves at the operating point. If the increments in v_c are small, but not unity, the vertical separation of the two curves has to be divided by this increment (δv_c) in order to determine k_v, and the result will be more accurate.

To analytically determine b_e and k_v, we must differentiate T_m in Equation 7.96 with respect to ω_m and v_c, separately. We get,

$$b_e = -\frac{\partial T_m}{\partial \omega_m} = \frac{1}{\omega_p}\left[\frac{v_{ref} + v_c}{2}\right]^2 \frac{\partial k(S)}{\partial S} - \frac{1}{\omega_p}\left[\frac{v_{ref} - v_c}{2}\right]^2 \frac{\partial k(2 - S)}{\partial S}, \tag{7.97}$$

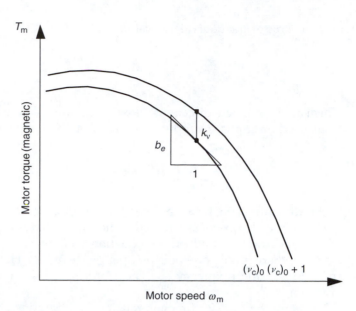

FIGURE 7.49
Graphic determination of transfer
function parameters for an induction
motor.

where $[\partial k(S)/\partial S]$ is given by Equation 7.93. To determine $[\partial k(2 - S)/\partial S]$, we note that

$$\frac{\partial k(2 - S)}{\partial S} = \frac{\partial k(2 - S)}{\partial(2 - S)} \frac{d(2 - S)}{dS} = -\frac{\partial k(2 - S)}{\partial(2 - S)} = -\frac{\partial k(S)}{\partial S}\bigg|_{S=2-S}. \qquad (7.98)$$

In other words, $[\partial k(2 - S)/\partial S]$ is obtained by first replacing S by $2-S$ in the right-hand side of Equation 7.93 and then reversing the sign of the result. Finally,

$$k_v = \frac{\partial T_m}{\partial v_c} = \frac{1}{2}k(S)[v_{ref} + v_c] + \frac{1}{2}k(2 - S)[v_{ref} - v_c]. \qquad (7.99)$$

∎

Induction motors have the advantages of brushless operation, low maintenance, ruggedness, and low cost. They are naturally suitable for constant-speed and continuous operation applications. With modern drive systems, they are able to function well in variable-speed and servo applications as well. Applications of induction motors include household appliances, industrial instrumentation, traction devices (e.g., ground transit vehicles), machine tools (e.g., lathes and milling machines), heavy-duty factory equipment (e.g., steel rolling mills, conveyors, and centrifuges), and equipment in large buildings (e.g., elevator drives, compressors, fans, and HVAC systems).

7.7.7 Single-Phase AC Motors

The multiphase (polyphase) ac motors are normally employed in moderate- to high-power applications (e.g., more than 5 hp). In low-power applications (e.g., motors used in household appliances such as refrigerators, dishwashers, food processors, and hair dryers; tools such as saws, lawn mowers, and drills), single-phase ac motors are commonly used, for they have the advantages of simplicity and low cost.

The stator of a single-phase motor has only one set of drive windings (with two or more stator poles) excited by a single-phase ac supply. If the rotor is running close to the frequency of the line ac, this single phase can maintain the motor torque, operating as an induction motor. But a single phase is obviously not capable of starting the motor. To overcome this problem, a second coil that is out of phase from the first coil is used during the starting period and is turned off automatically once the operating speed is attained. The phase difference is obtained either through a difference in inductance for a given resistance in the two coils or by including a capacitor in the second coil circuit.

7.8 Synchronous Motors

Phase-locked servos and stepper motors can be considered synchronous motors because they run in synchronism with an external command signal (a pulse train) under normal operating conditions. The rotor of a synchronous ac motor rotates in synchronism with the rotating magnetic field generated by the stator windings. The generation principle of this rotating field is identical to that in an induction motor, as explained before. Unlike an induction motor, however, the rotor windings of a synchronous motor are energized by an external dc source. The rotor magnetic poles generated in this manner lock themselves with the rotating magnetic field generated by the stator and rotate at the same speed (synchronous speed). For this reason, synchronous motors are particularly suited for constant-speed applications under variable-load conditions. Synchronous motors with permanent magnet (e.g., samarium cobalt) rotors are also commercially available.

A schematic representation of the stator–rotor pair of a synchronous motor is shown in Figure 7.50. The dc voltage, which is required to energize the rotor windings, may come from

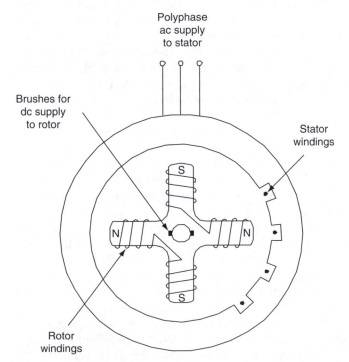

FIGURE 7.50
Schematic diagram of a stator–rotor configuration of a synchronous motor.

several sources. An independent dc supply, an external ac supply and a rectifier, or a dc generator driven by the synchronous motor itself, are three ways of generating the dc signal.

One major drawback of the synchronous ac motor is that an auxiliary starter is required to start the motor and bring its speed close to the synchronous speed. The reason for this is that in synchronous motors, the starting torque is virtually zero. To understand this, consider the starting conditions. The rotor is at rest and the stator field is rotating (at the synchronous speed). Consequently, there is 100% slip ($S = 1$). When, for example, an N pole of the rotating field in the stator is approaching an S pole in the rotor, the magnetic force tends to turn the rotor in the direction opposite to the rotating field. When the same N pole of the rotating field has just passed the rotor S pole, the magnetic force tends to pull the rotor in the same direction as the rotating field. These opposite interactions balance out, producing a zero net torque on the rotor. One method of starting a synchronous motor is by using a small dc motor. Once the synchronous motor reaches the synchronous speed, the dc motor is operated as a dc generator to supply power to the rotor windings. Alternatively, a small induction motor may be used to start the synchronous motor. A more desirable arrangement, which employs the principle of induction motor, is to include several sets of induction–motor-type rotor windings (cage-type or wound-type) in the synchronous motor rotor itself. In all these cases, the supply voltage to the rotor windings of the synchronous motor is disconnected during the starting conditions and is turned on only when the motor speed is close to the synchronous speed.

7.8.1 Control of a Synchronous Motor

Under normal operating conditions, the speed of a synchronous motor is completely determined by the frequency of the ac supply to the stator windings, because the motor speed is equal to the speed ω_f of the rotating field (see Equation 7.70). Hence, speed control can be achieved by the variable-frequency control method as described for an induction motor. In some applications of ac motors (both induction and synchronous types), clutch devices that link the motor to the driven load are used to achieve variable-speed control (e.g., using an eddy current clutch system that produces a variable coupling force through the eddy currents generated in the clutch). These dissipative techniques are quite wasteful and can considerably degrade the motor efficiency. Furthermore, heat removal methods would be needed to avoid thermal problems. Hence, they are not recommended for high-power applications where motor efficiency is a prime consideration, and for continous operation.

Note that unless a permanent magnet rotor is used, a synchronous motor would require a slip-ring and brush mechanism to supply the dc voltage to its rotor windings. This is a drawback that is not present in an induction motor.

The steady-state speed–torque curve of a synchronous motor is a straight line passing through the value of synchronous speed and parallel to the torque axis. But with proper control (e.g., frequency control), an ac motor can function as a servomotor. Conventionally, a servomotor has a linear torque–speed relationship, which can be realized by an ac servomotor with a suitable drive system. Applications of synchronous ac motors include steel rolling mills, rotary cement kilns, conveyors, hoists, process compressors, recirculation pumps in hydroelectric power plants, and, more recently, servomotors and robotics. Synchronous motors are particularly suitable in high-speed, high-power, and continous-operation applications where dc motors might not be appropriate. A synchronous motor can operate with a larger air gap between the rotor and the stator in comparison with an induction motor. This is an advantage for synchronous motors from the mechanical design point of view (e.g., bearing tolerances and rotor deflections due to thermal, static, and dynamic loads). Furthermore, rotor losses are smaller for synchronous motors than for induction motors.

7.9 Linear Actuators

Linear actuator stages are common in industrial motion applications. They may be governed by the same principles as the rotary actuators, but employing linear arrangements for the stator and the moving element, or a rotary motor with a rotary or linear motion transmission unit. Solenoids are typically on/off (or push/pull)-type linear actuators, and are commonly used in relays, valve actuators, switches, and a variety of other applications. Some useful types of linear actuators are presented in the following section.

7.9.1 Solenoid

The solenoid is a common rectilinear actuator, which consists of a coil and a soft iron core. When the coil is activated by a dc signal, the soft iron core becomes magnetized. This electromagnet can serve as an on/off (push/pull) actuator, for example, to move a ferromagnetic element (moving pole or plunger). The moving element is the load, which is typically restrained by a light spring and a damping element.

Solenoids are rugged and inexpensive devices. Common applications of solenoids include valve actuators, mechanical switches, relays, and other two-state positioning systems. An example of a relay is shown in Figure 7.51.

A relay of this type may be used to turn on and off devices such as motors, heaters, and valves in industrial systems. They may be controlled by a programmable logic controller (PLC). A time-delay relay provides a delayed on/off action with an adjustable time delay, as necessary in some process applications.

The percentage on time with respect to the total on/off period is the duty cycle of a solenoid (also see Section 7.4.3). A solenoid needs a sufficiently large current to move a load. There is a limit to the resulting magnetic force because the coil can saturate. In order to avoid associated problems, the ratings of the solenoid should match the needs of the load. There is another performance consideration. For long duty cycles, it is necessary

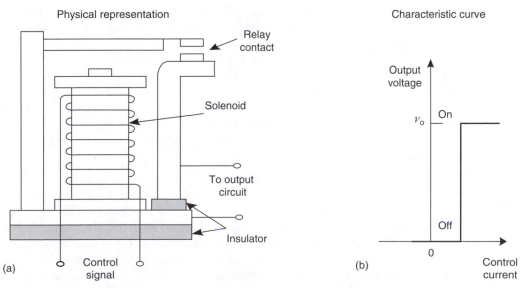

FIGURE 7.51
A solenoid operated relay. (a) Physical components. (b) Characteristic curve.

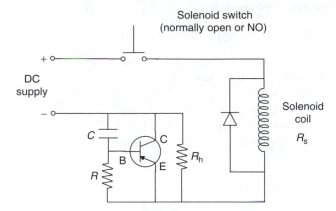

FIGURE 7.52
A hold-in circuit for a solenoid.

to maintain a current through the solenoid coil for a correspondingly long period. If the initial activating current of the solenoid is maintained over a long period, it heats up the coil and creates thermal problems. Apart from the loss of energy, this situation is undesirable because of safety issues, reduction in the coil life, and the need to have special means for cooling. A common solution is to incorporate a hold-in circuit, which reduces the current through the solenoid coil shortly after it is activated. A simple hold-in circuit is shown in Figure 7.52.

The resistance R_h is sufficiently large and comparable to resistance R_s of the solenoid coil. Initially, the capacitor C is fully discharged. Then the transistor is on (i.e., forward biased) and is able to conduct from the emitter (E) to collector (C). When the switch, which is normally open (denoted by NO), is turned on (i.e., closed), the dc supply sends a current through the solenoid coil (R_s), and the circuit is completed through the transistor. Since the transistor offers only a low resistance, the resulting current is large enough to actuate the solenoid. As the current flows through the circuit (while the switch is closed), the capacitor C becomes fully charged. The transistor becomes reverse biased due to the resulting voltage of the capacitor. This turns off the transistor. Then the circuit is completed not through the transistor but through the hold-in resistor R_h. As a result, the current through the solenoid drops by a factor of $R_s/(R_s + R_h)$. This lower current is adequate to maintain the state of the solenoid without overheating it.

A *rotary solenoid* provides a rotary push/pull motion. Its principle of operation is the same as that of a linear solenoid. Another type of solenoid is the proportional solenoid. It is able to produce a rectilinear motion in proportion to the current through the coil. Accordingly it acts as a linear motor. Proportional solenoids are particularly useful as valve actuators in fluid power systems; for example, as actuators for spool valves in hydraulic piston–cylinder devices (rectilinear actuators) and valve actuators for hydraulic motors (rotary actuators).

7.9.2 Linear Motors

It is possible to obtain a rectilinear motion from a rotary electromechanical actuator (motor) by employing an auxiliary kinematic mechanism (motion transmission device) such as a cam and follower, a belt and pulley, a rack and pinion, or a lead screw and nut (see Chapter 8). These devices inherently have problems of friction and backlash. Furthermore, they add inertia and flexibility to the driven load, thereby generating undesirable resonances and motion errors. Proper matching of the transmission inertia and the load inertia is essential.

Particularly, the transmission inertia should be less than the load inertia, when referred to one side of the transmission mechanism. Furthermore, extra energy is needed to operate the system against the inertia of the transmission mechanism.

For improved performance, direct rectilinear electromechanical actuators are desirable. These actuators operate according to the same principle as their rotary counterparts, except that flat stators and rectilinearly moving elements (in place of rotors) are employed. They come in different types:

1. Stepper linear actuators
2. DC linear actuators
3. AC linear actuators
4. Fluid (hydraulic and pneumatic) pistons and cylinders

In Chapter 6, we have indicated the principle of operation of a linear stepper motor (see Problem 6.16). Fluid pistons and cylinders are discussed later in the present chapter (see Section 7.10). Linear electric motors are also termed electric cylinders and are suitable as high-precision linear stages of motion applications. For example, a dc brushless linear motor operates similar to a rotary brushless motor and using a similar drive amplifier. Advanced rare earth magnets are used for the moving member, providing high force/mass ratio. The stator takes the form of a U-channel within which the moving member slides. Linear (sliding) bearings are standard. Since magnetic bearings can interfere with the force generating magnetic flux, air bearings are used in more sophisticated applications. The stator has the forcer coil for generating the drive magnetic field and Hall effect sensors for commutation. Since conductive material creates eddy-current problems, reinforced ceramic epoxy structures are used for the stator channel by leading manufacturers of linear motors. Applications of linear motors include traction devices, liquid-metal pumps, multiaxis positioning tables, Cartesian robots, conveyor mechanisms, and servovalve actuators.

7.10 Hydraulic Actuators

The ferromagnetic material in an electric motor saturates at some level of magnetic flux density (and the electric current, which generates the magnetic field). This limits the torque/mass ratio obtainable from an electric motor. Hydraulic actuators use the hydraulic power of a pressurized liquid. Since high pressures (on the order of 5000 psi) can be used, hydraulic actuators are capable of providing very high forces (and torques) at very high power levels simultaneously to several actuating locations in a flexible manner. The force limit of a hydraulic actuator can be an order of magnitude larger than that of an electromagnetic actuator. This results in higher torque/mass ratios than those available from electric motors, particularly at high levels of torque and power. This is a principal advantage of hydraulic actuators. The actuator mass considered here is the mass of the final actuating element, not including auxiliary devices such as those needed to pressurize and store the fluid. Another advantage of a hydraulic actuator is that it is quite stiff when viewed from the side of the load. This is because a hydraulic medium is mechanically stiffer than an electromagnetic medium. Consequently, the control gains required in a high-power hydraulic control system would be significantly less than the gains required in a comparable electromagnetic (motor) control system. Note that the stiffness of an actuator may be measured by the slope of the speed–torque (force) curve, and is representative of the speed of response (or, bandwidth). There are other advantages of fluid power systems. Electric motors generate heat. In continuous operation, then, the thermal problems can be serious, and special means of heat removal will be necessary. In a fluid power system, however, heat generated at the load can be quickly transferred to another location away from the load, by the hydraulic fluid itself, and effectively removed by means of a heat exchanger. Another advantage of fluid power systems is that they are self-lubricating and as a result, the friction in valves, cylinders, pumps, hydraulic motors, and other system components will be low and will not require external lubrication. Safety considerations will also be less because, for example, there is no possibility of spark generations as in motors with brush mechanisms. There are several disadvantages as well. Fluid power systems are more nonlinear than electrical actuator systems. Reasons for this include valve nonlinearities, fluid friction, compressibility, thermal effects, and generally nonlinear constitutive relations. Leakage can create problems. Fluid power systems tend to be noisier than electric motors. Synchronization of multi-actuator operations may be more difficult as well. Moreover, when the necessary accessories are included, fluid power systems are by and large more expensive and less portable than electrical actuator systems.

Fluid power systems with analog control devices have been in use in engineering applications since the 1940s. Smaller, more sophisticated, and less costly control hardware and microprocessor-based controllers were developed in the 1980s, making fluid power control systems as sophisticated, precise, cost-effective, and versatile as electromechanical control systems. Now, miniature fluid power systems with advanced digital control and electronics are used in numerous applications, directly competing with advanced dc and ac motion control systems. In addition, logic devices based on fluidics or fluid logic devices are preferred over digital electronics in some types of industrial applications. Applications of fluid power systems include vehicle steering and braking systems, active suspension systems, material handling devices, and industrial mechanical manipulators such as hoists, industrial robots, rolling mills, heavy-duty presses, actuators for aircraft control surfaces (ailerons, rudder, and elevators), ship steering and control devices, excavators, actuators for opening and closing of bridge spans, tunnel boring machines, food processing machines, reaction injection molding (RIM) machines, dynamic testing machines and heavy-duty shakers for structures and components, machine tools, ship building, and dynamic props, stage backgrounds, and structures in theatres and auditoriums.

7.10.1 Components of a Hydraulic Control System

A schematic diagram of a basic hydraulic control system is shown in Figure 7.53a. A view of a practical fluid power system is shown in Figure 7.53b. The hydraulic fluid (oil) is pressurized using a pump, which is driven by an ac motor. Typical fluids used are mineral oils or oil in water emulsions. These fluids have the desirable properties of self-lubrication, corrosion resistance, good thermal properties and fire resistance, environmental friendliness, and low compressibility (high stiffness for good bandwidth). Note that the motor converts electrical power into mechanical power, and the pump converts mechanical power into fluid power. In terms of through and across variable pairs, these power conversions can be expressed as

$$(i, v) \xrightarrow{\eta_m} (T, \omega) \xrightarrow{\eta_h} (Q, P)$$

in the usual notation. The conversion efficiency η_m of a motor is typically very high (over 90%), whereas the efficiency η_h of a hydraulic pump is not as good (about 60%), mainly

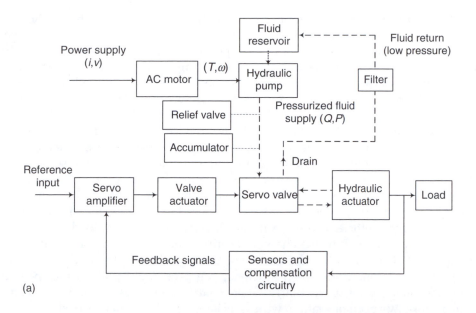

(a)

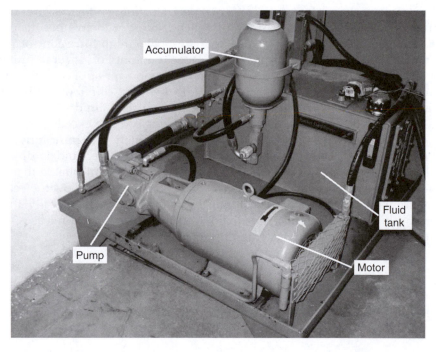

(b)

FIGURE 7.53
(a) Schematic diagram of a hydraulic control system. (b) An industrial fluid power system.

because of dissipation, leakage, and compressibility effects. Depending on the pump capacity, flow rates in the range of 1,000 to 50,000 gal/min (Note: 1 gal/min = 3.8 L/min) and pressures from 500 to 5000 psi (Note: 1 kPa = 0.145 psi) can be obtained. The pressure of the fluid from the pump is regulated and stabilized by a relief valve and an accumulator. A hydraulic valve provides a controlled supply of fluid into the actuator, controlling both the flow rate (including direction) and the pressure. In feedback control, this valve uses response signals (motion) sensed from the load, to achieve the desired response—hence the name servovalve. Usually, the servovalve is driven by an electric valve actuator, such as a torque motor or a proportional solenoid, which in turn is driven by the output from a servo amplifier. The servo amplifier receives a reference input command (corresponding to the desired position of the load) as well as a measured response of the load (in feedback). Compensation circuitry may be used in both feedback and forward paths to modify the signals so as to obtain the desired control action. The hydraulic actuator (typically a piston–cylinder device for rectilinear motions or a hydraulic motor for rotary motions) converts fluid power back into mechanical power, which is available to perform useful tasks (i.e., to drive a load). Note that some power in the fluid is lost at this stage. The low-pressure fluid at the drain of the hydraulic servovalve is filtered and returned to the reservoir, and is available to the pump.

One might argue that since the power that is required to drive the load is mechanical, it would be much more efficient to use a motor directly to drive that load. This issue has been addressed previously in this chapter. There are good reasons for using hydraulic power, however. For example, ac motors are usually difficult to control, particularly under variable-load conditions. Their efficiency can drop rapidly when the speed deviates from the rated speed, particularly when voltage control is used. They need gear mechanisms for low-speed operation, with associated problems such as backlash, friction, vibration, and mechanical loading effects. Special coupling devices are also needed. Hydraulic devices usually filter out high-frequency noise, which is not the case with ac motors. Thus, hydraulic systems are ideal for high-power, high-force control applications. In high-power applications, a single high-capacity pump or several pumps can be employed to pressurize the fluid. Furthermore, in low-power applications, several servo valve and actuator systems can be operated to perform different control tasks in a distributed control environment, using the same pressurized fluid supply. In this sense, hydraulic systems are very flexible. Hydraulic systems provide excellent speed–force (or torque) capability, variable over a wide range of speeds without significantly affecting the power-conversion efficiency, because the excess high-pressure fluid is diverted to the return line. Consequently, hydraulic actuators are far more controllable than ac motors. As noted before, hydraulic actuators also have an advantage over electromagnetic actuators from the point of view of heat transfer characteristics. Specifically, the hydraulic fluid promptly carries away any heat that is generated locally and releases it through a heat exchanger at a location away from the actuator.

7.10.2 Hydraulic Pumps and Motors

The objective of a hydraulic pump is to provide pressurized oil to a hydraulic actuator. Three common types of hydraulic pumps are

1. Vane pump
2. Gear pump
3. Axial piston pump

The pump type used in a hydraulic control system is not very significant, except for the pump capacity, in terms of the control functions of the system. But since hydraulic motors can be interpreted as pumps operating in the reverse direction, it is instructional to outline the operation of these three types of pumps.

A sliding-type vane pump is shown schematically in Figure 7.54. The vanes slide in the interior of the housing as they rotate with the rotor of the pump. They can move within

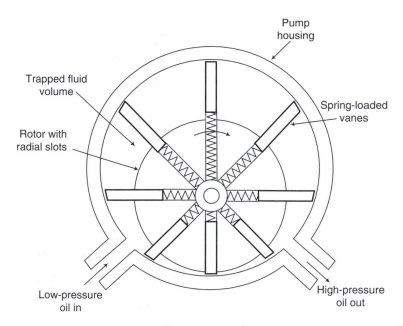

FIGURE 7.54
A hydraulic vane pump.

radial slots on the rotor, thereby maintaining full contact between the vanes and the housing. Springs or the pressurized hydraulic fluid itself may be used for maintaining this contact. The rotor is eccentrically mounted inside the housing. The fluid is drawn in at the inlet port as a result of the increasing volume between vane pairs as they rotate, in the first half of a rotation cycle. The oil volume trapped between two vanes is eventually compressed because of the decreasing volume of the vane compartment, in the second half of the rotation cycle. A pressure rise results from pushing the liquid volume into the high-pressure side and not allowing it to return to the low-pressure side of the pump, even when there is no significant compressibility in the liquid, when it moves from the low-pressure side to the high-pressure side. The typical operating pressure (at the outlet port) of these devices is about 2000 psi (13.8 MPa). The output pressure can be varied by adjusting the rotor eccentricity, because this alters the change in the compartment volume during a cycle. A disadvantage of any rotating device with eccentricity is the centrifugal force that is generated even while rotating at constant speed. Dynamic balancing is needed to reduce this problem.

The operation of an external gear hydraulic pump (or, simply a gear pump) is illustrated in Figure 7.55. The two identical gears are externally meshed. The inlet port is facing the gear enmeshing (retracting) region. Fluid is drawn in and trapped between the pairs of teeth in each gear, in rotation. This volume of fluid is transported around by the two gear wheels into the gear meshing region, at the pump outlet. Here, it undergoes an increase in pressure, as in the vane pump, as a result of forcing the fluid into the high-pressure side. Only moderate to low pressures can be realized by gear pumps (about 1000 psi or 7 MPa, maximum), because the volume changes that take place in the enmeshing and meshing regions are small (unlike in the vane pumps) and because fluid leakage between teeth and housing can be considerable. Gear pumps are robust and low-cost devices, however, and they are probably the most commonly used hydraulic pumps.

A schematic diagram of an axial piston hydraulic pump is shown in Figure 7.56. The chamber barrel is rigidly attached to the drive shaft. The two pistons themselves rotate with the chamber barrel, but since the end shoes of the pistons slide inside a slanted

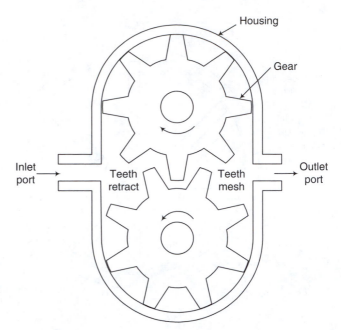

FIGURE 7.55
A hydraulic gear pump.

(skewed) slot, which is stationary, the pistons simultaneously undergo a reciprocating motion as well in the axial direction. As the chamber opening reaches the inlet port of the pump housing, fluid is drawn in because of the increasing volume between the piston

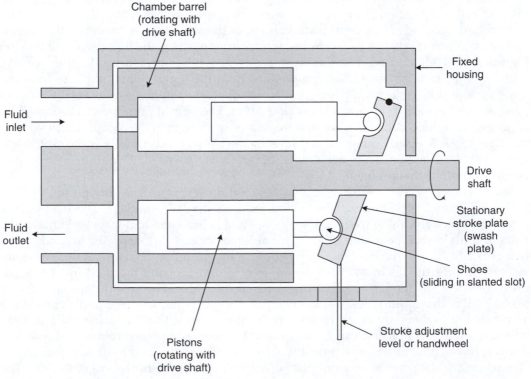

FIGURE 7.56
An axial piston hydraulic pump.

head and the chamber. This fluid is trapped and transported to the outlet port while undergoing compression as a result of the decreasing volume inside the chamber due to the axial motion of the piston. Fluid pressure increases in this process. High outlet pressures (4000 psi or 27.6 MPa, or more) can be achieved using piston pumps. As shown in Figure 7.56, the piston stroke can be increased by increasing the inclination angle of the stroke plate (slot). This, in turn, increases the pressure ratio of the pump. A lever mechanism is usually available to adjust the piston stroke. Piston pumps are relatively expensive.

The efficiency of a hydraulic pump is given by the ratio of the output fluid power to the motor mechanical power; thus,

$$\eta_p = \frac{PQ}{\omega T},$$
(7.100)

where P is the pressure increase in the fluid, Q is the fluid flow rate, ω is the rotating speed of the pump, and T is the drive torque to the pump.

7.10.3 Hydraulic Valves

Fluid valves can perform three basic functions:

1. Change the flow direction
2. Change the flow rate
3. Change the fluid pressure

The valves that accomplish the first two functions are termed flow-control valves. The valves that regulate the fluid pressure are termed pressure-control valves. A simple relief valve regulates pressure, whereas the poppet valve, gate valve, and globe valve are on/off flow-control valves. Some examples are shown in Figure 7.57. The directional valve (or check valve) shown in Figure 7.57a allows the fluid flow in one direction and blocks it in the opposite direction. The spring provides sufficient force for the ball to return to the seat when there is no fluid flow. It does not need to sustain any fluid pressure, and hence its stiffness is relatively low. A check valve falls into the category of flow control valves. Figure 7.57b shows a poppet valve. It is normally in the closed position, with the ball completely seated to block the flow. When the plunger is pushed down, the ball moves with it, allowing fluid flow through the seat opening. This on/off valve is bidirectional, and may be used to permit fluid flow in either direction. The relief valve shown in Figure 7.57c is in the closed condition under normal conditions. The spring force, which closes the valve (by seating the ball) is adjustable. When the fluid pressure (in a container or a pipe to which the valve is connected) rises above a certain value, as governed by the spring force, the valve opens thereby letting the fluid out through vent, which may be recirculated in the system. In this manner, the pressure of the system is maintained at a nearly constant level. Typically, an accumulator is used in conjunction with a relief valve, to take up undesirable pressure fluctuations and to stabilize the system. Valves are classified by the number of flow paths present under operating conditions. For example, a four-way valve has four ways in which flow can enter and leave the valve. In high-power fluid systems, two valve stages consisting of a pilot valve and a main valve may be used. Here, the pilot valve is a low-capacity, low-power valve, which operates the higher-capacity main valve.

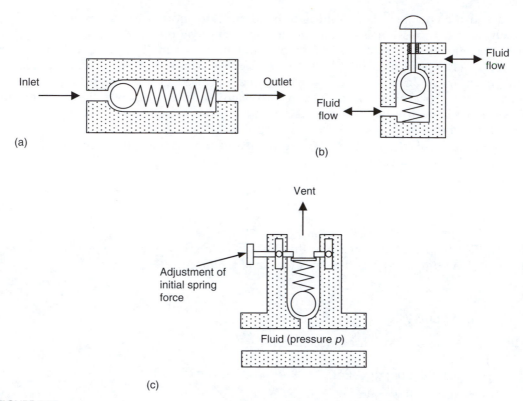

FIGURE 7.57
(a) A check valve (directional valve). (b) A poppet valve (an on/off valve). (c) A relief valve (a pressure-regulating valve).

7.10.3.1 Spool Valve

Spool valves are used extensively in hydraulic servo systems. A schematic diagram of a four-way spool valve is shown in Figure 7.58a. This is commonly called a servovalve because motion feedback is used by it to control the motion of a hydraulic actuator. The moving unit of the valve is called the spool. It consists of a spool rod and one or more expanded regions (or lobes), which are called lands. Input displacement (U) applied to the spool rod, using an actuator (torque motor or proportional solenoid), regulates the flow rate (Q) to the main hydraulic actuator as well as the corresponding pressure difference (P) available to the actuator. If the land length is larger than the port width (Figure 7.58b), it is an overlapped land. This introduces a dead zone in the neighborhood of the central position of the spool, resulting in decreased sensitivity and increased stability problems. Since it is virtually impossible to exactly match the land size with the port width, the underlapped land configuration (Figure 7.58c) is commonly employed. In this case, there is a leakage flow, even in the fully closed position, which decreases the efficiency and increases the steady-state error of the hydraulic control system. For accurate operation of the valve, the leakage should not be excessive. The direct flow at various ports of the valve and the leakage flows between the lands and the valve housing should be included in a realistic analysis of a spool valve. For small displacements δU about an operating point, the following linearized equations can be written. Since the flow rate Q_2 into the actuator increases as U increases and it decreases as P_2 increases, we have

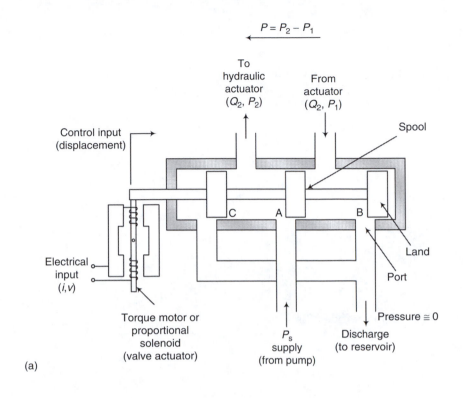

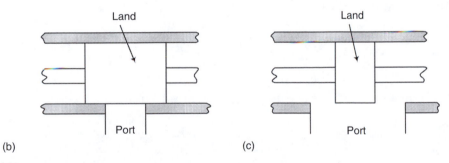

FIGURE 7.58
(a) A four-way spool valve. (b) An overlapped land. (c) An underlapped land.

$$\delta Q_2 = k_q \delta U - k_c' \delta P_2. \tag{7.101}$$

Similarly, since the flow rate Q_1 from the actuator increases with both U and P_1, we have

$$\delta Q_1 = k_q \delta U + k_c' \delta P_1. \tag{7.102}$$

The gains k_q and k_c' will be defined later.

In fact, if we disregard the compressibility of the fluid, $\delta Q_1 = \delta Q_2$, assuming that the hydraulic piston (actuator) is double-acting, with equal piston areas on the two sides of the actuator piston. We consider the general case where $Q_1 \neq Q_2$. Note, however, that the inlet port and the outlet port are assumed to have identical characteristics (hence, the associated coefficients in Equation 7.101 and Equation 7.102 are identical). By adding Equation 7.101 and Equation 7.102 and defining an average flow rate

$$Q = \frac{Q_1 + Q_2}{2} \tag{7.103}$$

and an equivalent flow–pressure coefficient

$$k_c = \frac{k_c'}{2} \tag{7.104}$$

we get

$$\delta Q = k_q \delta U - k_c \delta P, \tag{7.105}$$

where the flow gain is

$$k_q = \left(\frac{\partial Q}{\partial U}\right)_P \tag{7.106}$$

and the flow–pressure coefficient is

$$k_c = -\left(\frac{\partial Q}{\partial P}\right)_U. \tag{7.107}$$

Note further that the pressure sensitivity is

$$k_p = \left(\frac{\partial P}{\partial U}\right)_Q = \frac{k_q}{k_c}. \tag{7.108}$$

To obtain Equation 7.108, we use the well-known result from calculus:

$$\delta Q = \left(\frac{\partial Q}{\partial U}\right)_P \delta U + \left(\frac{\partial Q}{\partial P}\right)_U \delta P.$$

Since $\delta P/\delta U \to \partial P/\partial U$ as $\delta Q \to 0$, we have

$$\left(\frac{\partial P}{\partial U}\right)_Q = -\left(\frac{\partial Q}{\partial U}\right)_P \Big/ \left(\frac{\partial Q}{\partial P}\right)_U. \tag{7.109}$$

Equation 7.108 directly follows from Equation 7.109.

A valve can be actuated by several methods; for example, manual operation, the use of mechanical linkages connected to the drive load, and the use of electromechanical actuators such as solenoids and torque motors (or force motors). Regular solenoids are suitable for on/off control applications, and proportional solenoids and torque motors are used in continuous control. For precise control applications, electromechanical actuation of the valve (with feedback for servo operation) is preferred.

Large valve displacements can saturate a valve because of the nonlinear nature of the flow relations at the valve ports. Several valve stages may be used to overcome this saturation problem, when controlling heavy loads. In this case, the spool motion of the first stage (pilot stage) is the input motion. It actuates the spool of the second stage, which acts as a hydraulic amplifier. The fluid supply to the main hydraulic actuator, which drives the load, is regulated by the final stage of a multistage valve.

7.10.3.2 *Steady-State Valve Characteristics*

Although the linearized valve Equation 7.105 is used in the analysis of hydraulic control systems, it should be noted that the flow equations of a valve are quite nonlinear. Consequently, the valve constants k_q and k_c change with the operating point. Valve constants can be determined either by experimental measurements or by using an accurate nonlinear model. Now, we establish a reasonably accurate nonlinear relationship relating the (average) flow rate Q through the main hydraulic actuator and the pressure difference (load pressure) P provided to the hydraulic actuator.

Assume identical rectangular ports at the supply and discharge points in Figure 7.58a. When the valve lands are in the neutral (central) position, we set $U = 0$. We assume that the lands perfectly match the ports (i.e., no dead zone or leakage flows due to clearances). The positive direction of U is taken as shown in Figure 7.58a. For this positive configuration, the flow directions are also indicated in the figure. The flow equations at ports A and B are

$$Q_2 = Ubc_d \sqrt{\frac{2(P_s - P_2)}{\rho}}, \tag{7.110}$$

$$Q_1 = Ubc_d \sqrt{\frac{2P_1}{\rho}}, \tag{7.111}$$

where b is the land width, c_d is the discharge coefficient at each port, ρ is the density of the hydraulic fluid, and P_s is the supply pressure of the hydraulic fluid.

Note that in Equation 7.111 the pressure at the discharge end is taken to be zero. For steady-state operation, we use

$$Q_1 = Q_2 = Q. \tag{7.112}$$

Now, squaring Equation 7.110 and Equation 7.111 and adding, we get

$$2Q^2 = 2(Ubc_d)^2 \frac{(P_s - P)}{\rho},$$

where the pressure difference supplied to the hydraulic actuator is denoted by

$$P = P_2 - P_1. \tag{7.113}$$

Consequently,

$$Q = Ubc_d \sqrt{\frac{P_s - P}{\rho}} \quad \text{for } U > 0. \tag{7.114}$$

When $U < 0$, the flow direction reverses; furthermore, port A is now associated with P_1 (not P_2) and port C is associated with P_2. It follows that Equation 7.114 still holds, except that P_2-P_1 is replaced by P_1-P_2. Hence,

$$Q = Ubc_d\sqrt{\frac{P_s + P}{\rho}} \quad \text{for } U < 0. \tag{7.115}$$

Combining Equation 7.114 and Equation 7.115, we have

$$Q = Ubc_d\sqrt{\frac{P_s - P\mathrm{sgn}(U)}{\rho}}. \tag{7.116}$$

This can be written in the nondimensional form

$$\frac{Q}{Q_{max}} = \frac{U}{U_{max}}\sqrt{1 - \frac{P}{P_s}\mathrm{sgn}\left(\frac{U}{U_{max}}\right)}, \tag{7.117}$$

where $U_{max} = $ maximum valve opening (>0), and

$$Q_{max} = U_{max}bc_d\sqrt{\frac{P_s}{\rho}}. \tag{7.118}$$

Equation 7.117 is plotted in Figure 7.59. As with the speed–torque curve for a motor, it is possible to obtain the valve constants k_q and k_c defined by Equation 7.106 and Equation 7.107, from the curves given in Figure 7.59 for various operating points. For better accuracy, however, experimentally determined valve characteristic curves should be used.

7.10.4 Hydraulic Primary Actuators

Rotary hydraulic actuators (hydraulic motors) operate much like the hydraulic pumps discussed earlier, except that the direction flow is reversed and the mechanical power is delivered by the shaft, rather than taken in. High-pressure fluid enters the actuator. As it passes through the hydraulic motor, the fluid power is used up in turning the rotor, and the pressure is dropped. The low-pressure fluid leaves the motor en route to the reservoir. One of the more efficient rotary hydraulic actuators is the axial piston motor, quite similar in construction to the axial piston pump shown in Figure 7.56.

The most common type of rectilinear hydraulic actuator, however, is the hydraulic ram (or piston–cylinder actuator). A schematic diagram of such a device is shown in Figure 7.60. This is a double-acting actuator because the fluid pressure acts on both sides of the piston. If the fluid pressure is present only on one side of the piston, it is termed a single-acting actuator. Single-acting piston–cylinder (ram) actuators are also in common use for their simplicity and the simplicity of the other control components such as servovalves that are needed, although they have the disadvantage of asymmetry. The

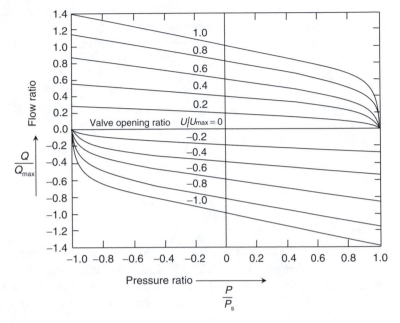

FIGURE 7.59
Steady-state characteristics of a four-way spool valve.

fluid flow at the ports of a hydraulic actuator is regulated typically by a spool valve. This valve may be operated by a pilot valve (e.g., a flapper valve).

To obtain the equations for the actuator shown in Figure 7.60, we note that the flow rate Q into a chamber depends primarily on two factors:

1. Increase in chamber volume
2. Increase in pressure (compressibility effect of the fluid)

When a piston of area A moves through a distance Y, the flow rate due to the increase in chamber volume is $\pm A\dot{Y}$. Now, with an increase in pressure δP, the volume of a given

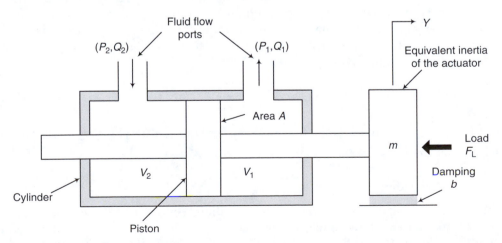

FIGURE 7.60
Double-acting piston–cylinder hydraulic actuator.

fluid mass would decrease by the amount $[-(\partial V/\partial P)\delta P]$. As a result, an equal volume of new fluid would enter the chamber. The corresponding rate of flow is $[-(\partial V/\partial P)(dP/dt)]$. Since the bulk modulus (isothermal, or at constant temperature) is given by

$$\beta = -V\frac{\partial P}{\partial V} \tag{7.119}$$

the rate of flow due to the rate of pressure change is given by $[(V/\beta)(dP/dt)]$. Using these facts, the fluid conservation (i.e., flow continuity) equations for the two sides of the actuator chamber in Figure 7.60 can be written as

$$Q_2 = A\frac{dY}{dt} + \frac{V_2}{\beta}\frac{dP_2}{dt}, \tag{7.120}$$

$$Q_1 = A\frac{dY}{dt} - \frac{V_1}{\beta}\frac{dP_1}{dt}. \tag{7.121}$$

For a realistic analysis, leakage flow rate terms (for leakage between piston and cylinder, and between piston rod and cylinder) should be included in Equation 7.120 and Equation 7.121. For a linear analysis, these leakage flow rates can be taken as proportional to the pressure difference across the leakage path. Note, further, that V_1 and V_2 can be expressed in terms of Y, as follows:

$$V_1 + V_2 = V_o, \tag{7.122}$$

$$V_1 - V_2 = V_o' + 2AY, \tag{7.123}$$

where V_o and V_o' are constant volumes, which depend on the cylinder capacity and on the piston position when $Y = 0$, respectively. Now, for incremental changes about the operating point $V_1 = V_2 = V$, Equation 7.120 and Equation 7.121 can be written as

$$\delta Q_2 = A\frac{d\delta Y}{dt} + \frac{V}{\beta}\frac{d\delta P_2}{dt}, \tag{7.124}$$

$$\delta Q_1 = A\frac{d\delta Y}{dt} - \frac{V}{\beta}\frac{d\delta P_1}{dt}. \tag{7.125}$$

Note that the total-value Equation 7.120 and Equation 7.121 are already linear for constant V. However, since the valve equation is nonlinear, and since V is not a constant, we should use the incremental-value Equation 7.124 and Equation 7.125 instead of the total-value equations, in a linear model. Adding Equation 7.124 and Equation 7.125 and dividing by 2, we get the hydraulic actuator equation

$$\delta Q = A\frac{d\delta Y}{dt} + \frac{V}{2\beta}\frac{d\delta P}{dt}, \tag{7.126}$$

where $Q = \dfrac{Q_1 + Q_2}{2}$ is the average flow into the actuator and $P = P_2 - P_1$ is the pressure difference on the piston of the actuator.

7.10.5 Load Equation

So far, we have obtained the linearized valve equation (Equation 7.105) and the linearized actuator equation (Equation 7.126). It remains to determine the load equation, which depends on the nature of the load that is driven by the hydraulic actuator. We may represent the load by a load force F_L, as shown in Figure 7.60. Note that F_L is a dynamic term, which may represent such effects as flexibility, inertia, and the dissipative effects of the load. In addition, the inertia of the moving parts of the actuator is modeled as a mass m, and the energy dissipation effects associated with these moving parts are represented by an equivalent viscous damping constant b. Accordingly, Newton's second law gives

$$m\frac{d^2Y}{dt^2} + b\frac{dY}{dt} = A(P_2 - P_1) - F_L. \tag{7.127}$$

This equation is also linear already. Again, since the valve equation is nonlinear, to be consistent, we should consider incremental motions δY about an operating point. Consequently, we have

$$m\frac{d^2\delta Y}{dt^2} + b\frac{d\delta Y}{dt} = A\delta P - \delta F_L, \tag{7.128}$$

where, as before,

$$P = P_2 - P_1.$$

If the active areas on the two sides of the piston are not equal, a net imbalance force would exist. This could lead to unstable response under some conditions.

7.11 Hydraulic Control Systems

The main components of a hydraulic control system are

1. A servovalve
2. A hydraulic actuator
3. A load
4. Feedback control elements

We have obtained linear equations for the first three components as Equation 7.105, Equation 7.126, and Equation 7.128. Now we rewrite these equations, denoting the incremental variables about an operating point by lowercase letters.

$$\text{Valve: } q = k_q u - k_c p \tag{7.129}$$

$$\text{Hydraulic actuator: } q = A\frac{dy}{dt} + \frac{V}{2\beta}\frac{dp}{dt} \tag{7.130}$$

$$\text{Load: } m\frac{d^2y}{dt^2} + b\frac{dy}{dt} = Ap - f_L. \tag{7.131}$$

The feedback elements depend on the specific feedback control method that is employed. We will revisit this aspect of a hydraulic control system later. Equation 7.129 through Equation 7.131 can be represented by the block diagram shown in Figure 7.61a. This is an open-loop control system because no external feedback elements have been used together

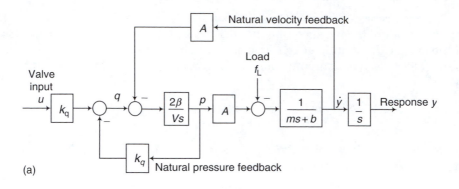

(a)

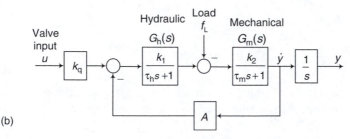

(b)

FIGURE 7.61
(a) Block diagram for an open-loop hydraulic control system. (b) An equivalent block diagram.

with response sensing. Note, however, the presence of a natural pressure feedback path and a natural velocity feedback path, which are inherent to the dynamics of the open-loop system.

The block diagram can be reduced to the equivalent form shown in Figure 7.61b. To obtain this equivalent representation, combine the first two summing junctions and then obtain the equivalent transfer function for the pressure feedback loop. This equivalent transfer function can be obtained using the relationship for reducing a feedback control system:

$$G_h = \frac{G}{1 + GH},\tag{7.132}$$

where G is the forward transfer function and H is the feedback transfer function. In the present case,

$$G = \frac{2\beta}{Vs}$$

and

$$H = k_c.$$

Hence,

$$G_h = \frac{k_1}{\tau_h s + 1},\tag{7.133}$$

where the pressure gain parameter is

$$k_1 = \frac{1}{k_c} \qquad (7.134)$$

and the hydraulic time constant is

$$\tau_h = \frac{V}{2\beta k_c}. \qquad (7.135)$$

Note that the pressure gain k_1 is a measure of the load pressure p generated for a given flow rate q into the hydraulic actuator. The smaller the pressure coefficient k_c, the larger the pressure gain, as is clear from Equation 7.107. The hydraulic time constant increases with the volume of the actuator fluid chamber and decreases with the bulk modulus of the hydraulic fluid. This is to be expected because the hydraulic time constant depends on the compressibility of the hydraulic fluid.

The mechanical transfer function of the hydraulic actuator is represented by

$$G_m = \frac{k_2}{\tau_m s + 1}, \qquad (7.136)$$

where the mechanical time constant is given by

$$\tau_m = \frac{m}{b} \qquad (7.137)$$

and $k_2 = 1/b$. Typically, the mechanical time constant is the dominant time constant, since it is usually larger than the hydraulic time constant.

Example 7.16

A model of the automatic gage control (AGC) system of a steel rolling mill is shown in Figure 7.62. The rollers are pressed using a single-acting hydraulic actuator with valve displacement u. The rollers are displaced through y, thereby pressing the steel that is rolled. For a given y, the rolling force F is completely known from the steel parameters.

1. Identify the inputs and the controlled variable in this control system.
2. In terms of the variables and system parameters indicated in Figure 7.62, write dynamic equations for the system, including valve nonlinearities.
3. What is the order of the system? Identify the response variables.
4. Draw a block diagram for the system, clearly indicating the hydraulic actuator with valve, the mechanical structure of the mill, inputs, and the controlled variable.
5. What variables would you measure (and feedback through suitable controllers) in order to improve the performance of the control system, using feedback control?

Solution

Part 1:
Valve displacement u and rolling force F are inputs. Roll displacement y is the controlled variable (output, response).

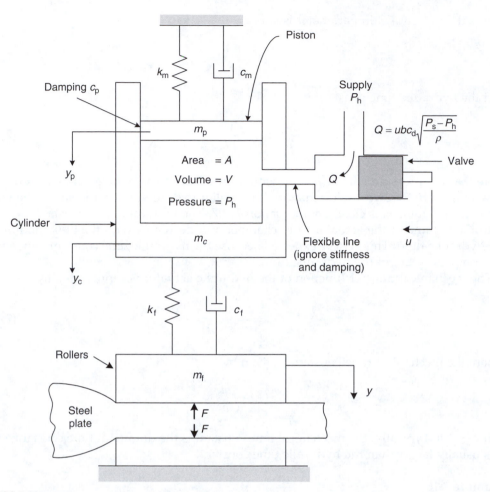

FIGURE 7.62
Automatic gage control (AGC) system of a steel rolling mill.

Part 2:
The mechanical-dynamic equations are

$$m_p\ddot{y}_p = -k_m y_p - c_m \dot{y}_p - c_p(\dot{y}_p - \dot{y}_c) - AP_h, \tag{i}$$

$$m_c\ddot{y}_c = -k_r(y_c - y) - c_r(\dot{y}_c - \dot{y}) - c_p(\dot{y}_c - \dot{y}_p) + AP_h, \tag{ii}$$

$$m_r\ddot{y} = -k_r(y - y_c) - c_r(\dot{y} - \dot{y}_c) - F. \tag{iii}$$

Note that the static forces balance and the displacements are measured from the corresponding equilibrium configuration, so that gravity terms do not enter into the equations.

The hydraulic actuator equation is derived as follows. For the valve, with the usual notation, the flow rate is given by

$$Q = buc_d\sqrt{\frac{P_s - P_h}{\rho}}.$$

For the piston and cylinder,

$$Q = A(\dot{y}_c - \dot{y}_p) + \frac{V}{\beta}\frac{dP_h}{dt}.$$

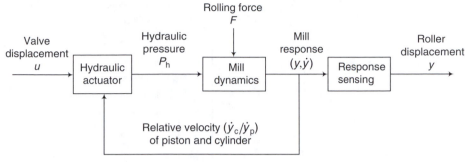

FIGURE 7.63
Block diagram for the hydraulic control system of a steel rolling mill.

Hence,

$$\frac{V}{\beta}\frac{dP_h}{dt} = A(\dot{y}_c - \dot{y}_p) + buc_d\sqrt{\frac{P_s - P_h}{\rho}}. \tag{iv}$$

Part 3:
There are three second-order differential Equation (i) through Equation (iii) and one first-order differential Equation (iv). Hence, the system is seventh order. The response variables are the displacements y_p, y_c, y and the pressure P_h.

Part 4:
A block diagram for the hydraulic control system of the steel rolling mill is shown in Figure 7.63.

Part 5:
The hydraulic pressure P_h and the roller displacement y are the two response variables, which can be conveniently measured and used in feedback control. The rolling force F may be measured and fed forward, but this is somewhat difficult in practice.

Example 7.17
A single-stage pressure control valve is shown in Figure 7.64. The purpose of the valve is to keep the load pressure P_L constant. Volume rates of flow, pressures, and the volumes of fluid subjected to those pressures are indicated in the figure. The mass of the spool and appurtenances is m, the damping constant of the damping force acting on the moving parts is b, and the effective bulk modulus of oil is β. The accumulator volume is V_a. The flow into the valve chamber (volume V_c) is through an orifice. This flow may be taken as proportional to the pressure drop across the orifice, the constant of proportionality being denoted by k_o. A compressive spring of stiffness k restricts the spool motion. The initial spring force is set by adjusting the initial compression y_o of the spring.

1. Identify the reference input, the primary output, and a disturbance input for the valve system.
2. By making linearization assumptions and introducing any additional parameters that might be necessary, write equations to describe the system dynamics.
3. Set up a block diagram for the system, showing various transfer functions.

Solution

Part 1:
1. Input setting $= y_o$
2. Primary response (controlled variable) $= P_L$
3. Disturbance input $= Q_L$

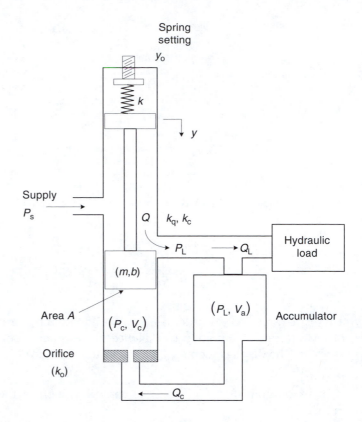

FIGURE 7.64
A single-stage pressure control valve.

Part 2:
Suppose that the valve displacement y is measured from the static equilibrium position of the system. The equation of motion for the valve spool device is

$$m\ddot{y} = -b\dot{y} - k(y - y_o) + A(P_s - P_c). \tag{i}$$

The flow through the chamber orifice is given by

$$Q_c = k_o(P_L - P_c) = -A\frac{dy}{dt} + \frac{V_c}{\beta}\frac{dP_c}{dt}. \tag{ii}$$

The outflow Q from the spool port increases with y and decreases with the pressure drop $(P_L - P_s)$. Hence, the linearized flow equation is

$$Q = k_q y - k_c(P_L - P_s).$$

Note that k_q and k_c are positive constants, defined previously by Equation 7.106 and Equation 7.107.

The accumulator equation is

$$Q - Q_c - Q_L = \frac{V_a}{\beta}\frac{dP_L}{dt}.$$

Substituting for Q and Q_c, we have

$$k_q y - k_c(P_L - P_s) - k_o(P_L - P_c) - Q_L = \frac{V_a}{\beta}\frac{dP_L}{dt}$$

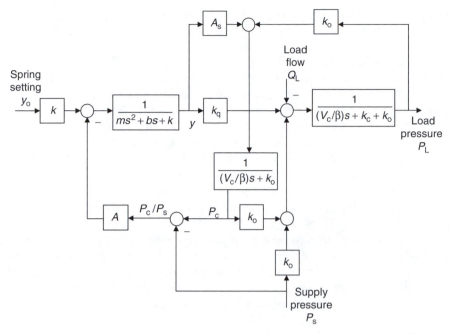

FIGURE 7.65
Block diagram for the single-stage pressure control valve.

or

$$k_q y - (k_c + k_o)P_L + (k_c P_s + k_o P_c) - Q_L = \frac{V_a}{\beta} \frac{dP_L}{dt}. \tag{iii}$$

The equations of motion are Equation (i) through Equation (iii).

Part 3:
Using Equation (i) through Equation (iii), the block diagram shown in Figure 7.65 can be obtained. Note in particular the "natural" feedback path of load pressure P_L. This feedback is responsible for the pressure control characteristic of the valve.

7.11.1 Feedback Control

In Figure 7.61a, we have identified two natural feedback paths that are inherent in the dynamics of the open-loop hydraulic control system. In Figure 7.61b, we have shown the time constants associated with these natural feedback modules. Specifically, we observe the following:

1. A pressure feedback path and an associated hydraulic time constant τ_h
2. A velocity feedback path and an associated mechanical time constant τ_m

The hydraulic time constant is determined by the compressibility of the fluid. The larger the bulk modulus of the fluid, the smaller the compressibility. This results in a smaller hydraulic time constant. Furthermore, τ_h increases with the volume of the fluid in the actuator chamber; hence, this time constant is related to the capacitance of the fluid as well. The mechanical time constant has its origin in the inertia and the energy dissipation (damping) in the moving parts of the actuator. As expected, the actuator becomes more sluggish as the inertia of the moving parts increases, resulting in an increased mechanical time constant.

These natural feedback paths usually provide a stabilizing effect to a hydraulic control system, but they are not adequate for satisfactory operation of the system. Specifically, this system, with natural feedback paths alone, does not represent a feedback control system. In particular note that the position of the actuator is provided through an integrator (see Figure 7.61). In open-loop operation, the position response steadily grows and displays an unstable behavior, in the presence of a slightest disturbance. Furthermore, the speed of response, which usually conflicts with stability, has to be adequate for proper performance. Consequently, it is necessary to include feedback control into the system. This is accomplished by measuring the response variables, and modifying the system inputs using them, according to some control law.

Schematic representation of a computer-controlled hydraulic system is shown in Figure 7.66. In addition to the motion (both position and speed) of the mechanical load, it is desirable to sense the pressures on the two sides of the piston of the hydraulic actuator, for feedback control. There are numerous laws of feedback control, which may be programmed into the control computer. Many of the conventional methods implement a combination of the following three basic control actions:

1. Proportional control (P)
2. Derivative control (D)
3. Integral control (I)

In proportional control, the measured response (or response error) is used directly in the control action. In derivative control, the measured response (or the response error) is differentiated before it is used in the control action. Similarly, in integral control, the response error is integrated and used in the control action. Modification of the measured responses to obtain the control signal is done in many ways, including electronic, digital, and mechanical means. For example, an analog hardware unit (termed a compensator or

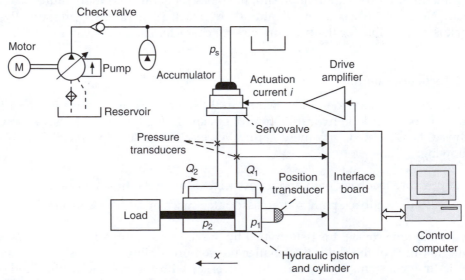

FIGURE 7.66
A computer-controlled hydraulic system.

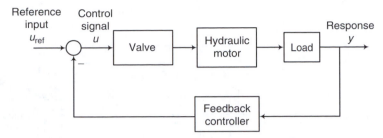

FIGURE 7.67
A closed-loop hydraulic control system.

controller), which consists of an electronic circuitry may be employed for this purpose. Alternatively, the measured signals, if they are analog, may be digitized and subsequently modified in a required manner through digital processing (multiplication, differentiation, integration, addition, etc.). This is the method used in digital control; either hardware control or software control may be used. The method represented in Figure 7.66 is the software approach.

Consider the feedback (closed-loop) hydraulic control system shown by the block diagram in Figure 7.67. In this case, a general controller is located in the feedback path. Then, a control law may be written as

$$u = u_{ref} - f(y),$$ (7.138a)

where $f(y)$ denotes the modifications made to the measured output y in order to form the control (error) signal u. The reference input u_{ref} is specified. Alternatively, if the controller is located in the forward path, as usual, the control law may be given by

$$u = f(u_{ref} - y).$$ (7.138b)

Mechanical components may be employed to obtain a robust control action.

Example 7.18
A mechanical linkage is employed as the feedback device for a servovalve of a hydraulic actuator. The arrangement is illustrated in Figure 7.68a. The reference input is u_{ref}, the input to the servovalve is u, and the displacement (response) of the actuator piston is y. A coupling element is used to join one end of the linkage to the piston rod. The displacement at this location of the linkage is x.

Show that rigid coupling gives proportional feedback action (Figure 7.68b). Now, if a viscous damper (damping constant b) is used as the coupling element and if a spring (stiffness k) is used to externally restrain the coupling end of the linkage (Figure 7.68c), show that the resulting feedback action is a lead compensation. Further, if the damper and the spring are interchanged (Figure 7.68d), what is the resulting feedback control action?

Solution
For all three cases of coupling, the relationship between u_{ref}, u, and x is the same. To derive this, we introduce the variable θ to denote the clockwise rotation of the linkage. With the linkage dimensions h_1 and h_2 defined as shown in Figure 7.68a, we have

$$u = u_{ref} + h_1\theta,$$
$$x = u_{ref} - h_2\theta.$$

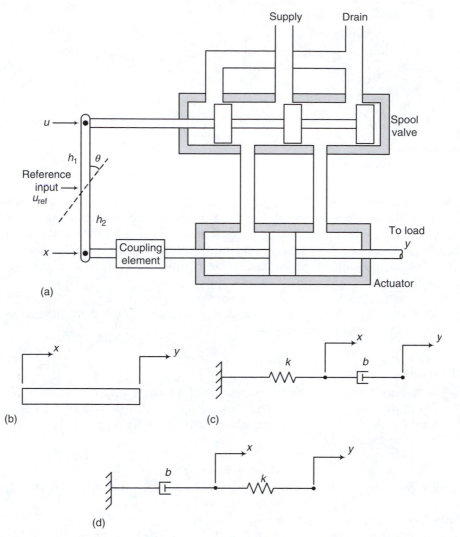

FIGURE 7.68
(a) A servovalve and actuator with mechanical feedback. (b) Rigid coupling (proportional feedback). (c) Damper–spring coupling (lead compensator). (d) Spring–damper coupling (lag compensator).

Now, by eliminating θ, we get

$$u = (r + 1)u_{\text{ref}} - rx,\tag{i}$$

where

$$r = h_1/h_2.\tag{ii}$$

For rigid coupling (Figure 7.68b),

$$y = x.$$

Hence, from Equation (i), we have

$$u = (r + 1)u_{\text{ref}} - ry.\tag{7.139}$$

Clearly, this is a proportional feedback control law.

Next, for the coupling arrangement shown in Figure 7.68c, by equating forces in the spring and the damper, we get

$$kx = b(\dot{y} - \dot{x}). \qquad \text{(iii)}$$

Introducing the Laplace variable s, we have the transfer-function relationship corresponding to Equation (iii):

$$x = \frac{bs}{bs + k}y. \qquad \text{(iv)}$$

By substituting Equation (iv) in Equation (i), we get

$$u = (r + 1)u_{\text{ref}} - \frac{rbs}{bs + k}y. \qquad (7.140)$$

Note that the feedback transfer function

$$G_c(s) = \frac{rbs}{bs + k} \qquad (7.141)$$

is a lead compensator, because the numerator provides a pure derivative action, which leads the denominator.

Finally, for the coupling arrangement shown in Figure 7.68d, we have

$$b\dot{x} = k(y - x). \qquad \text{(v)}$$

The corresponding transfer-function relationship is

$$x = \frac{k}{bs + k}y. \qquad \text{(vi)}$$

By substituting Equation (vi) in Equation (i), we get the transfer-function relationship for the feedback controller as

$$u = (r + 1)u_{\text{ref}} - \frac{rk}{bs + k}y. \qquad (7.142)$$

Note that in this case, the feedback transfer function is

$$G_c(s) = \frac{rk}{bs + k}. \qquad (7.143)$$

This is clearly a lag compensator because the denominator dynamics of the transfer function provide the lag action and the numerator has no dynamics (independent of s).

∎

Fluid power systems in general and hydraulic systems in particular are nonlinear. Nonlinearities have such origins as nonlinear physical relations of the fluid flow, compressibility, nonlinear valve characteristics, friction in the actuator (at the piston rings, which slide inside the cylinder) and the valves, unequal piston areas on the two sides of the actuator piston, and leakage. As a result accurate modeling of a fluid power system will be difficult, and a linear model will not represent the correct situation except near a small operating region. This situation may be addressed by using an accurate nonlinear model or a series of linear models for different operating regions. In either case, linear control laws (e.g., proportional, integral, and derivative (PID) actions) may not be adequate. This situation can be further exacerbated by factors such as parameter variations, unknown disturbances, and noise.

Many advanced control techniques have been applied to fluid power systems, in view of the limitations of such classical control techniques as PID. In one approach, an observer is used to estimate velocity and friction in the actuator, and a controller is designed to compensate for friction. Adaptive control is another advanced approach used in hydraulic control systems. In model-referenced adaptive control, the controller pushes the behavior of the hydraulic system towards a reference model. The reference model is designed to display the desired behavior of the physical system. Frequency-domain control techniques such as H-infinity control (H_∞ control) and quantitative feedback theory (QFT), where the system transfer function is shaped to realize a desired performance, have been studied. They are linear control techniques, which may not work perfectly when applied to a nonlinear system. Impedance control has been studied as well, with respect to hydraulic control systems. In impedance control, the objective is to realize a desired impedance function (Note: impedance = force/velocity, in the frequency domain) at the output of the control system, by manipulating the controller. These advanced techniques are beyond the scope of the present introductory treatment.

7.11.2 Constant-Flow Systems

So far, we have discussed only valve-controlled hydraulic actuators. There are two types of valve-controlled systems:

1. Constant-pressure systems
2. Constant-flow systems

Since there are four flow paths for a four-way spool valve, an analogy can be drawn between a spool valve–controlled hydraulic actuator and a Wheatstone bridge circuit (see Section 2.10), as shown in Figure 7.69. Each arm of the bridge corresponds to a flow path. As usual, P denotes pressure, which is an across variable analogous to voltage; and Q denotes the volume flow rate, which is a through variable analogous to current. The four fluid resistors R_i represent the resistances experienced by the fluid flow in the four paths of the valve. Note that these are variable resistors whose variation is governed by the spool movement (and hence the current of the valve actuator). When the spool moves to one side of the neutral (center) position, two of the resistors (say, R_2 and R_4) change due to the port opening and the remaining two resistors represent the leakage resistances (see Figure 7.58). The reverse is true when the spool moves in the opposite direction from the neutral position. The flow through the actuator is represented by a load resistance R_L, which is connected across the bridge.

In our discussion so far, we have considered only the constant-pressure system, in which the supply pressure P_s to the servovalve is maintained constant, but the corresponding supply flow rate Q_s is variable. This system is analogous to a constant-voltage bridge (see Chapter 2). In a constant-flow system, the supply flow Q_s is kept constant, but the corresponding pressure P_s is variable. This system is analogous to a constant-current bridge. Constant-flow operation requires a constant-flow pump, which may be more economical than a variable-flow pump. However, it is easier to maintain a constant pressure level by using a pressure regulator and an accumulator. As a result, constant pressure systems are more commonly used in practical applications.

Valve-controlled hydraulic actuators are the most common type used in industrial applications. They are particularly useful when more than one actuator is powered by

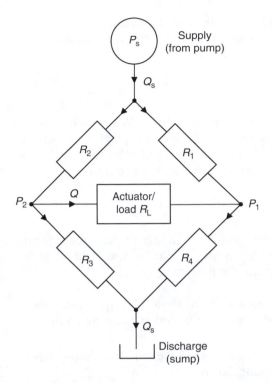

FIGURE 7.69
The bridge circuit representation of a four-way valve and an actuator load.

the same hydraulic supply. Pump-controlled actuators are gaining popularity, and are outlined next.

7.11.3 Pump-Controlled Hydraulic Actuators

Pump-controlled hydraulic drives are suitable when only one actuator is needed to drive a process. A typical configuration of a pump-controlled hydraulic-drive system is shown in Figure 7.70. A variable-flow pump is driven by an electric motor (typically, an ac motor). The pump feeds a hydraulic motor, which in turn drives the load. Control is provided by the flow control of the pump. This may be accomplished in several ways, for example, by controlling the pump stroke (see Figure 7.56) or by controlling the pump speed using a frequency-controlled ac motor. Typical hydraulic drives of this type can provide positioning errors less than $1°$ at torques in the range 25 to 250 N.m.

7.11.4 Hydraulic Accumulators

Since hydraulic fluids are quite incompressible, one way to increase the hydraulic time constant is to use an accumulator. An accumulator is a tank, which can hold excessive fluid during pressure surges and release this fluid to the system when the pressure slacks.

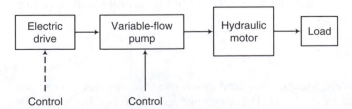

FIGURE 7.70
Configuration of a pump-controlled hydraulic-drive system.

In this manner, pressure fluctuations can be filtered out from the hydraulic system and the pressure can be stabilized. There are two common types of hydraulic accumulators:

1. Gas-charged accumulators
2. Spring-loaded accumulators

In a gas-charged accumulator, the top half of the tank is filled with air. When high-pressure liquid enters the tank, the air compresses, making room for the incoming liquid. In a spring-loaded accumulator, a movable piston, restrained from the top of the tank by a spring, is used in place of air. The operation of these two types of accumulators is quite similar.

7.11.5   Pneumatic Control Systems

Pneumatic control systems operate in a manner similar to hydraulic control systems. Pneumatic pumps, servovalves, and actuators are quite similar in design to their hydraulic counterparts. The basic differences include the following:

1. The working fluid is air, which is far more compressible than hydraulic oils. Hence, thermal effects and compressibility should be included in any meaningful analysis.
2. The outlet of the actuator and the inlet of the pump are open to the atmosphere (no reservoir tank is needed for the working fluid).

By connecting the pump (hydraulic or pneumatic) to an accumulator, the flow into the servovalve can be stabilized and the excess energy can be stored for later use. This minimizes undesirable pressure pulses, vibration, and fatigue loading. Hydraulic systems are stiffer and usually employed in heavy-duty control tasks, whereas pneumatic systems are particularly suitable for medium to low-duty tasks (supply pressures in the range of 500 kPa to 1 MPa). Pneumatic systems are more nonlinear and less accurate than hydraulic systems. Since the working fluid is air and since regulated high-pressure air lines are available in most industrial facilities and laboratories, pneumatic systems tend to be more economical than hydraulic systems. In addition, pneumatic systems are more environment-friendly and cleaner, and the fluid leakage does not cause a hazardous condition. However, they lack the self-lubricating property of hydraulic fluids. Furthermore, atmospheric air has to be filtered and any excess moisture removed before compressing. Heat generated in the compressor has to be removed as well.

  Both hydraulic and pneumatic control loops might be present in the same control system. For example, in a manufacturing work cell, hydraulic control can be used for parts transfer, positioning, and machining operations, and pneumatic control can be used for tool change, parts grasping, switching, ejecting, and single-action cutting operations. In a fish-processing machine, servo-controlled hydraulic actuators have been used for accurately positioning the cutter whereas pneumatic devices have been used for grasping and chopping of fish. We will not extend our analysis of hydraulic systems to include air as the working fluid. The reader may consult a book on pneumatic control for information on pneumatic actuators and valves.

7.11.6   Flapper Valves

Flapper valves, which are relatively inexpensive and operate at low-power levels, are commonly used in pneumatic control systems. This does not rule them out for hydraulic

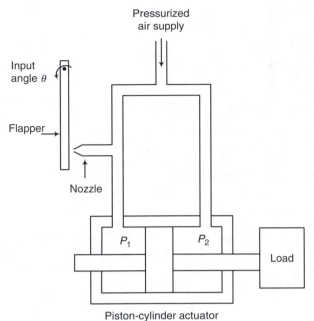

Piston-cylinder actuator

FIGURE 7.71
A pneumatic flapper valve system.

control applications, however, where they are popular in pilot valve stages. A schematic diagram of a single-jet flapper valve used in a piston–cylinder actuator is shown in Figure 7.71. If the nozzle is completely blocked by the flapper, the two pressures P_1 and P_2 will be equal, balancing the piston. As the clearance between the flapper and the nozzle increases, the pressure P_1 drops, thus creating an imbalance force on the piston of the actuator. For small displacements, a linear relationship between the flapper clearance and the imbalance force can be assumed.

Note that the operation of a flapper valve requires fluid leakage at the nozzle. This does not create problems in a pneumatic system. In a hydraulic system, however, this not only wastes power but also wastes hydraulic oil and creates a possible hazard, unless a collecting tank and a return line to the oil reservoir are employed. For more stable operation, double-jet flapper valves should be employed. In this case, the flapper is mounted symmetrically between two jets. The pressure drop is still highly sensitive to flapper motion, potentially leading to instability. To reduce instability problems, pressure feedback, using a bellows unit, can be employed.

A two-stage servovalve with a flapper stage and a spool stage is shown in Figure 7.72. Actuation of the torque motor moves the flapper. This changes the pressure in the two nozzles of the flapper in opposite directions. The resulting pressure difference is applied across the spool, which is moved as a result, which in turn moves the actuator as in the case of a single-stage spool valve. In the system shown in Figure 7.72, there is a feedback mechanism as well between the two stages of valve. Specifically, as the spool moves due to the flapper movement caused by the torque motor, the spool carries the flexible end of the flapper in the opposite direction to the original movement. This creates a back pressure in the opposite direction. Hence, this valve system is said to possess force feedback (more accurately, pressure feedback).

In general, a multistage servovalve uses several servovalves in series to drive a hydraulic actuator. The output of the first stage becomes the input to the second stage. As noted before, a common combination is between a hydraulic flapper valve and a

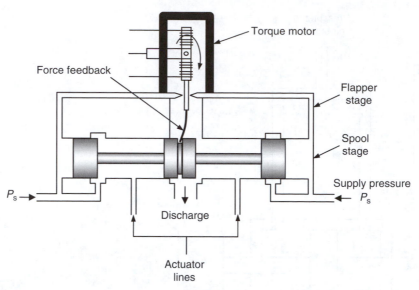

FIGURE 7.72
A two-stage servovalve with pressure feedback.

hydraulic spool valve, operating in series. A multistage servovalve is analogous to a multistage amplifier. The advantages of multistage servovalves are

1. A single-stage servovalve saturates under large displacements (loads). This may be overcome by using several stages, with each stage operating in its linear region. Hence, a large operating range (load variations) is possible without introducing excessive nonlinearities, particularly saturation.

2. Each stage filters out high-frequency noise, giving a lower overall noise-to-signal ratio.

The disadvantages are

1. They cost more and are more complex than single-stage servovalves.

2. Because of series connection of several stages, failure of one stage brings about failure of the overall system (a reliability problem).

3. Multiple stages decrease the overall bandwidth of the system (i.e., lower speed of response).

Example 7.19
Draw a schematic diagram to illustrate the incorporation of pressure feedback, using bellows, in a flapper-valve pneumatic control system. Describe the operation of this feedback control scheme, giving the advantages and disadvantages of this method of control.

Solution
One possible arrangement for external pressure feedback in a flapper valve is shown in Figure 7.73. Its operation can be explained as follows: If pressure P_1 drops, the bellows contract, thereby moving the flapper closer to the nozzle, thus increasing P_1. Hence, the

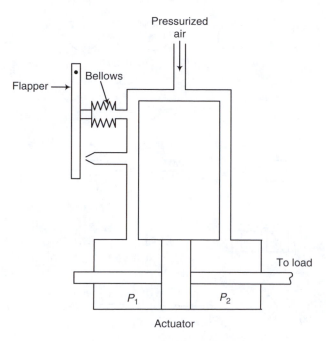

FIGURE 7.73
External pressure feedback for a flapper valve, using bellows.

bellows acts as a mechanical feedback device, which tends to regulate pressure disturbances. The advantages of such a device are:

1. It is a simple, robust, low-cost mechanical device.
2. It provides mechanical feedback control of pressure variations.

The disadvantages are:

1. It can result in a slow (i.e., low-bandwidth) system, if the inertia of the bellows is excessive.
2. It introduces a time delay, which can have a destabilizing effect, particularly at high frequencies.

7.11.7   Hydraulic Circuits

A typical hydraulic control system consists of several components such as pumps, motors, valves, piston–cylinder actuators, and accumulators, which are interconnected through piping. It is convenient to represent each component with a standard graphic symbol. The overall system can be represented by a hydraulic circuit diagram where the symbols for various components are joined by lines to denote flow paths. Circuit representations of some of the many hydraulic components are shown in Figure 7.74. A few explanatory comments would be appropriate. The inward solid pointers in the motor symbols indicate that a hydraulic motor receives hydraulic energy. Similarly, the outward pointers in the pump symbols show that a hydraulic pump gives out hydraulic energy. In general, the arrows inside a symbol show fluid flow paths. The external spring and arrow in the relief valve symbol shows that the unit is adjustable and spring restrained. There are three basic types of hydraulic line symbols. A solid line indicates a primary hydraulic

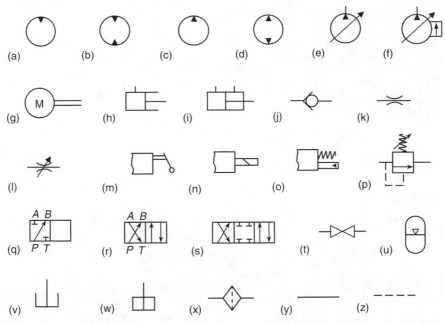

FIGURE 7.74

Typical graphic symbols used in hydraulic circuit diagrams. (a) Motor. (b) Reversible motor. (c) Pump. (d) Reversible pump. (e) Variable displacement pump. (f) Pressure-compensated variable displacement pump. (g) Electric motor. (h) Single-acting cylinder. (i) Double-acting cylinder. (j) Ball-and-seat check valve. (k) Fixed orifice. (l) Variable flow orifice. (m) Manual valve. (n) Solenoid-actuated valve. (o) Spring-centered pilot-controlled valve. (p) Relief valve (adjustable and pressure-operated). (q) Two-way spool valve. (r) Four-way spool valve. (s) Three-position four-way valve. (t) Manual shut-off valve. (u) Accumulator. (v) Vented reservoir. (w) Pressurized reservoir. (x) Filter. (y) Main fluid line. (z) Pilot line.

flow. A broken line with long dashes is a pilot line, which indicates the control of a component. For example, the broken line in the relief valve symbol indicates that the valve is controlled by pressure. A broken line with short dashes represents a drain line or leakage flow. In the spool valve symbols, P denotes the supply port (with pressure P_s) and T denotes the discharge port to the reservoir (with gage zero pressure). Finally, note that ports A and B of a four-way spool valve are connected to the two ports of a double-acting hydraulic cylinder (see Figure 7.58a).

7.12　Fluidics

The term fluidics is derived probably from fluid logic or perhaps fluid electronics. In fluidic control systems, the basic functions such as sensing, signal conditioning, and control are accomplished by the interaction of streams of fluid (liquid or gas). Unlike in mechanical systems, no moving parts are used in fluidic devices to accomplish these tasks. Of course, when sensing a mechanical motion by a fluidic sensor there will be a direct interaction with the moving object that is sensed. In addition, when actuating a mechanical load or valve using a fluidic device, there will be a direct interaction with a mechanical motion. These motions of the input devices and output devices should not be interpreted as mechanical motions within a fluidic device, but rather, mechanical motions external to the fluid interactions therein.

The fluidics technology was first introduced by the U.S. Army engineers in 1959 as a possible replacement for electronics in some control systems. The concepts themselves are quite old, and perhaps originated through electrical–hydraulic analogies where pressure is an across variable analogous to voltage, and flow rate is a through variable analogous to current. Since electronic circuitry is widely used for sensing, signal conditioning, and control tasks in hydraulic and pneumatic control systems, it was thought that the need for conversion between fluid flow and electrical variables could be avoided if fluidic devices were used for these tasks in such control systems, thereby bringing about certain economic and system performance benefits. Furthermore, fluidic devices are considered to have high reliability and can be operated in hostile environments (e.g., corrosive, radioactive, shock and vibration, high temperature) more satisfactorily than electronic devices. However, due to rapid advances in digital electronics with associated gains in performance and versatility, and reduction in cost, the anticipated acceptance of fluidics was not actually materialized in the 1960s and 1970s. Some renewed interest in fluidics was experienced in the 1980s, with applications primarily in the aircraft, aerospace, manufacturing, and process control industries. This section provides a brief introduction to the subject of fluidics. In view of its analogy to electronics and the use in mechanical control, fluidics is a topic that is quite relevant to the subject of control engineering.

7.12.1 Fluidic Components

Since fluidics was intended as a substitute for electronics, particularly in hydraulic and pneumatic control systems (i.e., in fluid power control systems), it is not surprising that much effort has gone into the development of fluidic devices that are analogous to electronic devices. Naturally, two types of fluidic components have been developed:

1. Analog fluidic components for analog systems
2. Digital fluidic components for logic circuits

Examples of analog components, which have been developed for fluidic systems are fluidic position sensor, fluidic rate sensor, fluidic accelerometer, fluidic temperature sensor, fluidic oscillator, fluidic resistor, vortex amplifier, jet-deflection amplifier, wall-attachment amplifier, fluidic summing amplifier, fluidic actuating amplifier, and fluidic modulator. A description of all such devices is beyond the scope of this book. Instead, we describe one or two representative devices in order to introduce the nature of fluidic components.

7.12.1.1 Logic Components

Examples of digital fluidic components are switches, flip-flops, and logic gates. Complex logic systems can be assembled by interconnecting these basic elements. As an example, consider the fluidic AND gate shown in Figure 7.75. The control inputs u_1 and u_2 represent the presence or absence of the high-speed fluid streams applied to the corresponding ports of the device. When only one control stream is present it passes through the drain channel aligned with it, due to the entrainment capability of the stream. When both control streams are present, there will be an interaction between the two streams, thereby producing a sizeable output stream y. Hence, there is an "AND" relationship between the output y and the inputs u_1 and u_2.

Operation of the fluid logic components depends on the wall-attachment phenomenon (or coanda effect). According to this phenomenon, a jet of fluid applied toward a wall

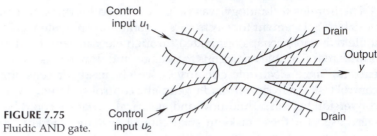

FIGURE 7.75
Fluidic AND gate.

tends to attach itself to the wall. If two walls are present, the jet will be attached to one of the walls depending on the conditions at the exit of the jet and the angles, which the walls make with the jet. Hence a switching action (i.e., attachment to one wall or the other) is created. The corresponding switching state can be considered a digital output.

A measure of the capability of a digital device is its fan-out. This is the number of similar devices that can be driven (or controlled) by the same digital component. Fluidic components have a reasonably high fan-out capability.

7.12.1.2 Fluidic Motion Sensors

A fluidic displacement sensor can be developed by using a mechanical vane to split a stream of incoming flow into two output streams. This is shown in Figure 7.76a. When the vane is centrally located, the displacement is zero ($\theta = 0$). Under these conditions, the pressure is the same at both output streams with the differential pressure $p_2 - p_1$ remaining zero. When the vane is not symmetrically located (corresponding to a nonzero displacement), the output pressures will be unequal. The differential pressure $p_2 - p_1$ provides both magnitude and direction of the displacement θ.

A fluidic angular speed sensor is shown in Figure 7.76b. The nozzle of the input stream is rotated at angular speed ω, which is to be measured. When $\omega = 0$, the stream travels straight in the axial direction of the input stream. When $\omega \neq 0$, the fluid particles emitting

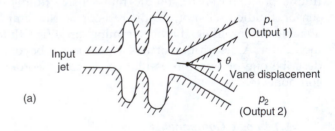

(a)

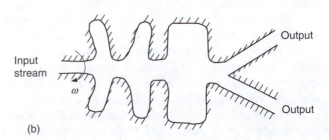

FIGURE 7.76
Fluidic motion sensors. (a) Angular displacement sensor. (b) Laminar angular speed sensor.

(b)

from the nozzle have a transverse speed (due to rotation) as well as an axial speed due to jet flow. Hence the fluid particles are deflected from the original (axial) path. This deflection is the cause of a pressure change at the output. Hence the output pressure change can be used as a measure of the angular speed.

There are many other ways to sense speed using a fluidic device. One type of fluidic angular speed sensor uses the vortex flow principle. In this sensor, the angular speed of the input device (object) is imparted on the fluid entering a vortex chamber at the periphery. In this manner a tangential speed is applied to the fluid particles, which move radially from the periphery toward the center of the vortex chamber. The resulting vortex flow will be such that the tangential speed becomes larger as the particles approach the center of the chamber (this follows by the conservation of momentum). Consequently, a pressure drop is experienced at the output (i.e., center of the chamber). The higher the angular speed imparted to the incoming fluid, the larger the pressure drop at the output. Hence the output pressure drop can be used as a measure of angular speed.

An angular speed sensor that is particularly useful in pneumatic systems is the wobble-plate sensor. This is a flapper valve–type device with two differential nozzles facing a wobble plate. The supply pressure is maintained constant. There are two output ports corresponding to the two nozzles. As the wobble plate rotates, the proximity of the plate to each nozzle changes periodically. The resulting fluctuation in the differential pressure can be used as a measure of the plate speed.

7.12.1.3 Fluidic Amplifiers

Fluidic amplifiers are used to apply a gain in pressure, flow, or power to a fluidic circuit. The corresponding amplifiers are analogous to voltage, current, and power amplifiers used in electronic circuits (see Chapter 2).

Many designs of fluidic amplifiers are available. Consider the jet-deflection amplifier shown in Figure 7.77. When the control input pressures p_{u1} and p_{u2} are equal, the supply stream passes through the amplifier with a symmetric flow. In this case, the output pressures p_{y1} and p_{y2} are equal. When $p_{u1} \neq p_{u2}$, the fluid stream is deflected to one side due to the nonzero differential pressure $\Delta p_u = p_{u1} - p_{u2}$. As a result, a nonzero differential pressure $\Delta p_y = p_{y2}; p_{y1}$ is created at the output. The pressure gain of the amplifier is given by

$$K_\mathrm{p} = \frac{\Delta p_y}{\Delta p_u}. \tag{7.144}$$

The gain K_p will be constant in a small operating range.

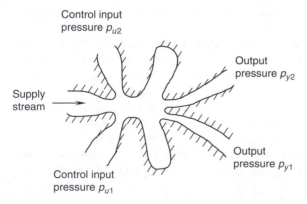

Control input
pressure p_{u2}

Output
pressure p_{y2}

Supply
stream

Control input
pressure p_{u1}

Output
pressure p_{y1}

FIGURE 7.77
Jet deflection amplifier.

7.12.2 Fluidic Control Systems

A fluidic control system is a control system that employs fluidic components to perform one or more functions such as sensing, signal conditioning, and control. The actuator is a symmetric configuration of a hydraulic piston–cylinder (ram) device. It is controlled using a pair of spool valves. Control signals to the spool valves are generated by appropriate fluidic components. Specifically, position of the load is measured using a fluidic displacement sensor, and the speed of the load is measured using a fluidic speed sensor. These signals are properly conditioned using fluidic amplifiers, compared with a reference signal using a fluidic summing amplifier, and the error signal is used through a fluidic interface amplifier to operate the actuator valve. This type of fluidic feedback control systems is useful in flight control, to operate the control surfaces (ailerons, rudders, and elevators) of an aircraft, particularly as a backup (e.g., for emergency maneuvering).

7.12.2.1 *Interfacing Considerations*

The performance of some of the early designs of fluidic control systems was disappointing because the overall control system did not function as expected whereas the individual fluidic components separately would function very well. Primary reason for this was the dynamic interactions between components and associated loading and impedance matching problems. Early designs of fluidic components, amplifiers in particular, did not have sufficient input impedances (see Chapter 2 for definitions of impedance parameters). Furthermore, output impedances were found to be higher than what was desired in order to minimize dynamic interaction problems. Much research and development effort has gone into improving the impedance characteristics of fluidic devices. Moreover, leakage problems are unavoidable when separate fluidic components are connected together using transmission lines. Modular laminated construction of fluidic systems has minimized these problems.

7.12.2.2 *Modular Laminated Construction*

Just as the integrated construction of electronic circuits has revolutionized the electronic technology, the modular laminated (or integrated) construction of fluidic systems has made a significant impact on the fluidic technology. The first generation fluidic components were machined out of metal blocks or moldings. Precise duplication and quality control in mass production were difficult with these types of components. Furthermore, the components were undesirably bulky. Since individual components were joined using flexible tubing, leakage at joints presented serious problems. Since the design of a fluidic control system is often a trial-and-error process of trying out different components, system design was costly, time-consuming, and tiresome, particularly because of component costs and assembly difficulties.

Many of these problems are eliminated or reduced with the modern day modular design of fluidic systems using component laminates. Individual fluidic components are precisely manufactured as thin laminates using a sophisticated stamping process. The system assembly is done by simply stacking and bonding together of various laminates (e.g., sensors, amplifiers, oscillators, resistors, modulators, vents, exhausts, drains, and gaskets) to form the required fluidic circuit. In the design stage, the stacks are clamped together without permanently bonding, and are tested. Then, design modifications can be implemented simply and quickly by replacing one or more of the laminates. Once an acceptable design is obtained, the stack is permanently bonded.

7.12.3 Applications of Fluidics

Fluidic components and systems have the advantages of small size and no moving parts, over conventional mechanical systems, which typically use bulky gear systems, clutches, linkages, cables, and chains. Furthermore, fluidic systems are highly reliable and are preferable to electronic systems, in hostile environments of explosives, chemicals, high temperature, radiation, shock, vibration, and electromagnetic interference (EMI). For these reasons, fluidic control systems have received a renewed interest in aircraft and aerospace applications, particularly as backup systems. In hydraulic and pneumatic control applications, the use of fluidics in place of electronics avoids the need for conversion between hydraulic or pneumatic signals and electrical signals, which could result in substantial cost benefits and reduction in the physical size.

Present-day fluidic components can provide high input impedances, low output impedances, and high gains comparable to typical electronic components (see Chapter 2). These fluidic components can provide bandwidths in the kHz range. Good dynamic range and resolution capabilities are available as well.

In addition to aircraft and aerospace flight control applications, fluidic devices are used in ground transit vehicles, heavy-duty machinery, machine tools, industrial robots, medical equipment, and process control systems. Specific examples include valve control for hydraulic or pneumatic actuated robotic joints and end effectors, windshield-wiper and windshield-washer controls for automobiles, respirator and artificial heart pump control, backup control devices for aircraft control surfaces, controllers for pneumatic power tools, control of food-packaging (e.g., bottle filling) devices, counting and timing devices for household appliances, braking systems, and spacecraft sensors. Fluidics will not replace electronics in a majority of control systems. But, there are significant advantages in using fluidics in some critical applications.

Problems

7.1 What factors generally govern (a) the electrical time constant (b) the mechanical time constant of a motor? Compare typical values for these parameters and discuss how they affect the motor response.

7.2 Write an expression for the back e.m.f. of a dc motor. Show that the armature circuit of a dc motor may be modeled by the equation

$$v_a = i_a R_a + k\phi\omega_m,$$

where v_a is the armature supply voltage, i_a is the armature current, R_a is the armature resistance, ϕ is the field flux, ω_m is the motor speed, and k is a motor constant.

Suppose that $v_a = 20$ V DC. At standstill, $i_a = 20$ A. When running at a speed of 500 rpm, the armature current was found to be 15 A. If the speed is increased to 1000 rpm while maintaining the field flux constant, determine the corresponding armature current.

7.3 In equivalent circuits for dc motors, iron losses (e.g., eddy current loss) in the stator are usually neglected. A way to include these effects is shown in Figure P7.3.

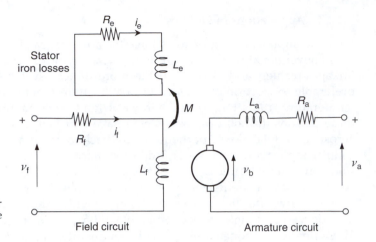

FIGURE P7.3
Equivalent circuit for a separately excited dc motor with iron losses in the
stator.

Iron losses in the stator poles are represented by a circuit with resistance R_e and
self-inductance L_e. The mutual inductance between the field circuit and the iron loss
circuit is denoted by M. It can be shown that

$$M = k\sqrt{L_f L_e},$$

where L_f is the self-inductance in the field circuit and k denotes a coupling constant.
With perfect coupling (no flux leakage between the two circuits), we have $k = 1$. But
usually, k is less than 1. The circuit equations are

$$v_f = R_f i_f + L_f \frac{di_f}{dt} - M \frac{di_e}{dt},$$

$$0 = R_e i_e + L_e \frac{di_e}{dt} - M \frac{di_f}{dt}.$$

The parameters and variables are defined in Figure P7.3. Obtain the transfer function
for i_f/v_f. Discuss the case $k = 1$ in reference to this transfer function. In particular,
show that the transfer function has a phase lag effect.

7.4 Explain the operation of a brushless dc motor. How does it compare with the
principle of operation of a stepper motor?

7.5 Give the steady-state torque–speed relations for a dc motor with the following three
types of connections for the armature and filed windings:

1. A shunt-wound motor

2. A series-wound motor

3. A compound-wound motor. with $R_{f1} = R_{f2} = 10\ \Omega$

The following parameter values are given:

$R_a = 5\ \Omega,\ R_f = 20\ \Omega,\ k = 1\ \text{N.m/A}^2$, and for a compound-wound motor, $R_{f1} = R_{f2} = 10\ \Omega$.
Note that

$$T_m = k i_f i_a.$$

Assume that the supply voltage is 115 V. Plot the steady-state torque–speed curves for these types of winding arrangements.

Using these curves, compare the steady-state performance of the three types of motors.

7.6 What is the electrical damping constant of a dc motor? Determine expressions for this constant for the three types of dc motor winding arrangements mentioned in Problem 7.5. In which case is this a constant value? Explain how the electrical damping constant could be experimentally determined. How is the dominant time constant of a dc motor influenced by the electrical damping constant? Discuss ways to decrease the motor time constant.

7.7 Explain why the transfer function representation for a separately excited and armature-controlled dc motor is more accurate than that of a field-controlled motor and still more accurate than those of shunt-wound, series-wound, or compound-wound dc motors. Give a transfer function relation (using Laplace variable s) for a dc motor where the incremental speed $\delta\omega_m$ is the output, the incremental winding excitation voltage δv_c is the control input, and the incremental load torque δT_L on the motor is a disturbance input. Assume that the parameters of the motor model are determined from experimental speed–torque curves for constant excitation voltage.

7.8 Explain the differences between full-wave circuits and half-wave circuits in thyristor (SCR) control of dc motors. For the SCR-drive circuit, shown in Figure 7.29a, sketch the armature current time history.

7.9 Using sketches, describe how pulse-width modulation (PWM) effectively varies the average value of the modulated signal. Explain how one could obtain

1. A zero average
2. A positive average
3. A negative average

by PWM. Indicate how PWM is useful in the control of dc motors. List the advantages and disadvantages of PWM.

7.10 In the chopper circuit shown in Figure 7.29a, suppose that $L_o = 100$ mH and $v_{ref} = 200$ V. If the worst-case amplitude of the armature current is to be limited to 1 A, determine the minimum chopper frequency.

7.11 Figure P7.11 shows a schematic arrangement for driving a dc motor using a linear amplifier. The amplifier is powered by a dc power supply of regulated voltage v_s. Under a particular condition, suppose that the linear amplifier drives the motor at voltage v_m and current i. Assume that the current drawn from the power supply is also i. Give an expression for the efficiency at which the linear amplifier is operating under these conditions. If $v_s = 50$ V, $v_m = 20$ V, and $i = 5$ A, estimate the efficiency of operation of the linear amplifier.

7.12 For a dc motor, the starting torque and the no-load speed are known, which are denoted by T_s and ω_o, respectively. The rotor inertia is J. Determine an expression for the dominant time constant of the motor.

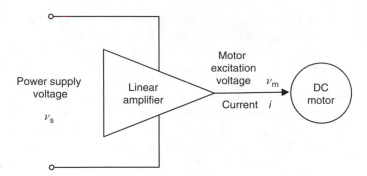

FIGURE P7.11
A linear amplifier for a dc motor.

7.13 A schematic diagram for the servo control loop of one joint of a robotic manipulator is given in Figure P7.13.

The motion command for each joint of the robot is generated by the controller of the robot in accordance with the required trajectory. An optical (incremental) encoder is used for both position and velocity feedback in each servo loop. Note that for a six-degree-of-freedom robot, there will be six such servo loops. Describe the function of each hardware component shown in the figure and explain the operation of the servo loop.

After several months of operation the motor of one joint of the robot was found to be faulty. An enthusiastic engineer quickly replaced the motor with an identical one without realizing that the encoder of the new motor was different. In particular, the original encoder generated 200 pulses/rev, whereas the new encoder generated 720 pulses/rev. When the robot was operated, the engineer noticed an erratic and unstable behavior at the repaired joint. Discuss reasons for this malfunction and suggest a way to correct the situation.

7.14 Consider the block diagram in Figure 7.16, which represents a dc motor, for armature control, with the usual notation. Suppose that the load driven by the

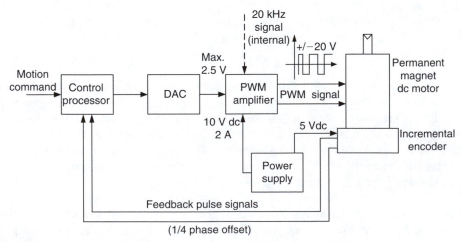

FIGURE P7.13
A servo loop of a robot.

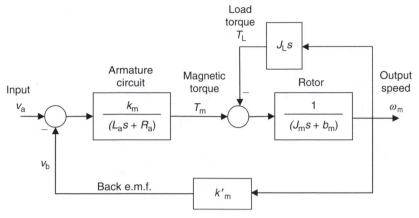

FIGURE P7.14
An armature-controlled dc motor with an inertial load.

    motor is a pure inertia element (e.g., a wheel or a robot arm) of moment of inertia J_L, which is directly and rigidly attached to the motor rotor.

1. Show that, in this case, the motor block diagram may be given as in Figure P7.14. Obtain an expression for the transfer function $\omega_m/v_a = G_m(s)$ for the motor with the inertial load, in terms of the parameters given in Figure P7.14.

2. Now neglect the leakage inductance L_a. Then, show that the transfer function given in 1 can be expressed as $G_m(s) = k/(\tau s + 1)$. Give expressions for τ and k in terms of the given system parameters.

3. Suppose that the motor (with the inertial load) is to be controlled using position plus velocity feedback. The block diagram of the corresponding control system is given in Figure 7.22, where the motor transfer function $G_m(s) = k/(\tau s + 1)$. Determine the transfer function of the (closed-loop) control system $G_{CL}(s) = \theta_m/\theta_d$ in terms of the given system parameters (k, k_p, τ, τ_v). Note that θ_m is the angle of rotation of the motor with inertial load and θ_d is the desired angle of rotation.

7.15 In the joint actuators of robotic manipulators, it is necessary to minimize backlash. Discuss the reasons for this. Conventional techniques for reducing backlash in gear drives include preloading, the use of bronze bearings that automatically compensate for wear, and the use of high-strength steel and other alloys that can be machined accurately and that have minimal wear problems. Discuss the shortcomings of some of the conventional methods of backlash reduction. Discuss the operation of a joint actuator unit that has virtually no backlash problems.

7.16 The moment of inertia of the rotor of a motor (or any other rotating machine) can be determined by a run-down test. With this method, the motor is first brought up to an acceptable speed and then quickly turned off. The motor speed vs. time curve is obtained during the run-down period that follows. A typical run-down curve is shown in Figure P7.16. Note that the motor decelerates because of its resisting torque T_r during this period. The slope of the run-down curve is determined at a suitable (operating) value of speed ($\bar{\omega}_m$) in Figure P7.16. Next, the motor is brought up to this speed ($\bar{\omega}_m$), and the torque ($\bar{T}_r$) that is needed to maintain the motor steady at this speed is obtained (either by direct measurement of torque, or by

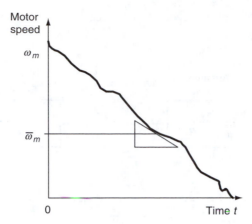

FIGURE P7.16
Data from a run-down test on an electric motor.

computing using field current measurement and known value for the torque constant, which is available in the data sheet of the motor). Explain how the rotor inertia J_m may be determined from this information.

7.17   In some types of (indirect-drive) robotic manipulators, joint motors are located away from the joints, and torques are transmitted to the joints through transmission devices such as gears, chains, cables, and timing belts. In some other types of (i.e., direct-drive) manipulators, joint motors are located at the joints themselves, the rotor is integral with one link, and the stator is integral with the joining link. Discuss advantages and disadvantages of these two designs.

7.18   In brushless motors, commutation is achieved by switching on the stator phases at the correct rotor positions (e.g., at the points of intersection of the static torque curves corresponding to the phases, for achieving maximum average static torque). We have noted that the switching points can be determined by measuring the rotor position using an incremental encoder. Incremental encoders are delicate, cannot operate at high temperatures, costly, and increase the size and cost of the motor package. In addition, precise mounting is required for proper operation. The generated signal may be subjected to electromagnetic interference (EMI) depending on the means of signal transmission. Since we need only to know the switching points (i.e., continuous measurement of rotor position is not necessary), and since these points are uniquely determined by the stator magnetic field distribution, a simpler and cost-effective alternative to an encoder for detecting the switching points would be to use Hall effect sensors. Specifically, Hall effect sensors are located at switching points around the stator (a sensor ring), and a magnet assembly is located around the rotor (in fact, the magnetic poles of the rotor can serve this purpose, without needing an additional set of poles). As the rotor rotates, a magnetic pole on the rotor triggers as appropriate Hall effect sensor, thereby generating a switching signal (pulse) for commutation at the proper rotor position. Microelectronic switching circuit (or switching transistor) is actuated by the corresponding pulse. Since Hall effect sensors have several disadvantages—such as hysteresis (and associated nonsymmetry of the sensor signal), low-operating temperature ratings (e.g., 125°C), thermal-drift problems, and noise due to stray magnetic fields and EMI—it may be more desirable to use fiber optic sensors for brushless commutation. Describe how the fiberoptic method of motor commutation works.

7.19 A brushless dc motor and a suitable drive unit are to be chosen for a continuous-drive application. The load has a moment of inertia 0.016 kg.m^2, and faces a constant resisting torque of 35.0 N.m (excluding the inertia torque) throughout the operation. The application involves accelerating the load from rest to a speed of 250 rpm in 0.2 s, maintaining it at this period for extended periods, and then decelerating to rest in 0.2 s. A gear unit with step-down gear ratio 4 is to be used with the motor. Estimate a suitable value for the moment of inertia of the motor rotor, for a fairly optimal design. Gear efficiency is known to be 0.8. Determine a value for continuous torque and a corresponding value for operating speed based on which the selection of a motor and a drive unit can be made.

7.20 Compare dc motors with ac motors in general terms. In particular, consider mechanical robustness, cost, size, maintainability, speed control capability, and possibility of implementing complex control schemes.

7.21 Compare frequency control with voltage control in induction motor control, giving advantages and disadvantages. The steady-state slip–torque relationship of an induction motor is given by

$$T_m = \frac{aSv_f^2}{[1 + (S/S_b)^2]}$$

with the parameter values $a = 1 \times 10^{-3}$ N.m/V^2 and $S_b = 0.25$. If the line voltage $v_f = 241$ V, calculate the breakdown torque. If the motor has two-pole pairs per phase and if the line frequency is 60 Hz, what is the synchronous speed (in rpm)? What is the speed corresponding to the breakdown torque? If the motor drives an external load, which is modeled as a viscous damper of damping constant $b = 0.03$ N.m/rad/s, determine the operating point of the system. Now, if the supply voltage is dropped to 163 V through voltage control, what is the new operating point? Is this a stable operating point?

7.22 Consider the induction motor in Problem 7.21. Suppose that the line voltage $v_f = 200$ V and the line frequency is 60 Hz. The motor is rigidly connected to an inertial load. The combined moment of inertia of the rotor and load is $J_{eq} = 5$ kg.m^2. The combined damping constant is $b_{eq} = 0.1$ N.m/rad/s. If the system starts from rest, determine, by computer simulation, the speed time history $\omega_L(t)$ of the load (and motor rotor) (Note: assume that the motor is a torque source, with torque represented by the steady-state speed–torque relationship).

7.23 (a) The equation of the rotor circuit of an induction motor (per phase) is given by (see Figure 7.40a)

$$i_r = \frac{Sv}{(R_r + jS\omega_p L_r)} = \frac{v}{(R_r/S + j\omega_p L_r)},$$
$$Z = R_r/S + j\omega_p L_r.$$

which corresponds to an impedance (i.e., voltage/current, in the frequency domain).

Show that this may be expressed as the sum of two impedance components:

$$Z = [R_r + j\omega_p SL_r] + [(1/S - 1)R_r + j\omega_p(1 - S)L_r].$$

For a line frequency of ω_p, this result is equivalent to the circuit shown in Figure 7.40c. Note that the first component of impedance corresponds to the rotor electrical loss and the second component corresponds to the useful mechanical power.

(b) Consider the characteristic shape of the speed vs. torque curve of an induction motor. Typically, the starting torque T_s is less than the maximum torque T_{max}, which occurs at a nonzero speed. Explain the main reason for this.

7.24 Prepare a table to compare and contrast the following types of motors:

1. Conventional dc motor with brushes
2. Brushless torque motor (dc)
3. Stepper motor
4. Induction motor
5. AC synchronous motor

In your table, include terms such as power capability, speed controllability, speed regulation, linearity, operating bandwidth, starting torque, power supply requirements, commutation requirements, and power dissipation. Discuss a practical method for reversing the direction of rotation in each of these types of motors.

7.25 Chopper circuits are used to chop a dc voltage so that a dc pulse signal results. This type of signal is used in the control of dc motors, by pulse-width modulation (PWM), because the pulse width for a given pulse frequency determines the mean voltage of the pulse signal. Inverter circuits are used to generate an ac voltage from a dc voltage. The switching (triggering) frequency of the inverter determines the frequency of the resulting ac signal. The inverter circuit method is used in the frequency control of ac motors. Both types of circuits use thyristor elements for switching. Either discrete circuit elements or integrated circuit (monolithic) chips may be developed for this purpose. Indicate how an ac signal may be generated by using a chopper circuit and a high-pass filter.

7.26 Show that the root-mean-square (rms) value of a rectangular wave can be changed by phase-shifting it and adding to the original signal. What is its applicability in the control of induction motors?

7.27 The direction of the rotating magnetic field in an induction motor (or any other type of ac motor) can be reversed by changing the supply sequence of the phases to the stator poles. This is termed phase switching. An induction motor can be decelerated quickly in this manner. This is known as plugging an induction motor. The slip vs. torque relationship of an induction motor may be expressed as

$$T_m = k(S)v_f^2.$$

Show that the same relationship holds under plugged conditions, except that $k(S)$ has to be replaced by $-k(2-S)$. Sketch the curves $k(S)$, $k(2-S)$, and $-k(2-S)$, from

$S=0$ to $S=2$. Using these curves, indicate the nature of the torque acting on the rotor during plugging (Note: $k(S) = (aS)/[1 + (S/S_b)^2]$).

7.28 What is a servomotor? AC servomotors that can provide torques in the order of 100 N.m at 3000 rpm are commercially available (Note: 1 N.m $=141.6$ oz. in.). Describe the operation of an ac servomotor that uses a two-phase induction motor. A block diagram for an ac servomotor is shown in Figure P7.28. Describe the purpose of each component in the system and explain the operation of the overall system. What are the advantages of using an ac amplifier after the inverter circuit in comparison to using a dc amplifier before the inverter circuit?

7.29 Consider the two-phase induction motor discussed in Example 7.15. Show that the motor torque T_m is a linear function of the control voltage v_c when $k(2-S) = k(S)$. How many values of speed (or slip) satisfy this condition? Determine these values.

7.30 A magnetically levitated rail vehicle uses the principle of induction motor for traction. Magnetic levitation is used for suspension of the vehicle slightly above the emergency guide rails. Explain the operation of the traction system of this vehicle, particularly identifying the stator location and the rotor location. What kinds of sensors would be needed for the control systems for traction and levitation? What type of control strategy would you recommend for the vehicle control?

7.31 What are common techniques for controlling
1. DC motors?
2. AC motors?

Compare these methods with respect to speed controllability.

7.32 Describe the operation of a single-phase ac motor. List several applications of this common actuator. Is it possible to realize three-phase operation using a single-phase ac supply? Explain your answer.

7.33 Speed control of motors (ac motors as well as dc motors) can be accomplished by using solid-state–switching circuitry. In one such method, a solid-state relay is

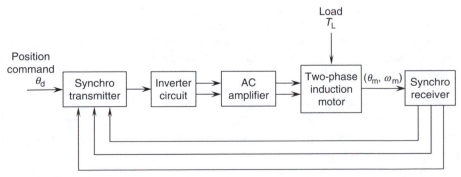

FIGURE P7.28
AC servomotor using a two-phase induction motor and a synchro transformer.

activated using a switching signal generated by a microprocessor so as to turn on and off at high speed, the power into the motor drive circuit. Speed of the motor can be measured using a sensor such as tachometer or optical encoder. This signal is read by the microprocessor and is used to modify the switching signal so as to correct the motor speed. Using a schematic diagram, describe the hardware needed to implement this control scheme. Explain the operation of the control system.

7.34   In some applications, it is necessary to apply a force without creating a motion. Discuss one such application. Discuss how an induction motor could be used in such an application. What are the possible problems arising from this approach?

7.35   The harmonic drive principle can be integrated with an electric motor in a particular manner in order to generate a high-torque gear motor. Suppose that the flexispline of the harmonic drive (see Chapter 8) is made of an electromagnetic material, as the rotor of a motor. Instead of the mechanical wave generator, suppose that a rotating magnetic field is generated around the fixed spline. The magnetic attraction causes the tooth engagement between the flexispline and the fixed spline. What type of motor principle may be used in the design of this actuator? Give an expression for the motor speed. How would one control the motor speed in this case?

7.36   List three types of hydraulic pumps and compare their performance specifications. A position servo system uses a hydraulic servo along with a synchro transformer as the feedback sensor. Draw a schematic diagram and describe the operation of the control system.

7.37   Giving typical applications and performance characteristics (bandwidth, load capacity, controllability, etc.), compare and contrast dc servos, ac servos, hydraulic servos, and pneumatic servos.

7.38   What is a multistage servovalve? Describe its operation. What are advantages of using several valve stages?

7.39   Discuss the origins of the hydraulic time constant in a hydraulic control system that consists of a four-way spool valve and a double-acting cylinder actuator. Indicate the significance of this time constant. Show that the dimensions (units) of the right-hand-side expression in the equation $\tau_h = V/2\beta k_c$ are [time]. Note: This has to be true because it represents a time constant (hydraulic).

7.40   Sometimes either a PWM ac signal or a dc signal with a superimposed constant frequency ac signal (or dither) is used to drive the valve actuator (torque motor) of a hydraulic actuator. What is the main reason for this superimposition of an oscillatory component into the drive signal? Discuss the advantages and disadvantages of this approach.

7.41   Compare and contrast valve-controlled hydraulic systems with pump-controlled hydraulic systems. Using a schematic diagram, explain the operation of a pump-controlled hydraulic motor. What are its advantages and disadvantages over a frequency-controlled ac servo?

7.42   Explain why accumulators are used in hydraulic systems. Sketch two types of hydraulic accumulators and describe their operation.

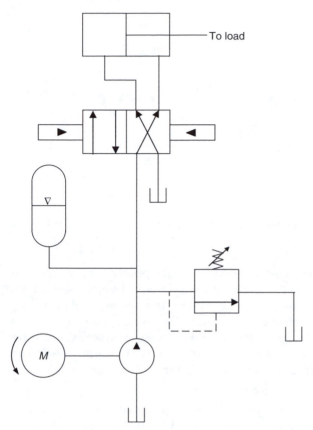

FIGURE P7.43
A hydraulic circuit diagram.

7.43 Identify and explain the components of the hydraulic system given by the circuit diagram in Figure P7.43. Describe the operation of the overall system.

7.44 If the load on the hydraulic actuator shown in Figure 7.60 consists of a rigid mass restrained by a spring, with the other end of the spring connected to a rigid wall, write equations of motion for the system. Draw a block diagram for the resulting complete system, including a four-way spool valve, and give the transfer function that corresponds to each block.

7.45 Suppose that the coupling of the feedback linkage shown in Figure 7.68c is modified as shown in Figure P7.45. What is the transfer function of the controller? Show that this feedback controller is a lead compensator.

7.46 The sketch in Figure P7.46 shows a half-sectional view of a flow control valve, which is intended to keep the flow to a hydraulic load constant regardless of variations of the load pressure P_3 (disturbance input).

1. Briefly discuss the physical operation of the valve, noting that the flow will be constant if the pressure drop across the fixed area orifice is constant.

2. Write the equations that govern the dynamics of the unit. The mass, the damping constant, and the spring constant of the valve are denoted by m, b, and k, respectively. The volume of oil under pressure P_2 is V, and the bulk modulus of the oil is β. Make the usual linearizing assumptions.

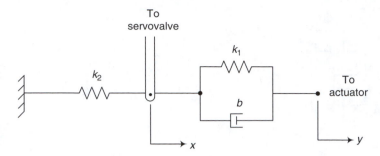

FIGURE P7.45
A mechanical coupling with lead action for a hydraulic servovalve.

   3. Set up a block diagram for the system from which the dynamics and stability of the valve could be studied.

7.47 A schematic diagram of a pump stroke-regulated hydraulic power supply is shown in Figure P7.47. The system uses a three-way pressure control valve of the type described in the text (see Figure 7.64). This valve controls a spring-loaded piston, which in turn regulates the pump stroke by adjusting the swash plate angle of the pump. The load pressure P_L is to be regulated. This pressure can be set by adjusting the preload x_o of the spring in the pressure control valve (y_o in Figure 7.64). The load flow Q_L enters into the hydraulic system as a disturbance input.

   1. Briefly describe the operation of the control system.

   2. Write the equations for the system dynamics, assuming that the pump stroke mechanism and the piston inertia can be represented by an equivalent mass m_p moving through x_p. The corresponding spring constant and damping constant are k_p and b_p, respectively. The piston area is A_p. The mass, spring constant, and damping constant of the valve are m, k, and b, respectively. The valve area is A_v and the valve spool movement is x_v. The volume of oil under pressure P_L is V_t, and the volume of oil under pressure P_p is V_o (volume of oil in the cylinder chamber). The bulk modulus of the oil is β.

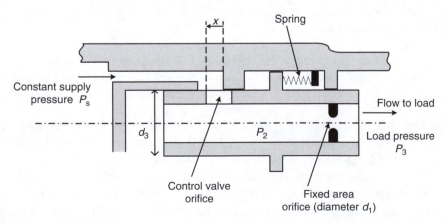

FIGURE P7.46
A flow control valve.

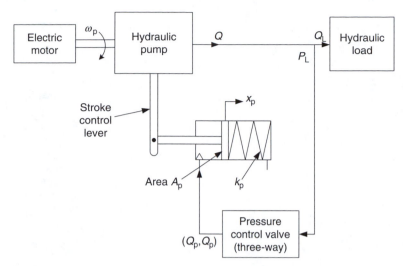

FIGURE P7.47
A pump stroke-regulated hydraulic power supply.

3. Draw a block diagram for the system from which the behavior of the system could be investigated. Indicate the inputs and outputs.

4. If Q_p is relatively negligible, indicate which control loops can be omitted from the block diagram. Hence, derive an expression for the transfer function $x_p(s)/x_v(s)$ in terms of the system parameters.

7.48 A schematic diagram of a solenoid-actuated flow control valve is shown in Figure P7.48a. The downward motion x of the valve rod is resisted by a spring of stiffness k. The mass of the valve rod assembly (all moving parts) is m, and the associated equivalent viscous damping constant is b. The voltage supply to the valve actuator (proportional solenoid) is denoted by v_i. For a given voltage v_i, the solenoid force is a nonlinear (decreasing) function of the valve position x. This steady-state variation of the solenoid force (downward) and the resistive spring force (upward), with respect to the valve displacement, are shown in Figure P7.48b. Assuming that the inlet pressure and the outlet pressure of the fluid flow are constants, the flow rate will be determined by the valve position x. Hence, the objective of the valve actuator would be to set x using v_i.

1. Show that for a given input voltage v_i, the resulting equilibrium position (x) of the valve is always stable.

2. Describe how the relationship between v_i and x could be obtained

   (a) Under quasistatic conditions

   (b) Under dynamic conditions

7.49 What are the advantages and disadvantages of pneumatic actuators in comparison with electric motors in process control applications? A pneumatic rack-and-pinion actuator is an on/off device that is used as a rotary valve actuator. A piston or diaphragm in the actuator is moved by allowing compressed air into the valve chamber. This rectilinear motion is converted into rotary motion through a rack-and-pinion device in the actuator. Single-acting types with spring return and

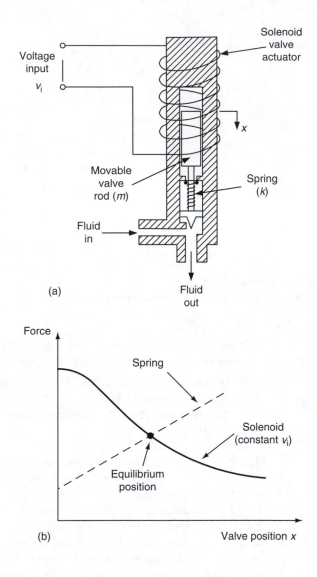

FIGURE P7.48
(a) A solenoid-actuated flow control valve.
(b) Steady-state characteristics of the valve.

double-acting types are commercially available. Using a sketch, explain the operation of a piston-type single-acting rack-and-pinion actuator with a spring-restrained piston. Could the force rating, sensitivity, and robustness of the device be improved by using two pistons and racks coupled with the same pinion? Explain.

7.50 Consider a pneumatic speed sensor that consists of a wobble plate and a nozzle arranged like a pneumatic flapper valve. The wobble plate is rigidly mounted at the end of a rotating shaft, so that the plane of the plate is inclined to the shaft axis. Using a sketch, explain the principle of operation of the wobble plate pneumatic speed sensor.

7.51 A two-axis hydraulic positioning mechanism is used to position the cutter of an industrial fish-cutting machine. The cutter blade is pneumatically operated. The hydraulic circuit of the positioning mechanism is given in Figure P7.51. Since the two hydraulic axes are independent, the governing equations are similar. State the nonlinear servovalve equations, hydraulic cylinder (actuator) equations, and

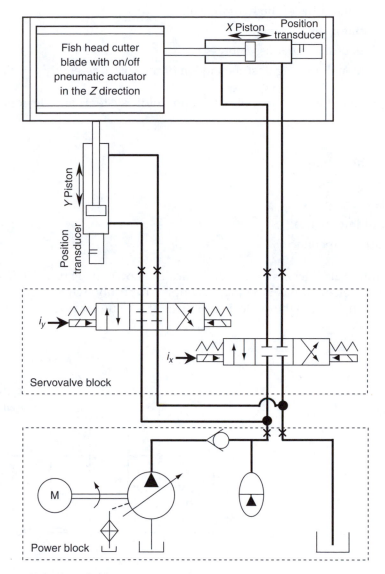

FIGURE P7.51

Two-axis hydraulic positioning system of an industrial fish cutter.

the mechanical load (cutter assembly) equations for the system. Use the following notation:

x_v = Servovalve displacement

K = Valve gain (nonlinear)

P_s = Supply pressure

P_1 = Head-side pressure of the cylinder, with area A_1 and flow Q_1

P_2 = Rod-side pressure of the cylinder, with area A_2 and flow Q_2

V_h = Hydraulic volume in the cylinder chamber

β = Bulk modulus of the hydraulic oil

x = Actuator displacement

M = Mass of the cutter assembly

F_f = Frictional force against the motion of the cutter assembly

7.52 Define the following terms in relation to fluidic systems and devices:

1. Fan-in
2. Fan-out
3. Switching speed
4. Transport time
5. Load sensitivity
6. Input impedance
7. Output impedance

7.53 A fluidic pulse generator is schematically shown in Figure P7.53a. The fluidic switching element has a regulated supply. Typically, when the input u_2 is larger than the input u_1 to the switching element, the resulting pressure differential turns

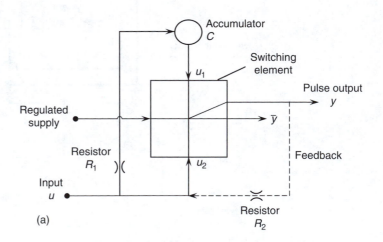

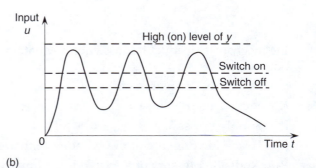

FIGURE P7.53

(a) A fluidic pulse generator. (b) An input signal.

on the output y to its high level, shown by a dotted line in Figure P7.53b. When $u_1 > u_2$, the output y will be turned off (and its complement output $\bar{y}$ will be turned on). There is a hysteresis band for this switching process. The pulse generator consists of a switching element, a fluid capacitor (accumulator) C, and a fluid restrictor (resistor) R_1. In the feedback configuration, there is also a feedback path through a second resistor R_2 as shown by the dotted line in Figure P7.53a.

First consider the open-loop configuration (without the feedback path) of the pulse generator. When the input (pressure) signal u is applied, u_2 immediately rises to the value of u, but u_1 rises to the value of u only after a time delay, due to the presence of transport lag and a finite time constant (modeled using R_1 and C). Hence, the output y will be turned on (to its high level) at the switch on level of u. Subsequently, u_1 reaches u. Now if u falls to the switch off level, so will u_2. However, due to the accumulator C, the level of u_1 will be maintained for sometime. Accordingly, the output y will be switched off (i.e., y will be zero). Sketch the shape of the output signal y for the input u given in Figure P7.53b.

Now consider the pulse generator with the feedback path through R_2. How will the output y change in this case, for the same input u?

An application of the pulse generator (with feedback) is in the pharmaceutical packaging industry. For example, consider the filling of a liquid drug into bottles and capping them. A packaging line consisting primarily of fluidic devices and fluid power devices can be designed for this purpose. Suppose that fluidic proximity sensors, fluidic amplifiers, hydraulic/pneumatic valves and actuators, and other auxiliary components (including resistors and logic elements) are available. Using a sketch, briefly describe a fluidic system that can accomplish the task.

8

Mechanical Transmission Components

When an actuator is chosen to drive a load, it is important to make sure that the two components are properly matched. In other words, the actuator must have the capability to drive the load precisely at the necessary speeds and possibly under transient motion conditions. One may have to employ a transmission device such as a gear to achieve this matching. Furthermore, the nature of the actuator motion may have to be modified to obtain the required load motion. For example, the rotatory (i.e., angular) motion of an actuator may have to be converted into a translatory (i.e., rectilinear) motion for moving a load. A transmission device can accomplish this function as well. A mechanical component can play a variety of crucial roles, which may include structural support or load bearing, mobility, transmission of motion and power or energy, and actuation and manipulation. The mechanical system has to be designed (integral with electronics, controls, etc.) to satisfy such desirable characteristics as light weight, high strength, high speed, low noise and vibration, long design life, fewer moving parts, high reliability, low-cost production and distribution, and infrequent and low-cost maintenance. Clearly, the requirements can be conflicting and there is a need for design optimization. Mechanical transmission devices play an important role in this regard. This concluding chapter studies several popular types of mechanical components and transmission devices.

8.1 Mechanical Components

Common mechanical components may be classified into some useful groups, as follows:

1. Load bearing/structural components (strength and surface properties)
2. Fasteners (strength)
3. Dynamic isolation components (both motion and force transmissibility)
4. Transmission components (motion conversion, load transmission)
5. Mechanical actuators (force or torque generation)
6. Mechanical controllers (controlled energy dissipation, controlled motion)

In each category we have indicated within parentheses the main property or attribute that is characteristic of the function of that category.

In load bearing or structural components, the main function is to provide structural support. In this context, mechanical strength and surface properties (e.g., hardness, wear resistance, and friction) of the component are crucial. The component may be rigid or flexible and stationary or moving. Examples of load bearing and structural components include bearings, springs, shafts, beams, columns, flanges, and similar load-bearing structures.

Fasteners are closely related to load bearing or structural components. The purpose of a fastener is to join two mechanical components or to mount/attach a component. Here as well, the primary property of importance is the mechanical strength. Mechanical flexibility may play a part as well, in some types of fasteners. Examples are bolts and nuts, locks and keys, screws, rivets, and spring retainers. Welding, bracing, gluing, cementing, and soldering are processes of fastening and will fall into the same category.

Dynamic isolation components perform the main task of isolating a system from another system (or environment) with respect to motion and forces. These involve the filtering or shielding of motions and forces or torques from a mechanical device such as a machine. Hence, motion transmissibility and force transmissibility are the key considerations in these components. Springs, dampers, and inertia elements may form isolation elements. Shock and vibration mounts for machinery, inertia blocks, and the suspension systems of vehicles are examples of isolation dynamic components.

Transmission components may be related to isolation components in principle, but their functions are rather different. The main purpose of a transmission component is the conversion of motion (in magnitude and from). In the process, the force or torque of the input member is also converted in magnitude and form. In fact, in some applications the modification of the force or torque may be the primary requirement of the transmission component. In any event, load (force or torque) transmission is an integral consideration together with motion transmission. Examples of transmission components are gears, lead screws and nuts (or power screws), racks and pinions, cams and followers, chains and sprockets, belts and pulleys (or drums), differentials, kinematic linkages, flexible couplings, and fluid transmissions.

Mechanical actuators are used to generate forces (and torques) for various applications. Specifically they are force sources or torque sources. Common actuators are electromagnetic in form (i.e., electric motors) and not purely mechanical. Since the magnetic forces are mechanical forces, which generate mechanical torques, electric motors may be considered as electromechanical devices. Other types of actuators that use fluids for generating the required effort may be considered as well in the category of mechanical actuators. In any event, load (force or torque) transmission is an integral consideration together with motion transmission. Examples are hydraulic pistons and cylinders (rams), hydraulic motors, their pneumatic counterparts, and thermal power units (prime movers) such as steam or gas turbines. Of particular interest in industrial applications are the electromechanical actuators and hydraulic and pneumatic actuators.

Mechanical controllers perform the task of modifying the dynamic response (motion and force or torque) of a system in a desired manner. Purely mechanical controllers typically carry out this task by controlled dissipation of energy. These are not as common as electrical or electronic controllers and hydraulic or pneumatic controllers. In fact, hydraulic or pneumatic servovalves may be treated in the category of purely mechanical controllers. Dynamic isolation components consisting of inertia, flexibility, and dissipation, may be considered as passive controllers. Examples are vibration dampers and dynamic absorbers. Furthermore, mechanical controllers are closely related to transmission components and mechanical actuators. Other examples of mechanical controllers are clutches and brakes.

In selecting a mechanical component for an application, many engineering aspects have to be considered. The foremost are the capability and performance of the component with respect to the design requirements (or specifications) of the system. For example, motion and torque specifications, flexibility and deflection limits, strength characteristics including stress–strain behavior, failure modes and limits, fatigue life, surface and material properties (e.g., friction, nonmagnetic, noncorrosive), operating range, and design life will be important. Other factors such as size, shape, cost, and commercial availability can be quite crucial as well.

Mechanical components

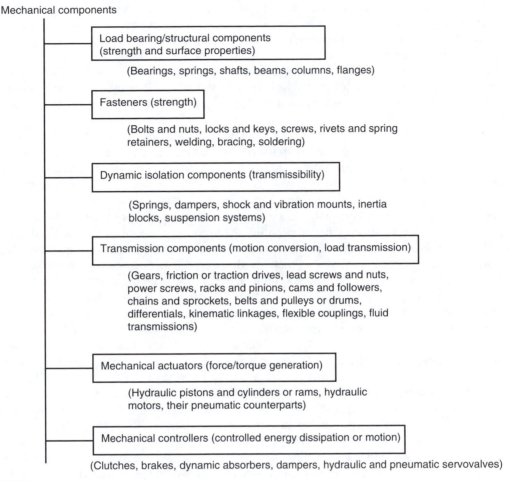

FIGURE 8.1
Classification of mechanical components.

The foregoing classification of mechanical components is summarized in Figure 8.1. It is not within the scope of the present chapter to study all the types of mechanical components that are summarized here. Rather, we select for further analysis a few important mechanical components that are particularly useful in practical systems.

8.2 Transmission Components

Transmission devices are indispensable in electromechanical system applications. We undertake to discuss a few representative transmission devices here. It should be cautioned that in the present treatment, a transmission is isolated and treated as a separate unit. In an actual application, however, a transmission device works as an integral unit with other components, particularly the actuator, the electronic drive unit, and the mechanical load that is manipulated. Hence, design or selection of a transmission should involve an integrated treatment of all interacting components.

Perhaps the most common transmission device is a gearbox. In its simplest form, a gearbox consists of two gear wheels, which contain teeth of identical pitch (tooth separation) and of unequal wheel diameter. The two wheels are meshed (i.e., the teeth are engaged) at one location. This device changes the rotational speed by a specific ratio (gear ratio) as dictated by the ratio of the diameters (or radii) of the two gear wheels. In particular, by stepping down the speed (in which case the diameter of the output gear is larger than that of the input gear), the output torque can be increased. Larger gear ratios can be realized by employing more than one pair of meshed gear wheels. Gear transmissions are used in a variety of applications including automotive, industrial-drive and robotics. Specific gear designs range from conventional spur gears to harmonic drives, as discussed later in the present chapter.

Gear drives have several disadvantages. In particular, they exhibit backlash because the tooth width is smaller than the tooth space of the mating gear. Some degree of backlash is necessary for proper meshing. Otherwise jamming will occur. Unfortunately, backlash is a nonlinearity, which can cause irregular and noisy operation with brief intervals of zero torque transmission. It can lead to rapid wear and tear and even instability. The degree of backlash can be reduced by using proper profiles (shapes) for the gear teeth. Backlash can be eliminated through the use of spring-loaded gears. Sophisticated feedback control may be used as well to reduce the effects of gear backlash.

Conventional gear transmissions, such as those used in automobiles with standard gearboxes, contain several gear stages. The gear ratio can be changed by disengaging the drive-gear wheel (pinion) from a driven wheel of one gear stage and engaging it with another wheel of a different number of teeth (different diameter) of another gear stage, while the power source (input) is disconnected by means of a clutch. Such a gearbox provides only a few fixed gear ratios. The advantages of a standard gearbox include relative simplicity of design and the ease with which it can be adapted to operate over a reasonably wide range of speed ratios, albeit in a few discrete increments of large steps. There are many disadvantages: Since each gear ratio is provided by a separate gear stage, the size, weight, and complexity (and associated cost, wear, and unreliability) of the transmission increase directly with the number of gear ratios provided. In addition, the drive source has to be disconnected by a clutch during the shifting of gears; the speed transitions are generally not smooth, and the operation is noisy. There is also dissipation of power during the transmission steps, and wear and damage can be caused by the actions of inexperienced operators. These shortcomings can be reduced or eliminated if the transmission is able to vary the speed ratio continuously rather than in a stepped manner. Further, the output speed and corresponding torque can be matched to the load requirements closely and continuously for a fixed input power. This results in more efficient and smooth operation, and many other related advantages. A continuously variable transmission (CVT), which has these desirable characteristics, will be discussed later in this chapter. First, we discuss a power screw, which is a converter of angular motion into rectilinear motion.

8.3 Lead Screw and Nut

A lead-screw drive is a transmission component, which converts rotary motion into rectilinear motion. Lead screws, power screws, and ball screws are rather synonymous. Lead screw and nut units are used in numerous applications including positioning tables, machine tools, gantry and bridge systems, automated manipulators, and valve actuators. Figure 8.2 shows the main components of a lead-screw unit. The screw is rotated by a motor, and as a result, the nut assembly moves along the axis of the screw. The support block, which is attached to the nut, provides the means for supporting the device that has to be moved using the lead-screw drive. The screw holes that are drilled on the support block

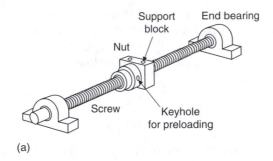

(a)

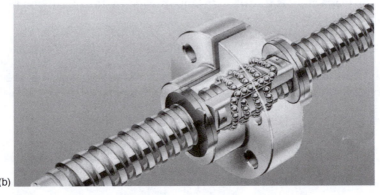

(b)

FIGURE 8.2
(a) A lead-screw and nut unit. (b) A commercial ball-screw unit. (Deutsche Star GmbH. With permission.)

may be used for this purpose. Since there can be backlash between the screw and the nut as a result of the assembly clearance or wear and tear, a keyhole is provided in the nut to apply a preload through some form of a clamping arrangement that is designed into the nut. The end bearings support the moving load. Typically, these are ball bearings that can carry axial loads as well by means of an angular-contact thrust bearing design.

The basic equation for operation of a lead-screw drive is obtained now. As shown in Figure 8.3, suppose that a torque T_R is provided by the screw at (and reacted by) the nut. This is the net torque after deducting the inertia torque (due to inertia of the motor rotor and the lead screw) and the frictional torque of the bearings, from the motor (magnetic) torque. Torque T_R is not completely available to move the load that is supported on the nut. The reason is the energy dissipation (friction) at the screw and nut interface. Suppose that the net force available from the nut to drive the load in the axial direction is F. Denote the screw rotation by θ and the rectilinear motion of the nut by x.

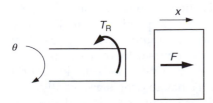

FIGURE 8.3
Effort and motion transmission at the screw and nut interface.

When the screw is rotated (say, by a motor) through a small angle $\delta\theta$, the nut, which is restrained from rotating because of the guides along which the support block moves, will move through a small distance δx along the axial direction. The work done by the screw is $T_R\delta\theta$ and the work done in moving the nut (with its load) is $F\delta x$. The lead-screw efficiency e is given by

$$e = \frac{F\delta x}{T_R\delta\theta}. \tag{8.1}$$

Now, geometric "compatibility" of the device gives $r\delta\theta = \delta x$, where the transmission parameter of the lead screw is r (axial distance moved per one radian of screw rotation). The lead l of the lead screw is the axial distance moved by the nut in one revolution of the screw, and it satisfies

$$l = 2\pi r. \tag{8.2}$$

In general, the lead is not the same as the pitch p of the screw, which is the axial distance between two adjacent threads. For a screw with n threads,

$$l = np. \tag{8.3}$$

By substituting the definition of r in Equation 8.1, we have

$$F = \frac{e}{r}T_R = \frac{2\pi e}{l}T_R. \tag{8.4}$$

This result is the representative equation of a lead screw, which may be used in the design and selection of components in a lead-screw drive system.

For a screw of mean diameter d, the helix angle α is given by

$$\tan\alpha = \frac{l}{\pi d} = \frac{2r}{d}. \tag{8.5}$$

Assuming square threads, we obtain a simplified equation for the efficiency of the screw in terms of the coefficient of friction μ. First, for a screw of 100% efficiency ($e = 1$), from Equation 8.4, a torque T_R at the nut can support an axial force (load) of T_R/r. The corresponding frictional force F_f is $\mu T_R/r$. The torque required to overcome this frictional force is $T_f = F_f d/2$. Hence, the frictional torque is given by

$$T_f = \frac{\mu d}{2r}T_R. \tag{8.6}$$

The screw efficiency is

$$e = \frac{T_R - T_f}{T_R} = 1 - \frac{\mu d}{2r} = 1 - \frac{\mu}{\tan\alpha}. \tag{8.7}$$

For threads that are not square (e.g., for slanted threads such as Acme threads, Buttress threads, and modified square threads), Equation 8.7 has to be appropriately modified.

It is clear from Equation 8.6 that the efficiency of a lead-screw unit can be increased by decreasing the friction and increasing the helix angle. Of course, there are limits. For example, typically the efficiency will not increase by increasing the helix angle beyond $30°$. In fact, a helix angle of $50°$ or more will cause the efficiency to drop significantly. The friction can be decreased by proper choice of material for screw and nut and through

TABLE 8.1

Some Useful Values for Coefficient of Friction

Material	Coefficient of Friction
Steel (dry)	0.2
Steel (lubricated)	0.15
Bronze	0.10
Plastic	0.10

surface treatments, particularly lubrication. Typical values for the coefficient of friction (for identical mating material) are given in Table 8.1. Note that the static (starting) friction will be higher (as much as 30%) than the dynamic (operating) friction. An ingenious way to reduce friction is using a nut with a helical track of balls instead of threads. In this case, the mating between the screw and the nut is not through threads but through ball bearings. Such a lead-screw unit is termed a *ball screw* (see Figure 8.2b). A screw efficiency of 90% or greater is possible with a ball screw unit.

In the driving mode of a lead screw, the frictional torque acts in the opposite direction to (and has to be overcome by) the driving torque. In the free mode where the load is not driven by an external torque from the screw, it is likely that the load will try to back-drive the screw (say, due to gravitational load). Then, however, the frictional torque will change direction and the back motion has to overcome it. If the back-driving torque is less than the frictional torque, motion will not be possible and the screw is said to be self-locking.

Example 8.1

A lead-screw unit is used to drive an object (a load) up an incline of angle θ, as shown in Figure 8.4. Under quasistatic conditions (i.e., neglecting inertial loads) determine the drive torque needed by the motor to operate the device. The total mass of the moving unit (load, nut, and fixtures) is m. The efficiency of the lead screw is e and the lead is l. Assume that the axial load (thrust) due to gravity is taken up entirely by the nut. (In practice, a significant part of the axial load is supported by the end bearings, which have the thrust-bearing capability.)

Solution

The effective load that has to be acted upon by the net torque (after allowing for friction) in this example is

$$F = mg \sin \theta.$$

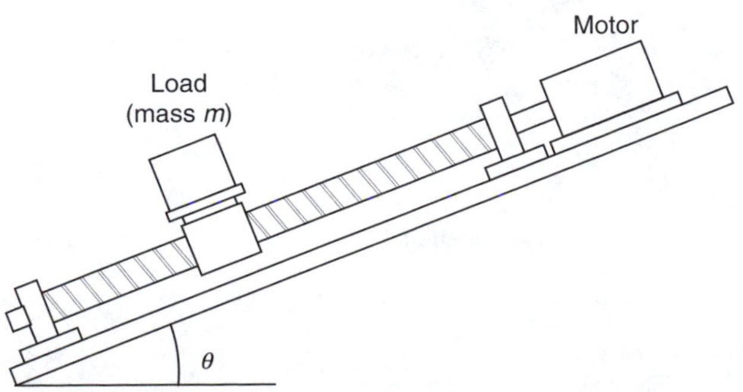

FIGURE 8.4
A lead-screw unit driving an object up an incline.

Substitute into Equation 8.4. The required torque at the nut is

$$T_R = \frac{mgr}{e} \sin\theta = \frac{mgl}{2\pi e} \sin\theta. \tag{8.8}$$

8.4   Harmonic Drives

Usually, motors run efficiently at high speeds. Yet in many practical applications, low speeds and high toques are needed. A straightforward way to reduce the effective speed and increase the output torque of a motor is to employ a gear system with high gear reduction. Gear transmission has several disadvantages, however. For example, backlash in gears would be unacceptable in high-precision applications. Frictional loss of torque, wear problems, noise, and the need for lubrication must also be considered. Furthermore, the mass of the gear system consumes energy from the actuator (motor) and reduces the overall torque-to-mass ratio and the useful bandwidth of the actuator.

A harmonic drive is a special type of transmission device that provides very large speed reductions (e.g., 200:1) without backlash problems. In addition, a harmonic drive is comparatively much lighter than a standard gearbox. The harmonic drive is often integrated with conventional motors to provide very high torques, particularly in direct-drive and servo applications. The principle of operation of a harmonic drive is shown in Figure 8.5. The rigid circular spline of the drive is the outer gear and it has internal teeth. An annular flexispline has external teeth that can mesh with the internal teeth of the rigid spline in a limited region when pressed in the radial direction. The external radius of the

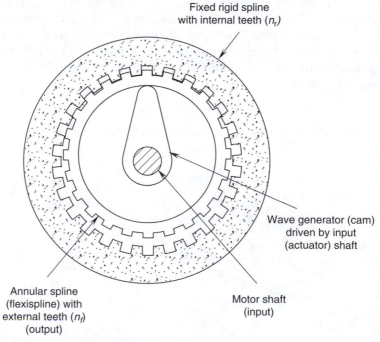

FIGURE 8.5
The principle of operation of a harmonic drive.

flexispline is slightly smaller than the internal radius of the rigid spline. As its name implies, the flexispline undergoes some elastic deformation during the meshing process. This results in a tight tooth engagement (meshing) without any clearance between the meshed teeth, and hence the motion is backlash free.

In the design shown in Figure 8.5, the rigid spline is fixed and may also serve as the housing of the harmonic drive. The rotation of the flexispline is the output of the drive; hence, it is connected to the driven mechanical load. The input shaft (motor shaft) drives the wave generator (represented by a cam in Figure 8.5). The wave generator motion brings about controlled backlash-free meshing between the rigid spline and the flexispline.

Suppose that n_r is the number of teeth (internal) in the rigid spline and n_f is the number of teeth (external) in the flexispline. It follows that the tooth pitch of the rigid spline $= \dfrac{2\pi}{n_r}$ (radians), and the tooth pitch of the flexispline $= \dfrac{2\pi}{n_f}$ (radians).

Further, suppose that n_r is slightly smaller than n_f. Then, during a single tooth engagement, the flexispline rotates through $(2\pi/n_r - 2\pi/n_f)$ radians in the direction of rotation of the wave generator. During one full rotation of the wave generator, there will be a total of n_r tooth engagements in the rigid spline (which is stationary in this design). Hence, the rotation of the flexispline during one rotation of the wave generator (around the rigid spline) is

$$n_r \left(\frac{2\pi}{n_r} - \frac{2\pi}{n_f} \right) = \frac{2\pi}{n_f}(n_f - n_r).$$

It follows that the gear reduction ratio (r:1) representing the ratio: input speed/output speed is given by

$$r = \frac{n_f}{n_f - n_r}. \tag{8.9a}$$

We can see that by making n_r very close to n_f, high gear reductions can be obtained from a harmonic drive. Furthermore, since the efficiency of a harmonic drive is given by

$$\text{Efficiency } e = \frac{\text{output power}}{\text{input power}}, \tag{8.10}$$

we have

$$\text{Output torque} = \frac{e n_f}{(n_f - n_r)} \times \text{input torque}. \tag{8.11}$$

This result illustrates the torque amplification capability of a harmonic drive.

An inherent shortcoming of the harmonic drive sketched in Figure 8.5 is that the motion of the output device (flexispline) is eccentric (or epicyclic). This problem is not serious when the eccentricity is small (which is the case for typical harmonic drives) and is further reduced because of the flexibility of the flexispline. For improved performance, however, this epicyclic rotation has to be reconverted into a concentric rotation. This may be accomplished by various means, including flexible coupling and pin-slot transmissions. The output device of a pin-slot transmission is a flange that has pins arranged on the circumference of a circle centered at the axis of the output shaft. The input to the pin-slot transmission is the flexispline motion, which is transmitted through a set of

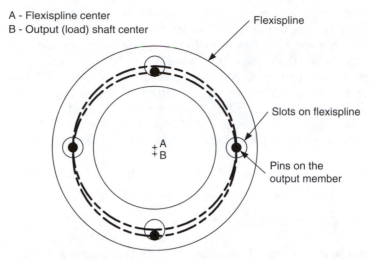

FIGURE 8.6
The principle of a pin-slot transmission.

holes on the flexispline. The pin diameter is smaller than the hole diameter, and the associated clearance is adequate to take up the eccentricity in the flexispline motion. This principle is shown schematically in Figure 8.6. Alternatively, pins could be attached to the flexispline and the slots on the output flange. The eccentricity problem can be eliminated altogether by using a double-ended cam in place of the single-ended cam as the wave generator in Figure 8.5. With this new arrangement, meshing takes place at two diametrically opposite ends simultaneously, and the flexispline is deformed elliptically in doing this. The center of rotation of the flexispline now coincides with the center of the input shaft. This double-mesh design is more robust and is quite common in industrial harmonic drives.

Other designs of harmonic drive are possible. For example, if $n_f < n_r$ then r in Equation 8.9 will be negative and the flexispline will rotate in the opposite direction to the wave generator (input shaft). Additionally, as indicated in the example below, the flexispline may be fixed and the rigid spline may serve as the output (rotating) member.

Traction drives (or friction drives) employ frictional coupling to eliminate backlash and overloading problems. These are not harmonic drives. In a traction drive, the drive member (input roller) is frictionally engaged with the driven member (output roller). The disadvantages of traction drives include indeterminacy of the speed ratio under slipping (overload) conditions, thermal problems and efficiency reduction due to energy dissipation, and large size and weight for a specified speed ratio.

Example 8.2
An alternative design of a harmonic drive is sketched in Figure 8.7a. In this design, the flexispline is fixed. It loosely fits inside the rigid spline and is pressed against the internal teeth of the rigid spline at diametrically opposite locations. Tooth meshing occurs at these two locations only. The rigid spline is the output member of the harmonic drive (See Figure 8.7b).

1. Show that the speed reduction ratio is given by

$$r = \frac{\omega_i}{\omega_f} = \frac{n_r}{(n_r - n_f)}. \tag{8.9b}$$

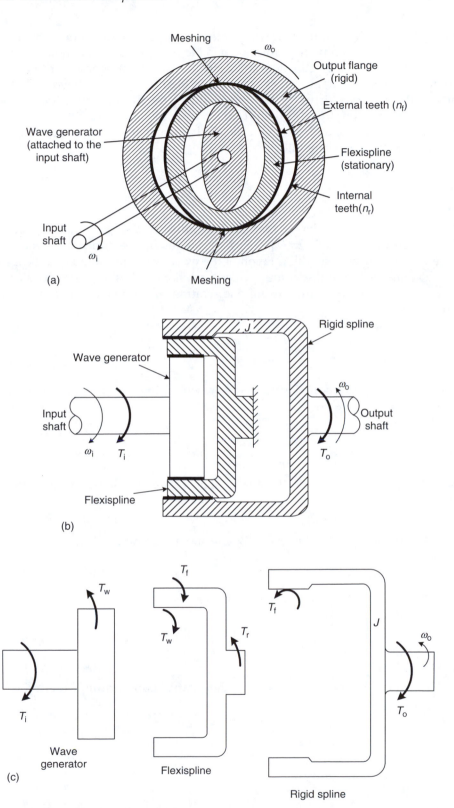

FIGURE 8.7
(a) An alternative design of harmonic drive. (b) Torque and speed transmission of the harmonic drive. (c) Free-body diagrams.

Note that if $n_f > n_r$ the output shaft will rotate in the opposite direction to the input shaft. Now consider the free-body diagram shown in Figure 8.7c. The axial moment of inertia of the rigid spline is J. Neglecting the inertia of the wave generator, write approximate equations for the system. The variables shown in Figure 8.7c are defined as T_i, the torque applied on the harmonic drive by the input shaft; T_o, the torque transmitted to the driven load by the output shaft (rigid spline); T_f, the torque transmitted by the flexispline to the rigid spline; T_r, the reaction torque on the flexispline at the fixture; and T_w, the torque transmitted by the wave generator.

Solution

Part 1

Suppose that n_r is slightly larger than n_f. Then, during a single tooth engagement, the rigid spline rotates through $(2\pi/n_f - 2\pi/n_r)$ radians in the direction of rotation of the wave generator. During one full rotation of the wave generator, there will be a total of n_f tooth engagements in the flexispline (which is stationary in the present design). Hence, the rotation of the rigid spline during one rotation of the wave generator (around the flexispline) is

$$n_f \left(\frac{2\pi}{n_f} - \frac{2\pi}{n_r} \right) = \frac{2\pi}{n_r} (n_r - n_f).$$

It follows that the gear reduction ratio (r:1) representing the ratio: input speed/output speed is given by

$$r = \frac{n_r}{n_r - n_f}. \tag{8.9c}$$

It should be clear that if $n_f > n_r$ the output shaft rotates in the opposite direction to the input shaft.

Part 2

Equations of motion for the three components are as follows:

1. Wave generator

   Here, since inertia is neglected, we have

   $$T_i - T_w = 0. \tag{8.12a}$$

2. Flexispline

   Here, since the component is fixed, the "static" equilibrium condition is

   $$T_w + T_f - T_r = 0. \tag{8.12b}$$

3. Rigid Spline

   Newton's second law gives

   $$T_f - T_o = J \frac{d\omega_o}{dt}. \tag{8.12c}$$

8.5 Continuously Variable Transmission

A continuously-variable transmission (CVT) is a transmission device whose gear ratio (speed ratio) can be changed continuously—i.e., by infinitesimal increments or having infinitesimal resolution—over its design range. Because of perceived practical advantages of a CVT over a conventional fixed-gear-ratio transmission, there has been significant interest in the development of a CVT that can be particularly competitive in automotive applications. For example, in the Van Doorne belt, a belt-and-pulley arrangement is used and the speed ratio is varied by adjusting the effective diameter of the pulleys in a continuous manner. The mechanism that changes the pulley diameter is not straightforward. Further, belt life and geometry are practical limitations.

An early automotive application of a CVT used the friction-drive principle. This used a pair of friction disks, with one rolling on the face of the other. By changing the relative position of the disks, the output speed can be changed for a constant input speed. All friction drives have the advantage of overload protection, but the performance depends on the frictional properties of the disks, and deteriorates with age. Thermal problems, power loss, and component wear can be significant. In addition, the range of speed ratios will depend on the disk dimension, which can be a limiting factor in applications with geometric constraints.

The infinitely variable transmission (IVT), developed by Epilogies Inc. (Los Gatos, CA), is different in principle to the other types of CVTs mentioned. The IVT achieves the variation in speed ratio by first converting the input rotation to a reciprocating motion using a planetary assembly of several components (a planetary plate, four epicyclic shafts with crank arms, an overrunning clutch called a mechanical diode, etc.), then adjusting the effective output speed by varying the offset of an index plate with respect to the input shaft, recovering the effective rotation of the output shaft through a differential-gear assembly. One obvious disadvantage of this design is the large number of components and moving parts that are needed.

Now we describe an innovative design of a CVT that has many advantages over existing CVTs. In particular, this CVT uses simple and conventional components such as racks and a pinion, and is easy to manufacture and operate. It has few moving parts and, as a result, has high mechanical efficiency and needs less maintenance than conventional designs.

8.5.1 Principle of Operation

Consider the rack-and-pinion arrangement shown in Figure 8.8a. The pinion (radius r) rotates at an angular speed (ω) about a fixed axis (P). If the rack is not constrained in some manner, its kinematics will be indeterminate. For example, as in a conventional drive arrangement, if the direction of the rack is fixed, it moves at a rectilinear speed of ωr with zero angular speed. Instead, suppose that the rack is placed in a housing and is only allowed a rectilinear (sliding) lateral movement relative to the housing and that the housing itself is free to rotate about an axis parallel to the pinion axis, at O. Let the offset between the two axes (OP) be denoted by e.

It should be clear that if the pinion is turned, the housing (along with the rack) also turns. Suppose that the resulting angular speed of the housing (and the rack) is Ω. Let us determine an expression for Ω in terms of ω. The rack must move at rectilinear speed v relative to the housing. The operation of the CVT is governed by the kinematic arrangement of Figure 8.8, with ω as the input speed, Ω as the output speed, and offset e as the parameter that is varied to achieve the variable speed ratio. Note that perfect meshing between the rack and the pinion is assumed and backlash is neglected. Dimensions such as r are given with regard to the pitch line of the rack and the pitch circle of the pinion.

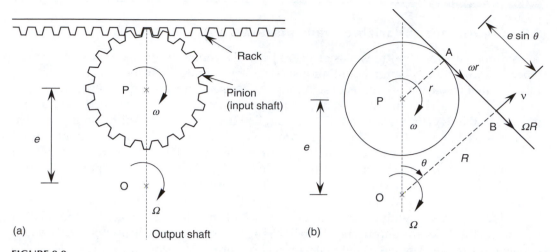

FIGURE 8.8
The kinematic configuration of the pinion and a meshed rack. (a) A reference configuration. (b) A general configuration.

Suppose that Figure 8.8a represents the reference configuration of the kinematic system. Now consider a general configuration as shown in Figure 8.8b. Here, the output shaft has rotated through angle θ from the reference configuration. Note that this rotation is equal to the rotation of the housing (with which the racks rotate). Hence, the angle θ can also be represented by the rotation of the line drawn perpendicular to a rack from the center of rotation O of the output shaft, as shown in Figure 8.8b. This line intersects the rack at point B, which is the middle point of the rack. Point A is a general point of meshing. Note that A and B coincide in the reference configuration (Figure 8.8a). The velocity of point B has two components—the component perpendicular to AB and the component along AB. Since the rack (with its housing) rotates about O at angular speed Ω, the component of velocity of B along AB is ΩR. This component has to be equal to the velocity of A along AB, because the rack (AB) is rigid and does not stretch. The latter velocity is given by ωr. It follows that

$$\omega r = \Omega R. \tag{8.13}$$

From geometry (see Figure 8.8b),

$$R = r + e\cos\theta. \tag{8.14}$$

By substituting Equation 8.14 in Equation 8.13, we get the speed ratio (p) of the transmission as

$$p = \frac{\omega}{\Omega} = 1 + \frac{e}{r}\cos\theta. \tag{8.15}$$

From Equation 8.15, it is clear that the kinematic arrangement shown in Figure 8.8 can serve as a gear transmission. It is also obvious, however, that if only one rack is made to continuously mesh around the pinion, the speed ratio p will simply vary sinusoidally about an average value of unity. This, then, will not be a very useful arrangement for a CVT. If, instead, the angle of mesh is limited to a fraction of the cycle, say from $\theta = -\pi/4$ to $+\pi/4$, and at the end of this duration another rack is engaged with the pinion to repeat the same motion while the first rack is moved around a cam without meshing with the pinion, then the speed reduction p can be maintained at an average value greater than unity. Furthermore, with such a system the speed ratio can be continuously changed by varying the offset parameter e. This is the basis of the two-slider CVT.

8.5.2 Two-Slider CVT

A graphic representation of a CVT that operates according to the kinematic principles described earlier is shown in Figure 8.9, a two-slider arrangement (U.S. Patent No. 4,800,768). Specifically, each slider unit consists of two parallel racks. The spacing of the racks (w) is greater than the diameter of the pinion. The meshing of a rack with the pinion is maintained by means of a suitably profiled cam, as shown. The two-slider units are placed orthogonally. It follows that each rack engages with the pinion at $\theta = -\pi/4$ and disengages at $\theta = +\pi/4$, according to the nomenclature given in Figure 8.8.

We note from Equation 8.15 that the speed ratio fluctuates periodically over periods of $\pi/2$ of the output-shaft rotation. For example, Figure 8.10 shows the variation in the output speed of the transmission for a constant input speed of 1.0 rad/s and an offset ratio of $e/r = 2.0$. It can be easily verified that the average speed ratio p is given by

$$\bar{p} = 1 + \frac{2\sqrt{2}}{\pi}\frac{e}{r}. \tag{8.16}$$

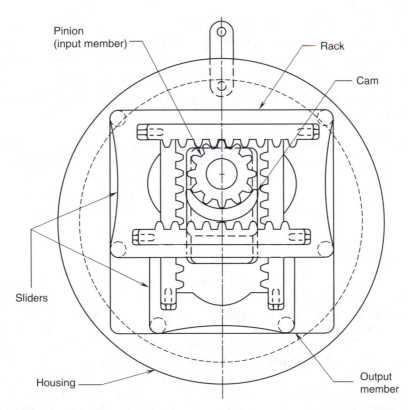

FIGURE 8.9
A drawing of a two-slider CVT.

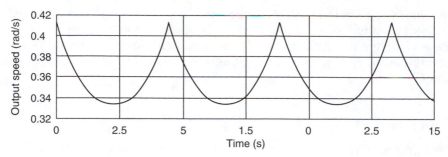

FIGURE 8.10
The response of the two-slider CVT for an input speed of 1.0 rad/s.

Note that $2\sqrt{2}/\pi \approx 9$. Also, the maximum value of speed ratio p occurs at $\theta = 0$ and the minimum value of p occurs at $\theta = \pm\pi/4$.

Offset ratio $e/r = 2.0$.

In summary, we can make the following observations regarding the present design of the CVT:

1. Speed ratio p (input shaft speed/output shaft speed) is not constant and changes with the shaft rotation.

2. The minimum speed ratio (p_{min}) occurs at the engaging and disengaging instants of a rack. The maximum speed ratio (p_{max}) occurs at halfway between these two points.

3. The maximum deviation from the average speed ratio is approximately 0.2 e/r and this occurs at the engaging and disengaging points.

4. Speed ratio increases linearly with e/r and hence the speed ratio of the transmission can be adjusted by changing the shaft-to-shaft offset e.

5. The larger the speed ratio, the larger the deviation from the average value (see items 3 and 4 above).

It has been indicated that the speed ratio of the transmission depends linearly on the offset ratio: (the offset between the output shaft and the input pinion)/(pinion radius). Figure 8.11 shows the variation in the average speed ratio p with respect to the offset ratio. Note that a continuous variation of the speed reduction in the range of more than 1 to 7 can be achieved by continuously varying the offset ratio e/r from 0 to 7.

8.5.3 A Three-Slider CVT

A three-slider CVT has been designed by us with the objective of reducing the fluctuations in the output speed and torque (Figure 8.12). The three-slider system consists of three rectangular pairs of racks (instead of two pairs), which slide along their slotted guideways, similar to the two-slider system. The main difference in the three-slider system is that each rack engages with the pinion for only 60° in a cycle of 360°. Hence, the fluctuating (sinusoidal) component of the speed ratio varies over an angle of 60°, in comparison with a 90° angle in the two-slider CVT. As a result, the fluctuations in the speed ratio will be less in the three-slider CVT. The six racks will engage and disengage sequentially during transmission. The cam profile of the three-slider system will be different from that of the two-slider system as well.

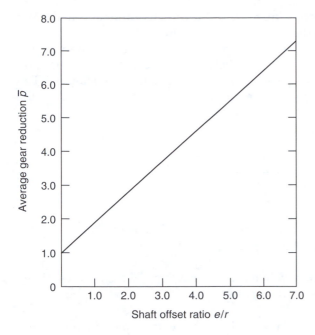

FIGURE 8.11
Average gear reduction curve for the two-slider CVT.

The speed reduction ratio of the three-slider CVT (for θ between $-\pi/6$ and $\pi/6$) is given by

$$p = \frac{\omega}{\Omega} = 1 + \frac{e}{r} \cos\theta. \qquad (8.17)$$

If we neglect inertia, elastic effects, and power dissipation (friction), the torque ratio of the transmission is given by the same equation. An advantage of the CVT is its ability to

FIGURE 8.12
A three-slider CVT.

continuously change the torque ratio according to output torque requirements and input torque (source) conditions. An obvious disadvantage in high-precision applications is the fluctuation in speed and torque ratios. This is not crucial in moderate-to-low-precision applications such as bicycles, golf carts, snowmobiles, hydraulic cement mixers, and generators. As a comparison, the percentage speed fluctuation of the two-slider CVT at an offset ratio of 6.0 (average speed ratio of approximately 6.5) is 18%, whereas for the three-slider CVT it is less than 8%.

Problems

8.1 In a lead-screw unit, the coefficient of friction μ was found to be greater than $\tan\alpha$, where α is the helix angle. Discuss the implications of this condition.

8.2 The nut of a lead-screw unit may have means of preloading, which can eliminate backlash. What are disadvantages of preloading?

8.3 A load is moved in a vertical direction using a lead-screw drive, as shown in Figure P8.3. The following variables and parameters are given: T, the motor torque; J, the overall moment of inertia of the motor rotor and the lead screw; m, the overall mass of the load and the nut; a, the upward acceleration of the load; r, the transmission ratio: (rectilinear motion)/(angular motion) of the lead screw; and e, the fractional efficiency of the lead screw.

Show that

$$T = \left(J + \frac{mr^2}{e}\right)\frac{a}{r} + \frac{r}{e}mg.$$

In a particular application, the system parameters are $m = 500$ kg, $J = 0.25$ kg.m^2, and the screw lead is 5.0 mm. In view of the static friction, the starting efficiency is

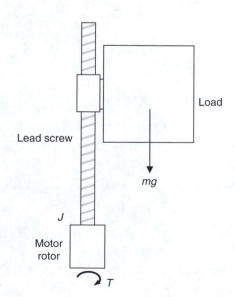

FIGURE P8.3
Moving a vertical load using a lead-screw drive.

50% and the operating efficiency is 65%. Determine the torque required to start the load and then move it upward at an acceleration of 3.0 m/s². What is the torque required to move the load downward at the same acceleration? Show that in either case much of the torque is used in accelerating the rotor (J). Note that in view of this observation it is advisable to pick a motor rotor and a lead screw with the least moment of inertia.

8.4 Consider the planetary gear unit shown in Figure P8.4. The pinion (pitch-circle radius r_p) is the input gear and it rotates at angular velocity ω_i. If the outer gear is fixed, determine the angular velocities of the planetary gear (pitch-circle radius r_g) and the connecting arm. Note that the pitch-circle radius of the outer gear is $r_p + 2r_g$.

8.5 List some advantages and shortcomings of conventional gear drives in speed transmission applications. Indicate ways to overcome or reduce some of the shortcomings.

8.6 A motor of torque T and moment of inertia J_m is used to drive an inertial load of moment of inertia J_L through an ideal (loss-free) gear of motor-to-load speed ratio r:1, as shown in Figure P8.6. Obtain an expression for the angular acceleration $\ddot{\theta}_g$ of the load. Neglect the flexibility of the connecting shaft. Note that the gear inertia may be incorporated into the terms J_m and J_L.

8.7 In mechanical drive units, it is important to minimize backlash. Discuss the reasons for this. Conventional techniques for reducing backlash in gear drives include preloading (or spring loading), the use of bronze bearings that automatically compensate for tooth wear, and the use of high-strength steel and other alloys that can be machined accurately to obtain tooth profiles of low backlash and that have minimal wear problems. Discuss the shortcomings of some of the conventional

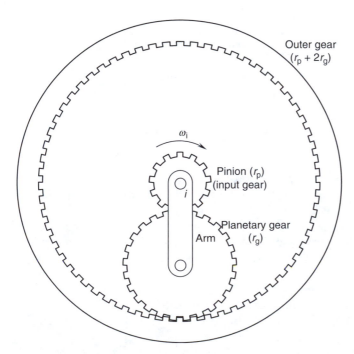

FIGURE P8.4
A planetary gear unit.

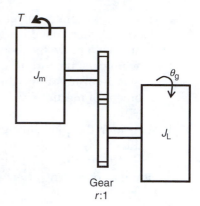

FIGURE P8.6

An inertial load driven by a motor through a gear transmission.

Gear
r:1

methods of backlash reduction. Discuss the operation of a drive unit that has virtually no backlash problems.

8.8 In some types of (indirect-drive) robotic manipulators, joint motors are located away from the joints and torques are transmitted to the joints through transmission devices such as gears, chains, cables, and timing belts. In some other types of (direct-drive) manipulators, joint motors are located at the joints themselves, with the rotor on one link and the stator on the joining link. Discuss the advantages and disadvantages of these two designs.

8.9 In the harmonic drive configuration shown in Figure 8.5, the outer rigid spline is fixed (stationary), the wave generator is the input member, and the flexispline is the output member. Five other possible combinations of harmonic drive configurations are tabulated below. In each case, obtain an expression for the gear ratio in terms of the gear ratio of the standard arrangement (shown in Figure 8.5) and comment on the drive operation.

Case	Rigid Spline	Wave Generator	Flexispline
1	Fixed	Output	Input
2	Output	Input	Fixed
3	Input	Output	Fixed
4	Output	Fixed	Input
5	Input	Fixed	Output

8.10 Figure P8.10 shows a picture of an induction motor connected to a flexible shaft through a flexible coupling. Using this arrangement, the motor may be used to drive a load that is not closely located and also not oriented in a coaxial manner with respect to the motor. The purpose of the flexible shaft is quite obvious in such an arrangement. Indicate the purpose of the flexible coupling. Could a flexible coupling be used with a rigid shaft instead of a flexible shaft?

8.11 Backlash is a nonlinearity, which is often displayed by robots that have gear transmissions. Indicate why it is difficult to compensate for backlash by using sensing and feedback control. What are the preferred ways to eliminate backlash in robots?

FIGURE P8.10
An induction motor linked to flexible shaft through a flexible coupling.

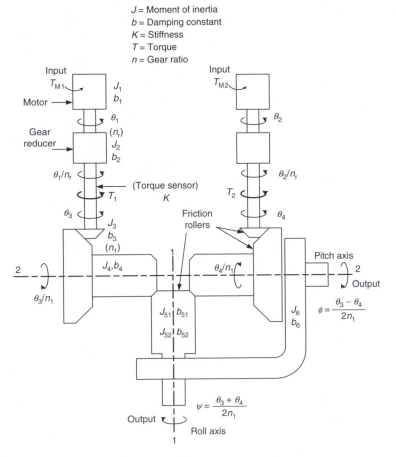

FIGURE P8.12
A traction-drive joint.

8.12   Friction drives (traction drives), which use rollers that make frictional contact, have been used as transmission devices. One possible application is for joint drives in robotic manipulators that typically use gear transmissions. An advantage of friction roller drives is the absence of backlash. Another advantage is finer motion resolution in comparison with gear drives.

a. Give two other possible advantages and several disadvantages of friction roller drives.

b. A schematic representation of the NASA traction drive joint is shown in Figure P8.12. Write dynamic equations for this model for evaluating its behavior.

Bibliography and Further Reading

This book has relied on many publications, directly and indirectly, in its development and evolution. Many of these publications are based on the work of the author and his co-workers. Also, there are some excellent books the reader may refer to for further information and knowledge. Some selected publications are listed below.

Books

Auslander, D.M. and Kempf, C.J., *Mechatronics Mechanical System Interfacing*, Prentice Hall, Upper Saddle River, NJ, 1996.

Barney, G.C., *Intelligent Instrumentation*, Prentice-Hall, Englewood Cliffs, N.J., 1985.

Beckwith, T.G., Buck, N.L., and Marangani, R.D., *Mechanical Measurement*, 3rd ed., Addison-Wesley, Reading, MA, 1982.

Bolton, W., *Mechatronics*, 2nd ed., Longman, Essex, England, 1999.

Chen, B.M., Lee, T.H., and Venkataramenan, V., *Hard Disk Drive Servo Systems*, Springer-Verlag, London, England, 2002.

Crandall, S.H., Karnopp, D.C., Kurtz, E.F. Jr., and Pridmore-Brown, D.C., *Dynamics of Mechanical and Electromechanical Systems*, McGraw-Hill, New York, 1968.

Dally, J.W., Riley, W.F., and Mcconnell, K.G., *Instrumentation for Engineering Measurements*, Wiley, New York, 1984.

De Silva, C.W., *Dynamic Testing and Seismic Qualification Practice*, Lexington Books, Lexington, MA, 1983.

De Silva, C.W., *Intelligent Control–Fuzzy Logic Applications*, CRC Press, Boca Raton, FL, 1995.

De Silva, C.W., Ed., *Intelligent Machines: Myths and Realities*, Taylor & Francis, CRC Press, BocaRaton, FL, 2000.

De Silva, C.W., *Mechatronics—An Integrated Approach*, Taylor & Francis, CRC Press, Boca Raton, FL, 2005.

De Silva, C.W., *Vibration Fundamentals and Practice*, 2nd ed., Taylor & Francis, CRC Press, Boca Raton, FL, 2007.

Doebelin, E.O., *Measurement Systems*, 3rd ed., McGraw-Hill, New York, 1983.

Gibson, J.E. and Tuteur, F.B., *Control System Components*, McGraw-Hill, New York, 1958.

Herceg, E.E., *Handbook of Measurement and Control*, Schaevitz Engineering, Pennsauken, NJ, 1972.

Histand, M.B. and Alciatore, D.G., *Introduction to Mechatronics and Measurement Systems*, WCB McGraw-Hill, New York, NY, 1999.

Hordeski, M.F., *The Design of Microprocessor, Sensor, and Control Systems*, Reston, VA, 1985.

Jain, L. and de Silva, C.W., Eds, *Intelligent Adaptive Control: Industrial Applications*, CRC Press, Boca Raton, FL, 1999.

Johnson, C.D., *Microprocessor-Based Process Control*, Prentice-Hall, Englewood Cliffs, NJ, 1984.

Karray, F. and de Silva, C.W., *Soft Computing Techniques and Their Applications*, Pearson, UK, 2004.

Necsulescu, D., *Mechatronics*, Prentice Hall, Upper Saddle River, NJ, 2002.

Potvin, J., *Applied Process Control Instrumentation*, Reston, VA, 1985.

Shetty, D. and Kolk, R.A., *Mechatronics System Design*, PWS Publishing Co., Boston, MA, 1997.

Tan, K.K., Lee, T.H., Dou, H., and Huang, S., *Precision Motion Control*, Springer-Verlag, London, England, 2001.

Sections of Books

Cao, Y. and de Silva, C.W., Adaptive control of a manipulator with flexible joints using neural networks, *Mechatronics and Machine Vision: Future Trends*, John Billingsley, Ed., Research Studies Press LTD, Baldock, Hertfordshire, England, pp. 71–78, 2003.

De Silva, C.W., Considerations of hierarchical fuzzy control, *Theoretical Aspects of Fuzzy Control*, H.T. Nguyen, M. Sugeno, R.M. Tong, and R.R. Yager, Eds, John Wiley & Sons, New York, 1995.

De Silva, C.W., Intelligent restructuring of automated production systems, *Fuzzy Logic and Its Applications to Engineering, Information Sciences, and Intelligent Systems*, Z. Bien and K.C. Min, Eds, Kluwer Academic Publishers, Boston, MA, 1995

De Silva, C.W., Electronic components, *Encyclopedia of Electrical and Electronics Engineering*, J.G. Webster, Ed., John Wiley & Sons, Inc., New York, Vol. 6, pp. 577–594, 1999.

De Silva, C.W., Sensors for control, *Encyclopedia of Physical Science and Technology*, 3rd ed., R.A. Meyers, Ed., Academic Press, San Diego, CA, Vol. 14, pp. 609–650, 2001.

De Silva, C.W., Wong, K.H., and Modi, V.J., Development of a novel multi-module manipulator system: dynamic model and prototype design, *Mechatronics and Machine Vision: Current Practice*, R.S. Bradbeer and J. Billingsley, Eds, Research Studies Press Ltd, Hertfordshire, Baldock, England, pp. 161–168, 2002.

Zhang, J. and de Silva, C.W., Intelligent hierarchical control of a space-based deployable manipulator, *Mechatronics and Machine Vision: Future Trends*, John Billingsley, Ed., Research Studies Press LTD., Baldock, Hertfordshire, England, pp. 61–70, 2003.

Journal Papers

Cao, Y. and de Silva, C.W., Dynamic modeling and neural-network adaptive control of a deployable manipulator system, *J. Guid Control Dyn.*, Vol. 29(1), pp. 192–195, 2006.

Cao, Y. and de Silva, C.W., Supervised switching control of a deployable manipulator system, *Int. J. Control Intell. Syst.*, Vol. 34(2), pp. 153–165, 2006.

Chen, Y., Wang, X.G., Sun, C., Devine, F., and de Silva, C.W., Active vibration control with state feedback in woodcutting, *J. Vib. Control*, Vol. 9(6), pp. 645–664, 2003.

Croft, E.A., de Silva, C.W., and Kurnianto, S., Sensor technology integration in an intelligent machine for herring roe grading, *IEEE ASME Trans. Mechatron.*, Vol. 1(3), pp. 204–215, September 1996.

De Silva, C.W., Design of PPD controllers for position servos, *J. Dyn. Syst. Meas. Control*, Trans. ASME, Vol. 112(3), pp. 519–523, September 1990.

De Silva, C.W., An analytical framework for knowledge-based tuning of servo controllers, *Int. J. Eng. Appl. Artif. Intell.*, Vol. 4(3), pp. 177–189, 1991.

De Silva, C.W., Trajectory design for robotic manipulators in space applications, *J. Guid. Control Dyn.*, Vol. 14(3), pp. 670–674, June 1991.

De Silva, C.W., A criterion for knowledge base decoupling in fuzzy-logic control systems, *IEEE Trans. Syst. Man Cybern.*, Vol. 24(10), pp. 1548–1552, October 1994.

De Silva, C.W., Applications of fuzzy logic in the control of robotic manipulators, *Fuzzy Sets Syst.*, Vol. 70(2–3), pp. 223–234, 1995.

De Silva, C.W., Intelligent control of robotic systems, *Int. J. Robot. Auton. Syst.*, Vol. 21, pp. 221–237, 1997.

De Silva, C.W., The role of soft computing in intelligent machines, *Phil. Trans. R. Soc. A*, UK, Vol. 361(1809), pp. 1749–1780, 2003 (By Invitation).

De Silva, C.W., Sensory information acquisition for monitoring and control of intelligent mechatronic systems, *Int. J. Inf. Acquisition*, Vol. 1(1), pp. 89–99, March 2004.

De Silva, C.W., Control system methodology for vibration suppression in machinery and structures, *J. Struct. Eng.*, Vol. 33(1), pp. 1–12, 2006.

De Silva, C.W. and Gu, J., An intelligent system for dynamic sharing of workcell components in process automation, *Int. J. Eng. Appl. Artif. Intell.*, Vol. 7(5), pp. 571–586, 1994.

De Silva, C.W. and Lee, T.H., Knowledge-based intelligent control, *Meas. Control*, Vol. 28(2), pp. 102–113, April 1994.

De Silva, C.W. and Lee, T.H., Fuzzy logic in process control, *Meas. Control*, Vol. 28(3), pp. 114–124, June 1994.

De Silva, C.W. and Wickramarachchi, N., An innovative machine for automated cutting of fish, *IEEE ASME Trans. Mechatron.*, Vol. 2(2), pp. 86–98, 1997.

De Silva, C.W., Price, T.E., and Kanade, T., A torque sensor for direct-drive manipulators, *J. Eng. Ind.*, Trans. ASME, Vol. 109(2), pp. 122–127, May 1987.

De Silva, C.W., Singh, M., and Zaldonis, J., Improvement of response spectrum specifications in dynamic testing, *J. Eng. Ind.*, Trans. ASME, Vol. 112, No. 4, pp. 384–387, November 1990.

De Silva, C.W., Schultz, M., and Dolejsi, E., Kinematic analysis and design of a continuously-variable transmission, *Mech. Mach. Theory*, Vol. 29(1), pp. 149–167, January 1994.

De Silva, C.W., Gamage, L.B., and Gosine, R.G., An intelligent firmness sensor for an automated herring roe Grader, *Int. J. Intell. Autom. Soft Comput.*, Vol. 1(1), pp. 99–114, 1995.

Goulet, J.F., de Silva, C.W., Modi, V.J., and Misra, A.K., Hierarchical knowledge-based control of a deployable orbiting manipulator, *Acta Astron.*, Vol. 50(3), pp. 139–148, 2002.

Gu, J.S. and de Silva, C.W., Development and implementation of a real-time open-architecture control system for industrial robot systems, *Eng. Appl. Artif. Intell., Int. J. Intell. Real-Time Autom.*, Vol. 17(1), pp. 469–483, 2004

Hu, B.G., Gosine, R.G., Cao, L.X., and de Silva, C.W., Application of a fuzzy classification technique in computer grading of fish products, *IEEE Trans. Fuzzy Syst.*, Vol. 6(1), pp. 144–152, February 1998.

Jain, A., de Silva, C.W., and Wu, Q.M.J., Intelligent fusion of sensor data for product quality assessment in a fish-cutting machine, *Int. J. Control Intell. Syst.*, Vol. 32(2), pp. 89–98, 2004.

Lee, T.H., Yue, P.K., and de Silva, C.W., Neural networks improve control, *Meas. Control*, Vol. 28(4), pp. 148–153, September 1994.

Lee, M.F.R, de Silva, C.W., Croft, E.A., and Wu, Q.M.J., Machine vision system for curved surface inspection, *Mach. Vis. Appl.*, Vol. 12, pp. 177–188, 2000.

Omar, F.K. and de Silva, C.W., Optimal portion control of natural objects with application in automated cannery processing of fish, *J. Food Eng.*, Vol. 46, pp. 31–41, 2000.

Rahbari, R. and de Silva, C.W., Comparison of two inference methods for p-type fuzzy logic control through experimental investigation using a hydraulic manipulator, *Int. J. Eng. Appl. Artif. Intell.*, Vol. 14(6), pp. 763–784, 2001.

Rahbari, R., Leach, B.W., Dillon, J., and de Silva, C.W., Expert system for an INS/DGPS integrated navigator installed in a Bell 206 helicopter, *Eng. Appl. Artif. Intell., Int. J. Intell. Real-Time Autom.*, Vol. 18(3), pp. 353–361, 2005.

Stanley, K., Wu, Q.M.J., de Silva, C.W., and Gruver, W., Modular neural-visual servoing with image compression input, *J. Intell. Fuzzy Syst.*, Vol. 10(1), pp. 1–11, 2001.

Tafazoli, S., de Silva, C.W., and Lawrence, P.D., Tracking control of an electrohydraulic manipulator in the presence of friction, *IEEE Trans. Control Syst. Technol.*, Vol. 6(3), pp. 401–411, May 1998.

Tang, P.L. and de Silva, C.W., Compensation for transmission delays in an ethernet-based control network using variable-horizon predictive control, *IEEE Trans. Control Syst. Technol.*, Vol. 14(4), pp. 707–718, 2006.

Tang, P.L., Poo, A.N., and de Silva, C.W., Knowledge-based extension of model-referenced adaptive control with application to an industrial process, *J. Intell. Fuzzy Syst.*, Vol. 10(3–4), pp. 159–183, 2001.

Tang, P.L., de Silva, C.W., and Poo, A.N., Intelligent adaptive control of an industrial fish cutting machine, *Trans. S. Afr. Inst. Elect. Eng.*, Vol. 93(2), pp. 60–72, June 2002.

Wu, Q.M. and de Silva, C.W., Dynamic switching of fuzzy resolution in knowledge-based self-tuning control, *J. Intell. Fuzzy Syst.*, Vol. 4(1), pp. 75–87, 1996.

Wu, Q.M.J., Lee, M.F.R., and de Silva, C.W., An imaging system with structured lighting for on-line generic sensing of three-dimensional objects, *Sens. Rev.*, Vol. 22(1), pp. 46–50, 2002.

Yan, G.K.C., de Silva, C.W., and Wang, G.X., Experimental modeling and intelligent control of a wood-drying kiln, *Int. J. Adapt. Control Signal Process.*, Vol. 15, pp. 787–814, 2001.

Zhou, Y. and de Silva, C.W., Adaptive control of an industrial robot retrofitted with an open-architecture controller, *J. Dyn. Syst. Meas. Control*, Trans. ASME, Vol. 118(1), pp. 143–150, March 1996.

Other Journal Publications

De Silva, C.W., Process variables, *Meas. Control*, Vol. 18(2), pp. 149–156, April 1984.

De Silva, C.W., Stability analysis, *Meas. Control*, Vol. 19(6), pp. 118–124, December 1985.

De Silva, C.W., Motor controllers, *Meas. Control*, Vol. 20(1), pp. 271–274, February 1986.

De Silva, C.W., Torque measurement, *Meas. Control*, Vol. 20(1), pp. 302–303, February 1986.

De Silva, C.W., Counters/frequency tachometers, *Meas. Control*, Vol. 20(2), pp. 201, April 1986.

De Silva, C.W., Mass/force measurement, load cells, *Meas. Control*, Vol. 20(5), pp. 230–231, October 1986.

De Silva, C.W., Control systems modeling terminology, *Meas. Control*, Vol. 21(3), pp. 125, June 1987.

De Silva, C.W., Linear algebra for modeling dynamic systems, *Meas. Control*, Vol. 21(4), pp. 165–171, September 1987.

De Silva, C.W., State models, *Meas. Control*, Vol. 21(5), pp. 117–121, October 1987.

De Silva, C.W., Linear graphs, *Meas. Control*, Vol. 21(6), pp. 102–107, December 1987.

De Silva, C.W., Through and across variables, *Meas. Control*, Vol. 22(1), pp. 120–123, February 1988.

De Silva, C.W., State models from linear graphs, *Meas. Control*, Vol. 22(2), pp. 122–127, April 1988.

De Silva, C.W., Bond graphs, *Meas. Control*, Vol. 22(3), pp. 130–141, June 1988.

De Silva, C.W., Transfer function models, *Meas. Control*, Vol. 22(4), pp. 157–169, September 1988.

De Silva, C.W., Frequency domain models, *Meas. Control*, Vol. 22(5), pp. 131–141, October 1988.

De Silva, C.W., Dynamics and control of robots, *Meas. Control*, Vol. 29(3), pp. 87–92, June 1995.

De Silva, C.W., Robot dynamics, *Meas. Control*, Vol. 29(4), pp. 133–141, September 1995.

De Silva, C.W., Control principles of robotics: kinematics and kinetics, *Meas. Control*, Vol. 29(5), pp. 77–83, October 1995.

De Silva, C.W., Lagrangian and Newton–Euler methods, *Meas. Control*, Vol. 29(6), pp. 69–73, December 1995.

De Silva, C.W., Trajectory design and optimization, *Meas. Control*, Vol. 30(1), pp. 109–113, February 1996.

De Silva, C.W., End-effector trajectory generation, *Meas. Control*, Vol. 30(2), pp. 101–109, April 1996.

De Silva, C.W., Feedforward nonlinearity compensation, *Meas. Control*, Vol. 30(3), pp. 90–95, June 1996.

De Silva, C.W., Application of nonlinearity compensation, *Meas. Control*, Vol. 30(4), pp. 125–135, September 1996.

De Silva, C.W., Robot control architectures, *Meas. Control*, Vol. 30(5), pp. 85–92, October 1996.

De Silva, C.W., Robot workcells and applications, *Meas. Control*, Vol. 30(6), pp. 77–86, December 1996.

De Silva, C.W., Control system instrumentation, *Meas. Control*, Vol. 33(1), pp. 69–82, February 1999.

De Silva, C.W., Frequency models and bandwidth, *Meas. Control*, Vol. 33(2), pp. 75–88, April 1999.

De Silva, C.W., Component interconnection and matching, *Meas. Control*, Vol. 33(3), pp. 61–75, June 1999.

De Silva, C.W., Performance specification, *Meas. Control*, Vol. 33(4), pp. 85–105, September 1999.

De Silva, C.W., Instrument error analysis: Part A, *Meas. Control*, Vol. 33(5), pp. 53–62, October 1999.

De Silva, C.W., Instrument error analysis: Part B, *Meas. Control*, Vol. 34(1), pp. 49–65, February 2000.

De Silva, C.W., Potentiometric transducer, *Meas. Control*, Vol. 34(2), pp. 49–61, April 2000.

De Silva, C.W., Differential transformers, *Meas. Control*, Vol. 34(3), pp. 49–60, June 2000.

De Silva, C.W., Inductive motion transducers, *Meas. Control*, Vol. 34(4), pp. 65–76, September 2000.

De Silva, C.W., Capacitive and piezoelectric sensors, *Meas. Control*, Vol. 34(5), pp. 49–63, October 2000.

De Silva, C.W., Miscellaneous sensors, *Meas. Control*, Vol. 34(6), pp. 49–58, December 2000.

De Silva, C.W. and Aronson, M.H., Response analysis by transform, *Meas. Control*, Vol. 18(3), pp. 149–154, June 1984.

De Silva, C.W. and Aronson, M.H., On-off and proportional control, *Meas. Control*, Vol. 18(4), pp. 165–170, September 1984.

De Silva, C.W. and Aronson, M.H., Reset and rate control, *Meas. Control*, Vol. 18(5), pp. 133–145, October 1984.

De Silva, C.W. and Aronson, M.H., Oscillation and stability criteria, *Meas. Control*, Vol. 19(1), pp. 125–131, February 1985.

De Silva, C.W. and Aronson, M.H., Transfer functions, poles, and zeros, *Meas. Control*, Vol. 19(2), pp. 134–137, April 1985.

De Silva, C.W. and Aronson, M.H., The S-domain and plane, *Meas. Control*, Vol. 19(3), pp. 134–139, June 1985.

De Silva, C.W. and Aronson, M.H., Lead and lag networks, *Meas. Control*, Vol. 19(4), pp. 182–186, September 1985.

De Silva, C.W., and Aronson, M.H., Response of common networks, *Meas. Control*, Vol. 19(5), pp. 125–129, October 1985.

De Silva, C.W. and Gamage, L.B., Introduction to C programming, *Meas. Control*, Vol. 25(5), pp. 85–90, October 1991.

De Silva, C.W. and Gamage, L.B., Data types and operators, *Meas. Control*, Vol. 25(6), pp. 66–71, December 1991.

De Silva, C.W. and Gamage, L.B., Control structures, *Meas. Control*, Vol. 26(1), pp. 85–91, February 1992.

De Silva, C.W. and Gamage, L.B., Pointers and other operators, *Meas. Control*, Vol. 26(2), pp. 70–75, April 1992.

De Silva, C.W. and Gamage, L.B., Arrays, *Meas. Control*, Vol. 26(3), pp. 78–83, June 1992.

De Silva, C.W. and Gamage, L.B., Functions, *Meas. Control*, Vol. 26(4), pp. 120–129, September 1992.

De Silva, C.W. and Gamage, L.B., Constructed data types, *Meas. Control*, Vol. 26(5), pp. 80–86, October 1992.

De Silva, C.W. and Gamage, L.B., Storage classes, *Meas. Control*, Vol. 26(6), pp. 72–77, December 1992.

De Silva, C.W. and Gamage, L.B., Processing, compiling, linking, *Meas. Control*, Vol. 27(1), pp. 93–97, February 1993.

De Silva, C.W. and Gamage, L.B., File handling and input/output, *Meas. Control*, Vol. 27(2), pp. 77–83, April 1993.

De Silva, C.W. and Gamage, L.B., Some useful features of C, *Meas. Control*, Vol. 27(3), pp. 86–90, June 1993.

De Silva, C.W. and Gamage, L.B., Programming language comparison, *Meas. Control*, Vol. 24(4), pp. 118–127, September 1993.

Answers to Numerical Problems

Chapter 1

1.10 1 ms; 11×10^{-6} Hz

Chapter 2

2.9 -7.5 V; -14 V
2.12 31.8 Hz; 5 μs
2.22 3600 rpm; 15 balls
2.26 1%

Chapter 3

3.3 **(a)** 2 mm; **(b)** 2500 N/m^2; **(c)** 74.0 dB
3.6 25.0
3.10 **(iv)** 0.09; **(v)** 6.25×10^4 to 25.0×10^4 N/m
3.13 16 Hz; 200 Hz; 50 Hz
3.16 **(b)** 0.99%
3.19 7.5%
3.21 **(b)** $e_m = \pm 1.1\%$; $e_l = \pm 11\%$; $e_r = \pm 1.2\%$; $e_\alpha = \pm 1.0\%$
3.23 **(ii)** $e_V = \pm 13.0\%$;
3.25 **(b) (ii)**$e_Q = \pm 1\%$; $e_s = \pm 1.2\%$; $e_f = \pm 3.1\%$
3.27 $e_w = \pm 1.7\%$; $e_s = \pm 0.8\%$; $e_y = \pm 1.7\%$
3.28 1.32%

Chapter 4

4.11 0.1%
4.17 1000 Hz
4.32 100 s
4.34 **(a)** 1.62%; **(b)** 170 Hz
4.37 9.5 kΩ
4.38 15.5 s
4.47 2.5%
4.49 144.0; 5.9%

Chapter 5

5.4 $\pm 0.088°$
5.9 >2
5.14 12 bits; 12 tracks; 4,096 sectors
5.15 0.0072°
5.22 **(c)** 20.0 m/s
5.24 20,000 hr

Chapter 6

6.1 90°
6.3 60°

6.6 6 teeth/pole; 8 stator poles; 50 rotor teeth
6.7 **(i)** 15°; **(ii)** 7.5°; **(iii)** 15°
6.8 5 phases; 20 stator poles; 0.72°step; 100 rotor teeth
6.9 **(a)** 30°; **(b)** 15°
6.10 7.2° tooth pitch; 1.8° step; 200 steps/rev; 2 poles/phase; max. 6 stator teeth/pole
6.17 2 ms; 50 steps/s
6.21 **(b)** 5°
6.22 168.65 rad/s
6.27 100, 50, 25, 20, 10, 5, 4, 2, and 1 steps/s
6.34 0.716 T_{max}
6.35 310 SM; rack-and-pinion with $r \leq 0.0318$ m/rad
6.36 Model 2
6.37 101 SM
6.38 310 SM; 3.0 mm
6.39 **(b) (i)** 310 SM; **(ii)** 8; **(iii)** 0.225°

Chapter 7

7.2 10 A
7.10 0.5 kHz
7.11 40%
7.19 1000 rpm, 12.12 N.m
7.21 7.26 N.m; 1800 rpm; 1350 rpm; 5 N.m, 1620 rpm; 25 N.m, 810 rpm (unstable)

Chapter 8

8.3 949.8 N.m; 937.8 N.m

Index